RENEWALS 458-4574
DATE DUE

MAR

THE MORAL AND POLITICAL WRITINGS OF MAHATMA GANDHI

VOLUME I

MOHANDAS KARAMCHAND GANDHI
(*2 October 1869—30 January 1948*)

THE MORAL AND POLITICAL WRITINGS OF MAHATMA GANDHI

VOLUME I

Civilization, Politics, and Religion

Edited by
RAGHAVAN IYER

CLARENDON PRESS · OXFORD
1986

Oxford University Press, Walton Street, Oxford OX2 6DP
Oxford New York Toronto
Delhi Bombay Calcutta Madras Karachi
Kuala Lumpur Singapore Hong Kong Tokyo
Nairobi Dar es Salaam Cape Town
Melbourne Auckland

and associated companies in
Beirut Berlin Ibadan Nicosia

Oxford is a trade mark of Oxford University Press

Published in the United States
by Oxford University Press, New York

The Writings of M.K. Gandhi © Navajivan Trust,
Ahmedabad-380 014, India, 1986

Compilation © Raghavan N. Iyer 1986

All rights reserved. No part of this publication may be reproduced, stored in a retrieval system, or transmitted, in any form or by any means, electronic, mechanical, photocopying, recording, or otherwise, without the prior permission of Oxford University Press

British Library Cataloguing in Publication Data
Gandhi, M.K.
The moral and political writings of Mahatma Gandhi.
Vol. 1: Civilization, politics and religion
1. Gandhi, M.K. 2. Statesmen—India—Biography
I. Title II. Iyer, Raghavan
954.03'5'0924 DS481.G3
ISBN 0-19-824754-0

Library of Congress Cataloging in Publication Data
Gandhi, Mahatma, 1869-1948.
The moral and political writings of Mahatma Gandhi.
Bibliography: p.
Includes index.
Contents: v. 1. Civilization, politics, and religion.
I. Iyer, Raghaven Narasimhan. II. Title.
DS481.G3A2 1985 954.03'5'0924 84-27158
ISBN 0-19-824754-0 (v. 1)

Set and Printed in Great Britain by
Butler and Tanner Ltd, Frome and London

Library
University of Texas
at San Antonio

To
Shri S. S. Aiyer
Shrimati Thangammal
Shri L. N. Iyer
Shrimati Lakshmi Iyer

Plan of the Series

Volume One
Civilization, Politics, and Religion

- I   Introduction
- II   Gandhi on Himself and His Mission
- III Influences and Books Read
- IV *Hind Swaraj*, Modern Civilization, and Moral Progress
- V   Politics and Religion
- VI Religion
- VII God

Volume Two
Truth and Non-Violence

- I   Introduction
- II   Human Nature, Perfectibility, and History
- III Principles and Vows
- IV Conscience, Heroism, and Humility
- V   *Satya*—Absolute and Relative Truth
- VI *Ahimsa*—the Scope and Power of Non-Violence
- VII *Ashram*—Experiments with Truth

Volume Three
Non-Violent Resistance and Social Transformation

- I   Introduction
- II   *Satyagraha*—Non-Violent Resistance
- III *Swaraj*—Freedom and Self-Rule
- IV *Swadeshi*—Self-Reliance
- V   *Sarvodaya*—Non-Violent Social Transformation

Preface

Despite the vast amount of proliferating literature on Mahatma Gandhi, there has as yet been no accessible and coherent record of his essential writings. During his lifetime he wrote weekly articles for the journals he edited—*Indian Opinion, Young India, Harijan,* and *Navajivan.* He was also unusually conscientious about replying to all of his correspondents in South Africa, England, India, and elsewhere, writing as many as seventy letters a day for over four decades. The enormous quantity of his correspondence is the main reason why his *Collected Works* (initiated by the Indian Government soon after his death and now drawing to its completion) has run into ninety volumes. His actual books were few, short, and somewhat inconclusive—*Hind Swaraj, The Story of My Experiments With Truth, Satyagraha in South Africa,* and *Ashram Observances in Action,* together with small tracts on the *Bhagavad Gita,* the Constructive Programme, and on health. His unfinished autobiography and several popular biographies remain the chief—and rather misleading—sources of public knowledge about the personality and impact of Gandhi. While a significant number of mostly perfunctory or fragmentary anthologies have been published, they have largely obscured the richness of his thought.

On the suggestion of G. D. H. Cole, John Plamenatz, and others at Oxford, I began my research into Gandhi's unpublished writings in 1956 when hardly two volumes of the *Collected Works* had appeared. Fortunately, Professor K. Swaminathan was persuaded by Vinoba Bhave to undertake the editorship and production of the *Collected Works.* Professor Swaminathan willingly assumed this onerous task and recently completed it with extraordinary patience, scrupulousness, and care. With his generous help, I was able to consult the corpus of materials in his office and in several libraries. This enabled me to complete *The Moral and Political Thought of Mahatma Gandhi,* published by Oxford University Press in 1973.

Since then, it has become evident that Gandhi has been most inadequately represented by the older anthologies. In attempting to rescue Gandhi's essential writings from the wealth of detail (and ephemera) in the *Collected Works*, I realized that there was no way of doing justice to the subtlety and scope of Gandhi's thought in less than three volumes of materials drawn from the entire corpus. It became necessary to sift every volume of the *Collected Works* and apply the most stringent criteria to produce a comprehensive, balanced, and accessible collection. While the materials selected rely on the definitive versions of the *Collected Works*, there are minor modifications. The entries are given apt titles, and the original titles provided by Gandhi or his associates are cited at the end of each entry. Footnotes have been kept to a bare minimum so that the reader is unencumbered by extraneous detail in reading the texts. The selections drawn from the entire lifework of Gandhi are organized under self-explanatory heads. They show the refinement of his thought over many years and also the underlying consistency of his commitments and perspectives.

This three-volume collection could help a variety of people in different countries—including India itself—to form a fuller and more just appreciation of Gandhi's significant and compelling contribution to the twentieth century and to the future.

R. N. I.

2 October 1983

Acknowledgements

I am grateful to the Navajivan Press, for permission to use materials from *The Collected Works of Mahatma Gandhi* (90 volumes); to Professor K. Swaminathan, for valuable suggestions concerning this three-volume edition; to Professors Kilian Coster and Elton Hall, for their generous help in preparing this volume; to Ruth Ahlroth and Paula Kelly, for their help in preparing the materials for publication; and to the editorial staff of Oxford University Press.

Contents

Mohandas Karamchand Gandhi		Frontispiece
Abbreviations		xix

I	Introduction	1
II	Gandhi on Himself and His Mission	13

§ 1. *Gandhi on Himself*   13

1. The Ochre Robe   13
2. Striving after *Moksha*   14
3. The Kingdom of Heaven   15
4. Becoming Fearless   19
5. Trying to See Light   20
6. One Step Enough for Me   21
7. Voluntary Retirement   24
8. The Royal Road   25
9. Freedom from Dogma   26
10. The Spirit, Not the Letter   28
11. 'Sacred Cow and Ferocious Tiger'   28
12. Truth from Whatever Source   29
13. Meeting Kindred Spirits   30
14. No False Modesty   32
15. No Miraculous Powers   33
16. My Sons   34
17. The Call to Lead   34
18. Out of My Ashes   35
19. The One-Sided and the All-Sided   35
20. Thinker and Doer   36
21. Words and Meanings   37
22. My Life Is My Message   37
23. Working Silently   38
24. Action Is My Domain   39

§ 2. *Saint or Politician?*   41

25. Saint or Politician?   41

§ 3. *My Mission* 45
    26. Identification with Labour 45
    27. Service of Humanity 48
    28. Cave in the Heart 55
    29. Brotherhood of Man 58
    30. Universal Message 60

§ 4. *Isms* 60
    31. Beyond Isms and Sects 60
    32. Isms and Followers 63
    33. Propounder of No Ism 64

III   Influences and Books Read   66

§ 1. *Narasinh Mehta* 66
    34. The *Vaishnava* Ideal 66
    35. *Vaishnava Dharma* 68

§ 2. *Theosophy* 72
    36. Theosophy, Renunciation, and Atheism 72
    37. Theosophy and True Religion 75

§ 3. *The* Bhagavad Gita 77
    38. The *Gita* and Non-Violence 77
    39. Fundamental Principles 84
    40. The Gospel of Service 89
    41. Self-Surrender 90
    42. Non-Attachment 92
    43. The Spirit of the *Gita* 93
    44. The Synthesis of the *Gita* 94
    45. Problem-Solving 96
    46. A Spiritual Dictionary 97
    47. Need for Patience 99

§ 4. *Socrates and Thoreau* 100
    48. The Wisdom of Socrates 100
    49. Reading Thoreau 102
    50. Thoreau on Civil Disobedience 102

§5.	*Ruskin*	103
	51. Ruskin on Education	103
§6.	*Mazzini*	105
	52. Patriotism of Mazzini	105
§7.	*Tolstoy*	107
	53. Tolstoy on Simplicity	107
	54. Tolstoy on Non-Retaliation	108
	55. The Teaching of Tolstoy	111
	56. Tolstoy on Non-Violence	112
	57. Tolstoy's Greatest Contribution	113
	58. Tolstoy's Influence	114
	59. Tolstoy on Self-Control	114
§8.	*Naoroji*	123
	60. Dadabhai Naoroji, MP	123
	61. The Simplicity of Naoroji	124
§9.	*Gokhale*	126
	62. The Message of Gokhale	126
	63. The Legacy of Gokhale	128
	64. Gokhale's Ideal of Service	135
§10.	*Rajchandra*	139
	65. The Culture of the Heart	139
	66. The Path to *Moksha*	140
	67. Contact with Rajchandra	143
	68. Freedom from Attachment	145
	69. Vigilance in Work	146
	70. Rajchandra on Religion	150
	71. Rajchandra's Faith in *Ahimsa*	153
§11.	*Books Read*	154
	72. Esoteric Christianity	154
	73. Sufi Mystics	155
	74. Gujarati and Hindi Books	156
	75. Books Read in Gaol (1922)	157
	76. Books Read in Gaol (1923)	167
	77. Studies in Gaol	178
	78. Islam, English History, and the *Mahabharata*	184

79.	Mystics, Sikhs, and the *Gita*	188
80.	Books Read in Gaol (1932)	190
81.	'The Hound of Heaven'	194

§ 12. *Reading and Reflection* — 194

82.	Peace of Mind	194
83.	Self-Restraint in Reading	196
84.	Assimilation of Reading	196
85.	Reading and Thinking	196
86.	Reading Without Reflection	197
87.	Discovery of Meaning	198

IV *Hind Swaraj*, Modern Civilization, and Moral Progress — 199

§ 1. Hind Swaraj — 199

88.	*Hind Swaraj*	199

§ 2. *On* Hind Swaraj — 271

89.	Imperfections in *Hind Swaraj*	271
90.	Knowing the Fundamentals	273
91.	Request to Tolstoy	274
92.	Panic over *Hind Swaraj*	275
93.	Aim of *Hind Swaraj*	276
94.	Means to Regeneration	278
95.	Higher Simplicity and Renunciation	278
96.	Return to Common Sense	280
97.	Relevance of *Hind Swaraj*	283
98.	Drawing the Line	284
99.	Emphasis on Essentials	285

§ 3. *Civilization* — 288

100.	Destructive Materialism	288
101.	Instability of Civilization	290
102.	Modern Civilization and Ancient Wisdom	291
103.	Civilization and Conscience	297
104.	Faith in Indian Civilization	302
105.	Ancient Glory and Present Inertia	304
106.	Cowardice and Hypocrisy in India	305
107.	Civilization and Education	309

	108. Science and Civilization	310
	109. Industrialism and India	315
	110. The Scientific Spirit	317
	111. Democratic *Swaraj*	320
§4.	*East and West*	324
	112. Union of East and West	324
	113. Test of Civilization	326
	114. Everything on Its Merits	329
§5.	*Modern Civilization*	329
	115. Esoteric Christianity and Modern Civilization	329
	116. Violence and Civilization	330
	117. High-Handedness and Hypocrisy	333
	118. Nature and Civilization	334
	119. Meshes of Civilization	336
	120. Hypocrisy of Civilization	340
	121. Glamour of Civilization	342
	122. Appeal to the Masses	348
	123. Control of Excesses	351
	124. Peace from Within	352
	125. The Hubbub of Cities	353
	126. Simplicity and Artificiality	354
§6.	*Progress Moral and Material*	355
	127. Economic and Moral Progress	355
	128. Moral Growth and Material Prosperity	363

V Politics and Religion 370

§1.	*Politics and Religion*	370
	129. Equanimity in Politics	370
	130. Soul-Power and Physical Force	371
	131. Spiritualizing Political Life	374
	132. Shadow and Substance in Politics	376
	133. 'Truth Even unto the Wicked'	379
	134. Men and Systems	380
	135. Voluntary Poverty	381

	136.	Religion and Irreligion in Politics	391
	137.	Towards Universal Brotherhood	391
	138.	Religion and the State	395
§ 2.	*Political Power*		396
	139.	The Crown of Thorns	396
	140.	Seeking Office	397
	141.	Political Power	398
	142.	Power and Representation	401
	143.	Silent Service	407
§ 3.	*Politics and Society*		408
	144.	Inclusiveness of the Political	408
	145.	Tricks and Trading	409
	146.	The Sanction of Public Opinion	410
	147.	Political Harangues and Camouflage	412
	148.	Political Power and Social Reform	413
	149.	True Politics and Power Politics	415
	150.	The Constructive Programme	421
	151.	Compromise on Non-Essentials	444
	152.	Politics and Social Work	445

VI Religion 447

§ 1. *Religion, Mysticism, and Society* 447

153. Religious Reform 447
154. Religion and Culture 450
155. The Soul and Its Source 451
156. Religion at Its Highest 453
157. True Spiritual Transformation 454
158. Religion and Art 458
159. The Unfailing Test 459
160. Ever-Growing Inwardness 460
161. Religion and Social Service 460
162. Limited Thinking and Limitless Service 461
163. The Heart and the Ego 463
164. Man Without Roots 464

§ 2. *Spiritual Teachers and Religious Traditions* 465

165. The Keynote of Hinduism 465

166.	The Elasticity of Hinduism	468
167.	The Trials of Hinduism	470
168.	Golden Sayings of Islam	475
169.	The Buddhist Reformation	476
170.	Interpretations of the Koran	477
171.	Buddha and the Masses	479
172.	The Sikh *Gurus*	483
173.	The Tyranny of Custom	485
174.	Hinduism a Living Organism	487
175.	The Process of Purification	490
176.	Buddha's Great Renunciation	491
177.	Teachers of Mankind	496
178.	The Sermon on the Mount	498
179.	Buddha on True Sacrifice	502
180.	Which Jesus?	506
181.	An All-Embracing Religion	507
182.	Buddha and *Ahimsa*	507
183.	Jesus the Anointed	509
184.	Jesus and Non-Co-operation	510
185.	The Message of *Ahimsa*	511
186.	The Fearless Prophet	512
187.	Hinduism Ever Evolving	513
188.	Salvation for All	515
189.	Mystics of Islam	516
190.	Beyond Theology	516
191.	The Example of Jesus	518
192.	Jesus on the Cross	520
193.	The Kingdom of God	520
194.	Jesus an Asiatic	521
195.	The Meaning of Islam	521

§ 3. *Religious Toleration and Equality* 523

196.	Christianity and Other Religions	523
197.	Hindu-Muslim Unity	525
198.	The Meaning of Brotherhood	527
199.	Tolerance and Faith	534
200.	Love and Mutuality	535
201.	Fellowship and Toleration	536
202.	Conversion and Humility	539
203.	Change of Religion	540

204.	Equality of Religions	542
205.	The Root of All Religions	546

§ 4. *Prayer and Devotion* 547

206.	Yearning of the Heart	547
207.	Cleansing Through Prayer	549
208.	The Servant of God	550
209.	The Sacred Alliance Between God and Man	552
210.	Inward Communion	555
211.	A Glimpse of God	558
212.	A Cry of the Heart	562
213.	Potency of Prayer	563
214.	Self-Examination	564

VII God 566

215.	The Omnipresence of God	566
216.	Defining the Indefinable	569
217.	The Sum Total of *Karma*	573
218.	Service of God	575
219.	God Is Formless	577
220.	Real Worship	577
221.	Living Faith	578
222.	Belief in God	579
223.	The Why of the First Cause	580
224.	Fusion of Thought, Word, and Deed	581
225.	Nearness to God	583
226.	God Eludes All Measure	586
227.	Spiritual Sight	587
228.	Realization of God	587
229.	God as the Knower	588
230.	Acceptance of God	589
231.	The Law and the Law-Giver	589
232.	Grace Through Silence	590
233.	The Transcendent and the Immanent	591

Chronology	593
Bibliography	606
Glossary	612
Index	615

Abbreviations

CWMG	*The Collected Works of Mahatma Gandhi* (ninety volumes), Publications Division, Government of India.
CW	Archives of the Office of the Collected Works, New Delhi.
G.	Originally written or spoken in Gujarati.
GN	Gandhi Memorial Museum and Library, New Delhi.
H.	Originally written or spoken in Hindi.
Hu.	Originally written or spoken in Hindustani.
MMU	Mobile Microfilm Unit, Gandhi Smarak Nidhi and Sangrahalaya, New Delhi.
SN	Sabarmati Sangrahalaya, Ahmedabad.
SWMG	*Speeches and Writings of Mahatma Gandhi*, Natesan, Madras.

I
Introduction

Mohandas Karamchand Gandhi was too modest to be comfortable with the title of 'Mahatma', and too candid to be readily understood by his contemporaries. Throughout his life he saw himself and his ideas distorted or oversimplified by others. Patiently, he kept on affirming and amplifying his ideals so that those who cared might comprehend. Politically, he sought to touch people's hearts so as to awaken their faith both in themselves and in his abiding vision of social transformation. At the same time, he was able to sidestep those contentious pundits who prefer verbal combat to patient assimilation or courageous experimentation. Through his remarkable capacity for self-criticism, his freedom from the complex reactions of others, and his firm insistence on essentials, he nurtured an enormous strength and moral toughness. Revered as a saint and reviled as a demagogue, Gandhi made so powerful an impact that we are still not ready to assess it. If he has already suffered the fate he was anxious to avoid—being lionized at a safe distance, only to be overlooked in daily practice—he none the less left mankind a challenging, and even haunting, image of the nobility of self-conquest. This memory will grace the corridors of history for centuries to come. It will long serve to disturb the complacency, and to question the unspoken assumptions, of modern life.

Within the tangled worlds of both politics and religion Gandhi moved freely; challenging sacrosanct dogmas about the limits of the possible, he explored daringly simple alternatives. Owing to his early experience of the meretricious glamour of modern civilization, he could at once declare that its influence was insidious, and deny that it was inescapable. Rather than retreat into stoical aloofness, he lived insistently in the world to show that even an imperfect individual could strive to purify politics and exemplify true religion—thereby restoring the lost meaning of humanity. By holding out at all times for the highest potential in every person, he raised the tone and refined the quality of human interaction.

An unsuspecting reader might be rather surprised at the range of Gandhi's writings. Although he recognized the power of the written word (his collected works fill ninety large volumes), he wrote no extensive treatises, devised no definitive theories, and refused to cultivate a written style in the usual sense of that word. A remarkably pellucid thinker, he was always a man of action, a *karma yogin* devoted to the moral transfiguration of mankind. For himself he asserted, 'Action is my domain, and what I understand, according to my lights, to be my duty, and what comes my way, I do. All my action is actuated by the spirit of service.'[1]

As a thinker, Gandhi was more resilient than rigorous. Having laid down the foundations of his thought during the pioneering days of his campaigns in South Africa, he elaborated upon its diverse applications as problems arose in his eventful life. With his superb sense of occasion and his assured faith that God provides what is needed by the aspiring soul, he used the enquiries of correspondents, speaking engagements, and the demands of day-to-day business to set the pace and scope of his pronouncements. Convinced that he should never take the next step until he was ready, Gandhi preferred to lead when persuaded, without claiming any messianic mantle. He would not be prompted or pushed; instead he waited for his inner voice to show the way, and often halted large-scale movements because that voice was silent. On one such occasion, when many were clamouring for his counsel, Gandhi simply explained his reticence by saying: 'I am trying to see light out of darkness.'[2] He was unerring in perceiving opportunities without becoming an opportunist, serving as an effective leader without recourse to expediency.

Gandhi was more inclined to underrate than to overstress the significance of his written words, largely because of his deep distaste for fathering a sectarian cult. Just as he disdained the title of 'Mahatma', he also disowned the notion of anything like 'Gandhism'. Leaving *The Story of My Experiments*

[1] *Harijan*, 3 Mar. 1946.
[2] *Amrita Bazar Patrika*, 7 Nov. 1924.

With Truth to stand as his sole account of himself, he unwittingly invited readers to imagine him as an unusually honest, but self-absorbed, individual. In his pathbreaking social experiments, Gandhi saw himself as an ethical scientist conducting an incomplete laboratory study of an imperfect specimen. He was, he stressed, an ordinary man who evolved by setting himself extraordinary, seemingly impossible, standards. As he wrote in more than one place, *The Story of My Experiments With Truth* was never intended to serve as an autobiography. It originated, rather, as a series of short notes on his life, written in gaol during the twenties, and subsequently issued in book form. By themselves, these fragments portray a deeply sensitive personality, but they do not, of course, touch upon the last twenty-five years of his life. As this three-volume series testifies, a thoughtful reader can gain a more rounded perspective of Gandhi by consulting his wide-ranging correspondence, his significant speeches, and his weekly essays.

Gandhi's moral and political insights grew out of a coherent set of concepts, the nuances of which he explored over six decades. Even the claim that he was a man of action rather than of introspection could be misleading. Gandhi worked from within outwardly. Through praying each day, repeatedly consulting his 'inner voice', probing his own motives, he would reach general conclusions. Then, after carefully considering the views of others, he would decide upon a course of action. This elusive and indefinable process, which he called 'heart churning', itself arose out of his unwavering conviction that constructive thought and timely action are inseparable. If skill in action can clarify and correct thought, soul-searching deliberation can purify action. Gandhi stressed fidelity to the greater good even when it remained hidden from view, together with the perseverance that springs from trust. Maintaining such faith was for Gandhi true *bhakti*. He also demonstrated that this practice need involve neither indecisiveness nor ineptitude in worldly matters. A keen alertness to detail can, he showed, be accompanied by a cultivated disinterest in immediate results. Upon a basis of unalterable conviction, one can confidently refine thoughts and redirect action. For Gandhi, this bedrock was spiritual truth gained through intense search and deep meditation; a developed art of fundamental commitment to *satya* and *ahimsa*, a moral dedication to self-chosen vows and sacrificial action.

Gandhi did not think that all human beings are alike, but he did fervently believe that all humanity originates in the same transcendental godhead. Recognizing that he could not define that sacred source, he found in *satya* or truth its best expression. God is Truth, and Truth is God. Since every human being can know and exemplify some truth—and indeed cannot live otherwise—every human being participates in the Divine. From this conviction, one is compelled to affirm universal brotherhood while attempting to enact it through authentic tolerance, mutual respect, and ceaseless civility. If Truth is God, man, who cannot exist without some inward truth, must at some level be sincere. Each individual enjoys both the ability and the sacred obligation to grow in Truth whilst acknowledging disagreements.

Gandhi could say without exaggeration that his all-absorbing goal in life was to seek and to serve God as Truth. Longing to obtain *moksha*, spiritual freedom, he maintained that it could not be won through great learning or preaching, but only through renunciation and self-control (*tapascharya*). Self-control was to be won through action, and the course of action to which Gandhi gave his life was the service of the downtrodden. Service of humanity alone could generate the disinterested self-control essential to spiritual emancipation. Through the selfless embodiment of *ahimsa* and *satyagraha*, Gandhi believed theophilanthropists could ameliorate human misery whilst freeing themselves from worldly hopes and fears. Freedom, he felt, lies in *anasakti*, selfless service. He was certain that he could never be a votary of principles which depended for their existence upon mundane politics or external support. While even social work is impossible without politics, political work must ever be judged in terms of social and moral progress, which are in turn inseparable from spiritual regeneration.

Gandhi viewed civilization as that which assists moral excellence, moving individuals and society to truth and nonviolence. True civilization aids self-realization and nurtures universal brotherhood. Gandhi decried modern civilization because he felt that it is less an instrument for soul-growth than a supposed end in itself. Its vaunted intellectual and technological achievements deflect it from any authentic concern with moral welfare. Its 'isms' and social structures, sciences and machines, are not evil in themselves—though in a true civilization many of them would not exist—but they actively participate in the contagion of corruption that pervades it. Modern civilization is diseased in the Socratic sense because it blinds the soul and eclipses the truth. It is, as Tolstoy also thought, bondage masquerading as freedom.

Gandhi contended that the earth has enough resources to provide for human need, but not human greed. He held, therefore, that every man, woman, and child would eat adequately, clothe and shelter themselves comfortably, if there were a greater sharing of wealth in all parts of the world. Spurning equally the insatiable acquisitiveness of capitalism and the mechanistic materialism of communism, Gandhi condemned the very basis of modern civilization. In his notion of authentic civility, a sense of spiritual and social obligation is fused with a spontaneous sense of natural reciprocity. He further upheld the belief, steadily undermined since the eighteenth century, that social institutions and political actions are by no means exempt from ethics. For social institutions are, he felt, the visible expression of moral values that mould the minds of individuals. It is therefore impossible to alter institutions without first affecting those values. Since modern civilization is one complex tissue of intertwined evils, no plan of partial and gradual reform from within the system can produce a lasting remedy. Gandhi sought to destroy systems, not persons; but he argued that the 'soulless system' had to be destroyed without its reformers themselves becoming soulless.

Holding that one should repudiate wrongs without reviling wrongdoers, Gandhi could not bring himself to condemn the British for their mistakes and even their misdeeds in India. They too, he felt, were the hapless victims of a commercial civilization. The theme of *Hind Swaraj* was not just the moral inadequacy and extravagant pretensions of modern civilization, but its treacherously deceptive self-destructiveness. 'This civilization is irreligion', he concluded, 'and it has taken such a hold on the people of Europe that those who are in it appear to be half mad.'³ Yet, he added, 'it is not the British that are responsible for the misfortunes of India but we who have succumbed to modern civilization.'⁴ For Gandhi, the villain is hypocritical materialism, the judge is he who frees himself from the collective hallucination, and the executioner is the Moral Law (Karma) which inexorably readjusts equilibrium throughout the cosmos.⁵

Gandhi did not preserve his feeling for common humanity by remaining conveniently apart from it. He knew poverty and squalor at first hand; he knew too the desperate violence found in those who have lived on the edge of starvation. Yet he could still extol the Indian peasant with ringing authority:

The moment you talk to them and they begin to speak, you will find wisdom drops from their lips. Behind the crude exterior you will find a deep reservoir of spirituality . . . . In the case of the Indian villager, an age-old culture is hidden under an encrustment of crudeness. Take away the encrustation, remove his chronic poverty and his illiteracy and you have the finest specimen of what a cultured, cultivated, free citizen should be.⁶

Gandhi's longing to transform contemporary civilization was mirrored in his political thought and action. No more than civilization is politics an end in itself. Gandhi invoked Indian tradition in rejecting the modern dichotomy between religion and politics, but he went much further than most classical Indian thinkers in dispensing entirely with notions of *raison d'état* and in hoping to counter the propensity of politics to become corrupt. Even if all wished to shed their pretensions and nurture the 'enlightened anarchy' of an ideal world community, politics would be necessary since human beings differ in their perspectives, needs, and desires. Accepting, then, that politics cannot simply be abolished, Gandhi sought to purify politics by showing that its sovereign principle is neither coercive nor manipulative power, but moral and social progress.

---

[3] *Hind Swaraj*, ch. IV.
[4] Preface to second Gujarati edition of *Hind Swaraj*, May 1914.
[5] For a fuller treatment, see *The Moral and Political Thought of Mahatma Gandhi*, chs. 2 and 3.
[6] *Harijan*, 28 Jan. 1939.

Gandhi rejected collectivist theories of both State and society. He argued that only the individual could exercise conscience, and, therefore, morally legitimate power. Refusing to hold political office himself or to endorse those compatriots who did, he saw power as a by-product of social activity at the family and community level. Through *satyagraha* he sought to introduce religious values into politics by extending the rule of domestic life into the political arena. Ascribing the underlying continuity of mankind to the sacrificial exercise of soulforce within families, he was convinced that the same energies could be brought to bear self-consciously in the larger sphere of life. For the *satyagrahi*, the individual committed to Truth, the only power that can be legitimately exercised is the capacity to suffer for the errors of others and on behalf of the welfare of all—whether it be the family, the nation, or the world.

The individual is therefore always to be treated as an end in himself, while social institutions are always to be treated as corrigible means to some greater end. The *satyagrahi* should be active in politics if he can stand firmly for social justice and initiate constructive change. Where he cannot, he must practise non-co-operation. One can at least refuse to participate in evils that one cannot directly alter, even if the *satyagrahi* soon finds that he can alter more than he previously supposed. Far from denying the existence of conflicts of interest, Gandhi evolved *ahimsa* so as to resolve such conflicts by limiting, if not wholly removing, their *himsa* (violence). Gandhi further advocated voluntary poverty as an essential prerequisite for any social or political worker who wished to remain untainted by the wasteful greed of power politics. He even maintained that possessions are anti-social: it is not enough to continue possessing goods in practice under the sincere illusion that one has given them up in spirit. Possessions, he believed, should be held in trust at the disposal of those who need them. Furthermore, those who trusted the community to provide for essential needs could come to experience true freedom.

Firmly believing in the fundamental unity of life, he rejected any distinction between public and private, between secular and sacred, and ultimately, between politics and religion. Religion, for Gandhi, signifies a spiritual commitment which is total but intensely personal, and which pervades every aspect of life. Gandhi was always concerned more with religious values than with beliefs; more with the fundamental ethics that he saw as common to all religions than with formal allegiance to received dogmas which hinder, rather than aid, religious experience. He staunchly refused to associate religion with sectarianism of any kind. 'Isms', he thought, appeal only to the immature; through religion he sought nothing less than the Truth itself. In his vision, each soul resembles a drop of water from the ocean of divinity, fallen into a muddy pool. To experience consanguinity with God it must cleanse itself of the mud. Whatever its tenets, assumptions, or practices, every true religion holds out this hope of self-regeneration. All true religions are therefore equal in Gandhi's estimation. He regularly advised enquirers to discover the true meanings of the faiths they were born into under Karma. The seeker pledged to Truth must, however, abstain from proselytizing others. He should rather encourage, or inspire, others to elevate the inner and outer practice of their own faiths. Different religions and sects emerge only because no tradition and no individual can be the exclusive receptacle for boundless Truth.

Gandhi found no difficulty in accepting his own religion, while also acknowledging that he was at heart a Christian, a Jain, a Muslim, and a Buddhist. He thought that accepting the Bible did not require rejecting the Koran, just because one scripture speaks more directly to an individual than another. The *Bhagavad Gita* was Gandhi's 'spiritual dictionary',[7] but his continued recourse to it did not negate any other sacred texts. He thought that the *Bhagavad Gita* was the most accessible text in the Indian tradition. As it affirmed that God represents perfect Truth, and that imperfect man, whatever his path, can follow its precepts and come closer to God, the *Gita* has universal application. Gandhi felt that enduring help could come only from within, from what one learns through *tapascharya*.

For Gandhi, religions and religious concepts grow through human experience just as individuals mature morally, socially, and spiritually. No religion can claim to be complete in time. No formulation is final. He could thus say, without condescension, that Hinduism included Jainism and Buddhism, while

---

[7] Preface to *Gitapadarthakosha*, *Harijanbandhu*, 25 Oct. 1936.

freely criticizing Hindu sectarian disagreements and dogmatism; he praised Islamic brotherhood, while decrying the intransigence of some Muslim zealots; he upheld Christianity as a 'blazing path of *bhakti yoga*' and the Sermon on the Mount as a model, while dismissing most theology because it invidiously tends to explain away what should be taken to heart and applied. Gandhi's radical reinterpretation of Hindu values in the light of the message of the Buddha was a constructive, though belated, response to the ethical impact of the early Buddhist Reformation on decadent India.

Given such beliefs, religion is ultimately priestless, because the capacity for prayer lies latent within human nature. Prayer and all devotion (*bhakti*) are, for Gandhi, a kind of petition. The noblest and purest petition is that one should become outwardly what one is inwardly—that one's thoughts, words, and deeds should ever more fully express the soul's core of truth and non-violence. Prayer is to God as thought is to Truth, but since God and Truth are beyond all limiting conceptions, they cannot accommodate egotistic petitions. Prayer is truly an intense supplication towards one's inmost ineffable nature, the source of one's being and strength, the touchstone of one's active life. Just as politics and religion should endeavour to reduce the gap between theory and practice, so too prayer must narrow the gulf between one's real being and one's manifest appearance.

Gandhi's heartfelt reverence for all religions and for their spiritual founders and exemplars, together with his restraint in attributing to any of them uttermost divine perfection, arose from his concept of Deity. God is alien to no human being, not even the atheist who risks sundering himself from his own source. 'To deny God', Gandhi believed, 'is like committing suicide.'[8] Since the divine is reflected within every individual as his inalienable core of Truth, God will appear in as many forms and formulations as there are possibilities of human thought. There are, at least, as many definitions of God as there are individuals, and God transcends them all. Beyond the boundaries of reason and imagination, God is ineffable, indescribable, without form or characteristic. Gandhi thought that the concepts and images used to express the divine, in-

[8] *Mahadevbhaini Diary*, Vol. 1, p. 82.

cluding his own formulations, were at best derived from glimpses of immense but partial truths. As aids, these images may assist human growth; but as dogmas, they tend to breed sectarianism and violence. As aids, they may foster the universal religion of duty and detachment (*dharma* and *vairagya*); but as dogmas, they tend to reinforce a harsh insistence upon rights and privileges. For Gandhi, all conceptions of God are merely means to be used in the service of Truth.

Gandhi knew that his ideas and ideals were difficult to instantiate precisely because of their inherent simplicity. He recognized, therefore, that he could only clarify and illustrate them to all who sought his counsel. Those others would, through *tapas*, have to assimilate and apply them for themselves. But the hero and villain jostle in every soul. The morally sensitive individual must learn to detect self-deception with firmness and forbearance, mellowness and maturity. He must come to know the obscuration of light within before he can ferret out evil at its roots. Eventually, 'a man with intense spirituality may without speech or a gesture touch the hearts of millions who have never seen him and whom he has never seen.'[9] Through meditation, man can attain a noetic plane on which thought becomes the primary and most potent mode of action. Gandhi unwaveringly affirmed that living this conviction would bring sacrificial suffering, as well as an inner joy which cannot be conveyed in words.

On his seventy-eighth birthday in 1947, when well-wishers showered him with lavish and affectionate greetings, Gandhi thought only of the violence and suffering of his recently independent and hastily partitioned homeland:

I am not vain enough to think that the divine purpose can only be fulfilled through me. It is as likely as not that a fitter instrument will be used to carry it out and that I was good enough to represent a weak nation, not a strong one. May it not be that a man purer, more courageous, more far-seeing, is wanted for the final purpose? Mine must be a state of complete resignation to the Divine Will .... If I had the impertinence openly to declare my wish to live 125 years, I must have the humility, under changed circumstances, openly to shed that wish .... In that state, I invoke the aid of the

[9] *Young India*, 22 Mar. 1928.

all-embracing Power to take me away from this 'vale of tears' rather than make me a helpless witness of the butchery by man become savage, whether he dares to call himself a Mussalman or Hindu or what not. Yet I cry, 'Not my will but Thine alone shall prevail.'[10]

By upholding vows, any person, Gandhi held, can align his conduct to the motionless centre of the wheel of life. But the individual must first adopt stern measures to control the mind in its everyday vagaries, monitoring or even selecting his every thought. Only in this way can one become single-minded and so incarnate one's beliefs in one's sphere of *dharma*. Gandhi felt that conscience is kept alive not by a preoccupation with intention, but by concern for rectitude of action. He deliberately shifted emphasis from the spiritual emancipation of the individual to the collective benefit of all.

Gandhi's fundamental convictions constitute a world-view of far-reaching dimensions. They cannot be proved, for 'truth is its own proof, and non-violence is its supreme fruit.'[11] But Gandhi never doubted that if these ideals were practised with sincerity and humility, aimed not at the applause of the world, but at the support of the soul, they would gradually prove to be self-validating, helping the individual, painfully but assuredly, to mature into a joyous state of spiritual freedom and self-mastery. It is awe-inspiring, but hardly surprising, that upon receiving his assassin's bullets, Gandhi made a final gesture of forgiveness and whispered, 'Hey Ram! Hey Ram!'

Gandhi did not wish to be considered an inspired prophet. His metaphysical presuppositions only deepened his disarming faith in a human solidarity that admits of no degree. He persisted in seeing himself as a somewhat unworthy exemplar of his exacting ideals. And yet, by his lifelong fidelity to his vows, Gandhi demonstrated the liberating and transforming power of any attempt to fuse metaphysics and conduct, theory and practice, through an enormous effort of the will. A few months before the assassination, Sarojini Naidu, the poetess who had played a leading role in the Salt March, tried to capture something of the enigma of Gandhi in the context of the twentieth century:

[10] D. G. Tendulkar, *Mahatma*, Vol. 8, pp. 144–5.
[11] *Navajivan*, 11 Oct. 1925.

With Christ he shares the great gospel that love is the fulfilling of the law. With the great Muhammad he shares the gospel of brotherhood of man, equality of man and oneness of man. With Lord Buddha he shares the great evangel that the duty of life is not self-seeking but to seek the truth, no matter at what sacrifice. With the great poets of the world, he shares the ecstasy of the vision that the future of man is great, that the future of man can never be destroyed, that all sin will destroy itself, but that love and humanity must endure, grow and reach the stars. Therefore, today, a broken world ruined by wars and hatred, a broken world seeking for a new civilization honours the name of Mahatma Gandhi.

In himself, he is nothing. There are men of learning, greater than his, and there are men of wealth and power, and men of fame, but who is there that combines in one frail body the supreme qualities of virtue enshrined in him: courage indomitable, faith invincible, and compassion that embraces the entire world? This transcendental love of humanity that recognizes no limitations of race, no barriers of country but gives to all, like a shining sun, the same abundance of love, understanding and service. Every day—today and yesterday and tomorrow—every day is the same story of the miracle of Gandhi in our own age.

Who said that the age of miracles is past? How should the age of miracles be past while there is such a superb example of embodied miracle in our midst? ... He was born like other men, he will die like other men, but unlike them he will live through the beautiful gospel he has enunciated, that hatred cannot be conquered by hatred, the sword cannot be conquered by the sword, that power cannot be exploited over the weak and the fallen, that the gospel of non-violence which is the most dynamic and the most creative gospel of power in the world, is the only true foundation of a new civilization, yet to be built.[12]

[12] D. G. Tendulkar, *Mahatma*, Vol. 8, p. 144.

II

Gandhi on Himself and His Mission

§ 1. Gandhi on Himself

1. THE OCHRE ROBE

[19 January 1921]

I am always eager to meet *sadhus*. When I visited the Kumbhmela[1] in Hardwar, I tried to go into all *akhadas*[2] of *sadhus* in search of a *sadhu* who would gladden my heart. I met every *sadhu* who had acquired some reputation but I must say I was disappointed. I am convinced that *sadhus* are an ornament to India and that, if the country lives, it will be thanks to them. But I see very little of the goodness of *sadhus* today. On the last day in Hardwar, I spent the whole night thinking what I could do so that *sadhus* in the country would be real *sadhus*. Finally, I took a hard vow.[3] I shall not say what it is, but many believe that it is a difficult one to keep. By God's grace, I have still kept it inviolate.

Some friends suggested to me that I should become a *sannyasi*. However, I have not become one. My conscience did not approve of such a step then and does not do so today. I am sure you will not believe that the reason for my not doing so is love of enjoyments. I am struggling to the best of my ability to conquer the desire for them. But in the very process of struggling, I see that I am not worthy of the ochre robe. I cannot say I always practise truth, non-violence and *brahmacharya* in action, speech and thought. Whether I want or no, I feel attachments and aversions, feel disturbed by desire; I try to control them with an effort of mind and succeed in repressing their physical manifestation. If I could practise them to perfection, I would be in possession today of all the supernatural powers they speak of; humble myself, the world would be at my feet and no one would ever want to laugh me out or treat me with contempt.

But I have not come here to persuade you to give up your dress. I would have failed in my duty if, in gratitude for the straightforwardness which I have found in the Swaminarayana sect and the love with which you have invited me here, I did not tell you what I feel. I submit to you, therefore, that you should bring credit to your outward garb of *sadhus* through the virtues of *sadhus*, and thus shine in lustre yourselves and shed lustre on the Swaminarayana sect.

Speech at Meeting of *Sadhus*, Vadtal (G.)
Navajivan, 23 Jan. 1921

[1] Gandhi visited this fair in 1915 and joined a volunteer corps for the service of the pilgrims.
[2] Centres for *sadhus* of particular sects.
[3] To have only five articles of food.

2. STRIVING AFTER *MOKSHA*

November 1, 1921

It is dawn. Your letter is lying before me. Why should you apologize for what you wrote?

There may be an element of egoism in my article or conduct without my being aware of it. The word *klesha* here should be interpreted in a different way, though I cannot say how. I feel pain when I see others suffer. It is the nature of compassion that whenever one is unable to relieve the suffering of others one suffers unbearably. Logic will not avail when one is describing the condition of one's mind. I have given a graphic description of my feelings. It may be that these feelings were not quite pure. However, my state of mind at the time of writing the article had also in it a yearning for *moksha*—my aspiration for *moksha* was by no means weak then—but the fact is that I am no more than an aspirant after *moksha*. But I am not yet fit for *moksha* in this life. My *tapascharya* is not intense enough. I can control my passions no doubt, but I have not yet become completely free from them. I can control the palate, but the tongue has not yet ceased relishing good food.

He who can restrain the senses is a man of self-control; but the man whose senses have become, through constant practice, incapable of enjoying their objects has transcended self-control, has in fact attained *moksha*. I would not be tempted to give up my striving after *moksha* even for the sake of *swaraj*, but that does not mean that I have attained *moksha*. You will, therefore, notice many defects in my language. For me, even the effort for attaining *swaraj* is a part of the effort for *moksha*. Writing this to you is also a part of the same effort. If I find it a hindrance in the path of *moksha*, my pen would drop down this very moment, such is my yearning for *moksha*. Nevertheless, the mind is like a drunken monkey; mere effort is not enough for controlling it. Our actions also must be coming in our way.

In my article 'Optimism' I have suggested a rule of conduct, viz., that we should cease to have any dealings with a person who breaks a promise. This is the sign of a detached person. If even next year I find that the atmosphere in India is unfavourable and still we go on harping the same tune, it would be an outrage. In that case, I must first acquire the necessary fitness. This is why silence has been considered the best speech. Whatever I may do will be, I am sure, quite natural to me, for I will say and act upon nothing but what I believe to be the truth.

But 'Enjoy what you get today, who ever has seen tomorrow?' has been accepted as their motto alike by libertines and the self-controlled.

May the new year bring happiness to you.

Letter to Mathuradas Trikumji (G.)
Bapuni Prasadi, pp. 38–9

3. THE KINGDOM OF HEAVEN

Pundit Ghasita Ram, 'President, All-India, Sub-Assistant Surgeons Association, Punjab Province, Amritsar', sent to the Editor some days ago an 'open letter' addressed to me. After omitting from it laudatory phrases and sentences of good wishes and after correcting obvious grammatical errors, the letter reads:

I am a *Brahman*, a doctor and an old man like you. It will not be out of courtesy if I in this triple capacity offer you a piece of counsel. If you see wisdom and truth in it and if it appeals to your common sense and sentiments, kindly take it to heart.

You have seen much of the world; you have read much of it. Consequently you possess a wonderful experience of it, but in this world of mortals none till now has been able to accomplish the task he has undertaken in his lifetime. Buddha in spite of his high morality could not convert the whole of India to Buddhism.

Shankaracharya in spite of his high intellectuality could not make all India Vedantist. Christ in spite of his high spirituality could not bring into the fold of Christianity the whole Jewish nation. I do not think, and I am not prepared to believe for a single moment about the accomplishment of your task. Still in face of these historical facts, if you believe in its accomplishment in your lifetime, then Sir, I venture to say that it is nothing but a dream.

This world is a place of trials, troubles and turmoils. The more a man sinks into it the more he is restless and, eventually, he loses his spiritual calmness and peace of mind. Consequently, the *mahatmas* of olden time kept themselves aloof from worldly worries, anxieties and cares and strove to gain perfect peace and true quality of mind and enjoyed an everlasting happiness and bliss thereby.

The Jail life has brought a great change in your life and vigour and the disease has reduced you much. Therefore, in the fitness of things, you may live a calm life and spin the thread of your remaining days in a certain solitary cave in the meditation of God, and in realization of your own self in perfect spiritual tranquillity and calmness, because your health will not allow you to bear the burden of the worldly cares any more. It shall not be out of place to mention that you are absolutely convinced of the goodwill, mercy and sympathy of good officers. That very system of European medicines and surgery, which you condemned more than once, has saved you from the jaws of monstrous death. The English officers helped you in time of your troubles and needs.

'A friend in need is a friend indeed.' Now it remains for you to show your true friendship and to become the true ally of the British Raj in gratitude for the safety of your life and your release from the jail. If anyhow you cannot do so by your words and deeds, then pray do not come into the arena of political activity, but still if your restless soul does not allow you to sit in rest, then in this *bhumi*, the motherland of great sages and saints, *rishis* and *munis*, take up the task of spiritualizing your brother Indians, teach them a lesson of true realization of the self. By doing so, instead of gaining this Kingdom of Earth, you will gain the Kingdom of Heaven.

In my opinion, the writer is desperately in earnest and merits an answer on that account if for nothing else. But it enables me, too, to clear up certain misapprehensions about my mission in life.

Let me first, however, dispose of the advice about my views on medicine. I have not *Indian Home Rule*[1] before me, but I recollect sufficient to be able to say that I have nothing to revise about the views set forth there. If I had written it for English readers and in English, I would have put the same thought in a manner that would have been more acceptable to the English ear. The original is in Gujarati and was written for the Gujarati readers of *Indian Opinion* of Natal. Moreover, what is written there has reference to an ideal state. It is a common error to think that condemnation of measures involves that of men. Medicine does often benumb the soul of the patient. It may, therefore, be considered evil, but not, therefore, necessarily medicine-men. I had precious medical friends when I wrote the book and did not hesitate to seek their advice in times of need. That was, as the writer implies, inconsistent with my belief regarding the use of medicine. Several friends have said to me the same thing in so many words. I plead guilty. But that is to admit that I am not a perfect man. Unfortunately for me, I am far from being perfect. I am an humble aspirant for perfection. I know my way to it also. But knowing the way is not reaching its end.

If I was perfect, if I had acquired full control over all my passions even in thought, I should be perfect in body. I am free to confess that daily I am obliged to expend a great amount of mental energy in acquiring control over my thoughts. When I have succeeded, if I ever do, think what a storehouse of energy would be set free for service. As I hold that appendicitis was a result of infirmity of thought or mind, so do I concede that my submission to the surgical operation was an additional infirmity of mind. If I was absolutely free of egoism, I would have resigned myself to the inevitable; but I wanted to live in the present body. Complete detachment is not a mechanical process. One has to grow to it by patient toil and prayer. As for gratitude, I have more than once publicly expressed my gratitude to Col. Maddock and his staff for the kindness with which they overwhelmed me. But there is no connection between the kind treatment I received from Col. Maddock and the system of Government I condemn. Col Maddock himself would think little of me, if I revised my views about Dyerism because he, Col. Maddock, was a competent surgeon and did his duty as such. Nor have I any cause to be thankful to the Government for providing me with best surgical assistance or for prematurely releasing me. The former they were bound to provide for every prisoner. The latter has embarrassed me. I knew my course in prison whether well or ill. Outside the prison-walls, although I am slowly regaining my health, I do not know with certainty how to shape my course.

Now for the central point of the letter. The confusion in the writer's mind has arisen because of his misconception of the work of the prophets he names and of an awkward comparison between them and me. I do not know that Buddha did not accomplish his task which was to reach *Nirvana*. Tradition has it that he did. Conversion of others was a by-product, if one may so describe a sacred function. The Gospels record it of Jesus that he testified on the Cross of his own work, 'It is finished.'[2] Nor has their work of love died after them. The truest part of it will live for ever. The two or three thousand years that have gone by since their ministry are but a speck in the vast time circle.

I do not consider myself worthy to be mentioned in the same breath with the race of prophets. I am an humble seeker after truth. I am impatient to realize myself, to attain *moksha* in this very existence. My national service is part of my training for freeing my soul from the bondage of flesh. Thus considered, my service may be regarded as purely selfish. I have no desire for the perishable kingdom of earth. I am striving for the Kingdom of Heaven which is *moksha*. To attain my end it is not necessary for me to seek the shelter of a cave. I carry one about me, if I would but know it. A cave-dweller can build castles in the air, whereas a dweller in a palace like Janak has no castles to build. The cave-dweller who hovers round the world on the wings of thought has no peace. Janak, though living in the midst of 'pomp and circumstance', may have peace that passeth understanding. For me the road to salvation lies through incessant toil in the service of my country and therethrough of humanity. I want to identify myself with everything that lives. In the language of the *Gita* I want to live at peace with both friend and foe. Though, therefore, a Mussalman or a Christian or a Hindu may despise me and hate me, I want to love him and serve him even as I would love my wife or son though they hate me. So my patriotism is for me a stage in my journey to the land of eternal freedom and peace. Thus it will be seen that for me there are no politics devoid of religion. They subserve religion. Politics bereft of religion are a death-trap because they kill the soul.

'My Mission'
Young India, 3 Apr. 1924

[1] Translation of *Hind Swaraj*.
[2] John 19: 30.

4. BECOMING FEARLESS

August 21, 1924

Dear Shri Ghanshyamdas,[1]

God has provided me with conscience-keepers. You, I think, are one of them. Some of my own children, some ladies and a few grown-ups like Jamnalalji[2] and yourself want to make me a perfect man. Regarding you thus, how could I be offended by your letter? In fact, I want you always to caution me in this manner.

You complain against three things: One, my absolving the Swaraj Party of the charge of corruption; two, my giving a testimonial to Suhrawardy; and, three, my trying to get Sarojini Devi elected as Congress President.

In the first place, it is one's duty to say only that which, after a painstaking inquiry, one has come to regard as the truth, even if the world considers it to be an error. In no other way can one become fearless. I cannot consider anything dearer to me than *moksha*. Yet even that *moksha* I would renounce if it were to conflict with truth and non-violence. In all these three things I only followed truth. When I said that, I had in mind what you had told me at Juhu. In the absence of any definite proof, it becomes my duty to treat the Swaraj Party as free from the guilt imputed to it. If there is any evidence that you can furnish I shall certainly examine it. I shall even make it public if you will allow me to do so; if not I shall know and keep it to myself and remain silent.

As for Suhrawardy, I only testified to his sagacity and I am having practical experience of it even now.

About Sarojini Devi your fears are, I think, uncalled for. I hold strongly that she has served India well and is still doing so. While I have done nothing so far for her presidentship, I do believe that if the others who have so far occupied that office were fit for it, she too is fit. Everybody is charmed with her enthusiasm. I can bear witness to her courage. I have noticed nothing blameworthy in her character.

Please do not conclude from this that I approve of everything done by her or anyone else.

God has filled this world with objects, animate and inanimate, good and bad. The wise man considers only the good and ignores the bad, even as the swan leaves the water and takes only the milk from a mixture of milk and water.[3]

Yours,
MOHANDAS GANDHI

Letter to G. D. Birla (H.)
CW 6030

[1] G. D. Birla (1894-1983); industrialist and mill owner; Chairman, Harijan Sevak Sangh.

[2] Jamnalal Bajaj (1889-1942); social worker and philanthropist; Treasurer of Indian National Congress for many years.

[3] Tulsidas, *Ramacharitamanasa*, 'Balakanda'.

5. TRYING TO SEE LIGHT

November 2, 1924

I have no message to give. What shall I say? I am thinking. I am trying to see light out of darkness.

Message to *Bengalee*[1]
Amrita Bazar Patrika, 7 Nov. 1924

[1] In reply to a telegram from Bipin Chandra Pal.

6. ONE STEP ENOUGH FOR ME

[21 December 1925]

Ten years have passed since I left South Africa. I have received hundreds of letters, and have replied to them. I have explained this matter a hundred times over in *Young India* and in *Navajivan*, and yet, when I come to the Wardha Ashram I am being asked the very same questions. This has revived old memories for me and distressed me very much. I do not say that such questions should not occur to anyone, but, if they do, people may approach Vinoba and have their doubts answered. However, the reason for my being distressed was that it has become a widespread disease to ask such questions. We should resist the temptation to ask them. Please understand my words properly. What I wish to say is that such questions may certainly occur to us, but, they should be kept back in one's mind.

When, thousands of years ago, the battle of Kurukshetra was fought, the doubts which occurred to Arjuna were answered by Shri Krishna in the *Gita*; but that battle of Kurukshetra is going on, will go on, for ever within us, the Prince of Yogis, Lord Krishna, the universal *atman* dwelling in the hearts of us all, will always be there to guide Arjuna, the human soul, and our Godward impulses represented by the Pandavas will always triumph over the demoniac impulses represented by the Kauravas. Till, however, that victory is won, we should have faith and let the battle go on, and be patient meanwhile. This does not mean that we should suppress our inner urge for fear of anyone; it means that if such an urge takes the form of the question 'Who created God?', we should curb it, tell ourselves that it is impious to ask such a question and have faith that the question will answer itself by and by.

This physical frame which God has given us is a prison, but it is also the door leading to deliverance and, if we wish that it should serve only that purpose, we should understand its limitations. We may well desire to clutch the stars in the heavens, but we should note that it is beyond our power to do so; for our soul is imprisoned in a cage, its wings, therefore, have been clipped and it cannot fly as high as it would. It can secure a great many occult powers, but it will fail in its aim of winning deliverance if it goes after such powers. Hence, the kind of abstract questions which were put to me the other day should be avoided—in the conviction that in the course of time the soul will become strong enough and know the answers to them.

Instead of discussing such abstract questions, we should follow the advice of the poet: 'Let us spend today to some purpose, for who knows what tomorrow will bring?' This line may seem to come from the pen of Charvak, who also says: 'Live in ease while you live, drink ghee even if you have to borrow money for it, for the body will never return to life after it is cremated.' But the line is not by Charvak. Its author was a devotee and, when he advised us to spend today profitably, he meant that we should discharge the duty which lies before us today. We do not know if we shall be alive tomorrow, though a little later he says that we shall be born again. This duty is what was explained by Vinoba the other day, 'ending the misery of all creatures that suffer', destroying the chain of ever-recurring birth and death. The only means for this is *bhakti*. An Englishman named Newman, a great devotee, wrote in a poem of his 'One step enough for me.'

This half line is the quintessence of all philosophy. That one step means patient, unswerving *bhakti*. If a sick person gets up and tries to walk down a staircase, he would feel giddy and fall. If we do not understand our limitations and try to get knowledge which is beyond us, we would not only not be able to digest it but would be sick with surfeit.

We should, therefore, cure ourselves of the disease of asking abstract questions, should attend to the immediate duty before us today and leave these questions for some other day. The couplet from a *bhajan* which was sung here today teaches us the very same thing, that instead of talking about *mukti* all the time we should spend our time in *bhakti*. Without *bhakti* there can be no deliverance. Only he, therefore, wins deliverance who is devoted to duty and fills his heart with love of God— he alone wins deliverance who never thinks about it.

Bhakti, moreover, does not imply ineptitude in practical affairs. That which produces such ineptitude cannot be called *bhakti*. It may, of course, be that, looking at the way we conduct our affairs, people will think of us as simpletons. A true devotee, though fully attentive to practical affairs, brings the spirit of *bhakti* into them. His conduct will always be in harmony with *dharma*. It is because Krishna acted in this manner that he is looked upon as the *Purnavatara*. A devotee finds no difficulty in attending to the practical affairs of life.

Ashrams like this one are established so that such a way of life in complete harmony with *dharma* may prevail everywhere. I have, therefore, always cherished the hope that these Ashrams will serve as instruments for raising the country and teaching and spreading true *dharma*. I do not worry whether that hope will be fulfilled in the present or after many generations—it is sufficient for us that we go on doing our duty along the path we have chalked out for ourselves. For this, we should strive to cultivate the qualities of both a *Brahmin*—truth and faith—and a *Kshatriya*—strength and non-violence. It is my faith that this Ashram will help its inmates to cultivate both these types of qualities. I do not suggest, of course, that other Ashrams cannot do that. I believe that this Ashram, at any rate, will do some good.

If we realize that truth and non-violence have a particular value for us and practise them in our lives, if we have the faith that there is no principle in this world which admits of an exception, we shall in the course of time understand the meaning of perfect truth and perfect non-violence. The peace which I have enjoyed here during the past ten days on observing that the inmates of the Ashram perform their duty in the spirit I have explained, I have enjoyed nowhere else, and you can very well imagine what my feelings must be now that I shall have to leave this peaceful atmosphere and go back to a world full of turmoil. But, as I told a friend, our study of the *Gita* would have been to no purpose if we get frightened of the turmoil in the world; we should get our peace not from the external environment, but from within us, and so I do not worry.

Speech at Wardha Ashram (G.)
Navajivan, 27 Dec. 1925

7. VOLUNTARY RETIREMENT

When I decided on voluntary retirement from active work, I excluded, whether out of ignorant attachment or fear, from the scope of my decision, though I personally wished it, a visit to Ahmedabad. If I make an exception in favour of it, I fear I might be prevented from doing the kind of service I wish to do by staying in the Ashram for one year. I encountered this danger only last week. The Ramakrishna Mission was celebrating its foundation day. I was invited to preside over the function. Since I had decided now to live in the Ashram all the time, how could I decline the invitation? If, on the other hand, I attended this function, why should I not attend many other functions on similar auspicious occasions which might be arranged in Ahmedabad? If I attended them, my purpose in retiring from active work in order to be at peace with myself would be defeated. Should Dr. Hariprasad ask me to give one day to every street in Ahmedabad and sweep it, I would certainly count that as fit work for me. If I undertook to do it, every day in the year would be occupied and then I would be where I was.

The friends who had come to invite me were convinced by this argument and gave me freedom. I expect similar consideration from every worker in the city. As the rest of the country is to forget me up to December 20, Ahmedabad should do so too. If permitted by Vallabhbhai, I want to be bold and include Ahmedabad too in the scope of my vow, so that I might have no temptation and no need to argue with anyone. Even if, however, Vallabhbhai cannot give me such freedom, I should like the citizens of Ahmedabad to spare me and not invite me to any function.

As I study the various activities of the Ashram and the work of the Spinning Association, I realize that, if I want to do complete justice to the Ashram, the Spinning Association, *Young India* and *Navajivan*, I shall have no time to spare for any other activity. If I can attend to this quiet work for a year, I am confident that my capacity for service will increase. I appeal to workers in Ahmedabad to understand my position and spare me during this year the necessity of having even to go to the city even on public work.

P. S. After I had written the note above I had a discussion with Vallabhbhai and he gave his consent to my including Ahmedabad, too, in my vow. He also believes that if I really wished to have peace, my retirement from public life should mean retirement to the Ashram. I cannot, therefore, leave the Ashram to attend any function or participate in any activity outside it, even in Ahmedabad. If some unforeseen contingency arises and if it becomes necessary for me to leave Ahmedabad and go somewhere else, for the sake of my health, these would certainly be treated as exceptional circumstances.

'Spare Me' (G.)
Navajivan, 10 Jan. 1926

8. THE ROYAL ROAD

Ashram, Sabarmati,
February 12, 1926

Dear Friend,

I have been receiving your letters regularly. Please do not think that you are unworthy to be my disciple. I regard myself as too imperfect to have any disciples. Do not for one moment think that those who are living with me at the Ashram I regard as my disciples. They are all co-workers with me. I am in the position of an elder to them. And I am an elder because I may be considered to be more experienced than they are and my experience is at their disposal equally with theirs. There is no secret either about the royal road I told you of. The royal road is the doing of one's appointed duty to the best of one's ability and the dedication of all service to God. Work done in this fashion always clears difficulties in front of us and shows us also whenever we err. You should certainly continue the union amongst your friends of the little circle you have mentioned and my advice will be always at your disposal.

I hope you are at peace with yourself and your neighbours and in the enjoyment of good health.

Yours sincerely,

Madame A. Mirbel
100, Rue Brale Maison
Lille
(France)

Letter to Antoinette Mirbel
SN 14096

9. FREEDOM FROM DOGMA

An American friend who subscribes herself as a lifelong friend of India writes:

As Hinduism is one of the prominent religions of the East, and as you have made a study of Christianity and Hinduism, and on the basis of that study have announced that you are a Hindu, I beg leave to ask of you if you will do me the favour to give me your reasons for that choice. Hindus and Christians alike realize that man's chief need is to know God and to worship Him in spirit and in truth. Believing that Christ was a revelation of God, Christians of America have sent to India thousands of their sons and daughters to tell the people of India about Christ. Will you in return kindly give us your interpretation of Hinduism and make a comparison of Hinduism with the teachings of Christ? I will be deeply grateful for this favour.

I have ventured at several missionary meetings to tell English and American missionaries that if they could have refrained from 'telling' India about Christ and had merely lived the life enjoined upon them by the Sermon on the Mount, India instead of suspecting them would have appreciated their living in the midst of her children and directly profited by their presence. Holding this view, I can 'tell' American friends nothing about Hinduism by way of 'return'. I do not believe in people telling others of their faith, especially with a view to conversion. Faith does not admit of telling. It has to be lived and then it becomes self-propagating.

Nor do I consider myself fit to interpret Hinduism except through my own life. And if I may not interpret Hinduism through my written word, I may not compare it with Christianity. The only thing it is possible for me therefore to do is to say, as briefly as I can, why I am a Hindu.

Believing as I do in the influence of heredity, being born in a Hindu family, I have remained a Hindu. I should reject it, if I found it inconsistent with my moral sense or my spiritual growth. On examination, I have found it to be the most tolerant of all religions known to me. Its freedom from dogma makes a forcible appeal to me inasmuch as it gives the votary the largest scope for self-expression. Not being an exclusive religion, it enables the followers of that faith not merely to respect all the other religions, but it also enables them to admire and assimilate whatever may be good in the other faiths. Non-violence is common to all religions, but it has found the highest expression and application in Hinduism. (I do not regard Jainism or Buddhism as separate from Hinduism.)

Hinduism believes in the oneness not of merely all human life but in the oneness of all that lives. Its worship of the cow is, in my opinion, its unique contribution to the evolution of humanitarianism. It is a practical application of the belief in the oneness and, therefore, sacredness of all life. The great belief in transmigration is a direct consequence of that belief. Finally the discovery of the law of *varnashrama* is a magnificent result of the ceaseless search for truth. I must not burden this article with definitions of the essentials sketched here, except to say that the present ideas of cow-worship and *varnashrama* are a caricature of what in my opinion the originals are. The curious may see the definitions of cow-worship and *varnashrama* in the previous numbers of *Young India*. I hope to have to say on *varnashrama* in the near future. In this all-too-brief a sketch I have mentioned what occur to me to be the outstanding features of Hinduism that keep me in its fold.

'Why I am a Hindu'
Young India, 20 Oct. 1927

10. THE SPIRIT, NOT THE LETTER

The Ashram,
Sabarmati,
January 18, 1928

Dear Friend,

I had your letter. I have kept some of your letters in my *Young India* file yet for use.

About the proposed *smriti*, I cannot yet see eye to eye with you. You often seem to emphasize the letter rather than the spirit. When I use the word 'inspired', I do not give it a technical meaning. When I feel 'inspired', you will find that nothing would deter me from giving a new *smriti* to Hinduism, and let me secretly tell you that I am aiming at such inspiration. Till then I must wait.

It gave me much pleasure to be able to see you face to face in Madras.

Yours sincerely,

Sjt. S. D. Nadkarni

Letter to S. D. Nadkarni
SN 13043

11. 'SACRED COW AND FEROCIOUS TIGER'

Satyagraha Ashram,
Sabarmati,
March 12, 1928

Dear Friend,

I was delighted to receive your long letter. I endorse every word of what you say about Mrs. Gandhi and the wretched incident I have related in the autobiographical chapters. Of course you have not imagined that I am in any way proud of recalling the brutality or that I am today capable of any such brutality. But I thought that if people recognize me as a gentle peace-loving man, they should also know that at one time I could be a positive beast even though at the same time I claimed to be a loving husband. It was not without good cause that a friend once described me as a combination of sacred cow and ferocious tiger.

It would have been a pity if you had burnt your beautiful letter as at one time you thought you should. You have certainly not appeared to me to be rude or ill-mannered but most natural and on that account lovable. I do indeed wish that I had come in closer contact with your dear brother, but I knew him enough to love him and to appreciate his sterling worth.

Yours sincerely,

Miss Jane Howard
'Rosemary'
50 Pandora Road
Malvern
Johannesburg
(Transvaal, S. Africa)

Letter to Jane Howard
SN 11967

12. TRUTH FROM WHATEVER SOURCE

The Ashram,
Sabarmati,
March 28, 1928

My dear C. R.,

I have your letter about the proposed European visit. I have myself no heart in it, nor have I any confidence in myself about making it successful; but an interview with Rolland still remains an attraction. All the reputation I enjoy in the West is borrowed from him and I feel that if I meet him face to face, there may be disillusionment on many points. It may be that we should come closer than we ever were. I do attach considerable importance to our knowing each other much better than we do.

I quite agree with you that there is nothing to gain from the health point of view. I might possibly suffer, and health is no consideration whatsoever in the proposed trip. From that point of view any hill station in India would be infinitely superior for me.

I feel also with you that the withdrawal of my presence is likely to unsettle things a bit especially in Bardoli. Foreign cloth boycott can certainly make no headway during my absence. But now that you are all gathering together at Calcutta, I would like you to discuss the proposed visit at the Council meeting. I am most anxious that I should not become exclusive and should be humble enough to arrive at truth no matter from what source it comes.

I am sorry about the defalcations, but I shall accept your warning not to disturb myself or discuss them.

I understand what you say about Ramachandran. I want you to write him a warm letter and go out of your way to draw him towards you. He is a kind of 'Chetty' also, for he did wonderfully well in the way of *khadi* at Jamia.

I must not forget one thing, though, about your reference to the defalcations. If the defaulter gives you Rs. 500 and tenders an apology for publication, you should be entirely satisfied. But this is an unconsidered opinion of a layman.

What do you say to my exploit in conducting an exclusively milk experiment? I do not want to be told you swooned at my saying it is a literally milk-and-water experiment.

Yours sincerely,

Letter to C. Rajagopalachari
SN 13123

13. MEETING KINDRED SPIRITS

It is not without deep sorrow that I am now able to announce that the much-talked-of visit of mine to Europe is not to come off this year at any rate. To those in Austria, Holland, England, Scotland, Denmark, Sweden, Germany and Russia who had sent me kind invitations I can only say that their disappointment will be no greater than mine.

Somehow or other I dread a visit to Europe and America. Not that I distrust the peoples of these great Continents any more than I distrust my own, but I distrust myself. I have no desire to go to the West in search of health or for sightseeing. I have no desire to deliver public speeches. I detest being lionized. I wonder if I shall ever again have the health to stand the awful strain of public speaking and public demonstrations. If God ever sent me to the West, I should go there to penetrate the hearts of the masses, to have quiet talks with the youth of the West and have the privilege of meeting kindred spirits—lovers of peace at any price save that of Truth.

But I feel that I have as yet no message to deliver personally to the West. I believe my message to be universal but as yet I feel that I can best deliver it through my work in my own country. If I can show visible success in India, the delivery of the message becomes complete. If I came to the conclusion that India had no use for my message, I should not care to go elsewhere in search of listeners even though I still retained faith in it. If, therefore, I ventured out of India, I should do so because I have faith, though I cannot demonstrate it to the satisfaction of all, that the message is being surely received by India be it ever so slowly.

Thus whilst I was hesitatingly carrying on the correspondence with friends who had invited me, I saw that there was need for me to go to Europe, if only to see M. Romain Rolland. Owing to my distrust of myself over a general visit, I wanted to make my visit to that wise man of the West the primary cause of my journey to Europe. I therefore referred my difficulty to him and asked him in the frankest manner possible whether he would let me make my desire to meet him the primary cause of my visit to Europe. In reply I have a noble letter from him through Mirabai[1] wherein he says that in the name of truth itself, he will not think of letting me go to Europe if a visit to him is to be the primary cause. He will not let me interrupt my labours here for the sake of our meeting. I read in his letter no false humility. I read in it a most genuine expression of truth. He knew when he wrote his reply that my desire to go to Europe to meet him was not for a mere courteous discussion but in the interest of the cause as dear to him as to me. But evidently he was too humble to bear the burden of calling me merely so that in furtherance of the common interest we might by mutual talks understand each other better. And I wanted him to shoulder that very burden, if he felt that truth required us to meet each other face to face. His reply therefore I have taken as a clear answer to my prayer. Apart from this visit, I felt within me no imperative call.

I have taken the public into my confidence as, against my wish, the fact that a visit to Europe during this season was under serious contemplation was published in the papers. I regret my decision but it seems to be the correct one. For whilst there is no urge within to go to Europe, there is an incessant call within for so much to do here. And now the death of my best comrade seems to keep me rooted to the Ashram.

But I may say to the many friends in Europe, that next year, if all is well and if they still will have me I shall try to undertake the postponed tour, under the strict limitations mentioned by me and this I shall do whether I am ready to deliver my message or not. To see my numerous friends face to face will be no small privilege. But let me conclude this personal explanation by saying that if ever I am privileged to visit the West, I shall go there without changing my dress or habits, save in so far as the climate may require a change and self-imposed restrictions may permit. My outward form is I hope an expression of the inward.

'To European Friends'
Young India, 26 Apr. 1928

[1] Miss Madeleine Slade (1892-1982).

14. NO FALSE MODESTY

The Ashram,
Sabarmati,
May 1, 1928

Dear Dr. Bidhan,

Your letter flatters me, but I must not succumb to my pride. Apart from the fact that as a non-co-operator I may have nothing to do with the University that is in any way connected with Government, I do not consider myself to be a fit and proper person to deliver Kamala lectures. I do not possess the literary attainment which Sir Ashutosh undoubtedly contemplated for the lecturers.

You are asking me to shoulder a responsibility which my shoulders cannot bear. I am keeping fairly fit. I am biding my time and you will find me leading the country in the field of politics when the country is ready. I have no false modesty about me. I am undoubtedly a politician in my own way, and I have a scheme for the country's freedom. But my time is not yet and may never come to me in this life. If it does not, I shall not shed a single tear. We are all in the hands of God. I therefore await His guidance.

Yours sincerely,

Letter to Dr B. C. Roy
SN 13210a

15. NO MIRACULOUS POWERS

Satyagraha Ashram, Sabarmati,
July 13, 1928

Dear Friend,

I have your letter. I don't know how the story about miraculous powers possessed by me has got abroad. I can only tell you that I am but an ordinary mortal susceptible to the same weakness, influences and the rest as every other human being and that I possess no extraordinary powers.

Yours sincerely,

Miss Barbara Bauer
Big Spring, Texas, U.S.A.

Letter to Barbara Bauer
SN 14349

16. MY SONS

Satyagraha Ashram, Sabarmati,
August 11, 1928

My dear Olive,

I was thankful to receive your letter giving me detailed news about yourself and your brave and wonderful work. I was thankful too to hear about Clement and Comber.

You want to know something about my boys. Harilal, the eldest, has become a rebel. He even drinks and makes himself merry and is honestly of opinion that I have gone astray in all I am doing. Manilal is in Phoenix keeping up *Indian Opinion*. He was married two years ago and took his wife with him. They are both happy. Ramdas and Devdas are with me helping me in my work. Ramdas was married a year ago. Devdas is still unmarried. I am conducting here a fairly big institution. The enclosed will give you its constitution and its composition.

When you write to the different members of the family please send my love to them all and accept it for yourself.

Yours sincerely,
M. K. GANDHI

Miss O. C. Doke
Kafulafuta, P. O. Naola, N. W. Rhodesia (South Africa)

Letter to Olive Doke
CW 9226

17. THE CALL TO LEAD

[1 November 1928]

I could still lead India. I shall only lead India when the nation comes to me to be led, when there is a national call.

I shall not go before then. I shall not go unless I am certain of my power over the masses. I could not lead India again until I realized that they are numerous enough to pursue a policy of non-violence, nor until I could control them. But I see nothing on the horizon at the moment. That would not make me at all anxious to take that position. Perhaps it will not be in my lifetime. It may be in the time of my successor.

I cannot name one at this moment. There must be one who could lead India today but I cannot name him. Truly I should be ashamed to remain inactive but it may be necessary in my lifetime. It may be there will come a man, but not now.

Interview with *Civil and Military Gazette*
Hindustan Times, 3 Nov. 1928

18. OUT OF MY ASHES

[Faizpur,
26 December 1936]

What new message can I give you at the age of 68? And where is the use of my giving you a message if you pass a resolution there of assassinating me or burning my effigy? Assassinating the body of course does not matter, for out of my ashes a thousand Gandhis will arise. But what if you assassinate or burn the principles I have lived for?

'Message to Students'
Harijan, 16 Jan. 1937

19. THE ONE-SIDED AND THE ALL-SIDED

Kohat,
[22/23 October 1938]

Chi. Brajkrishna,

I feel on reading your letter that this time at any rate you have to stay in Delhi. All the tasks suggested are a must for you.

I shall write to the Meerut people.

You may do whatever seems proper regarding S. Shall I write to F.?

I certainly gave you permission to live with me but take it that this desire is born of attachment. It would not do simply to assert that Ramana Maharshi and Aurobindo are one-sided while I am all-sided. One who is one-sided but understands his mission and pursues it has merit. One who claims to be all-sided but is only experimenting has even less worth than broken almond shells. Only God knows where I stand. I am an aspirant while they are known to be, and perhaps are, realized souls. Anyway their followers attribute to them full self-realization.

Blessings from
BAPU

Letter to Brajkrishna Chandiwala (H.)
GN 2459

20. THINKER AND DOER

Sevagram,
March 12, 1945

Chi. Chhaganlal,

If you can keep Purushottam, please do keep him. He has written a letter to Bapa. See it. I understand about his complaint against you. What can be done? It is for you to impress people by your work. Engaging yourself in all the activities will no doubt impress but will not the work suffer? Isn't that my lot? How have I managed so far? I am a thinker as much as I am a doer and can derive a kind of satisfaction by expressing original views. Many others are not able to do that. I shall be in Bombay in April. It will be good if Rama meets me there then.

Blessings from
BAPU

Chhaganlal Joshi
Harijan Seva Sangh
Rajkot

Letter to Chhaganlal Joshi (G.)
Pyarelal Papers

21. WORDS AND MEANINGS

Mahabaleshwar,
May 31, 1945

Chi. Kishorelal,

You work wonders. I have gone through your Preface or whatever you call it. It is all right. However, I do not intend to write in that way. I will not involve myself and my readers in the controversy. I will try to give the key for reading my writings. You have written on the basis of my writings, so it would perhaps be better if it is published as it is, under your name. I would, however, know about it only after I finish writing.

I understand about *paricharya*. It is like this. I have already written—haven't I?—that words like men need growth. As knowledge grows, the meaning of a word becomes wider, which it should. Why should we stick to the meanings of the critics? Even then, what you say seems correct from the point of view of language. The pity is that I am not a linguist and therefore whatever occurred to me on the spur of the moment, I wrote down. Enough, as it is time to go for a walk.

Blessings from
BAPU

Letter to Kishorelal G. Mashruwala (G.)
Pyarelal Papers

22. MY LIFE IS MY MESSAGE

[Mahabaleshwar,
On or before 30 May 1945]

Q. Gandhiji, is there any special message you would care to send to the Negro people of America?

A. My life is its own message. If it is not, then nothing I can now write will fulfil the purpose.

When asked to comment on the probable trend of the race relations, Mr. Gandhi said:

My faith burns brighter today, even brighter than it has in

the past; we are fast approaching a solution to troublesome race problems.

This he feels will be accomplished in spite of present-day discouraging symptoms. And he still feels that the best weapon for use by under-privileged peoples is non-violence.

Pointing to his recent statement made at the beginning of the San Francisco Conference, he indicated that India's freedom was closely identified with the welfare of all other under-privileged peoples. At that time he had said: 'The freedom of India will demonstrate to all exploited races of the earth that their freedom is very near and that in no case will they be exploited.'

Interview with Denton J. Brooks[1]
The Hindu, 15 June 1945

[1] Far Eastern correspondent of *Chicago Defender*, in which the interview appeared on 10 June 1945. Brooks reported: ' ... in the exclusive interview given to me last week ... Gandhiji was observing silence, with the exception of an hour after evening prayers.... I asked questions and he hurriedly jotted down his answers.'

23. WORKING SILENTLY

Nature Cure Clinic,
5 Todiwala Road, Poona,
October 28, 1945

Dear Madam,

I have just received your letter of 20th September, and I have devoured almost every line of the enclosure to your letter. Miss Slade whom we know here as Mirabai only, as she desires to be known, is in a basin of the Himalayas which she adores and loves. It is near Hardwar, a celebrated place of pilgrimage, through which the mighty Ganges flows down.

Your Green Cross scheme appeals to me forcibly, not that there is anything new in it for me. Your resolution is also brief and to the point, and therefore is a temptation and invitation for me to sign. But I must resist the temptation. The Green Cross Society will, I hope, forgive me for the resistance. It will do so readily if it appreciates the fact that some, among whom I count myself, render greater help by refraining from signing anything like your resolution but working silently and probably effectively.

Though I refrain from sending you my signature I will ask you, if you at all can, to keep me in touch with your activities from time to time. It may interest and even please you to know that I have enforced in my own life now for years your 'ten don'ts' and invited my neighbours to do likewise for I have long believed that there is a 'spirit in the wood', using the word 'wood' in a double sense.

Yours sincerely,

Mrs. M. H. Morrison
Hon. Secretary
The Green Cross Society
41 Asmuns Place, London N. W. 11

Letter to Mrs M. H. Morrison
Pyarelal Papers

24. ACTION IS MY DOMAIN

A friend suggests that I should resume writing my autobiography from the point where I left off and, further, that I should write a treatise on the science of *ahimsa*.

I never really wrote an autobiography. What I did write was a series of articles narrating my experiments with truth which were later published in book form. More than twenty years have elapsed since then. What I have done or pondered during this interval has not been recorded in chronological order. I would love to do so but have I the leisure? I have resumed the publication of *Harijan* in the present trying times as a matter of duty. It is with difficulty that I can cope with this work. How can I find time to bring the remainder of my experiments with truth up to date? But if it is God's will that I should write them, He will surely make my way clear.

To write a treatise on the science of *ahimsa* is beyond my powers. I am not built for academic writings. Action is my domain, and what I understand, according to my lights, to be my duty, and what comes my way, I do. All my action is actuated by the spirit of service. Let anyone who can systematize *ahimsa* into a science do so, if indeed it lends itself to such treatment. In the event of my inability, the correspondent has suggested three names in order of preference for this task: Shri Vinoba, Shri Kishorelal Mashruwala, Shri Kaka Kalelkar. The first named could do it, but I know he will not. Every hour of his is scheduled for his work and he would regard it as sacrilege to take a single moment therefrom for writing a *shastra*. I would agree with him. The world does not hunger for *shastras*. What it craves, and will always crave, is sincere action. He who can appease this hunger will not occupy his time in elaborating a *shastra*.

Shri Kishorelal has already written an independent treatise. If his health permits, I know he would like to write further. It may not be correct to call his work a *shastra*, but it may be said to be very near to one. In his present state of health, however, I do not think he can shoulder the burden, and I would be the last person to lay it on him. Like Shri Vinoba he too does not allow a moment of his time to be wasted. Much of it is given to help solve the personal problems of large circle of friends. The end of the day leaves him utterly exhausted.

Shri Kakasaheb, like Shri Thakkar, is an incorrigible nomad. Just now he has made the propagation and development of the national and provincial languages his special concern. Even if he wanted to divert a moment of his time to the writing of a *shastra*, I would try to prevent him from doing so.

From the above it may be concluded that there is no need at present for the treatise in question. Any such during my lifetime would necessarily be incomplete. If at all, it could only be written after my death. And even so let me give the warning that it would fail to give a complete exposition of *ahimsa*. No man has ever been able to describe God fully. The same holds true of *ahimsa*. I can give no guarantee that I will do or believe tomorrow what I do or hold to be true today. God alone is omniscient. Man in the flesh is essentially imperfect. He may be described as being made in the image of God, but he is far from being God. God is invisible, beyond the reach of the human eye. All that we can do, therefore, is to try to understand the words and actions of those whom we regard as men of God. Let them soak into our being and let us endeavour to translate them into action, but only so far as they appeal to the heart. Could any scientific treatise do more for us?

'Two Requests' (G.)
Harijan, 3 Mar. 1946

§ 2. Saint or Politician?

25. SAINT OR POLITICIAN?

A kind friend has sent me the following cutting from the April number of *East and West*:

Mr. Gandhi has the reputation of a saint but it seems that the politician in him often dominates his decisions. He has been making great use of *hartals* and there can be no gainsaying that under his direction *hartal* is becoming a powerful political weapon for uniting the educated and the uneducated on a single question of the day. The *hartal* is not without its disadvantages. It is teaching direct action, and direct action, however potent, does not work for unity. Is Mr. Gandhi quite sure that he is serving the highest behests of *ahimsa*, harmlessness? His proposal to commemorate the shooting at Jallianwala Bagh is not likely to promote concord. It is a tragic incident into which our Government was betrayed, but is the memory of its bitterness worth retaining? Can we not commemorate the event by raising a temple of peace, to help the widows and orphans to bless the souls of those who died without knowing why? The world is full of politicians and pettifoggers who, in the name of patriotism, poison the inner sweetness of man and, as a result, we have wars and feuds and such shameless slaughter as turned Jallianwala Bagh into a shambles. Shall we not now try for a larger symbiosis such as Buddha and Christ preached, and bring the world to breathe and prosper together? Mr. Gandhi seemed destined to be the apostle of such a movement, but circumstances are forcing him to seek the way of raising resistances and group unities. He may yet take up the larger mission of uniting the world.

I have given the whole of the quotation. As a rule I do not notice criticism of me or my methods except when thereby I acknowledge a mistake or enforce still further the principles criticized. I have a double reason for noticing the extract. For, not only do I hope further to elucidate the principles I hold dear, but I want to show my regard for the author of the criticism whom I know and whom I have admired for many years for the singular beauty of his character. The critic regrets to see in me a politician, whereas he expected me to be a saint. Now I think that the word 'saint' should be ruled out of present life. It is too sacred a word to be lightly applied to anybody, much less to one like myself who claims only to be a humble searcher after truth, knows his limitations, makes mistakes, never hesitates to admit them when he makes them, and frankly confesses that he, like a scientist, is making experiments about some of 'the eternal verities' of life, but cannot even claim to be a scientist because he can show no tangible proof of scientific accuracy in his methods or such tangible results of his experiments as modern science demands. But though by disclaiming sainthood I disappoint the critic's expectations, I would have him to give up his regrets by answering him that the politician in me has never dominated a single decision of mine, and if I seem to take part in politics, it is only because politics encircle us today like the coil of a snake from which one cannot get out, no matter how much one tries. I wish therefore to wrestle with the snake, as I have been doing, with more or less success, consciously since 1894, unconsciously, as I have now discovered, ever since reaching the years of discretion.

Quite selfishly, as I wish to live in peace in the midst of a bellowing storm howling round me, I have been experimenting with myself and my friends by introducing religion into politics. Let me explain what I mean by religion. It is not the Hindu religion, which I certainly prize above all other religions, but the religion which transcends Hinduism, which changes one's very nature, which binds one indissolubly to the truth within and which ever purifies. It is the permanent element in human nature which counts no cost too great in order to find full expression and which leaves the soul utterly restless until it has found itself, known its Maker and appreciated the true correspondence between the Maker and itself.

It was in that religious spirit that I came upon *hartal*. I wanted to show that it is not a knowledge of letters that would give India consciousness of herself, or that would bind the educated together. The *hartal* illuminated the whole of India as if by magic on the 6th of April, 1919. And had it not been for the interruption of the 10th of April, brought about by Satan whispering fear into the ears of a Government conscious of its own wrong and inciting to anger a people that were prepared for it by utter distrust of the Government, India would have risen to an unimaginable height. The *hartal* had not only been taken up by the great masses of people in a truly religious spirit but it was intended to be a prelude to a series of direct actions.

But my critic deplores direct action. For, he says, 'it does not work for unity.' I join issue with him. Never has anything been done on this earth without direct action. I rejected the word 'passive resistance', because of its insufficiency and its being interpreted as a weapon of the weak. It was direct action in South Africa which told and told so effectively that it converted General Smuts to sanity. He was in 1906 the most relentless opponent of Indian aspirations. In 1914 he took pride in doing tardy justice by removing from the statute-book of the Union a disgraceful measure which in 1909 he had told Lord Morley would be never removed, for he then said South Africa would never tolerate repeal of a measure which was twice passed by the Transvaal Legislature. But what is more, direct action sustained for eight years left behind it not only no bitterness, but the very Indians who put up such a stubborn fight against General Smuts, ranged themselves round his banner in 1915 and fought under him in East Africa. It was direct action in Champaran which removed an age-long grievance.

A meek submission when one is chafing under a disability or a grievance which one would gladly see removed, not only does not make for unity, but makes the weak party acid, angry and prepares him for an opportunity to explode. By allying myself with the weak party, by teaching him direct, firm, but harmless action, I make him feel strong and capable of defying the physical might. He feels braced for the struggle, regains confidence in himself and knowing that the remedy lies with himself, ceases to harbour the spirit of revenge and learns to be satisfied with a redress of the wrong he is seeking to remedy.

It is working along the same lines that I have ventured to suggest a memorial about Jallianwala Bagh. The writer in *East and West* has ascribed to me a proposal which has never once crossed my mind. He thinks that I want 'to commemorate the shooting at Jallianwala Bagh'. Nothing can be further from my thought than to perpetuate the memory of a black deed. I dare say that before we have come to our own we shall have a repetition of the tragedy and I will prepare the nation for it by treasuring the memory of the innocent dead. The widows and the orphans have been and are being helped, but we cannot 'bless the souls of those who died without knowing why', if we will not acquire the ground which has been hallowed by innocent blood and there erect a suitable memorial for them. It is not to serve, if I can help it, as a reminder of the foul deed but it shall serve as an encouragement to the nation that it is better to die helpless and unarmed and as victims rather than as tyrants. I would have the future generations remember that we who witnessed the innocent dying did not ungratefully refuse to cherish their memory. As Mrs. Jinnah truly remarked when she gave her mite to the fund, the memorial would at least give us an excuse for living. After all it will be the spirit in which the memorial is erected that will decide its character.

What was the 'larger symbiosis' that Buddha and Christ preached? Buddha fearlessly carried the war into the enemy's camp and brought down on its knees an arrogant priesthood. Christ drove out the money-changers from the temple of Jerusalem and drew down curses from Heaven upon the hypocrites and the Pharisees. Both were for intensely direct action. But even as Buddha and Christ chastised they showed unmistakable gentleness and love behind every act of theirs. They would not raise a finger against their enemies, but would gladly surrender themselves rather than the truth for which they lived. Buddha would have died resisting the priesthood, if the majesty of his love had not proved to be equal to the task of bending the priesthood. Christ died on the Cross with a crown of thorns on his head defying the might of a whole Empire. And if I raise resistances of a non-violent character I simply and humbly follow in the footsteps of the great teachers named by my critic.

Lastly, the writer of the paragraph quarrels with my 'grouping unities' and would have me take up 'the larger mission of uniting the world'. I once told him under a common roof that I was probably more cosmopolitan than he. I abide by that expression. Unless I group unities I shall never be able to unite the whole world. Tolstoy once said that if we would but get off the backs of our neighbours the world would be quite all right without any further help from us. And if we can only serve our immediate neighbours by ceasing to prey upon them, the circle of unities thus grouped in the right fashion will ever grow in circumference till at last it is co-terminus with that of the whole world. More than that it is not given to any man to try or achieve. *Yatha pinde, tatha brahmande*[1] is as true today as ages ago when it was first uttered by an unknown *rishi*.

'Neither a Saint nor a Politician'
Young India, 12 May 1920

[1] 'As with the body, so with the universe.'

§ 3. My Mission

26. IDENTIFICATION WITH LABOUR

[8 August 1925]

I have great pleasure in being able to visit these great steel works. I have been thinking of coming to this place ever since 1917, the year in which I was trying to serve the Champaran agriculturists. It was then that Sir Edward Gait told me that I ought not to leave Bihar without having seen these works. But man proposes and God disposes and with me God had disposed otherwise. I made many attempts to see this place.

As you know I am a labourer myself, I pride myself on calling myself a scavenger, weaver, spinner, farmer and what not, and I do not feel ashamed that some of these things I know but indifferently. It is a pleasure to me to identify myself with the labouring classes, because without labour we can do nothing. There is a great Latin saying of which the meaning is 'to labour is to pray', and one of the finest writers of Europe has said that a man is not entitled to eat unless he labours, and by labour he does not mean labour with the intellect, but labour with the hands. The same thought runs throughout Hindu religion. 'He who eats without labour eats sin, is verily a thief.' This is the literal meaning of a verse in *Bhagavad Gita*. I therefore pride myself on the fact that I can identify myself with labour throughout the world.

It was my ambition to see one of the greatest—if not the greatest—Indian enterprises in India, and study the conditions of work there. But none of my activities is one-sided, and as my religion begins and ends with truth and non-violence, my identification with labour does not conflict with my friendship with capital. And believe me, throughout my public service of 35 years, though I have been obliged to range myself seemingly against capital, capitalists have in the end regarded me as their true friend. And in all humility I may say that I have come here also as a friend of the capitalists—a friend of the Tatas. And here it would be ungrateful on my part if I do not give you a little anecdote about how my connection with the Tatas began.

In South Africa, when I was struggling along with the Indians there in the attempt to retain our self-respect and to vindicate our status, it was the late Sir Ratan Tata who first came forward with assistance. He wrote me a great letter and sent a princely donation,—a cheque for Rs. 25,000 and a promise in the letter to send more, if necessary. Ever since I have a vivid recollection of my relations with the Tatas and you can well imagine how pleasurable it has been for me to be with you, and you will believe me when I say that, when I part company with you tomorrow, I shall do so with a heavy heart, because I shall have to go away without having seen so many things, for it would be presumption on my part to say at the end of two days that I had really studied things here. I know well enough the magnitude of the task before one who wants to study this great enterprise.

My Mission

I wish to this great Indian firm all the prosperity that it deserves and to this great enterprise every success. And may I hope that the relations between this great house and labourers who work here under their care will be of the friendliest character? At Ahmedabad I have had much to do with the capitalists and workmen, and I have always said that my ideal is that capital and labour should supplement and help each other. They should be a great family living in unity and harmony, capital not only looking to the material welfare of the labourers but their moral welfare also,—capitalists being trustees for the welfare of the labouring classes under them.

I am told that though so many Europeans and Indians live here, their relations are of a happy character. I hope the information is literally true. It is the privilege of both of you to be associated in this great enterprise and it is possible for you to give India an object-lesson in amity and goodwill. You will, I hope, have best relations with one another not only under the roofs of the huge workshops you work in, but you will also carry your amity outside your workshops and both of you will realize that you have come to live and work here as brothers and sisters, never regarding another as inferior, or oneself as inferior. And if you succeed in doing that you will have a miniature *swaraj*.

I have said that I am a non-co-operator, I call myself a civil resister—and both words have come to possess a bad odour in the English language like so many other English words—but I non-co-operate in order that I may be able to co-operate. I cannot satisfy myself with false co-operation—anything inferior to 24 carats gold. My non-co-operation does not prevent me from being friendly even to Sir Michael O'Dwyer and General Dyer. It harms no one, it is non-co-operation with evil, with an evil system and not with the evil-doer.

My religion teaches me to love even an evil-doer, and my non-co-operation is but part of that religion. I am saying these things not to soothe the ears of any one—I have in my life never been guilty of saying things I did not mean—my nature is to go straight to the heart, and if often I fail in doing so for the time being, I know that truth will ultimately make itself heard and felt, as it has often done in my experience. The wish, therefore, that the relations between you should be of the friendliest character is a desire from the bottom of my heart. And it is my deep prayer that you may help in delivering India from evil and bondage and help her to give the message of peace to the outside world. For this meeting of Indians and Europeans in India must have or can be made to have a special meaning, and what can be better than that we two may live together so as to spread peace and goodwill on earth? May God grant that, in serving the Tatas, you will also serve India and will always realize that you are here for a much higher mission than merely working for an industrial enterprise.

Speech at Indian Association, Jamshedpur
Amrita Bazar Patrika, 14 Aug. 1925
Young India, 20 Aug. 1925

27. SERVICE OF HUMANITY

It is a privilege for me to enjoy the friendship of so many unknown American and European friends. It pleases me to note that the circle is ever widening, perhaps more especially in America. I had the pleasure of receiving a warm invitation about a year ago to visit that continent. The same invitation has now been repeated with redoubled strength and with the offer to pay all expenses. I was unable then as I am now, to respond to the kind invitation. To accept it is an easy enough task, but I must resist the temptation, for I feel that I can make no effective appeal to the people of that great continent unless I make my position good with the intellectuals of India.

I have not a shadow of doubt about the truth of my fundamental position. But I know that I am unable to carry with me the bulk of educated India. I can therefore gain no effective help for my country from the Americans and Europeans so long as I remain isolated from educated India. I do want to think in terms of the whole world. My patriotism includes the good of mankind in general. Therefore, my service of India includes the service of humanity. But I feel that I should be

My Mission

going out of my orbit if I left it for help from the West. I must be satisfied for the time being with such help as I can get from the West, speaking to it from my smaller Indian platform. If I go to America or to Europe, I must go in my strength, not in my weakness, which I feel today,—the weakness I mean, of my country. For the whole scheme for the liberation of India is based upon the development of internal strength. It is a plan of self-purification. The peoples of the West, therefore, can best help the Indian movement by setting apart specialists to study the inwardness of it.

Let the specialists come to India with an open mind and in a spirit of humility as befits a searcher after Truth. Then, perhaps, they will see the reality instead of a glorified edition that, in spite of all my desire to be absolutely truthful, I am likely to present if I went to America. I believe in thought-power more than in the power of the word, whether written or spoken. And if the movement that I seek to represent has vitality in it and has divine blessing upon it, it will permeate the whole world without my physical presence in its different parts. Anyway, at the present moment I see no light before me. I must patiently plod in India until I see my way clear for going outside the Indian border.

After pressing the invitation, the American friend puts a number of questions for my consideration. I welcome them and gladly take the opportunity of answering them through these columns. He says:

Whether you decide, now or later, to come here or not to come, I trust you will find the following questions worth considering. They have developed insistently in my mind for a long time.

His first question is:

Has the time arrived—or is it coming—when your best way to help India will be by moving the whole world—and especially England and America—to a new consciousness?

I have partly answered the question already. In my opinion the time has not yet arrived—it may come any day—for me to go out of India to move the whole world to a new consciousness. The process, however, is even now indirectly and unconsciously going on though slowly.

Are not the present-day interests of all mankind, everywhere, so inextricably interwoven that no single country, like India, can be moved far out of its present relationships to the others?

I do believe with the writer that no single country can remain in isolation for any length of time. The present plan for securing *swaraj* is not to attain a position of isolation but one of full self-realization and self-expression for the benefit of all. The present position of bondage and helplessness hurts not only India, not only England, but the whole world.

Is not your message and method essentially a world gospel—which will find its power in responsive souls, here and there, in many countries, who will thereby, gradually, remake the world?

If I can say so without arrogance and with due humility, my message and methods are indeed in their essentials for the whole world and it gives me keen satisfaction to know that it has already received a wonderful response in the hearts of a large and daily-growing number of men and women of the West.

If you demonstrate your message in the language only of the East and in terms only of Indian emergencies, is there not grave danger that inessentials will be confused with fundamentals—that some features which correspond only to extreme situations in India will be wrongly understood to be vital in the universal sense?

I am alive to the danger pointed out by the writer, but it seems to be inevitable. I am in the position of a scientist who is in the midst of a very incomplete experiment and who, therefore is unable to forecast large results and larger corollaries in a language capable of being understood. In the experimental stage, therefore, I must run the risk of the experiment being misunderstood as it has been, and probably still is, in many places.

My Mission

Ought you not to come to America (which in spite of all her faults is perhaps, potentially, the most spiritual of all living peoples) and tell the world what your message means in terms of Western, as well as Eastern, civilization?

People in general will understand my message through its results. The shortest way, therefore, perhaps of making it effectively heard is to let it speak for itself, at any rate for the time being.

For example, should the Western followers of your inspiration preach and practise the spinning-wheel?

It is certainly not necessary for the Western people to preach and practise the spinning-wheel unless they will do so out of sympathy or for discipline or with a view to applying their matchless inventive faculty to making the spinning-wheel a better instrument while retaining its essential characteristic as a cottage industry. But the message of the spinning-wheel is much wider than its circumference. Its message is one of simplicity, service of mankind, living so as not to hurt others, creating an indissoluble bond between the rich and the poor, capital and labour, the prince and the peasant. That larger message is naturally for all.

Is your condemnation of railroads, doctors, hospitals and other features of modern civilization essential and unalterable? Should we not, first, try to develop a spirit great enough to spiritualize the machinery and the organized, scientific and productive powers of modern life?

My condemnation of railroads, etc., whilst true where it stands, has little or no bearing on the present movement which disregards none of the institutions mentioned by the writer. In the present movement, I am neither attacking railroads nor hospitals; but in an ideal State they seem to me to have little or no place. The present movement is just the attempt the writer desires. Yet it is not an attempt to spiritualize the machinery—because that seems to me an impossible task—but to introduce, if it is at all possible, a human or the humane spirit among the men behind the machinery. Organization of machinery for the purpose of concentrating wealth and power in the hands of a few and for the exploitation of many I hold to be altogether wrong. Much of the organization of machinery of the present age is of that type.

The movement of the spinning-wheel is an organized attempt to displace machinery from that state of exclusiveness and exploitation and to place it in its proper state. Under my scheme, therefore, men in charge of machinery will think not of themselves or even of the nation to which they belong but of the whole human race. Thus Lancashire men will cease to use their machinery for exploiting India and other countries but, on the contrary, they will devise means of enabling India to convert in her own villages her cotton into cloth. Nor will Americans under my scheme seek to enrich themselves by exploiting the other races of the earth through their inventive skill.

Is it not possible, in conditions so favourable as America's, to clarify and advance the evolution of the best human consciousness into such purpose and power, courage and beneficence, as shall liberate the souls of India's millions—and of all men everywhere?

It is undoubtedly possible. Indeed, it is my hope that America will seek the evolution of the best human consciousness; but that time is perhaps not yet. Probably it will not be before India has found her own soul. Nothing will please me more than to find America and Europe making the difficult path of India as easy as it is possible for them to do. They can do so by withdrawing the temptations in India's way and by encouraging her in her attempt to revive her ancient industries in her own villages.

Why is it that people like myself, in every country, are grateful to you and eager to follow you? Is it not for two reasons chiefly: first; Because the next [sic] and basic need throughout the world is for a new spiritual consciousness—a realization, in the thought and feeling of average people, of the equal divinity of all human beings and the unity, brotherhood, of all; second, because you, more than any other widely known man, have this consciousness—together with the power to arouse it in others?

I can only hope that the writer's estimate is true.

It is a world need—is it not?—to which you have the best answer that God has vouchsafed to man? How can your mission be fulfilled in India alone? If my arm or leg could be vitalized to an extent far beyond the balance of my body, would that make for my general health—or even for the permanent best good of the one favoured member?

I am fully aware that my mission cannot be fulfilled in India alone, but I hope I am humble enough to recognize my limitations and to see that I must keep for the time being, to my restricted Indian platform till I know the result of the experiment in India itself. As I have already replied, I would like to see India free and strong so that she may offer herself as a willing and pure sacrifice for the betterment of the world. The individual, being pure, sacrifices himself for the family, the latter for the village, the village for the district, the district for the province, the province for the nation, the nation for all.

May I even submit,—with deep reverence for your message—that possibly your own vision and inspiration would benefit by adjustment to the world instead of only, or chiefly, to India?

I recognize the considerable force of the foregoing statement. It is not at all impossible that a visit to the West may give me not a wider outlook—for I have endeavoured to show that it is the widest possible but it may enable me to discover new methods of realizing the outlook. If such is my need, God will open the way for me.

Is the political form of government, in India or anywhere, so important as the average individual's soul-force—his courageous expression of the best inspiration he can derive from the divine spirit within and all about him?

The average individual's soul-force is any day the most important thing. The political form is but a concrete expression of that soul-force. I do not conceive the average individual's soul-force as distinguished and existing apart from the political form of government. Hence I believe that after all a people has the government which it deserves. In other words self-government can only come through self-effort.

Is not the basic need, everywhere, for the clarification and development of this soul-force in individuals—beginning, possibly, with a few people and spreading like a divine contagion to the many?

It is, indeed.

You teach, rightly, that the faithful development of such soul-force in India will assure India's freedom. Will it not everywhere shape all political, economic and international institutions including the issues of Peace or War? Can those forms of human civilization be made radically superior in India to the rest of the world—now, when all mankind are neighbours?

I have already answered this question in the preceding paragraphs. I have claimed in these pages before now that India's freedom must revolutionize the world's outlook upon Peace and War. Her impotence affects the whole of mankind.

You know, better than I or anyone, how all these questions should be answered. I chiefly seek to express my eager faith in your gospel, my hungry desire for your leadership in solving the urgent problems of America and of all mankind. Therefore, will you graciously remember that, if (or when) the time may come that India's progress in the directions you have so inspiringly outlined appears to pause— waiting for the Western world to come up alongside—then we of the West stand urging you to give us a few months of your time and your personal presence. My own feeling is that if you will call us and instruct us, we (your uncounted followers scattered obscurely over the wide earth) will join our lives to yours in the discovery and realization of a new and noble, worldwide Commonwealth of the Spirit in which man's age-old dreams of brotherhood, democracy, peace and soul progress shall characterize the daily life of average people—in India, England, America and everywhere.

I wish I had confidence in my leadership on the world platform. I have no false modesty about me. If I felt the call within, I would not wait a single second but straightway respond to an invitation so cordial as this. But with my limitations of which I am painfully conscious, I feel somehow that my experiment must be restricted to a fragment. What may be true of the fragment is likely to be true of the whole. It is true indeed that India's progress in the direction I desire seems to have come to a pause but I think that it only seems so. The little seed that was sown in 1920 has not perished. It is, I think, taking deep root. Presently it will come out as a stately tree. But if I am labouring under a delusion, I fear that no artificial stimulus that my visit to America may temporarily bring can revive it. I am pining for the assistance of the whole world. I see it coming. The urgent invitation is one of the many signs. But I know that we shall have to deserve it before it comes upon us like a mighty flood, a flood that cleanses and invigorates.

'To American Friends'
Young India, 17 Sept. 1925

28. CAVE IN THE HEART

There are many friends who point out 'my *dharma*' to me, and I am happy that they do it. The fact that they write to me freely is evidence of their love for me and their confidence that what they say will not hurt me. I have just received one such letter. The correspondents are well-known Gujarati workers and leaders in their respective districts. The reader will easily see that the letter has been prompted by their regard for me. I, therefore, give it here with a few omissions:

Though the letter has been written with a good motive and seems at first to be well-reasoned, I am unable to follow the advice of these friends.

Our holy books tell us in the plainest words that one's own *dharma* is to be preferred even though it may require no great virtue to follow it. Another's *dharma* may seem to have superior virtue in it but it is better even to lose one's life in following one's modest *dharma*. It is dangerous to follow another's *dharma*. Can I leave the field merely because today people do not accept my views? The idea of non-co-operation was conceived first by me. I did not know then how it would be received. I put into practice what I believed to be *dharma* and invited others to follow me. A good many were attracted by the idea. What difference does it make to me if it has now lost attraction for them? Should I give up my *dharma* because of that? If I do so, I would dishonour my ideal of service. My faith in the efficacy of non-co-operation remains what it was at the time of its inception.

The alternation of flood and ebb is a law of nature. Why should we be elated with pride at flood-tide and be filled with despair at ebb-tide? One who has no control over the helm may lose direction. My hands are firm on the helm and, therefore, I have no such fear.

People's love of *khadi* has increased rather than diminished. Blind worship has changed into intelligent love. On the whole, the quality of *khadi* being produced is daily improving and the demand for it is increasing. Among public activities which are going on independently of the Government, I think there is none which is as much alive as the Khadi Movement. This statement can be proved with the help of figures. It is a fact that spinning and carding have stopped in a few places; nevertheless, they are better organized today than they were at any time during the past four years.

The Hindu-Muslim problem is, today, like the lump of clay on the potter's wheel. The Almighty alone knows what kind of pot will emerge. However, having regard to the unparalleled popular awakening, the present developments, though painful, should surprise no one. All the dirt has come up to the surface and, therefore, it is all that we see. What today the Hindus and Muslims cannot be persuaded to do, they will be forced to do sooner or later by the pressure of circumstances. They have no choice but to become united, and so I am not worried on that score. If Fate has decreed that we should fight a few battles among ourselves, let us. This will not be the first instance of such fighting in the annals of the world. Brothers sometimes fight with one another, but unite again. When the era of peace dawns upon us, wars will seem barbaric. Today, however, fighting is looked upon as civilized.

Untouchability is on its last legs. Its soul is dead, what we see is only its skeleton.

We need not be filled with despair because our struggle for

swaraj has resulted in creating disunity among us. Such things have taken place among all nations which have become free. Our duty is only to take note of them and seek remedies for them; to lose heart in the face of them would be evidence of our cowardice.

A person who has accepted defeat in India, what can he give to America or bring thence for our country? People's admiration for me in America or Europe cannot blind me. We shall profit nothing by begging for help from the West. My returning with a certificate from the West would be a matter of shame for the country and me. I see no good reason at present which could induce me to go to Europe or America. Let no one believe that the leaders of those continents are simply pining to meet me or hear me. Whatever reputation I enjoy in those countries is among people whose voice would carry no weight at all with the general public. They, too, like me, seem to have nothing else to do and so build castles in the air and formulate plans for the betterment of the world. I shall retain their love so long as I remain devoted to truth and non-violence. The reader, however, should know that these persons do not hold the reins of power in the West. Whatever strength I have can best be demonstrated in our country. Hills seem beautiful from a distance. Immediately I leave India, I shall have stepped out of my true sphere, and such persons have no place anywhere in the world.

Even in Africa, I can do nothing now. I am, in this matter, in the same plight as Arjuna, who was robbed by a *kaba* though he held the same bow and arrow which he had wielded all his life. My Krishna would not be there by my side now. A soldier distinguishes himself in a fight which comes to him unsought. One who goes out in search of a fight would be a gambler. I may say that I have never once gambled in my life, and, fortunately for me, I lost even on the one occasion when I played for stakes.

If the leaders and the country have wearied of me, I should retire to the Himalayas. By the Himalayas I do not mean the Dhavalagiri Hills, I mean the Himalaya in my heart. It would be easy for me to find a cave in that Himalaya and live there. Even this I will not myself seek, it will come seeking me. The devotee does not himself go to God. If he did, he would not be able to bear His dazzling light. Hence God Himself comes down to His devotees and appears to them in the form in which they have adored Him. My God knows that I am impatiently awaiting His coming; a mere sign from Him will suffice for me. 'With a frail thread has Hari tied me, and I turn the way He pulls me.' So sang Mirabai. I am Mirabai's disciple, and so I may also sing this song after making the necessary change in the gender of the pronoun. I should always hold myself ready to be pulled by such a thread. That is why I am always drawing out thread and reminding my 'roving mind' that it should always be ready to set out for its homeland. It matters little whether that homeland be a cave in my heart or some unknown country. Whatever the place to which I retire, He will be present there and, therefore, I shall have nothing to fear.

I would immediately open shops for selling *khadi* in every *taluk*, if only the workers in each *taluk* would assure me that they would be able to sell a fixed quantity of *khadi*. People should write to the Khadi Association for detailed information.

'My *Dharma*' (G.)
Navajivan, 20 Dec. 1925

29. BROTHERHOOD OF MAN

[9 March 1929]

Mr. Chairman and Friends, I propose to speak to a certain length in Hindustani and as your address is worded in English, I propose to make a brief reply first in English and then say my say in Hindustani. I thank you for this warmth of reception and the kindly sentiments expressed in your address. I am not able just now to appropriate, much less to assimilate, all the compliments that you have paid me. But I could certainly claim two things of which you have made kind mention. The first thing is that my mission is not merely brotherhood of Indian humanity. My mission is not merely freedom of India, though today it undoubtedly engrosses practically the whole of my life and the whole of my time. But through realization of freedom of India I hope to realize and carry on the mission of brotherhood of man. My patriotism is not an exclusive thing. It is all-embracing and I should reject that patriotism which sought to mount upon the distress or the exploitation of other nationalities. The conception of my patriotism is nothing if it is not always in every case, without exception, consistent with the broadest good of humanity at large. Not only that but my religion and my patriotism derived from my religion embrace all life.

I want to realize brotherhood or identity not merely with the beings called human, but I want to realize identity with all life, even with such beings as crawl on earth. I want, if I don't give you a shock, to realize identity with even the crawling things upon earth, because we claim common descent from the same God, and that being so, all life in whatever form it appears must be essentially one. I can therefore safely claim all the credit that you may choose to give me in describing my mission of brotherhood of man. As a necessary corollary you may naturally mention, as you have kindly mentioned, untouchability. I have said times without number that untouchability is a serious blot on Hinduism and, I think, in the long run, in the race for life in which all the religions of the world are today engaged, either Hinduism has got to perish or untouchability has to be rooted out completely, so that the fundamental principle of Advaita Hinduism may be realized in practical life. Beyond these two things among those that you have mentioned in your address I am unable today to appropriate or assimilate anything. It will be time enough to pronounce a verdict upon my work after my eyes are closed and this tabernacle is consigned to the flames.

You have very kindly asked me to give the indigenous population of Burma some advice. I must own to you that I regard myself as altogether unfit for giving you any advice. My study of your great traditions is merely superficial. My study of your present-day problems is still more superficial, although I yield to none in my love and admiration for you to which I gave expression yesterday at two meetings. I wish I had all the facts before me. I wish that all the different parties in Burma could meet me and I could gain access to your hearts. My heart is there open to receive you, but it is for you to call and certainly that call will not be in vain. And if I find I have enough material before me to give you even provisional limited advice I should be at your disposal

Speech at Public Meeting, Rangoon
Amrita Bazar Patrika, 10 Mar. 1929
Young India, 4 Apr. 1929

30. UNIVERSAL MESSAGE

June 11, 1937

Dear Friend,

I thank you for your letter of the 20th May last. I have no message to give except this that there is no deliverance for any people on this earth or for all the people of this earth except through truth and non-violence in every walk of life without any exceptions. And this is based on an unbroken experience extending practically over half a century.

Yours sincerely,
M. K. GANDHI

Daniel Oliver, Esq.
Hammana
Lebanon, Syria

Letter to Daniel Oliver
Pyarelal Papers

§ 4. Isms

31. BEYOND ISMS AND SECTS

Q. Does not your identifying yourself with the Patna resolution of the Working Committee betray distrust in Lord Linlithgow, although you have professed to believe in his sincerity?

A. You have read into the resolution what is wholly unwarranted by the text. I do not doubt the Viceroy's sincerity. I have not known a Viceroy who has weighed his word as Lord Linlithgow does. It is a pleasure to have a talk with him. For he speaks with the greatest deliberation. His speech is, therefore, always brief and to the point. I adhere to the remark I made about our last meeting that, although we could not agree, we had come nearer each other. We might have gone on talking for a few days, but we would only have talked round the subject and repeatedly come to the same point of disagreement. I was under no handicap for I was speaking for myself. He was under a severe handicap. He was speaking under orders. He had no authority to go outside his instructions. And so we parted the best of friends. But so far as I am concerned, I expect many more meetings. The resolution makes the Congress position clear beyond doubt. It represents also my own position. If the British Government really mean full-hearted Dominion Status with the right to secede, then they can have no difficulty in accepting the Congress position. Unfortunately Lord Zetland's interview shows that it is not India which is to determine her future but Britain will do so for her. This is not even Dominion Status of any known variety. Once the British Government are sure that they can no longer hold India, all the difficulties that are now being put forth on their behalf will vanish like darkness before dawn. For they are all of their creation. They are inherent in exploitation. I hope you now see that there is no question of distrust of the Viceroy. Events had to move to where they are.

Q. You say that no such thing as Gandhism exists, and that what you stand for is nothing new. I am a Muslim. I see flashes of Islamic glory in Gandhism. As a student of theology I see the grandeur of Hinduism and the vigour of Christianity amply expounded in Gandhism. It includes also to a considerable extent the chaste philosophy of the entire East. I search the pages of India's past history, but your creed I do not find. Why, therefore, is it not new, and why may it not be termed Gandhism for those of us who believe in you and therefore it?

A. I have a horror of 'isms', especially when they are attached to proper names. Even if all that you say of me is true, it does not make a new sect. My effort is to avoid not only new sects but even to do away with old and superfluous ones. *Ahimsa* abhors sects. *Ahimsa* is a unifying force. It discovers unity in diversity. All that you say is derivable from *ahimsa*. To bring into being a new cult is repugnant to *ahimsa*, to the very experiment I am making. Thus you will, I hope, see that there is no room for 'Gandhism'.

Q. You say, 'It is degrading both for man and woman that woman should be called upon or induced to forsake the hearth and shoulder the rifle for the protection of that hearth. It is a reversion to barbarity and the beginning of the end.' But what about the millions of female labourers in fields, factories, etc? They are forced to forsake the hearth and become 'bread-winners'. Would you abolish the industrial system and revert to the stone age? Would that not be a reversion to barbarity and the beginning of the end? What is the new order that you envisage where the sin of making women work will be absent?

A. If millions of women are forced to forsake their hearth and become bread-winners, it is wrong, but not so wrong as shouldering the rifle. There is nothing inherently barbarous in labour. I see no barbarity in women voluntarily working on their fields whilst they are looking after their homes. In the new order of my imagination all will work according to their capacity for an adequate return for their labour. Women in the new order will be part-time workers, their primary function being to look after the home. Since I do not regard the rifle as a permanent feature in the new order, its use will be progressively restricted even so far as men are concerned. It will be tolerated as a necessary evil while it lasts. But I would not deliberately contaminate women with the evil.

Q. Why may not the illiterate masses be taught the Roman script? This would eliminate the existing controversy between Urdu and Hindi.

A. To teach the Roman script in the place of Hindi and Urdu would be like putting the cart before the horse. Our children have first to learn both Hindi and Urdu scripts. Difficult questions cannot be solved by ignoring them or suggesting apparently easy substitutes. So long as hearts are divided the Roman scripts will not cement them. It would be an additional burden. The learning of the two scripts is the best and the easiest way of at least solving the national language riddle. It opens Hindi and Urdu thought to both Hindu and Muslim boys and girls who will be the men and women of the future generation. The Roman script will be learnt at its proper time, i.e., when our boys and girls are taught the English language, as some undoubtedly will be.

Q. Congress clamours for unity, but the principles which must be followed to attain that unity, viz., Hindu-Muslim fellow-feeling, no caste distinctions, no hatred towards each other and towards foreigners, co-operative endeavour, all these are presented to audiences through the microphone but not acted upon. Tell me, what are the duties of a Congress member? I would love to join and will put forth all my energy to do my bit for the country.

A. You need not mind what others do or ought to do. Charity begins at home. Let yours begin with yourself. Abolish all caste and religious or race distinctions from your heart. Be true to everyone—Hindu, Muslim, Harijan, English, etc., as you are, I hope, to yourself, and you will find that so far as you are concerned your difficulty will be solved and your example will be copied by others. Be sure that you have banished all hate from your heart, and that you have no political or other objective in loving and serving your neighbour as if he was your own self.

Sevagram, March 12, 1940

'Question Box'
Harijan, 16 Mar. 1940

32. ISMS AND FOLLOWERS

July 16, 1945

Chi. Shanta,

I got your letter yesterday. I am writing this after the morning prayer. Address your reply at Sevagram.

You have remained as crazy after becoming a communist and a mother as you were as a child.

Which Ashram has boycotted you? Where is the Ashram? Who has boycotted you? Many communists have stayed with me. In the same way you also can stay. You know that Jayanti had stayed with me.

You should know that I have received many complaints,

but I have not acted on any of them. I am in correspondence with the Secretary. He had asked me for my permission to publish the letters and I have granted it. Whether he has published them or not I do not know.

The Working Committee has taken no step. There has been no time to consider the matter.

If Jawaharlalji goes against them, all the communists will have to sit up and think. For he has a soft corner for the Party but he will not tolerate anything unworthy. I myself have not been able to come to a final decision. I have received quite a few complaints. I have sent them to the Head Office.

You have written your letter without thinking. If you calm down and think before writing, you can help the communist cause.

You must learn to distinguish between communism and communists. Besides, Marx stands for one thing, Lenin for another and Stalin for a third. The followers of the last are again divided into two groups. Gandhi is one thing, Gandhism is another and Gandhi-ites are a third thing. There are always, and will remain, such differences. Immature people may identify themselves with one or the other group.

Blessings from
BAPU

Letter to Shanta Patel (G.)
CW 4287

33. PROPOUNDER OF NO ISM

Mussoorie,
June 7, 1946

Chi. Rameshwari,

I know Ratnamayidevi very well. I have no objection to your taking her in. I do not know myself who is a Gandhian. Gandhism is a meaningless word for me. An ism follows the propounder of a system. I am not one, hence I cannot be the cause for any ism. If an ism is built up it will not endure, and if it does it will not be Gandhism. This deserves to be properly understood.

I like your work, it is neat and clean. It is you who have built up the Balika Ashram, you conduct it and I would be pleased if Ratnamayidevi gave you full satisfaction.

Blessings from
BAPU

Letter to Rameshwari Nehru (H.)
CW 3110

III

Influences and Books Read

§ 1. Narasinh Mehta

34. THE *VAISHNAVA* IDEAL

A true *Vaishnava* is he
  Who is moved by others' sufferings;
Who helps people in distress,
  And feels no pride for having done so.
Respectful to everyone in the world,
  He speaks ill of none;
Is self-controlled in action, speech and thought—
  Twice-blessed the mother who bore such a one.
He has an equal-seeing eye, and is free from all craving,
  Another's wife is to him a mother;
His tongue utters no untruth,
  And never his hand touches another's wealth.
Moha and *maya* have no power over him,
  In his mind reigns abiding detachment;
He dances with rapture to Rama's name—
  No centre of pilgrimage but is present in his person.
A man he is without greed and cunning,
  And purged of anger and desire;
Offering reverence to such a one, says Narasainyo,[1]
  Will bring release to seventy-one generations of one's forbears.[2]

From the marks of a *Vaishnava* described by Narasinh Mehta we see that he is a man who
  1. is ever active in bringing relief to the distressed,
  2. takes no pride in doing so,
  3. is respectful to all,
  4. speaks ill of none,
  5. is self-controlled in speech,
  6. in action and
  7. in thought,

8. holds all in equal regard,
9. has renounced desires,
10. is loyal to one woman, his wife,
11. is ever truthful,
12. keeps the rule of non-stealing,
13. is beyond the reach of *maya*,
14. is, in consequence, free from all desire,
15. is ever absorbed in repeating Rama's name,
16. and, as a result, has been sanctified,
17. covets nothing,
18. is free from guile,
19. from the urge of desire and
20. from anger.

Here Narasinh, the best among the *Vaishnavas*, has given pride of place to non-violence. This means that a man who has no love in him is no *Vaishnava*. One who does not follow truth and has not acquired control over all his senses is not a *Vaishnava*. He teaches us in his *prabhatiyan* that one does not become a *Vaishnava* simply by studying the Vedas, by following the rules of *varnashram*, by wearing a string of basil seeds or the *tilak* mark. All these things can be the origin of sin. Even a hypocrite may wear a string of beads or put the *tilak* mark or study the Vedas or keep repeating Rama's name with his lips. But such a one cannot follow truth in his life, cannot, without giving up his hypocrisy, help people in distress or be self-controlled in speech, action and thought.

I invite everyone's attention to these principles, since I still continue to receive letters regarding *Antyajas*. The advice I receive from one and all is that, if I do not exclude *Antyajas* from the national schools, the movement for *swaraj* will end in smoke. If I have even a little of the true *Vaishnava* in me, God will also vouchsafe me the strength to reject the *swaraj* which may be won by abandoning the *Antyajas*.

The resolution, to the effect that the *Antyajas* cannot be excluded from any place which is open to members of other classes or communities, is not mine but that of the senate as a whole. I welcome the resolution. Had the senate not passed it, it would have been guilty of *adharma*.

The resolution lays down nothing new. One to the same effect is actually in operation in the existing schools. The

Congress, a body which the *Vaishnavas* respect, has also passed such a resolution. They have not opposed it. I realize, however, that they honour me by criticizing me for having a hand in a resolution of this kind. The point of their argument is that others may violate *dharma* but that I, especially, should not do so. This is very gratifying to me.

I have been endeavouring to show that *dharma* requires that we do not look upon *Antyajas* as untouchables. Old veils prevent us from seeing that we are guilty of *adharma* in acting to the contrary. Just as, through such veils, British rule cannot see its own Satanism, so also, thanks to them, some of us are unable to see the chains of slavery which bind us. I think it my duty to reason with such people patiently.

But I cannot stand hypocrisy and sophistry. I saw in *Gujarati* an account of a talk I had with Maharajshri,[3] as also the comments on it. I have been very much pained by both. I seldom comment on views expressed in newspapers. In fact I scarcely read papers. But the *Gujarati* is a widely read paper and it claims to present the *sanatan dharma* in its true nature. Hence I am pained whenever I find in it even the least element of unfairness. A friend has sent me a cutting giving the report of my talk with Maharajshri and the criticism on it. I see in both an attempt, deliberate or otherwise, to prove *adharma* to be *dharma*. I shall explain next time what this is.

'To *Vaishnavas*' (G.)
Navajivan, 5 Dec. 1920

[1] Narasinh Mehta (1414–79); saint-poet of Gujarat.
[2] This poem formed part of the daily prayers at the Ashram.
[3] Goswami Shri Gokulnathji Maharaj, religious head of the *Vaishnavas* in Bombay.

35. *VAISHNAVA DHARMA*

I drew attention to the virtues of the true *Vaishnava* which Narasinh Mehta, the great *Vaishnava*, has sung of in his poem, and I expressed my pain at the comments on my talk with Maharajshri.

I saw in them, not an attempt to determine the meaning of *dharma*, but only *duragraha* and attacks on me. May it not be said that I, too, am guilty of *duragraha* and attack others? Certainly, it may. It will be for the reader to judge whether this is true or not. At the very beginning of our talk, Maharajshri told me that, in the interpretation of *shastra*, reason had no scope. That itself pained me. In my view, that which reason cannot understand and which the heart does not accept can be no *shastra*, and I think that anyone who wants to follow *dharma* in its purity cannot but admit this principle. If we do not do this, we shall run the risk of violating our *dharma*. I have heard the *Gita* being interpreted to the effect that, if any of our relations are wicked, we may use force to reduce them to submission; indeed, that it is our *dharma* to do so. Because Rama killed Ravana, is it *dharma* to kill a person whom we regard as Ravana? The *Manusmriti* permits the eating of meat; is a *Vaishnava* free, therefore, to eat meat? I have had it from the lips of *shastris* and men claiming to be *sannyasis* that, when ill, one may take even beef. Where would I have been if, accepting all these interpretations of the Shastras, I had gone about destroying my relatives, had advised the killing of the British and eaten beef in my illness? On such occasions, I was saved because I accepted only that as *dharma* which my reason and my heart accepted as such, and I advise everyone to do the same.

For this reason, holy men of *tapascharya* have told us that those who study the Vedas but do not follow *dharma* in conduct are mere pedants; that they neither swim across themselves, nor help others to do so. So it is that I am never impressed by those who have the Vedas on their lips or have got the commentaries by heart and, instead of marvelling at their learning, cherish my little knowledge as of greater value.

These being my views, I was pained when Maharajshri enunciated the principle for determining the meaning of *shastra*, but I was also pleased by his frankness. Though he held, thus, that the Shastras were against me, his decision finally was that it would not be justice to exclude *Antyajas* from schools which were open to Muslims, Parsis, Christians, Jews and others. *Vaishnavas*, who give money for innumerable worldly activities and spend it even in gambling and similar things, could not take cover behind objections on religious grounds and refuse donations to national schools which admit *Antyajas* along with others. If they do not wish to send their children to schools attended by *Antyajas*, they should be under no pressure to do so. This was the pragmatic decision which Maharajshri gave.

However, the arguments of the *shastris* who surrounded Maharajshri made my heart sink. I saw no frankness in them, but only an obstinate clinging to their views. Shastri Vasantram has given us a sample of this in *Gujarati*.

I wish respectfully to tell him and the editor of *Gujarati* that a public worker's duty is not to follow the popular current but, should its direction be wrong, to lead it in the right direction.

I do not know the Shastras, have no experience, am obstinate—by asserting all this I cannot be disqualified from being a *Vaishnava*. So long as I hold that the test of being a *Vaishnava* lies in moral conduct and not in debating or the gift of clever speech or in determining the meaning of the Shastras, I do not wish to give up my claim.

To say that looking upon the practice of untouchability as sin is a Western idea is to make out a sin to be a virtue. Akha Bhagat[1] had had no Western education, but it was he who said in a poem, 'Like a superfluous limb is this notion of defilement by touch.' It is sheer fanaticism, and will lead to the degradation of *dharma*, to look upon efforts at eradicating our evils as deriving their inspiration from other religions and to cling on to those evils.

It has been argued that the practice of untouchability implies no contempt; the British put forward a similar argument in regard to their attitude to us. If they keep us away, call us 'natives', they mean no contempt. If they force us to occupy separate compartments in trains, it is entirely for 'convenience of hygiene' and there is no ill will in the procedure—such is their claim. I have seen *Vaishnavas* abusing and beating up *Antyajas* who happened to touch them unintentionally. It is sheer hypocrisy, or sin rather, to describe such conduct as *dharma*. There is arrogance in ordering *Antyajas* to turn their faces to the wall when a *Brahmin* is out in the street. It is baseness to offer them the remains on our plates or things which have become rotten. Such conduct has its source in the practice of untouchability.

I simply do not understand the contention that bathing and putting on clean clothes will not cleanse an *Antyaja*. Is it that an *Antyaja* carries filth in his heart, or that he was not born a human being? Are *Antyajas* lower even than animals?

I have seen a number of *Antyajas* who were of an open frank heart, were upright, were men of knowledge and lovers of God. I look upon such *Antyajas* as worthy of all reverence.

I can understand refusal to touch an *Antyaja* who is dirty or who has not bathed after carrying night-soil. But it is the very limit of *adharma* to refuse to touch *Antyajas* however clean. I have seen many who were not *Antyajas* but were extremely dirty. There are many Christians among those who carry night-soil. It is part of a physician's duty to help remove stools. We do not consider contact with these as sinful. But we despise our physicians who hold no degrees, committing a sin thereby and bringing a bad name to the *Vaishnava dharma*.

It seems, rather, that Shastri Vasantram and the editor of *Gujarati* have identified the practice of untouchability with *varnashram*. In my humble opinion, the latter is *dharma*, unalterable, universal and in harmony with Nature, as also a social arrangement; it is a pure outward form of Hinduism.

The practice of untouchability is a blot on Hinduism; it was probably introduced as a temporary expedient, during the period of decline. It is based on no principle of universal application and has no support in the Shastras. The verses which are cited in justification of it are interpolations; or, at any rate, there are differences of opinion about their meaning. There are no statements by *Vaishnavas* describing the practice of untouchability to be a part of *dharma*. It is disappearing day by day. It is not observed in trains, Government schools, centres of pilgrimage and courts. In mills and other big factories, people touch *Antyajas* with the utmost freedom. My request to *Vaishnavas* is that this contact with the *Antyajas* which in any case they tolerate, though looking upon it as sinful, they should accept deliberately and as an act of virtue. The *Gita* says the same thing: 'To the man who looks on all with an equal eye, a *Brahmin*, a dog and an *Antyaja*—all are the same.' 'Narsainyo' says in his poem that a *Vaishnava* should have the same eye for all. *Vaishnavas* cannot claim that they can maintain this attitude to *Antyajas* though looking upon them as people with whom all contact is forbidden.

'*Vaishnavas* and *Antyajas*' (G.)
Navajivan, 12 Dec. 1920

[1] Mystic poet of the seventeenth century.

§ 2. Theosophy

36. THEOSOPHY, RENUNCIATION, AND ATHEISM

Towards the end of my second year in England I came across two Theosophists, brothers, and both unmarried. They talked to me about the *Gita*. They were reading Sir Edwin Arnold's translation—*The Song Celestial*—and they invited me to read the original with them. I felt ashamed, as I had read the divine poem neither in Sanskrit nor in Gujarati. I was constrained to tell them that I had not read the *Gita*, but that I would gladly read it with them, and that though my knowledge of Sanskrit was meagre, still I hoped to be able to understand the original to the extent of telling where the translation failed to bring out the meaning. I began reading the *Gita* with them. The verses in the second chapter

> If one
> Ponders on objects of the sense, there springs
> Attraction; from attraction grows desire,
> Desire flames to fierce passion, passion breeds
> Recklessness; then the memory—all betrayed—
> Lets noble purpose go, and saps the mind,
> Till purpose, mind, and man are all undone.

made a deep impression on my mind, and they still ring in my ears. The book struck me as one of priceless worth. The impression has ever since been growing on me with the result that I regard it today as the book *par excellence* for the knowledge of Truth. It has afforded me invaluable help in my moments of gloom. I have read almost all the English transla-

tions of it, and I regard Sir Edwin Arnold's as the best. He
has been faithful to the text, and yet it does not read like a
translation. Though I read the *Gita* with these friends, I cannot
pretend to have studied it then. It was only after some
years that it became a book of daily reading.

The brothers also recommended *The Light of Asia* by Sir
Edwin Arnold, whom I knew till then as the author only of
The Song Celestial, and I read it with even greater interest than
I did the *Bhagavad Gita*. Once I had begun it I could not leave
off. They also took me on one occasion to the Blavatsky Lodge
and introduced me to Madame Blavatsky and Mrs. Besant.
The latter had just then joined the Theosophical Society, and
I was following with great interest the controversy about her
conversion. The friends advised me to join the Society, but
I politely declined saying, 'With my meagre knowledge of my
own religion I do not want to belong to any religious body.'
I recall having read, at the brothers' instance, Madame Blavatsky's
Key to Theosophy. This book stimulated in me the desire
to read books on Hinduism, and disabused me of the notion
fostered by the missionaries that Hinduism was rife with
superstition.

About the same time I met a good Christian from Manchester
in a vegetarian boarding house. He talked to me about
Christianity. I narrated to him my Rajkot recollections. He
was pained to hear them. He said, 'I am a vegetarian. I do
not drink. Many Christians are meat-eaters and drink, no
doubt; but neither meat-eating nor drinking is enjoined by
Scripture. Do please read the Bible.' I accepted his advice,
and he got me a copy. I have a faint recollection that he
himself used to sell copies of the Bible, and I purchased from
him an edition containing maps, concordance, and other aids.
I began reading it, but I could not possibly read through the
Old Testament. I read the book of Genesis, and the chapters
that followed invariably sent me to sleep. But just for the sake
of being able to say that I had read it, I plodded through the
other books with much difficulty and without the least interest
or understanding. I disliked reading the book of Numbers.

But the New Testament produced a different impression,
especially the Sermon on the Mount which went straight to
my heart. I compared it with the *Gita*. The verses, 'But I say

unto you, that ye resist not evil: but whosoever shall smite thee on thy right cheek, turn to him the other also. And if any man take away thy coat let him have thy cloak too', delighted me beyond measure and put me in mind of Shamal Bhatt's 'For a bowl of water, give a goodly meal', etc. My young mind tried to unify the teaching of the *Gita*, *The Light of Asia* and the Sermon on the Mount. That renunciation was the highest form of religion appealed to me greatly.

This reading whetted my appetite for studying the lives of other religious teachers. A friend recommended Carlyle's *Heroes and Hero-worship*. I read the chapter on the Hero as a prophet and learnt of the Prophet's greatness and bravery and austere living.

Beyond this acquaintance with religion I could not go at the moment, as reading for the examination left me scarcely any time for outside subjects. But I took mental note of the fact that I should read more religious books and acquaint myself with all the principal religions.

And how could I help knowing something of atheism too? Every Indian knew Bradlaugh's name and his so-called atheism. I read some book about it, the name of which I forget. It had no effect on me, for I had already crossed the Sahara of atheism. Mrs. Besant who was then very much in the limelight, had turned to theism from atheism, and that fact also strengthened my aversion to atheism. I had read her book *How I Became a Theosophist*.

It was about this time that Bradlaugh died. He was buried in the Woking Cemetery. I attended the funeral, as I believe every Indian residing in London did. A few clergymen also were present to do him the last honours. On our way back from the funeral we had to wait at the station for our train. A champion atheist from the crowd heckled one of these clergymen. 'Well, Sir, you believe in the existence of God?'

'I do,' said the good man in a low tone.

'You also agree that the circumference of the Earth is 28,000 miles, don't you?' said the atheist with a smile of self-assurance.

'Indeed.'

'Pray tell me then the size of your God and where he may be?'

'Well, if we but knew, He resides in the hearts of us both.'

'Now, now, don't take me to be a child,' said the champion with a triumphant look at us.

The clergyman assumed a humble silence.

This talk still further increased my prejudice against atheism.

'Acquaintance With Religions', *The Story of My Experiments With Truth*, Ch. XX (G.)
Navajivan, 18 Apr. 1926

37. THEOSOPHY AND TRUE RELIGION

Panchgani,
[17 July 1946]

GANDHIJI: Jaiprakash will have to join me. I will not pit myself against him. In 1942, I said I was sailing out on uncharted waters. I will not do it now. I did not know the people then. I know now what I can do and what I cannot.

LOUIS FISCHER: You did not know in 1942 that there would be violence?

G. Correct.

L. F. So if the Constituent Assembly fails you will not stage a civil disobedience campaign?

G. Not unless the Socialists and the Communists are subdued by that time.

L. F. That is not likely ...

G. I cannot think of civil disobedience when there is so much violence in the air in India. Today some caste Hindus are not playing the game by the untouchables.

L. F. By some caste Hindus you mean some Congressmen?

G. Not many Congressmen. But there are some who have not banished untouchability from their hearts. That is the tragedy ... The Muslims also feel they are wronged. In an orthodox Hindu house a Muslim will not be permitted to sit on the same carpet with a Hindu and have his meal. That is false religion. India is falsely religious. It must get true religion.

L. F. You have not succeeded with Congress?

G. No, I have not. I have failed. Something, however, has been accomplished. The Harijans are admitted to the temples in Madura and in many other holy places, and the caste Hindus worship in the same temples.

L. F. Some say Hindu-Muslim relations are better and some say they are worse.

G. Jinnah and other Muslim leaders were once members of Congress. They left it because they felt the pinch of Hindu patronizing. In the beginning, the leading Congressmen were Theosophists. Mrs. Annie Besant attracted me very much. Theosophy is the teaching of Madame Blavatsky. It is Hinduism at its best.... Cultivated Muslims joined the Theosophists. Later, Congress membership grew and with it the Hindu patronizing attitude. The Muslims are religious fanatics, but fanaticism cannot be answered with fanaticism. Bad manners irritate. Brilliant Muslims in Congress became disgusted. They did not find the brotherhood of man among the Hindus. They say Islam is the brotherhood of man. As a matter of fact, it is the brotherhood of Muslims. Theosophy is the brotherhood of man. Hindu separatism has played a part in creating the rift between Congress and the League. Jinnah is an evil genius. He believes he is a prophet.

L. F. He is a lawyer.

G. You do him an injustice. I give you the testimony of my eighteen days of talks with him in 1944. He really looks upon himself as the saviour of Islam....

L. F. But Jinnah is cold. He is a thin man. He pleads a case, he does not preach a cause.

G. I agree he is a thin man. But I don't consider him a fraud. He has cast a spell over the Muslim, who is a simple-minded man ...

L. F. Jinnah told me in 1942 you did not want independence ... He said you wanted Hindu rule.

G. He is utterly wrong. That is absurd. I am a Muslim, a Hindu, a Buddhist, a Christian, a Jew, a Parsi. He does not know me when he says I want Hindu rule. He is not speaking the truth. He is speaking like a pettifogging lawyer. Only a maniac resorts to such charges. I believe that the Muslim

League will go into the Assembly. But the Sikhs have refused. They are stiff-necked like the Jews ...

L. F. What did you learn from your eighteen days with Jinnah?

G. I learned that he was a maniac. A maniac leaves off his mania and becomes reasonable at times. I have never regretted my talks with him. I have never been too stubborn to learn. Every one of my failures has been a stepping-stone. I could not make any headway with Jinnah because he is a maniac, but many Muslims were disgusted with Jinnah for his behaviour during the talks ...

Report of Interview with Louis Fischer
Louis Fischer, *The Life of Mahatma Gandhi*, pp. 472-3

§ 3. The *Bhagavad Gita*

38. THE *GITA* AND NON-VIOLENCE

A friend puts the following question ...[1]

Such doubts will continue to arise. Those who have made some study [of the *Gita*] should try to resolve them to the best of their ability. I will try to do so, but I must say at the same time that in the last resort man acts according to the dictates of his heart. The heart takes precedence over the intellect. The principle is accepted first, and proof follows afterwards. Inspiration precedes the arguments with which we justify it. That is why it is said that the intellect is led by one's actions. Man discovers arguments in favour of what he wants to do or has done.

I can, therefore, understand that my interpretation of the *Gita* may not be acceptable to everyone. In these circumstances, I think it should suffice if I describe how I arrived at my interpretation of the *Gita* and explain the principles which I have followed in determining the meaning of *shastras*. 'My duty is to fight, and be unconcerned with the result. The enemies who deserve to die are dead already, my part is simply to be an instrument in killing them.'

I became acquainted with the *Gita* in 1889. I was twenty

years of age at that time. I had not yet fully understood the significance of non-violence as a principle of *dharma*. It was from Shamal Bhatt's couplet, 'Let him offer water, and a good meal to eat', that I had first learnt the principle of winning over even an enemy with love. Its truth had made a deep appeal to my heart, but the couplet had not suggested to me the principle of compassion for all creatures. I had even eaten meat before that time while I was still in India. I believed that it was one's duty to kill snakes and other such creatures. I remember having killed bed-bugs and other insects. I remember killing a scorpion once. Today I think that we should not kill even such poisonous creatures. In those days I believed that we would have to fit ourselves to fight the British. I used to murmur to myself the lines of the poem beginning, 'Is it any wonder that the British rule over us?' My eating meat was for the purpose of fitting myself for this fighting in future. These were the views I held before I left for England. It was my desire to keep, even at the cost of my life, the promises which I had given to my mother that saved me from eating meat and other sins. My love of truth has saved me in many difficult situations.

It was at this time that, coming into contact with two Englishmen, I was induced to read the *Gita*. I say 'induced' because I had no particular desire to read it. When these two friends asked me to read the *Gita* with them, I felt rather ashamed. The consciousness that I knew nothing about our holy books made me feel miserable. The reason, I think, was my vanity. I did not know Sanskrit well enough to be able to read the *Gita* without help. The two English friends, on their part, did not know Sanskrit at all. They gave me Sir Edwin Arnold's excellent translation of the poem. I went through the whole of it immediately and was fascinated by it. From that time till now, the last nineteen stanzas of Chapter II have ever remained engraved in my heart. For me, they contain the essence of *dharma*. They embody the highest knowledge. The principles enunciated in them are immutable. The intellect, too, is active in them in the highest degree, but it is intellect disciplined to high purpose. The knowledge which they contain is the fruit of experience.

This was my first introduction to the *Gita*. Since then, I

have read many other translations and commentaries and listened to many discourses but the impression made by that first reading persists. These stanzas are the key to the understanding of the *Gita*. I would even go so far as to advise people to reject statements in the poem which bear a meaning contrary to that of these nineteen stanzas. For a person who is humble there can be no question of rejecting anything. He will merely reason: 'It is the imperfection of my own intellect that today other stanzas seem to me inconsistent with these. In the course of time, I shall be able to see their consistency.' So he will tell himself and others, and leave the matter there.

For understanding the meaning of the Shastras, one must have a well-cultivated moral sensibility and experience in the practice of their truths. The injunction against *Sudras* studying the Vedas is not altogether unjustified. A *Sudra*, in other words a person without moral education, without sense and without knowledge, would completely misread the Shastras. No person, even if grown up in age, is qualified to understand difficult problems in Algebra without preparation. Before anyone can understand such problems, he must have studied the elements of the subject. How would '*Aham Brahmasmi*'[2] sound in the mouth of a lustful man? What meaning, or distorted meaning, would he not attach to it?

Hence, anyone who offers to interpret the Shastras must have observed the prescribed disciplines in his life. A mechanical observance of these disciplines is as futile as it is difficult. The Shastras regard it essential that one should have a *guru*. But *gurus* are rare in this age and, therefore, wise men of learning advise regular study of books in regional languages which are steeped in the spirit of devotion. Those, however, who are devoid of this spirit and lack even faith, are not qualified to explain the meaning of the Shastras. Learned men may please themselves and draw seemingly profound meanings from the Shastras, but what they offer is not the real sense of these. Only those who have experience in the practice of their truths can explain the real meaning of the Shastras.

There are, however, principles for the guidance of the common man too. Any interpretation of a *shastra* which is opposed to truth cannot be right. The Shastras are not meant for those who question the validity of the principle of truth itself, or, rather, the Shastras are no better than ordinary books for such a person. No one can meet him in argument. Anyone, on the other hand, who does not find the principle of non-violence in the Shastras is indeed in danger, but his case is not hopeless. Truth is a positive value, while non-violence is a negative value. Truth affirms. Non-violence forbids something which is real enough. Truth exists, untruth does not exist. Violence exists, non-violence does not. Even so, the highest *dharma* for us is that nothing but non-violence can be. Truth is its own proof, and non-violence is its supreme fruit. The latter is necessarily contained in the former. Since, however, it is not evident as truth is, one may try to discover the meaning of the Shastras without believing in it. But the spirit of non-violence alone will reveal to one the true meaning of the Shastras.

Tapascharya is certainly necessary for the realization of truth. Some sage who had realized truth revealed to the world the goddess of non-violence from amidst the prevailing violence, and said: 'Violence comes of illusion; it avails not. Non-violence alone is true.' Without non-violence, it is not possible to realize truth. The vows of *brahmacharya*, non-stealing and non-possession are of importance for the sake of non-violence, they help one to realize it in oneself. It is the life-breath of truth. Without it, man is a beast. The seeker after truth will discover all this very early in his quest, and then he will have no difficulty at any time in understanding the meaning of the Shastras.

The second rule to be followed in determining the meaning of a text in a *shastra* is that one should not stick to its letter, but try to understand its spirit, its meaning in the total context. Tulsidas's *Ramayana* is one of the greatest works because its spirit is that of purity, compassion and devotion to God. An evil fate awaits one who beats his wife because Tulsidas has said in his work that a *Sudra*, a dull-witted person, a beast and a woman merit chastisement. Rama not only never raised his hand against Sita, he did not even displease her at any time. Tulsidas merely stated a common belief. He could never have thought that there would be brutes who might beat their wives and justify their action by reference to his verse. May be Tulsidas himself, following the practice of his time, used to beat his wife; what even then? The practice does not cease to be reprehensible. In any case, his *Ramayana* was not composed to justify men beating their wives. It was composed to display the character of a perfect man, to tell us about Sita, the noblest among chaste and devoted wives, and to delineate the ideal devotion of Bharat. The support which the work seems to lend to evil customs should be ignored. Tulsidas did not compose his priceless work to teach geography. We should, therefore, reject any erroneous statements of a geographical character which we may find in it.

Let us now examine the *Gita*. Its subject-matter is simply the realization of *Brahman* and the means thereto; the battle is only the occasion for its teaching. One can say, if one likes, that the poet used it as an occasion because he did not look upon war as morally wrong. On reading the *Mahabharata*, I formed quite a different impression. Vyasa wrote his supremely beautiful epic to depict the futility of war. What did the Kauravas' defeat and the Pandavas' victory avail? How many among the victors survived? What was their fate? What was the end of Kunti, mother of the Pandavas? What trace is left today of the Yadava race?

Since the *Gita*'s subject is not description of the battle and justification of violence, it is perfectly wrong to give much importance to these. If, moreover, it is difficult to reconcile a few of the verses with the idea that the *Gita* advocates non-violence, it is still more difficult to reconcile the teaching of the work as a whole with the advocacy of violence.

When a poet composes his work, he does not have a clear conception of all its possible implications. It is the very beauty of a good poem that it is greater than its author. The truth which a poet utters in his moment of inspiration, we do not often see him following in his own life. Hence the lives of many poets are at variance with the teaching of their poems. That the overall teaching of the *Gita* is not violence but non-violence is evident from the argument which begins in Chapter II and ends in Chapter XVIII. The intervening chapters propound the same theme. Violence is simply not possible unless one is driven by anger, by ignorant love and by hatred. The *Gita*, on the other hand, wants us to be incapable of anger and attain to a state unaffected by the three *gunas*.[3] Such a person can never feel anger. I see even now the red eyes of Arjuna every time he aimed an arrow from his bow, drawing the string as far as his ear.

But, then, had Arjuna's obstinate refusal to fight anything to do with non-violence? In fact, he had fought often enough in the past. On the present occasion, his reason was suddenly clouded by ignorant attachment. He did not wish to kill his kinsmen. He did not say that he would not kill anyone even if he believed that person to be wicked. Shri Krishna is the Lord dwelling in everyone's heart. He understands the momentary darkening of Arjuna's reason. He, therefore, tells him: 'You have already committed violence. By talking now like a wise man, you will not learn non-violence. Having started on this course, you must finish the job.' If a passenger travelling in a train which is running at a speed of forty miles an hour suddenly feels aversion to travelling and jumps out of the train, he will have but committed suicide. He has not in truth realized the futility of travelling as such or of travelling by train. Arjuna was in a similar condition. Krishna, who believed in non-violence, could not have given Arjuna any advice other than what he did. But to conclude from this that the *Gita* teaches violence or justifies war is as unwarranted as to argue that, since violence in some form or other is inescapable for maintaining the body in existence, *dharma* lies only in violence. The man of discriminating intellect, on the other hand, teaches the duty of striving for deliverance from this body which exists through violence, the duty, that is, of striving for *moksha*.

But whom does Dhritarashtra represent, and likewise Duryodhana, Yudhishthira, or Arjuna? Whom does Krishna represent? Were they historical personages? Does the *Gita* relate their actual doings? Is it likely that Arjuna should suddenly, without warning, ask a question when the battle was about to commence, and that Krishna should recite the whole *Gita* in reply? And then, Arjuna, who had said that his ignorance had been dispelled, forgets what he was taught in the *Gita*, and Krishna is made to repeat his teaching in the *Anugita*.

Personally, I believe that Duryodhana and his supporters stand for the Satanic impulses in us, and Arjuna and others stand for God-ward impulses. The battle-field is our body.

The poet-seer, who knows from experience the problems of life, has given a faithful account of the conflict which is eternally going on within us. Shri Krishna is the Lord dwelling in everyone's heart who is ever murmuring His promptings in a pure *chitta* like a clock ticking in a room. If the clock of the *chitta* is not wound up with the key of self-purification, the in-dwelling Lord no doubt remains where He is, but the ticking is heard no more.

I do not wish to suggest that violence has no place at all in the teaching of the *Gita*. The *dharma* which it teaches does not mean that a person who has not yet awakened to the truth of non-violence may act like a coward. Anyone who fears others, accumulates possessions and indulges in sense-pleasures will certainly fight with violent means, but violence does not, for that reason, become justified as his *dharma*. There is only one *dharma*. Non-violence means *moksha*, and *moksha* means realizing *Satyanarayana*.[4] But this *dharma* does not under any circumstances countenance running away in fear. In this world which baffles our reason, violence there will then always be. The *Gita* shows the way which will lead us out of it, but it also says that we cannot escape it simply by running away from it like cowards. Anyone who prepares to run away would do better, instead, to kill and be killed.

If the verses cited by the correspondent cannot be understood even after this explanation, I cannot explain them. I am sure no one doubts that God, who is omnipotent, is, and must be, the Creator, the Preserver and the Destroyer of the Universe. He who creates has certainly the right to destroy. Even so, He does not kill, for He does nothing. God is so merciful He does not violate the law that every creature that is born will die one day. If He were to follow His fancies and whims, where should we be?

'Meaning of the *Gita*' (G.)
Navajivan, 11 Oct. 1925

[1] The letter is not translated here. The correspondent had argued that Chapters I and XI of the *Gita* did not seem to support the view that it taught non-violence.
[2] 'I am the *Brahman*, the Absolute', the central teaching of Advaita Vedanta.
[3] *Sattva* (purity), *rajas* (restlessness), and *tamas* (torpidity).
[4] God in the form of Truth.

39. FUNDAMENTAL PRINCIPLES

Bangalore,
July 2, 1927

Shri Santoji Maharaj,

I have carefully preserved your questions, and with God's grace I shall try to answer them now. I enclose the questions with this answer so that you may not have to take the trouble of recalling them and I may not have to copy them. I have numbered them serially, and there should be, therefore, no confusion.

1. Only he can interpret the *Gita* correctly who tries to follow its teaching in practice, and the correctness of the interpretation will be in proportion to his success in living according to the teaching. The *Gita* was not composed as a learned treatise. It may be a profound one, but in my view the realization of its profound quality depends on the depth of one's sincerity in putting its teaching into practice. I have read Lokamanya Tilak's and Shankaracharya's commentaries and tried to understand them as well as I could. I am not qualified to pronounce judgment on their learning. If we accept the point of view I have suggested, the question of expressing an opinion on their learning does not arise. The *Gita* is related to the Vedas and the Upanishads, for it gives the essence of both.

2. The *Gita* teaches that one should cultivate the state of *samatva* and explains with every manner of argument the means of doing so, namely, *bhakti* accompanied with *jnana*, that is, service of every living creature without thought of reward.

3. The godly heritage, according to the *Gita*, is that which helps one to attain self-realization. The sign of having acquired such heritage is the weakening of one's attachments and aversions, and the means of acquiring it is cultivation of *bhakti* for the Lord.

4. As far as I have been able to read and understand the teachings of our sacred writers of past times, I don't think there is any difference of view.

5. A *Brahmin* and a *Bhangi* are stung by a scorpion. I see

that a *vaid* has arrived to help the former, and another has been sent for. The *vaid* simply does not look towards the *Bhangi* who has been crying for help. Both the *vaid* and the *Brahmin* hear his cries. If the *Brahmin* has learnt to regard all beings with an equal eye, he would ask the *vaid* to treat the *Bhangi* first. If I were the *vaid*, I would run to the *Bhangi*, suck out the poison from the wound and apply to him whatever other remedies I knew. Having done that, I would offer my service to the *Brahmin*, if he required it, and then attend to my other work. To cultivate an equal eye towards all beings means to serve all people in the world with equal regard.

6. It is not true at all that the Lord teaches in the *Gita* that one may kill one's kinsmen. When Arjuna got ready to fight for what he believed to be a just cause, and was overcome with ignorant attachment and weakness, through which he made a distinction between kinsmen and others, the Lord cured him of his attachment and weakness. We can only guess what reply Krishna would have given if Arjuna had protested and said that he did not wish to kill at all, whether his opponents were kinsmen or others. It is my humble view, however, that the *Gita* was not composed to give a direct answer to that question.

7. I have felt that there is a greater spirit of universality and liberalism in the *sanatana* Vedic *dharma*.

8. The question, which is the chief religious work, one can answer only for oneself. For me it is the *Gita*. One distinction at any rate among such works is based on the subject they deal with, namely, those which lay down codes of conduct and others which state and explain the nature of Godhead. If the question refers to anything more than this, I have not understood it.

9. The differences between the rules of conduct enjoined by the various religions will change from age to age, and as knowledge and the spirit of liberalism grow such differences will diminish.

10. I think this question is involved in question 9. But I shall say this by way of further explanation. We should believe that the practices and modes of conduct which we find represented in the Koran, the Bible, the Vedas and other works were the best in those times and those lands. If our reason

cannot accept them in this age, it is our *dharma* to change them or abandon them altogether. Only fundamental principles are immutable.

11. In acting towards other people and understanding their faiths one should follow the principle of regarding others as oneself.

12. I think it almost impossible to decide which out of the many interpretations of religious works represents undiluted truth. That is why the *Gita* has convincingly argued that *samanvaya* is the best. God alone represents perfect truth. Imperfect man, therefore, should humbly believe that as one's truth is dear to one, so others' truths are bound to be dear to them. Hence everyone should follow his own path and others should not hinder him from doing so. People will then follow, of their own accord, that path which is found from experience to be the smoothest.

13. Till we come across a person of experience and of perfect purity in conduct, we should observe the rules of conduct and discipline enjoined in the religious work which we have accepted as our scripture, read it regularly, reflect over it and put its teachings into practice. Those who cannot do even this are ignorant people. Such of them as cannot observe purity in conduct have a protector only in God. I believe in the assurance given in the *Gita* that even people of this class are saved somehow. Words certainly have a meaning, but there are ebbs and floods in the meaning of words as if they had a life of their own.

14. As I understand the problem, without a belief in reincarnation it would be almost impossible to prove that the world is governed by justice. Moreover, one soul cannot have experience of the world within the span of one life, which is but a moment in a vast cycle of time. I can practically say that I have direct proof every moment of the truth of the belief in reincarnation.

15. Virtue and sin exist in the same sense that light and darkness, happiness and suffering, truth and untruth, do. However, just as there is an unknowable and indescribable Reality beyond the categories of existence and non-existence, so also there is something beyond virtue and sin of which this body can have no experience. The descriptions given in Bud-

dhist writings or in those of the Nyaya and the Sankhya schools of philosophy are not unalterable, but these too, can be understood and accepted from the respective viewpoints.

16. For the development of man's reasoning faculty, the right influences are certainly essential. To put it scientifically, every society solves the problem in every age for itself.

17. Violence means injuring a creature through bodily action or speech or in thought, with the intention of injuring it. Non-violence means not injuring any creature in this manner. The doctrine of non-violence propounded in *Vedanta* literature, as far as I have understood it, seems to me all right. But I cannot say myself whether I have understood the *Vedanta* teaching correctly, nor can I claim that my study of the *Vedanta* is deep.

18. For being able to observe *brahmacharya*, one must keep the mind, speech and body constantly engaged in morally pure activity. We can, therefore, say in a general way that a *brahmachari* should do the opposite of what householders given to the enjoyment of pleasures do. It is my experience that desires in the mind are closely connected with the kind of food one eats, but I know that such desires arise even when one's food is pure and small in quantity. We may, therefore, say that right food is a great help for the observance of *brahmacharya*, but that it is not all. The purest food is fruit, which has naturally ripened, eaten in solitude. I have no doubt about this at all. The chief thing is that, if the palate could be controlled, the observance of *brahmacharya* would become quite easy. We commit two errors in saying that knowledge has reference to the mind and food is connected with the body, which is a material object. The body of a living human being is not altogether inert. And the thing which we call mind and which learns from experience is as closely connected with the body as light is with the sun. A dead body is one from which the mind has gone; such a body does not eat or drink. Thus, it is the mind which eats through the body, and likewise it is through the body that really speaking it acquires knowledge.

19. The universal Soul in which all the souls exist is God. The living creature which does not know that universal Soul and looks upon itself as separate from other creatures is what we call *jiva*. That universal Soul, though dwelling in all, is not

directly experienced; that is its beauty, its miracle, its *maya*. The true end of human effort consists in crossing that *maya* and knowing that universal Soul, which is the one source of all. It is not a thing which can be experienced in a manner our reason can understand; how, then, can there be any means of experiencing it? But anyone who has the strength to forget the 'I' in him and make himself a cipher can have a glimpse of this universal Soul, though he cannot help someone else to have it too. Such a person is so dazzled by the mere glimpse, and so utterly spell-bound, that he merges in it. He feels no desire and no need to describe his supreme bliss to anyone.

20. I have somehow found something for myself by combining all the suggestions made by the authors of *shastras*. It is, therefore, quite difficult for me to say which path is acceptable. Shankara is dear to me, and so are Ramanuja, Madhva, Vallabha and others—I have relished delicacies from all, but have not been able to satisfy my hunger through what I got from any of them.

21. It may now be seen that the reply to this question is contained in the replies to the preceding questions. *Yajna*, *dana*, *tapas*, are obligatory duties, but that does not mean that the manner of performing them in this age should be the same as in ancient times. *Yajna*, *dana*, etc., are permanent principles. The social practices and the concrete forms through which they are put into practice may change from age to age and country to country. In my view, for example, the supreme *yajna* for this country and in this age is the spinning-wheel. The right gift which a seeker of *moksha* in this country and this age may make is to dedicate his all, body, intellect and possessions, to the service of the country. And, likewise, the right *tapas* for this country and this age consists in burning with agony at the suffering of countless untouchables and others who are starving for want of food or because of famines. Anyone who performs these three important duties certainly becomes purified and he may even have a vision of God's cosmic form which Arjuna had.

22. *Saguna*, *nirguna*, etc., belong to the speech of human beings trying to express their imperfect knowledge which is sheer ignorance. In truth God is beyond description. Even to call Him *nirguna* is an utterly vain attempt to describe Him.

Since, however, He is a slave of His devotees, He can be described not with a thousand but with an infinite number of epithets, all of which can be applied to him from the point of view of the respective devotees, and it is His supreme mercy that He tolerates them all. There is no error, therefore, in saying that He exists as all bodies, all sense-organs and all other things. We may thus confess our inability to describe Him.

23. I wish to say in all humility that my fasts and other ordeals are inspired by the desire to see God face to face. I fast in order that, even if I go without food altogether, I may have a glimpse of God such as I have described in these answers. But one cannot force oneself to undertake a fast. One must have fitness even for a fast. I am constantly striving for such fitness. But it is possible that I may not acquire it in this life and may even die ingloriously.

This completes the list of your questions. If you have any other questions to ask besides these, do ask them. You will notice a certain positiveness in some or all of the answers. Please do not put it down to presumption or pride. If I did not write what I have done, I would be guilty of untruth, for it would mean concealing my real beliefs through false humility. If, therefore, you see impropriety in my positiveness, kindly forgive me for that.

The revered Malaviyaji Maharaj is here. I have discussions with him on all manner of religious issues. I will tell him of your desire too.

Yours,
MOHANDAS GANDHI

Letter to Santosh Maharaj (G.)
SN 12323

40. THE GOSPEL OF SERVICE

[23 October 1927]

In declaring the *Gita* class open Mahatmaji advised the students to get up at 4 o'clock in the morning and regularly read the *Bhagavad Gita* daily. He was anxious that they should begin the study of the *Gita* in right earnest. If they could not read Sanskrit they could go in for a Tamil translation of the *Gita*, but not the English one, because the English rendering could not impart the true significance of the *Gita*. He said that the third chapter is an important one in the *Gita*.

The *Gita* contains the gospel of *karma* or work, the gospel of *bhakti* or devotion and the gospel of *jnana* or knowledge. Life should be a harmonious whole of these three. But the gospel of service is the basis of all, and what can be more necessary for those who want to serve the country than that they begin with the chapter enunciating the gospel of work? But you must approach it with the five necessary equipments, viz., *ahimsa* (non-violence), *satya* (truth), *brahmacharya* (celibacy), *aparigraha* (non-possession), and *asteya* (non-stealing). Then and then only will you be able to reach a correct interpretation of it. And then you will read it to discover in it *ahimsa* and not *himsa*, as so many nowadays try to do. Read it with the necessary equipment and I assure you you will have peace of which you were never aware before.

Speech to Students, Tiruppur
The Hindu, 25 Oct. 1927
Young India, 3 Nov. 1927

41. SELF-SURRENDER

January 7, 1928

This chapter[1] is as sweet as the manner in which it has been sung. It is balm for people like us who suffer from inner torments. We are all troubled by evil desires, and the Lord holds out an assurance here to all those who seek refuge in Him that He will free them from such desires. We can also see from this chapter that at the time when the *Gita* was composed, distinctions of high and low had already made their appearance in the *varnashrama* system and some had come to be regarded as lower than others. In truth, however, can we describe anyone as higher or lower than others? Let him who is completely free from evil desires point an accusing finger at another. All of us are equal in this regard, and this chapter points out the unfailing means of winning freedom from evil cravings, namely, total surrender to the Lord. We should not of course conclude from this that such surrender will by itself, without further effort on our part, purify us of our evil. If a person who is dragged by his senses, against his will, to objects of pleasures turns to God for help, with tears in his eyes, as he ceaselessly struggles against them, the Lord will certainly free him from the evil desires which trouble him.

Another thought also arises out of it, but we shall consider it tomorrow.

That thought is that this also explains the means of expiation of one's sins. Such expiation consists not in fasting, but in *bhakti*, in self-surrender. I fully understand the usefulness of fasting but there are also limits to it. Fasting can never expiate a sin; it may, on the contrary, serve to cover it up. A sinner is one who has committed a sin, but *papayoni* means one born of sin itself and therefore the most wicked of all sinners. We cannot say to whom this notion first occurred, but everyone is assured of freedom through total surrender to the Lord. Expiation of sin consists in *bhakti*. Fasting may seem necessary sometimes as a means of filling the heart with *bhakti*, and everyone may decide for himself when it is so. The real means, however, is *bhakti*, and *bhakti* means reducing oneself to a cipher, erasing the 'I'. If we can do this, no matter how many our sins in the past, they will not stand in the way of our freedom. The thoroughly wicked mentioned in this chapter are none else but ourselves. We, who commit all manner of sins in our hearts and move in the world as respectable people, are sinners, all of us, and in this chapter the Lord holds out an assurance to us.

Chapter XIV contains a description of the three *gunas* and Chapter XV a description of *Purushottama*. Thirty years ago I read Drummond's book in which he had established by means of several examples that the rules of the material world apply to the spiritual world also. We see it proved in this world which is made of the three *gunas*. There are not three *gunas* but many; those three are the major divisions of the many. He who transcends those three becomes one with *Purushottama*. None has been born in this world who could exist with only one *guna*. Even if a man possesses a high degree of the *sattvik guna*, the latter still includes something of the *gunas* of *tamas* and *rajas*. The example of water occurs to me. In the form of ice, it remains like a stone. But when it boils and becomes steam, it rises in the sky. As ice it lacks the capacity to rise, but as steam, it ascends higher and higher. Its highest power appears in the form of steam. And finally, ceasing to be steam, it becomes a cloud and benefits the world in the form of rain. However, if steam becomes ice, it lies dead and still. Ice too has its uses. Melted ice flows in the form of rivers. It also causes floods but we are not concerned with that. It is a proven fact that without the sun, even water cannot turn into steam. But that shows that it cannot do without someone else's help. The long and short of what I want to say is that steam points to a state of *moksha*. It benefits the world in its state of *moksha*. This is how we should understand the meaning of these two chapters.

Discourses on the *Gita* (G.)
Mahadev Desai's Diary (MSS)

[1] *Bhagavad Gita*, ch. IX.

42. NON-ATTACHMENT

It is more than forty years since I have been reading, pondering and following the *Gita*. Friends expressed a desire that I should put before the Gujaratis my understanding of that work. I embarked on translating it. From the point of view of scholarship my qualifications for attempting the translation would seem to be nil, but as one following its teaching I may be considered to be fairly well qualified. The translation has now been published. Many editions of the *Gita* carry the Sanskrit text also. It has been intentionally left out in this one. I would like it if all knew Sanskrit, but all will never learn Sanskrit. Moreover, many cheap editions with Sanskrit are available. I, therefore, decided to leave out the Sanskrit text and reduce both the size and the price. This edition has 19 pages of introduction and 187 pages of translation so that it can be easily carried in one's pocket. Ten thousand copies have been printed.

My aspiration is that every Gujarati will read this *Gita*, ponder it and practise its teaching. An easy way to ponder it is to try and understand the meaning without referring to the Sanskrit text and then to conduct oneself accordingly. For example, those who interpret the teaching of the *Gita* to mean that one ought to kill the wicked without making a distinction between one's kinsmen and others, should kill their parents or other relatives if they are wicked. In practice, they will not be able to do so. Naturally, then, it would occur to the reader that where destruction is prescribed the work contemplates some other form of destruction.

Almost every page of the *Gita* advises us not to make a distinction between our own people and others. How is this to be done? Reflection will lead us to the conclusion that we should perform all our acts in a spirit of non-attachment. In the very first chapter we find Arjuna facing the troublesome question of one's own people and others. In every chapter the *Gita* brings out how such a distinction is false and harmful. I have called the *Gita* 'Anasaktiyoga'. The interested reader will be able to learn from the work what that is, how non-attachment is to be cultivated, what its characteristics are, etc. Trying as I do to live according to the teaching of the *Gita*, I could not avoid the present struggle. As a friend says in his telegram to me, this is a holy war for me. It is a happy augury for me that this book is being published just as the holy war is entering its last phase in the form of the present struggle.

Bhagavad Gita (*Anasaktiyoga*) (G.)
Navajivan, 16 Mar. 1930

43. THE SPIRIT OF THE *GITA*

September 29, 1930

Dear Sister,

I have your two letters to answer. First, Satis Babu's question. The literal interpretation of the battle in Chapter II of the *Gita* may certainly be physical conflict. But I have not a shadow of doubt that the spirit of the *Gita* inclines us to interpret it as mental conflict. The moment this is doubted, the *Gita* will cease to be sacred scripture for me.

I hope you are well. After consulting Vinoba, do send the boys desirous of learning Sanskrit to Wardha. Chhotelal is now in jail. I am very glad to note even the slight improvement in Tarini. Give my blessings to Tarini, Charu, Arun and others. The *tapovanas* like those of ancient India can exist even now, it all depends upon our *tapascharya*. Certainly their form may be different. Our duty lies in going a step ahead of the achievements of our ancestors.

Blessings from
BAPU

Letter to Hemaprabha Das Gupta (H.)
GN 1671

44. THE SYNTHESIS OF THE *GITA*

March 31, 1932

As I had no news from you at all, at last I wrote to Taramati. Then unexpectedly I got your letter yesterday and felt happy. I do not worry about your loss of weight. Your health should remain good in other respects. From the point of view of climate, Nasik is excellent. Write to me regularly from wherever you may be. If writing to me comes in the way of your writing to others, ask one of those to whom you must write to convey your message to me. The understanding between the authorities here and me is that I can write to any of my fellow-prisoners. This should mean that, if the prisoner chooses to reply to me, his letter should not count among the number he is entitled to write.

... A prisoner ought not to even think of the outside world. *Karmayoga* for him would mean doing the duty which now has become his in jail.

In the verse[1] 'seek refuge in the attitude of detachment', etc., the word *buddhi* certainly does not mean the Samkhya, or the path of knowledge. Here it has the meaning which it bears in the phrase *yogabuddhi*, about which the Lord had promised to explain in the verse 'Thus have I set before thee the attitude of Knowledge; hear now the attitude of Action', etc. That is to say, if any comparison is to be made at all, it is between *karmayoga* and *bhaktiyoga*. But instead of making a comparison between *jnanayoga*, *karmayoga* and *bhaktiyoga*, the *Gita* aims at a synthesis of the three. To practise any one of them perfectly, one must also practise the other two. That is, the three are inseparable. There is some indication of the superiority of *karmayoga* from among the three, but only because there is less risk in it of being deceived. 'Here no effort undertaken is lost, no disaster befalls',[2] etc. I believe that we must first grasp the general purport of the book we are trying to study. In other words, our thinking about it should be independent.

As for rebirth, there is no room for doubt at all. Our body changes a little every day, so that in seven years the whole gets changed altogether. It seems to be the same because the form remains the same. If the body can be transformed every seven years, there is no reason whatever to think that what we know as death means complete annihilation. The *atman* is different from the body and it is certainly not destroyed when the body is destroyed. Then, all that takes place at death is a change of state. If a change of state is possible, why not a change of the body? However, we can have a full discussion of this subject only when we meet. Meanwhile, I certainly wish that you should be able to get rid of your doubts about certain conclusions. For this purpose, you may put me any questions that you wish.

I don't think that I can write a *smriti*. What I write or say has not been thought out in conformity with a system. I have only enough strength to enable me to meet every situation as it arises in the course of my quest for truth. That is to say, I am no *shastri*, and only a person learned enough to write a *shastra* can compose a *smriti*. This suggestion was first made to me by Kishorelal. I would certainly write such a treatise if I could. I would be fortunate, however, if I can use rightly whatever powers God has given me.

We three are quite well. The third is Mahadev.

Letter to Mathuradas Trikumji (G.)
Bapuni Prasadi, pp. 105–7

[1] *Bhagavad Gita*, II.49.
[2] Ibid. II.40.

45. PROBLEM-SOLVING

May 10, 1932

When deciding what one's duty is, many questions may arise. But when examining the teaching of the *Gita*, we have only to consider what Arjuna's question was. If a teacher goes beyond the scope of the question put to him, he would be a bad teacher, for the questioner's attention would be fixed on his question and he would be in no fit state of mind to listen to other things. The discussion may even be beyond his capacity and may, therefore, bore him. Further, the reply to his question is likely to get lost in the discussion about irrelevant issues, in the same way that plants of grains in a field sometimes get buried under the weeds which grow round them. From this point of view, Krishna's reply to Arjuna is perfect. And when we leave Chapter I and come to Chapter II, we find that it breathes the pure spirit of *ahimsa*. We should not, believing or claiming that Shri Krishna is the most perfect *Avatar* of God, expect that we can get from his words a simple reply to every question that may arise in our minds in the same manner that we can find the meaning of a word by looking up a dictionary. Even if we could find such replies, that would do us harm, for in that case nobody would strive for further advance or new discoveries. Man's reason would simply atrophy.

Thus, the people of every age should strive to solve the problems of that age by their own hard labour and *tapascharya*. We should, therefore, try to solve the problems which arise in the present age in regard to war and other issues with the help of the principles enunciated in a noble work like the *Gita*. But even such help will be negligible. We shall get real help only from what we learn through *tapascharya*. Books on Ayurveda describe the properties of innumerable herbs and plants. We may read those descriptions for guidance. But the knowledge so gained will be useless, may even be burdensome, if the properties attributed to those herbs and plants do not pass the test of actual experience. We have to seek solutions to the many difficult problems of life in exactly the same manner. If you still wish to ask me any questions on this subject, please do.

A Letter (G.)
Mahadevbhaini Diary, Vol. I, pp. 146-7

46. A SPIRITUAL DICTIONARY

Segaon, Wardha,
September 24, 1936

Kakasaheb has explained in his own foreword why this *kosha*, which was prepared twelve years ago and which has not come up to the required standard yet, is being published today.

This *kosha* is indeed necessary for those who are at all interested in the translation which has been published under my name. Maybe it will be useful to other students of the *Gita* as well. To them I would suggest that if they do not like the interpretations given in the *Padarthakosha* and prefer others, they should note these down in the book itself. By doing so, they will have a glossary of their own liking with very little effort. And if such students send me the interpretations of their choice, I shall be grateful.

The more I study the *Gita*, the more I am made aware of its uniqueness. For me, it is a spiritual dictionary. Whenever I am in doubt as to what I should or should not do, I fall back upon it, and so far it has never disappointed me. It is indeed a Kamadhenu. In the beginning, we recited one *sloka* a day, then two, then five, then a chapter a day, then a complete recitation in two weeks and, finally, in the last few years some of us have been able to do a complete recitation in seven days. And on particular days of the week the sound of particular chapters is heard around 4.30 a.m. Some—a very few—have learnt all the eighteen chapters by heart. The morning prayers include the chapters in the following order during the days of the week.

Friday I, II
Saturday III, IV, V
Sunday VI, VII, VIII
Monday IX, X, XI, XII
Tuesday XIII, XIV, XV
Wednesday XVI, XVII
Thursday XVIII.

Suffice it to say that there has been some deliberation behind this schedule. Our experience has been that this order facilitates understanding of the text.

It may well be asked why the recitation has to begin on a Friday. The reason for this is simply the following: For considerable length of time a complete recitation used to take fourteen days. In the Yeravda jail I thought of a seven-day recitation and the idea was put into effect on a Friday. In this way and since that day, the recitation week begins on a Friday.

There are two reasons why I mention the matter of complete recitation here: to show how far devotion to the *Gita* has taken some of us up to date and to indicate to the reader the way to encourage its study.

But we are not to rest content with merely reciting the *Gita*. It is a spiritual dictionary; it is a tremendous force for straightening the creases of the soul; it is the refuge of the afflicted; it is an awakener from stupor. Only those who have this faith can derive help from reciting the *Gita*. I do not in the least suggest that recitation of the *Gita* without an understanding of its meaning is by itself beneficial to a man. With sufficient effort a tame parrot can certainly be taught to memorize the *Gita*. But that is not going to bring the slightest merit either to the parrot or to its trainer.

The *Gita* is the living, life-giving, immortal mother. The mother who suckled us will one day disappear and leave us in the lurch. We see countless mothers failing to protect their children from peril. But one who seeks the support of Mother *Gita* saves himself from the greatest peril. She is ever wakeful and she never lets one down. But a mother will not give one nourishment unless one asks for it. Mother *Gita* too gives nothing without our asking for it. Before taking anyone under her wing she tests him severely, expects whole-hearted devotion. The devotion, too, should not be dry. She wants single-minded devotion. Hence, she plainly refuses to help those who are not prepared to surrender their all to her.

The student of physics gains an insight into his subject only when he is crazy after it. Those who obtain M.A. and B.A. degrees read day and night, spend money and put themselves to great physical strain. From amongst those who make such efforts, only a few are successful at the first attempt. Those who fail do not despair but try again and again and only rest when they have passed. And in the end—?

For those who wish to drink the nectar that is the *Gita*, much greater effort than this should be and is required. But how many want to partake of that nectar? Even of those who feel the need, how many are prepared to make a strenuous effort? We know that the number of people devoted to the *Gita* in the way I have suggested is negligible. Yet all agree that the *Gita* is the essence of all the Upanishads. No Hindu should remain ignorant of it. At present, however, all religion has depreciated in value. This is not the occasion to go into the reasons for this. In this note of appeal all that I have tried to do on the occasion of the publication of this volume is to draw the attention of seekers to the gem that is the *Gita* and to suggest how it can be put to good use. May it bear fruit.

MOHANDAS KARAMCHAND GANDHI

Preface (G.)
Gitapadarthakosha, pp. iii–vi
Harijanbandhu, 25 Oct. 1936

47. NEED FOR PATIENCE

[Before 22 January 1946]

Q. In the *Bhagavad Gita* comes a *sloka*: ... If everything goes according to fixed law, then if you pray to God, can He intervene and set aside the law for your sake?

A. But the meaning at the bottom is: 'Thy will be done, not mine.' God's law is never interrupted, but that very law says that every act has its result. In the prayer for the child, father surrenders his will to God. The law works as if it were a person, and since every action tells, this prayer has an unforseeable result. You should work out what I have written. The sum total of every act is a resultant. Draw a parallelogram of forces.

Q. *Jnana, karma* and *bhakti*—should not they all go together? Unless you know God, how can you have *bhakti*? You cannot even offer him your *karma*.

A. You must not reason like this. If you have no work and so-called *bhakti*, the result will be lop-sided. Do you know God to whom you pray? I don't. He is unknown to you and me.

Q. To whom shall we then pray?

A. To God, whom you do not know. We always do not know the person to whom we are praying.

Q. But the Shastras say that He is knowable.

A. Since He is knowable we search. It may take a billion years. Even if you do not believe, you must continue to pray. 'Help thou mine unbelief' is a verse from the Bible. But it is improper for you to ask such questions. For you there must be infinite patience and inward longing. Inward longing obviates all such questions. 'Have faith and you will be whole' is another chip from the Bible. *Gita* has many such parallel passages.

Q. When I look at nature around me, I say to myself there must be one Creator, one God and to Him I should pray.

A. That is reasoning. God is beyond reason. But I have nothing to say, if your reason is enough to sustain you.

'Answers to Questions'
GN 3230

§ 4. Socrates and Thoreau

48. THE WISDOM OF SOCRATES

February 29, 1932

There was once a wise man, named Socrates, who lived in Athens. His unconventional ideas, which, however, spread love of truth and goodness, displeased the authorities, and he was sentenced to death. In that country people were in those days sometimes sentenced to die by drinking poison. Like Mirabai, Socrates was asked to drink a cup of poison. Our purpose here is to discuss the substance of the concluding words of Socrates's defence at the time of his trial. We can all derive a moral from it. Let us call Socrates by the name Sukrit. The Arabs knew him as Sokrat.

This is what Sukrit said: 'It is my unshakeable faith that no harm comes to a good man either in this world or the next. God never forsakes good men and their friends. I also believe that nobody ever dies before his time. I don't look upon the sentence of death as a punishment. The time has come for me to die and be delivered from the sufferings of this life. That is why you have condemned me to die by drinking poison. I am sure that my good lies in that. I, therefore, bear no anger against my prosecutors or against those who have condemned me. They may not have meant well by me, but they can do me no harm either.

'I have one request to make to the council of elders. If my sons forsake the path of goodness and follow that of evil, if they become lovers of wealth, punish them as you have punished me. Punish them also if they become hypocrites and try to show themselves other than what they are. If you do so, my sons and I will believe that you have acted justly.'

This is a request of Sukrit in regard to his sons. The elders of the city who had assembled to pass judgment on him, did not know the law of non-violence. He, therefore, made the fore-going request and thereby warned his sons and told them what he expected of them, and also rebuked the elders gently for punishing him for doing good. By suggesting to his sons that they should follow in his footsteps, he told them in effect that the path which he had shown to the citizens of Athens was also meant for them, and even went to the extent of saying that, if they did not follow that path, they should be treated as deserving punishment.

I had decided to let this week be blank. But I could not reconcile myself to the idea. Glancing at the books, I saw Socrates's speech. I opened the book, intending to send something from it, and my eyes fell on the world-famous passage in it which I have summarized above.

BAPU

'Death, a Friend' (G.)
MMU/I and II (Microfilm)

49. READING THOREAU

Camp Hardoi,
October 12, 1929

Dear Friend,

I was agreeably surprised to receive your letter. Yes, indeed your book[1] which was the first English book I came across on vegetarianism was of immense help to me in steadying my faith in vegetarianism. My first introduction to Thoreau's writings was I think in 1907 or later when I was in the thick of passive resistance struggle. A friend sent me Thoreau's essay on civil disobedience. It left a deep impression upon me. I translated a portion of that essay for the readers of *Indian Opinion* in South Africa which I was then editing and I made copious extracts from that essay for that paper. That essay seemed to be so convincing and truthful that I felt the need of knowing more of Thoreau and I came across your life of him, his 'Walden' and other short essays all of which I read with great pleasure and equal profit.

Yours sincerely,

Henry S. Salt, Esq.
21 Cleveland Road
Brighton (England)

Letter to Henry S. Salt
SN 15663

[1] *A Plea for Vegetarianism.*

50. THOREAU ON CIVIL DISOBEDIENCE

Wardha,
September 10, 1935

Dear Kodanda Rao,

The statement that I had derived my idea of civil disobedience from the writings of Thoreau is wrong. The resistance to authority in South Africa was well advanced before I got the essay of Thoreau on civil disobedience. But the movement was then known as passive resistance. As it was incomplete I had coined the word *satyagraha* for the Gujarati readers. When I saw the title of Thoreau's great essay, I began the use of his phrase to explain our struggle to the English readers. But I found that even civil disobedience failed to convey the full meaning of the struggle. I therefore adopted the phrase civil resistance. Non-violence was always an integral part of our struggle.

As per your advice, a copy is going to Mr. Pearson. I hope you have done well. Mahadev is in Bombay just now.

Yours sincerely,
M.K. GANDHI

Sjt. Kodanda Rao

Letter to P. Kodanda Rao
GN 6280

§ 5. Ruskin

51. RUSKIN ON EDUCATION

March 28, 1932

John Ruskin was a great writer, teacher and religious thinker. He died about 1880.[1] I suppose most inmates of the Ashram know that one book[2] of his had a great effect on me and that it was this book which inspired me to introduce an important change in my life practically on the instant. He started in 1871 writing monthly letters addressed to factory workers. I had read praise of these letters in some article of Tolstoy, but I had not been able to secure them till now. I had brought with me from England a book about Ruskin's work and his efforts in the field of constructive activities. I read it here. This book also mentioned the letters referred to above. So I wrote to a woman disciple of Ruskin in England, who was none else but the author of that book. Being a poor woman, she could not send me the volumes of these letters. Through foolishness or false courtesy, I had not asked her to write to the Ashram for the money. This good woman sent my letter to a friend of hers who was comparatively in better circumstances. This friend was the editor of *The Spectator*. I had even met him while in England. He sent me the four volumes in which these letters had been published. I have been reading the first part. The thoughts expressed in these letters are beautiful and resemble some of our own ideas, so much so that an outsider would think that the ideas which I have set forth in my writings and which we try to put into practice in the Ashram, I had stolen from these letters of Ruskin. I hope readers will understand what is meant by 'stolen'. If an idea or ideal of life is borrowed from somebody but is presented as one's own conception, it is said to be stolen.

Ruskin has discussed many matters. Here I will mention only a few of his ideas. He says that it is a sheer error to suppose, as is generally done, that some education however little or however faulty is better than no literary education at all. It is his view that we should strive for real education alone. And then he says that every human being requires three things and three virtues. Anyone who fails to cultivate them does not know the secret of life. These six things should therefore form the basis of education. Every child, whether boy or girl, should learn the properties of pure air, clean water and clean earth, and should also learn how to keep air, water and earth pure or clean and know their benefits. Likewise, he has mentioned gratitude, hope and charity as the three virtues. Anybody who does not love truth and cannot recognize goodness or beauty lives in his own self-conceit and remains ignorant of spiritual joy. Similarly, he who has no hope, who has, in other words, no faith in divine justice, will never be cheerful in heart. And he who is without love, that is, lacks the spirit of *ahimsa*, who cannot look upon all living things as his kith and kin, will never know the secret of living.

Ruskin has explained these ideas at great length in his wonderful language. I hope I shall be able to write about them some time in a language which all the inmates of the Ashram can understand. Today I rest content with the brief precis given above. But I will say one thing, that what Ruskin has explained in his finished and cultivated prose with English readers in view, is practically the same ideas which we discuss in our rustic language and which we have been trying to put into practice. I am comparing here not two languages, but two writers. I cannot hope to equal Ruskin's mastery of language. But a time will certainly come when the love of our language will have become universal and we shall have writers like Ruskin who will have dedicated themselves heart and soul to it and will write as powerful Gujarati as the English of Ruskin.

'Some Reflections on Education' (G.)
MMU/II (Microfilm)

[1] Actually in 1900.
[2] *Unto This Last.*

§ 6. Mazzini

52. PATRIOTISM OF MAZZINI

Italy as a nation came into existence recently. Before 1870 Italy comprised a number of small principalities, each with its petty chief. Before 1870, she was like the India or Kathiawad of today. Though the people spoke the same language and had the same character, they all owed allegiance to different petty states. Today Italy is an independent European country and her people are regarded as a distinct nation. All this can be said to be the achievement of one man. And his name— Joseph Mazzini. Joseph Mazzini was born in Genoa on June 22, 1805. He was a man of such sterling character, so good-natured and so patriotic, that great preparations are being made throughout Europe to commemorate the centenary of his birth. For, although he dedicated his whole life to the service of Italy, he was so broadminded that he could be regarded a citizen of every country. It was his constant yearning that every nation should become great and live in unity.

Even at the early age of thirteen, Mazzini showed great intelligence. In spite of great scholarship that he evidenced, he gave up his books out of patriotism and undertook the study of law, and began using his legal knowledge gratuitously to help the poor. Then he joined a secret organization which was working for the unification of Italy. When the Italian chiefs learnt of this, they put him into prison. While still in prison, he continued to advance his plans for freeing his country. At last he had to leave Italy. He went to Marseilles and lived there. The Italian princes, however, using their influence, had him banished from that city. Though obliged to fly from place to place, he did not lose heart and kept on sending his writings secretly to Italy, which gradually influenced the minds of the people. He suffered a lot in the process. He had to run about in disguise to evade spies. Even his life was frequently in danger, but he did not care.

At last he went to England in 1837. He did not suffer so much there but had to live in extreme poverty. In England he came into contact with the great leaders of that country and sought their aid.

In 1848 Mazzini returned with Garibaldi to Italy, and set up the self-governing State of Italy. But it did not last long, thanks to the activities of crafty persons and though Mazzini had to flee the country once again, his influence did not fade. The seed of unity that he had sown endured and, though Mazzini remained in banishment, Italy became a single united kingdom in 1870. Victor Emmanuel became its king. Mazzini was gratified to see his country thus united. But as he was not permitted to enter the country, he used to go there in disguise. Once when the police went to arrest him, he opened the door for them as if he were an usher and gave them the slip.

This great man died in March, 1873. His foes had now become his friends. People had come to recognise his true worth. Eighty thousand people joined his funeral procession. He was buried at the highest spot in Genoa. Today Italy and the whole of Europe worship this man. In Italy he is considered one of the greatest of men. He was a pious and religious man, ever free from selfishness and pride. Poverty was for him an ornament. The sufferings of others he regarded as his own. There are very few instances in the world where a single man has brought about the uplift of his country by his strength of mind and his extreme devotion during his own lifetime. Such was the unique Mazzini.

'Joseph Mazzini—A Remarkable Career' (G.)
Indian Opinion, 22 July 1905

§ 7. Tolstoy

53. TOLSTOY ON SIMPLICITY

It is believed that, in the western world at any rate, there is no man so talented, learned and as ascetic as Count Tolstoy. Though he is now nearly eighty years old, he is quite healthy, industrious and mentally alert.

Tolstoy was born of a noble family in Russia. His parents had enormous wealth, which he inherited. He is himself a Russian nobleman, and has, in his youth, rendered very good service to his country by fighting gallantly in the Crimean War. In those days, like the other noblemen of his time, he used to enjoy all the pleasures of the world, kept mistresses, drank and was strongly addicted to smoking. However, when he saw the carnage and bloodshed during the war, his mind brimmed over with compassion. His ideas changed; he began a study of his own religion and read the Bible. He read the life of Jesus Christ which made a deep impression on his mind. Not satisfied with the then current Russian translation of the Bible, he studied Hebrew, the language in which it was originally written, and continued his researches into the Bible. It was also about this time that he discovered in himself a great talent for writing. He wrote a very effective book on the evil consequences of war. His fame spread throughout Europe. To improve the morals of the people he wrote several novels which can be equalled by few books in Europe. The views expressed by him in all these books were so very advanced that the Russian clergy were displeased with him, and he was excommunicated. Disregarding all this, he kept up his efforts and began to propagate his ideas. His writings had a great effect on his own mind. He gave up his wealth and took to a life of poverty. He has lived like a peasant for many years now and earns his needs by his own labour. He has given up all his vices, eats very simple food and has it in him no longer to hurt any living being by thought, word or deed. He spends all his time in good works and prayer. He believes that:

1. In this world men should not accumulate wealth;
2. no matter how much evil a person does to us, we should

always do good to him. Such is the Commandment of God, and also His Law;
3. no one should take part in fighting;
4. it is sinful to wield political power, as it leads to many of the evils in the world;
5. man is born to do his duty to his Creator; he should therefore pay more attention to his duties than to his rights;
6. agriculture is the true occupation of man. It is therefore contrary to divine law to establish large cities, to employ hundreds of thousands for minding machines in factories so that a few can wallow in riches by exploiting the helplessness and poverty of the many.

These views he has very beautifully supported by examples from various religions and other old texts. There are today thousands of men in Europe who have adopted Tolstoy's way of life. They have given up all their worldly goods and taken to a very simple life.

Tolstoy is still writing with great energy. Though himself a Russian, he has written many strong and bitter things against Russia concerning the Russo-Japanese War. He has addressed a very pungent and effective letter to the Czar in regard to the war. Selfish officers view him with bitterness, but they, and even the Czar, fear and respect him. Such is the power of his goodness and godly living that millions of peasants are ever ready to carry out his wish no sooner than it is spoken.

'Count Tolstoy' (G.)
Indian Opinion, 2 Sept. 1905

54. TOLSTOY ON NON-RETALIATION

S.S. Kildonan Castle,
November 18, 1909

The letter[1] translated below calls for an explanation.

Count Tolstoy is a Russian nobleman. He has had his full share of life's pleasures, and was once a valiant soldier. He has no equal among European writers. After much experience and study, he has come to the conclusion that the political policies generally followed in the world are quite wrong. The chief reason for that, according to him, is that we are vengeful, a habit unworthy of us and contrary to the tenets of all religions. He believes that to return injury for injury does harm both to ourselves and to our enemy. According to him, we should not retaliate against anyone who may injure us, but reward him with love instead. He is uncompromising in his loyalty to the principle of returning good for evil.

He does not mean by this that those who suffer must seek no redress. He believes rather that we invite suffering on ourselves through our own fault. An oppressor's efforts will be in vain if we refuse to submit to his tyranny. Generally, no one will kick me for the mere fun of it. There must be some deeper reason for his doing so. He will kick me to bend me to his will if I have been opposing him. If, in spite of the kicks, I refuse to carry out his orders, he will stop kicking me. It would make no difference to me whether he did so or not. What matters to me is the fact that his order is unjust. Slavery consists in submitting to an unjust order, not in suffering ourselves to be kicked. Real courage and humanity consist in not returning a kick for a kick. This is the core of Tolstoy's teaching.

The letter translated below was originally written in Russian. It was rendered into English by Tolstoy himself and sent to the editor of *Free Hindustan* in reply to a letter of his. This editor holds different views from Tolstoy's and hence he did not publish the letter. It reached my hands and a friend asked me whether or not it should be published. I liked the letter. What I saw was a copy of the original letter. I sent it to Tolstoy and sought his permission to publish it, asking him at the same time whether the letter was in fact written by him. His permission having been received, both the English version of the letter and a Gujarati translation are being published in *Indian Opinion*.

To me Tolstoy's letter is of great value. Anyone who has enjoyed the experience of the Transvaal struggle will perceive its value readily enough. A handful of Indian *satyagrahis* have pitted love or soul-force against the might of the Transvaal Government's guns. That is the central principle of Tolstoy's teaching, of the teaching of all religions. Khuda-Ishwar has endowed our soul with such strength that sheer brute force is of no avail against it. We have been employing that strength against the Transvaal Government not out of hatred or with a view to revenge, but merely in order to resist its unjust order.

But those who have not known what a happy experience *satyagraha* can be, who have been caught up in the toils of this huge sham of modern civilization, like moths flitting round a flame, will find no interest in Tolstoy's letter all at once. Such men should pause for a moment and reflect.

Tolstoy gives a simple answer to those Indians who appear impatient to drive the whites out of India. We are our own slaves, not of the British. This should be engraved in our minds. The whites cannot remain if we do not want them. If the idea is to drive them out with firearms, let every Indian consider what precious little profit Europe has found in these.

Everyone would be happy to see India free. But there are as many views as men on how that can be brought about. Tolstoy points out a simple way to such men.

Tolstoy has addressed this letter to a Hindu and that is why it cites thoughts from Hindu scriptures. Such thoughts, however, are to be found in the scriptures of every religion. They are such as will be acceptable to all, Hindus, Muslims and Parsis. Religious practices and dogmas may differ, but the principles of ethics must be the same in all religions. I therefore advise all readers to think of ethics.

No one should assume that I accept all the ideas of Tolstoy. I look upon him as one of my teachers. But I certainly do not agree with all his ideas. The central principle of his teaching is entirely acceptable to me, and it is set out in the letter given below.

In this letter, he has not spared the superstitions of any religion. That is, however, no reason why any proud follower of Hinduism or of any other religion should oppose his teaching. It should suffice for us that he accepts the fundamental principles of every religion. When irreligion poses as religion, as it so often does, even true religion suffers. Tolstoy points this out repeatedly. We must pay the utmost attention to his thought whatever the religion we belong to.

In translating, I have endeavoured to use the simplest possible Gujarati. I have been mindful of the fact that readers of

Indian Opinion prefer simple language. Moreover, I want Tolstoy's letter to be read by thousands of Gujarati Indians, and difficult language may prove tedious reading to such large numbers. Though all this has been kept in mind, slightly difficult words may have been occasionally used when simpler ones were not available, for which I apologize to the readers.

MOHANDAS KARAMCHAND GANDHI

Preface to Leo Tolstoy's 'Letter to a Hindoo' (G.)
Indian Opinion, 25 Dec. 1909

[1] Not reproduced here.

55. THE TEACHING OF TOLSTOY

The great Tolstoy has quit this corporeal frame at the ripe old age of 83. It is truer to say that 'he has quit this corporeal frame' than that 'he has died'. There can be no death for Tolstoy's soul. His name will ever remain immortal. Only his body, which was of dust, has returned to dust.

Tolstoy is known to the entire world; but not as a soldier, though once he was reputed to be an expert soldier; not as a great writer, though indeed he enjoys a great reputation as a writer; nor as a nobleman, though he owned immense wealth. It was as a good man that the world knew him. In India, we would have described him as a *maharshi* or *fakir*. He renounced his wealth, gave up a life of comfort to embrace that of a simple peasant. It was Tolstoy's great virtue that he himself put into practice what he preached. Hence thousands of men clung loyally to his words—his teaching.

We believe Tolstoy's teaching will win increasing appreciation with the passage of time. Its foundation was religion. Being a Christian, he believed that Christianity was the best religion. He did not, however, denounce any other religion. He said, on the contrary, that truth was undoubtedly present in all the religions. At the same time, he also pointed out that selfish priests, *Brahmins* and *Mullas* had distorted the teaching of Christianity and other religions and misled the people.

What Tolstoy believed with especial conviction was that in

essence all religions held soul-force to be superior to brute force and taught that evil should be requited with good, not evil. Evil is the negation of religion. Irreligion cannot be cured by irreligion, but only by religion. There is no room in religion for anything other than compassion. A man of religion will not wish ill even to his enemy. Therefore, if people always want to follow the path of religion, they must do nothing but good.

In his last days, this great man wrote a letter to Mr. Gandhi to acknowledge copies of *Indian Opinion* in which he expressed these same ideas. The letter is in Russian. We give in this issue a Gujarati translation of it, based on an English translation. The translation is worth reading. What he has said there about *satyagraha* deserves to be pondered over by all. According to him, the Transvaal struggle will leave its mark on the world. Everyone [he says] has much to learn from it. He extends encouragement to the *satyagrahis* and assures them of justice from God, if not from the rulers. The latter, being enamoured of their strength, will certainly not be pleased with *satyagraha*. Despite that, *satyagrahis* must have patience and continue to fight. Citing, further, the example of Russia, Tolstoy states that there, too, soldiers everyday turn their back upon their profession. He is convinced that, though this movement has had no tangible results in the present, it will assume a big form in the end and Russia will be free.

It is no small encouragement to us that we have the blessings of a great man like Tolstoy in our task. We publish his photograph in today's issue.

'The Late Lamented Tolstoy the Great' (G.)
Indian Opinion, 26 Nov. 1910

56. TOLSTOY ON NON-VIOLENCE

Ashram, Sabarmati,
March 11, 1926

Dear Friend,

I have your letter. I wish indeed that I could visit Europe and see so many of my unknown European friends. But for the time being, I feel I must not leave India. When I feel that the way is clear for me, I shall not hesitate to go to Europe. Till that time we must meet one another through correspondence. Nor will it be possible at the present moment to send Mr. Andrews or any other friends. Mr. Andrews is away in South Africa. He returns next month but the work here is already cut out for him and it will keep him for several months.

There is no doubt about it that Tolstoy's writings had a powerful effect on me. He strengthened my love of non-violence. He enabled me to see things more clearly than I had done before. His manner of putting this is all his own. At the same time I know that there were fundamental differences between us and though they will abide, they are of little consequence compared with so many things for which I shall feel ever grateful to him. My patriotism is patent enough; my love for India is ever growing but it is derived from my religion and is therefore in no sense exclusive.

Yours sincerely,

A Letter
SN 19353

57. TOLSTOY'S GREATEST CONTRIBUTION

The Ashram,
Sabarmati,
April 20, 1928

Dear Friend,

I have your letter. I cannot resist you, but I take you at your word. I send you a single sentence as follows:

Tolstoy's greatest contribution to life lies, in my opinion, in his ever attempting to reduce to practice his professions without counting the cost.

Thanks for your inquiry about my health. I appear to be keeping well at the present moment.

Yours sincerely,

Letter to John Haynes Holmes
SN 14287

58. TOLSTOY'S INFLUENCE

Satyagraha Ashram,
Sabarmati,
September 7, 1928

Dear Friend,

I have your letter. I have never quoted Tolstoy or any other author without acknowledgment in any of my writings. And I do not remember having often quoted authors in my writings; not because I would not, but because my reading is so poor and capacity for reproducing what I have read is still less.

The vow of celibacy was undoubtedly taken after I had acquired considerable acquaintance with Tolstoy's teachings. And, whilst it is as a general statement quite true that my life is based upon the teachings of the *Gita*, I would not be able to swear that Tolstoy's writings and teachings did not influence my decision about celibacy.

So much for your satisfaction. I hope at some date to deal with your vital questions in the pages of *Young India*.

Yours sincerely,

Dhan Gopal Mukerjee

Letter to Dhan Gopal Mukerjee
SN 14378

59. TOLSTOY ON SELF-CONTROL

[10 September 1928]

My present state of mind does not at all permit me to join in celebrating any day or festival. Some time ago a reader of *Navajivan* or *Young India* asked me a question: 'You have stated, writing about *shraddha*, that the right way of performing the *shraddha* of our elders, on their death-anniversary day, is to recall their virtues and make them our own. May I ask you, therefore, how you observe the *shraddha* days of your elders?' I used to observe these days when I was young, but I don't mind telling you that now I do not even remember the dates on which they fall. I do not recall to have observed any such day during the past many years. Such is my unhappy state of mind, or rather, you may say, my charming or, as some friends believe, profound, ignorance. I believe it is enough if we fix our attention every minute of the day on the task in hand, think about it and do it as methodically as we can. We thereby celebrate the death anniversary of our elders as also the memory of men like Tolstoy. If Dr. Hariprasad had not drawn me into the net, it is quite likely that I would have arranged no celebration in the Ashram on this day, the 10th; it is even likely that I would have forgotten the day altogether. I had letters three months ago from Aylmer Maude and others engaged in collecting Tolstoy's writings, requesting me to send an article on the occasion of this centenary celebration and to draw the country's attention to this date. You must have seen an abstract of Aylmer Maude's letter, or perhaps the whole of it, published in *Young India*. Afterwards I forgot all about this matter. This is an auspicious occasion for me, but I would not have felt sorry if I had discovered that I had forgotten it. All the same, I welcome the opportunity which members of the Youth Association have offered of celebrating this day in the Ashram.

I wish I could say that, like Dattatreya, I had accepted many persons in this world as my *gurus*, but I am not in that position. I have said, on the contrary, that I am still in search of a *guru* in religious matters. It is my belief, which grows stronger day by day, that one must have especial fitness to find a *guru*. A *guru* comes unsought to him who has it. I lack such fitness. I have described Gokhale as my political *guru*. He had satisfied all my expectations of a *guru* in that field. I never doubted or questioned the propriety of his views or instructions. I cannot say that of anyone as a *guru* in religious matters.

And yet, I would say that three men have had a very great influence on my life. Among them I give the first place to the poet Rajchandra, the second to Tolstoy and the third to Ruskin. If I had to choose between Tolstoy and Ruskin and if I knew more about the lives of both, I would not know to whom to give preference. At present, however, I give the place to Tolstoy. I have not read as much of Tolstoy's life as many others may have, and in fact I have not read very much of his writings either. Among his works the one which has had the greatest effect on me is *The Kingdom of God Is within You*. The title means that God's Kingdom is in our heart, that if we search for it outside we shall find it nowhere. I read the book forty years ago. At that time, I was sceptical about many things and sometimes entertained atheistic ideas. When I went to England, I was a votary of violence, I had faith in it and none in non-violence. After I read this book, that lack of faith in non-violence vanished. Later I read some of his other books, but I cannot describe what effect they had on me. I can only say what effect his life as a whole had on me.

I attach importance to two things in his life. He did what he preached. His simplicity was extraordinary; it was not merely outward; outward simplicity of course he had. Though he was born in an aristocratic family and had all the good things of life to enjoy, had at his disposal all that wealth and possessions could give a man, he changed the direction of his life's voyage in the prime of youth. Though he had enjoyed all the pleasures and tasted all the sweetness which life can offer, the moment he realized the futility of that way of life he turned his back on it, and he remained firm in his new convictions till the end of his life. I have, therefore, stated in some message I have sent that Tolstoy was the very embodiment of truth in this age. He strove uncompromisingly to follow truth as he saw it, making no attempt to conceal or dilute what he believed to be the truth. He stated what he felt to be the truth without caring whether it would hurt or please the people or whether it would be welcome to the mighty emperor. Tolstoy was a great advocate of non-violence in his age. I know of no author in the West who has written as much and as effectively for the cause of non-violence as Tolstoy has done. I may go even further and say that I know no one in India or elsewhere who has had as profound an understanding of the nature of non-violence as Tolstoy had and who has tried to follow it as sincerely as he did.

I feel unhappy about this state of affairs, I do not like it. India is *karmabhumi*. The sages and seers of this country have made the biggest discoveries in the sphere of non-violence. But we cannot live on inherited wealth. If we do not continue to add to it, we would be eating it away. The late Justice Ranade has cautioned us against this. We may complacently quote the

Vedas and Jain literature and talk profound things, or propound great principles and strike the world dumb, but people will not believe in our sincerity. Hence Ranade pointed it out as our duty that we should add to our inheritance. We should compare it with the writings of other religious thinkers and if, as a result of such comparison, we discover anything new or find new light shed on a subject, we should not reject it. We have, however, failed to do this. Our religious heads are always one-sided in their thinking. There is no harmony between their words and deeds. We do not have among us men who, like Tolstoy, would speak out the plain truth irrespective of whether or not that would please the people or the society in which they work. Such is the pitiable condition of this our land of non-violence. Our non-violence is an unworthy thing. We see its utmost limit in refraining somehow from destroying bugs, mosquitoes and fleas, or from killing birds and animals. We do not care if these creatures suffer, nor even if we partly contribute to their suffering. On the contrary, we think it a heinous sin if anyone releases or helps in releasing a creature that suffers. I have already written and explained that this is not non-violence, and I take this occasion, when I am speaking about Tolstoy, to repeat that that is not the meaning of non-violence. Non-violence means an ocean of compassion, it means shedding from us every trace of ill-will for others. It does not mean abjectness or timidity, or fleeing in fear. It means, on the contrary, firmness of mind and courage, a resolute spirit.

We do not see this non-violence in the educated classes in India. For them Tolstoy's life should be a source of inspiration. He strove hard to put into practice what he believed in, and never turned back from his chosen path. I do not believe that he did not find that stick. He himself said, of course, that he had failed to discover it, but that was his humility. I do not agree with his critics that he did not find that stick. I might perhaps agree if anyone asserted that he did not fully act upon the principle of non-violence of which he had had a glimpse. But, then, has there been anyone in this world who could act upon the principle of non-violence fully while he lived? I believe it impossible for one living in this body to observe non-violence to perfection. While the body endures, some degree of egotism is inescapable. We retain the body only so long as egotism persists. Bodily life, therefore, necessarily involves violence. Tolstoy himself said that anyone who believed that he had realized his ideal would be lost. From the moment he believed that, his fall would begin. The further we travel towards an ideal the further it recedes. As we advance in its search, we realize that we have one step after another to climb. No one can climb all the steps in one leap. This view does not imply cravenness of spirit or pessimism but certainly there is humility in it. Hence our sages and seers said that the state of *moksha* meant utter emptiness. He who aspires after *moksha* must develop a state of such emptiness. One cannot attain this without God's grace. That state of emptiness can only remain an ideal as long as one lives in this body. The moment Tolstoy saw this truth clearly, grasped it with his intellect and started on his journey towards the ideal, he had found the green stick. He could not describe it, but could have only said that he had found it. If, however, he had in fact said that he had found it, progress in life would have been over for him.

The seeming contradictions in Tolstoy's life are no blot on him or sign of his failure. They signify the failure of the observer. Emerson has said that a foolish consistency is the hobgoblin of little minds. We would be utterly lost if we tried to live and show that there was no contradiction in our lives. In trying to live in that manner, we would have to remember what we did yesterday and then harmonize our actions today with that; in trying to preserve such forced harmony, we would have to resort to untruth. The best way is to follow the truth as one sees it at the moment. If we are progressing from day to day, why should we worry if others see contradictions in us? In truth, what looks like contradiction is not contradiction, but progress. And so, what seems to be contradiction in Tolstoy's life is really not contradiction, but only an illusion in our minds. Only the man himself knows how much he struggles in the depth of his heart or what victories he wins in the war between Rama and Ravana. The spectator certainly cannot know that. If the person slips ever so little, the world will think that there was nothing in him; this, of course, is for the best. One should not condemn the world on that account and so the saints have said that we should rejoice when the world speaks ill of us, but tremble with fear when it praises us. The world cannot act otherwise than it does; it must censure where it sees evil. But, whenever we examine the life of a great man, we should bear in mind what I have explained. God is witness to the battles he may have fought in his heart and the victories he may have won. These are the only evidence of his failures and successes.

By saying this, I do not wish to suggest that you should cover up your weaknesses, or, when they are as big as hills, think that they are as small as grains of sand. What I have said is in regard to other people. We should look upon others' weaknesses, huge as the Himalayas, to be as small as mustard seeds and ours, as small as mustard seeds, to be as big as the Himalayas. When we become aware of the slightest lapse on our part or seem to have become guilty of untruth, intentionally or otherwise, we should feel as if we were burning, as if we were caught in flames. A snake bite or a scorpion sting is of little consequence; you will find many who can cure them. Is there anyone, however, who can cure us of the sting of untruth or violence? God alone can do that, and He will do it only if we strive in earnest. Hence, we should be vigilant against our weaknesses and magnify them to the utmost, so that, when the world censures us, we should not think that people were mean-minded and exaggerated our faults. If anyone pointed out a weakness in Tolstoy, though there could hardly be an occasion for anyone to do so for he was pitiless in his self-examination, he would magnify that weakness to fearful proportions. He would have seen his lapse and atoned for it in the manner he thought most appropriate before anyone had pointed it out to him. This is a sign of goodness, and I think, therefore, that he had found that stick.

Tolstoy drew people's attention to another thing through his writings and his life, and that is the idea of 'bread labour'. It was not his own discovery. Another author had mentioned it in a Russian Miscellany. Tolstoy made his name known to the world and also put before it his idea. The cause of the inequalities we see in the world, of the contrasts of wealth and poverty, lies in the fact that we have forgotten the law of life. That law is the law of 'bread labour'. On the authority of

Chapter III of the *Gita*, I call it *yajna*. The *Gita* says that he who eats without performing *yajna* is a thief and sinner. Tolstoy has said the same thing. We should not distort the meaning of 'bread labour' and forget the real idea. Its simple meaning is that he has no right to eat who does not bend his body and work. If every one of us did bodily labour to earn his food, we would not see the poverty which we find in the world. One idler is the cause of two persons starving, for his work has to be done by someone else. Tolstoy said that people came forward for philanthropic service, spent money for the purpose and earned titles as reward for their service, but he said it would be enough if, instead of all this, they did a little physical work and got off the backs of others. That is true indeed. In that lies humility. To do philanthropic service but refuse to give up one's luxuries is to act in the way described by Akha Bhagat, 'Stealing an anvil and gifting a needle'. Can we hope thereby to go up in a *viman* to heaven?

It is not that others have not said what Tolstoy said, but there was magic in Tolstoy's language, for he acted upon what he preached. He who was accustomed to the comforts of wealth started doing physical labour. He used to work on the farm or do other labour for eight hours a day. That does not mean that he gave up literary work. In fact, after he started doing physical labour his literary work came to have greater life in it. It was during spare time in this period of *yajna* that he wrote what he described as his most important work, *What Is Art?* Physical labour did not tell upon his health, and he believed that it sharpened his intellect. Students of his works will bear testimony that he was right.

If we wish to benefit from Tolstoy's life, we should learn these three things from it. I am addressing members of a youth association, and I wish to remind them that they have to choose between two paths in life: one of self-indulgence and the other of self-restraint. If you think that Tolstoy lived and died well, you will see that there is only one right path in life for all, especially for the young—and that is the path of self-control. That is particularly true in India. *Swaraj* is not something to be won from the Government. If you examine the causes of our degradation, you will see that we are more responsible for it than the Government. You will then see that the key to *swaraj* is in our hands, and not in England nor in Simla nor in Delhi. It is in your pocket and mine. Our lethargy is responsible for the delay in remedying the degradation and listlessness of our society. If we overcome that, there is no power on earth which can prevent us from raising ourselves and securing *swaraj*. We ourselves choose to lie helpless on the path and refuse to lift ourselves out of that condition.

I should like to tell the members of the Youth Association that this is a golden time for them, or from another point of view, a hard time, a time of trial, if I put it in a third way. It is not enough that they pass university examinations and secure degrees. They will have secured real degrees only when they pass the examination of life and stand the test of hardships and difficulties. This is a period of transition, a golden time for you. You have two paths before you: one leading to the north and another to the south, one to the east and another to the west. You have to choose between the two. You must consider which path you will choose. All kinds of winds—poisonous winds, in my view—are blowing into the country from the West. There are, of course, some beautiful currents too, like Tolstoy's life. But these do not blow with every ship that arrives! You may say 'with every ship' or 'every day', for every day a ship arrives in the Bombay or Calcutta port. Along with other foreign goods, foreign literature too arrives. Its ideas intoxicate people and draw them to the path of self-indulgence. I have no doubt about that. Do not be vain and believe that your thoughts, or what in your immaturity you have read in books and understood from them, are the only truth, that what is old is barbarous and uncivilized and that truth lies only in things newly discovered. If you suffer from such vanity, I don't think you will bring credit to your Association. If you have still not fulfilled my hope that you have learnt humility, culture, a sense of propriety and purity from Sarala Devi, do so in future.

Do not be puffed up because you have been praised for some good things you have done. Run away from praise, and don't think that you have done much. If you collected money for Bardoli, worked hard and sweated for the cause, if a few of you went to jail for it, I ask you, as a man of experience, 'Is it much that you have done?' Others may say that you have, but you should not rest satisfied with what you have done. You have to purify your inner life, and it is from your conscience that you have to obtain a real certificate. Truly speaking our *atman* too is generally asleep. It was said by Tilak Maharaj that in our languages we have no word corresponding to 'conscience'. We do not believe that everyone has a conscience; in the West they do. What conscience can an adulterous or dissolute man have? Tilak Maharaj, therefore, rejected the idea of conscience. Our seers and sages of old said that one must have an inner ear to hear the inner voice, that one must have the inner eye, and must cultivate self-control to acquire these. Hence, in Patanjali's treatise on *yoga*, the first step prescribed for the student of *yoga*, for one aspiring after self-realization, is the observance of the disciplines of *yama-niyama*. There is no path but that of self-control for you or me or others. Tolstoy showed this by leading a long life of self-control. I wish and pray to God that we should be able to see this as clearly as daylight, and should leave this meeting with a resolution that we shall learn the lesson of self-control from Tolstoy's life.

Let us resolve that we will never give up the pursuit of truth. To follow truth, the only right path in this world is that of non-violence. Non-violence means an ocean of love, whose vastness no one has ever been able to measure. If it fills us we would be so large-hearted that we would have room in it for the whole world. I know this is difficult to achieve, but not impossible. Thus we heard the poet say, in the prayer with which we commenced, that he would bow his head only to him who was free from attachment and aversion, who had overcome all desires and who was the perfect embodiment of non-violence, that is, love, whether he was named Shankar or Vishnu or Brahma or Indra or whether he was Buddha or Siddha. Such non-violence is not limited to refraining from killing disabled creatures. It may be *dharma* not to kill them, but love goes infinitely further than that. What does it profit a person that he saves the lives of disabled creatures, if he has had no vision of such love? In God's court, his work will have little value.

The third thing is bread labour—*yajna*. We earn the right to eat only by putting the body to hard work, by doing physical labour. *Yajna* means any work done for the service of others. It is not enough that we do physical labour; we should live only in order that we may serve others, and not that we may run after immoral and worldly pleasures. If a young man who has trained his body with rigorous exercise spends eight hours every day in such exercise, he is not doing 'bread labour'. I do not belittle your doing exercise and training your body; but such exercise does not constitute the *yajna* which Tolstoy has advised and which is described in Chapter III of the *Gita*. He who believes that this life is for *yajna*, for service, will day by day give up running after pleasures. True human effort consists in striving to realize this ideal. It does not matter if no human being has succeeded in doing that to perfection; let the ideal ever remain distant from us. We should walk and break stones, as Farhad did for Shirin, our Shirin being the ideal of non-violence. This certainly holds our little *swaraj*, but it holds everything else too.

Speech on Tolstoy's Birth Centenary, Ahmedabad Youth Association (G.)
Navajivan, 16 Sept. 1928

§8. Naoroji

60. DADABHAI NAOROJI, MP

The first Indian to become a member of the British Parliament was Mr. Dadabhai Naoroji. Born on September 4th, 1825, in the city of Bombay, he was educated at the Elphinstone School and College, and was, at the age of 29, made Professor of Mathematics and Natural Philosophy—being the first Indian to receive that honour. In 1855, Mr. Naoroji visited England as partner in the first Indian business to be established in that country. The University College, London, did him the honour of appointing him Professor of Gujarati; and one of the benefits gained for India by Mr. Naoroji was the admission of Indians to the Civil Service in 1870. He was made Prime Minister of Baroda in 1874, and a year later was elected a member of the Corporation and Municipal Council of Bombay, to which body he gave five years' valuable service. Mr. Naoroji was a member of the Bombay Legislative Council from 1885 to 1887. The Indian National Congress honoured him by electing him President in 1886, 1893, and again in 1906. Mr. Naoroji sat in the House of Commons from 1893 to 1895 as Liberal member for Central Finsbury, London, and he did good work for his country as member of the Royal Commission on Indian Expenditure, etc., and, in 1897, gave evidence before the Welby Commission.

From the very commencement of the British Committee of the Indian National Congress, he was a diligent member and hard worker. Among the publications from the pen of Mr. Dadabhai Naoroji are: *England's Duty to India*, *Admission of Educated Natives into the Indian Civil Service*, *Financial Administration of India*, and what is, perhaps, the best known of his many writings, *Poverty and Un-British Rule in India*. In 1906, the venerable Dadabhai journeyed to the Motherland to preside over the Indian National Congress, a task which was a tremendous strain upon even his iron constitution and indomitable spirit. Since the Calcutta Congress of 1906, Mr. Dadabhai has practically retired from public life, and in 1907 he went to reside at Varsova, a small fishing village in the Bombay Presidency where he still watches with a keen interest the progress of events in India which go to make or mar its future. Truly has he earned for himself the honoured title of THE GRAND OLD MAN OF INDIA.

'The Grand Old Man of India'
Indian Opinion, 3 Sept. 1910

61. THE SIMPLICITY OF NAOROJI

The birth anniversary of the Grand Old Man of India, Dadabhai Naoroji, fell on 4th September; but the National Women's Council arranged the function on the 30th August to suit my convenience since I had to be present in Poona on the 4th. Dadabhai led the life of a *rishi*. I have many sacred memories of him. This Grand Old Man of India was, and continues to be, one of the great men who have moulded my life. I think the memories that I recounted before the sisters are worth being reported to the readers.

I had the privilege to see Dadabhai in 1888 for the first time. A friend of my father's had given me a letter of introduction to him, and it is worth noting that this friend was not at all acquainted with Dadabhai. He, however, took it for granted that anyone from the public could write to such a saintly person. In England, I found that Dadabhai came in contact with all students. He was their leader and attended their gatherings. Ever since, I have seen his life flowing in the same rhythm till the end. I was in South Africa for twenty years, and exchanged hundreds of letters with Dadabhai during the period. I was astonished at the regularity with which his replies came. My letters used to be typed, but I do not remember any typed reply from him. The replies were all in his own hand, and moreover, as I came to know subsequently, he would himself make copies of his letters on a tissue-paper book. I could find that most of my letters were replied to by the return of post. Whenever I met him I tasted nothing but love and sweetness.

Dadabhai would talk to me exactly like a father to a son, and I have heard from others that their experience was the same as mine. The thought uppermost in his mind all the time was how India could rise and attain her freedom. My first acquaintance with the extent of Indian poverty was through Dadabhai's book;[1] I learnt from that book itself that about three crores of men in our country are half-starved. Today this number has increased. His simplicity was without limit. It so happened that someone criticized him in 1908. I found it extremely intolerable and yet I was unable to prove that it was wrong. I was troubled by many doubts.

I thought that it was sinful to entertain doubts about a great patriot like Dadabhai. Therefore I sought an appointment and went to see him with the consent of the critic. That was the first time I went to his private office. It was made up of a very small room with only two chairs. I entered. He asked me to sit in a vacant chair but I went and sat near his feet. He saw distress on my face and questioned me, asking me to speak out whatever weighed on my mind. With great hesitation I reported to him the criticisms of his detractors and said, 'I was troubled by doubts on hearing these things and, because I worship you, I consider it a sin to keep them back.' Smilingly, he asked me, 'What reply do I give you? Do you believe this thing?' His manner, his tone and the pain that was so apparent in his words, were enough for me. I said, 'I do not now want to hear anything more. I have no trace of a doubt left in me.' Even then he told me many things relating to this matter, which it is not necessary to recapitulate here. After this event I realized that Dadabhai was an Indian living in the simple style of a *fakir*. A *fakir*'s style does not imply that a man should not have even a farthing; but Dadabhai had forsaken the luxuries and standards which other people of his stratum were enjoying during those days.

I myself and many others like me have learnt the lessons of regularity, single-minded patriotism, simplicity, austerity and ceaseless work from this venerable man. At a time when criticism of the Government was considered sedition and hardly anyone dared to speak the truth, Dadabhai criticized the Government in the severest terms and boldly pointed out the shortcomings of the administration. I have absolutely no doubt that the people of India will remember Dadabhai affectionately as long as India endures as an entity in the world.

'Birth Anniversary of Dadabhai Naoroji' (G.)
Navajivan, 7 Sept. 1924

[1] *Poverty and Un-British Rule in India.*

§9. Gokhale

62. THE MESSAGE OF GOKHALE

[20 February 1915]

My one desire tonight is that my heart may reach your hearts and that there should be a real at-one-ment between us.

You have all learnt something about Tulsidas's *Ramayana*. The most stirring part is that about the companionship of the good. We should seek the company of those who have suffered and served and died. One such was Mr. Gokhale. He is dead, but his work is not dead, for his spirit lives.

The masses came to know of Gokhale's efficiency in work. All know Gokhale's life of action. But few know of his religious life. Truth was the spring of all his actions.

This was behind all his works, even his politics. This was the reason he founded the Servants of India Society, the ideal of which was to spiritualise the political as well as the social life of the nation.

It was fearlessness which ruled all the actions of his life. But as he was fearless he was also thorough. One of his favourite *shlokas* from the Shastras says: Real wisdom is not to begin a thing but to see the thing through to the end. This characteristic of thoroughness may be seen from this incident. He once had to speak to a large audience and he spent three days in order to prepare a short speech for this meeting and he asked me to write out a speech for him. I wrote out the speech. He took it and smiled his heavenly smile, discussed it with me and said, 'Give me something better, rewrite it.' For three days he worried over it. When the speech was given, it thrilled the whole audience. He delivered his speeches without notes, but he did so, because he was so thorough, that one might say he wrote his speeches with his own blood. As he was thorough and fearless, so he was gentle. He was human from top to toe in all his dealings. He was sometimes impatient, but he would ask forgiveness, coming forward with his smile, whether to a servant or a great man, saying, 'I know you will forgive me, won't you?'

He had a great struggle during the latter days of his life, a struggle with his conscience. He had to decide whether he should continue to take part in a struggle at the expense of his health. His conscience ruled every action of his life. He did not wear it on his sleeve, he wore it in his heart. Therefore he is living still, and may we all have the strength to carry out his last wish. His last word to those members of the Servants of India Society who were with him were: 'I do not want any memorial or any statue. I want only that men should love their country and serve it with their lives.' This is a message for the whole of India and not only for them. It was through service that he learnt to know his own nature and to know his country. His love for India was truthful and therefore he wanted nothing for India which he did not want for humanity also. It was not blind love, for his eyes were open to her faults and failings. If we can love India in the same way that he did, we have done well in coming to Shantiniketan to learn how to live our lives for India's sake. Copy the zeal which he showed in all he took up, the love that was the law of his life, the truthfulness which guided every action and the thoroughness which was characteristic of all his work.

Remember that our *shastras* teach us that these simple virtues are the stepping stones to the higher state of life, without which all our worship and works are useless.

I was in quest of a really truthful hero in India and I found him in Gokhale. His love and reverence for India were truly genuine. For serving his country, he completely eschewed all happiness and self-interest. Even while lying on his sick-bed, his mind was occupied in thinking about the welfare of India. A few days ago, when at night he was under the grip of a painful ailment, he called for some of us and began talking about the bright future of India, as envisaged by him. Doctors repeatedly advised him to retire from work but he would not listen to them. He said, 'None but death can separate me from work.' And death at last brought peaceful rest to him. May God bless his soul!

Speech at Shantiniketan on Gokhale's Death
The Ashram, June–July 1915

63. THE LEGACY OF GOKHALE

[Before 4 February 1916]

All that you do, and all you enjoy,
    What you offer in sacrifice, what you give in charity,
All austerities that you go through,
    Render up everything to me.[1]
When, smiling and playing my way through life,
    I see Hari revealed to me, a visible presence,

> Then shall I consider my life
> To have attained its true end.
> Muktanand's Lord, who sports with us,
> O Odha! He is the thread of our lives.

The counsel Shri Krishna addressed to Arjuna was, as it were, addressed by Lady India to Mahatma Gokhale and taken to heart by the latter, such was the manner of life of this great, departed soul. It is a fact known to everyone that all his activities, all his joys and the sacrifices he made, all the suffering he went through, were dedicated to Mother India.

The state of Odhava's mind with regard to Shri Krishna, as depicted by Muktanand, was that of the late Gokhale with regard to India.

What is the message of a life such as this? The Mahatma did not leave even this unsaid. When dying, he sent for the members of the Servants of India Society who were then present and told them: 'Do not occupy yourselves with writing my biography or spend your time in putting up my statues. If you are true servants of India, dedicate your lives to the fulfilment of our aims, to the service of India.' We know, too, what he felt in his heart about the meaning of that service. The Congress should of course be kept alive, the true condition of the country should be placed before the people through speeches and writings and efforts made to have education provided to every Indian. What was the aim behind all this? And how was it to be realized? In answering these questions, we get to know his point of view. Framing a constitution for the Servants of India Society, he laid down that the duty of its members would be to spiritualize political life in India. This embraces everything.

His was a religious life. My soul stands witness that, in all that he did, at all times, he acted wholly in the spirit of religion. Some twenty years ago, this Mahatma's sentiments sometimes appeared to be those of an atheist. He said once: 'I don't have Ranade's[2] faith. How I should like to have it!' Even at that time, however, I could see a religious bent of mind in his actions. It would not be wrong to say that his very doubt proceeded from such a bent of mind. He who lives in the manner of a *sadhu*, whose desires are simple, who is the image of truth, is full of humility, who represents the very essence of truth and has wholly renounced his ego, such a one is a holy soul, whether he knows it or not. Such a one was Mahatma Gokhale, as I could see from my twenty years' acquaintance with him.

In 1896, I discussed in India the question of indentured labour in Natal. At that time, I knew the Indian leaders only by name. This was the first occasion when I had contact with the leaders at Calcutta, Bombay, Poona and Madras. The late Gokhale was then known as a follower of Ranade. He had already at this time dedicated his life to the Fergusson College. I was a mere youth, with no experience. The bond which developed between us on the occasion of our very first meeting in Poona never came to exist between any other leader and me. Sure enough, all that I had heard about Mahatma Gokhale was confirmed by my own experience; but especially the effect which the soft expression on his lotus-like face had on me has still not vanished from my mind. I instantly recognized him as *dharma* incarnate. I had an audience with Shri Ranade, too, at that time, but I could get no glimpse into his heart. I could only see him as Gokhale's mentor. Whether it was that he was much senior to me in age and experience or that there were some other reasons, whatever the reason, I could not understand Shri Ranade as well as I could Gokhale.

After this contact with him in 1896, Gokhale's political life became my ideal. That very time he took possession of my heart as my *guru* in matters political. He edited the Quarterly of the Sarvajanik Sabha, he made Fergusson College illustrious by teaching there. He gave evidence before the Welby Commission[3] and demonstrated his true worth to India. He created so fine an impression on Lord Curzon[4] by his ability that the latter, who feared none, feared him. By his performance in the Central Legislative Assembly, he brought credit to India. At the risk of his life, he served on the Public Service Commission. He did all this and much more. Others have given a far better account of these things than I could hope to. Moreover, one cannot claim that his message, as I have understood it and defined it here, may be clearly deduced from these actions of his. I, therefore, propose to conclude this article by relating what I have myself known and what exemplifies his message.

The *satyagraha* struggle made so profound an impression on his mind that, though his health absolutely forbade it, he decided to pay a visit to South Africa. He went there in 1912. The Indians in South Africa gave him a right royal welcome. On the very next day after his arrival in Cape Town, there was a meeting in the local Town Hall. The Mayor was in the chair. Gokhale was in no condition to attend meetings and make speeches. But he left intact all the countless and taxing engagements that had been fixed. Following this decision, he attended the meeting in the Town Hall. At that very first appearance, he conquered the hearts of the whites in Cape Town. Everyone felt that a great soul was visiting South Africa. Mr. Merriman, a prominent leader in South Africa and a man of character and liberal views, had this to say when they met: 'Sir, a visit by a person like you brings a breath of fresh air into this land of ours.'

As the late Mr. Gokhale's tour progressed, this first impression became stronger. At every place the distinction between whites and Coloureds was forgotten for the moment. There were meetings in all places like the one held in Cape Town. The whites and Indians sat in the same rows at these meetings and, according equal honour to the late Mr. Gokhale, earned similar honour for themselves. There was a dinner in his honour in Johannesburg. It was attended by nearly three hundred prominent whites. The Mayor was in the chair. The whites in Johannesburg are not likely to be awed by anyone. If they have some multi-millionaires among them, they have also men who know people's worth. These vied with one another in shaking hands with Mr. Gokhale. There was only one reason for this. In his speeches, the audiences saw Gokhale's overflowing love for his motherland and a sense of fairness, at the same time. He wanted his country to be treated with the fullest respect and honour but did not want that any other country should be humiliated. If he was anxious to see all the rights of his countrymen preserved, he was equally anxious to see that the rights of others were not jeopardized in the process. Because of this, everyone felt a genuine sweetness in his utterances.

Mr. Gokhale believed that he delivered in Johannesburg his best speech in South Africa. It lasted more than three quarters of an hour but I never felt that anyone in the audience was bored. How did he make this speech? He started preparing for it six days in advance. He acquainted himself with the history of the question, as much as was necessary for his purpose, posted himself with the relevant figures and, sitting up late the preceding night, got ready with his language. The result was as I have stated. He satisfied both the whites and his own people.

I shall never forget while I live the pains he took to prepare himself for his meeting with Generals Botha and Smuts in Pretoria, the capital of South Africa. On the day before the interview, he closely examined Mr. Kallenbach and me. He got up at three o'clock and woke us up. He had finished with the literature he had been supplied and now wanted to cross-examine me, in order to make sure whether he was fully prepared. I told him politely that he need not have exerted himself so much, that we would fight it out if we obtained nothing then, but that we did not want him to be sacrificed for our sake. But how would a man, who had made it a rule to throw himself heart and soul into everything that he undertook, listen to my words? How shall I describe his manner of cross-examining me? How shall I praise his thoroughness? Such pains could have only one result. The cabinet promised Mr. Gokhale that a Bill conceding the *satyagrahis'* demands would be introduced in Parliament in the ensuing session and that the annual tax of £3 on the indentured labourers would be repealed.

The promises were not kept at the time mentioned. Did Mr. Gokhale hold his peace thereafter? Not for a moment. I am sure his exertions in 1913 to secure the fulfilment of the promises must have shortened his life by at least ten years. This is what his doctors believe. It is difficult to give an idea of the labour he went through in that year in rousing India and collecting funds. India was in an uproar about the issue of South Africa. The power which brought this about was Mr. Gokhale's. Lord Hardinge made a speech in Madras[5] which will go down in history. This too was owing to Gokhale. Those who were the nearest to him bear witness that, worrying himself over the South African issue, he became permanently bed-ridden. Even so, till the very last he refused to rest. He would receive, at midnight, telegrams from South Africa as long as letters. He attended to them immediately and drafted a reply on the instant. A telegram would be sent to Lord Hardinge the same moment and a statement for the Press prepared. In attending to the question, he delayed his meal and delayed going to bed, ignored the difference between day and night. Such single-minded and selfless devotion would be possible only to an elevated soul.

On the Hindu-Muslim question, too, his approach was ever the most religious. Once a man dressed as a *sadhu* went to see him, claiming to speak for Hindus. He would have the Muslims treated as inferior and the Hindus as superior. When Mr. Gokhale refused to play this game, he was accused of wanting in pride as a Hindu. Knitting his brows, he replied in a voice that pierced the heart: 'If Hinduism consists in doing what you say, I am not a Hindu. Please leave me.' One *sannyasi* left another and walked off.

Mr. Gokhale possessed in an eminent degree the quality of fearlessness. Among the qualities that make for the religious way of life, this occupies almost the first place. There was a reign of terror in Poona after the assassination of Lieutenant Rand. Mr. Gokhale was in England at that time. He made a famous speech there in defence of Poona. Some of the statements he made in that speech could not subsequently be proved. After some time, he returned to India. He apologized to the British troops against whom he had levelled charges. This action even displeased a section of the Indian people. Some persons advised the Mahatma to retire from public life. A few ignorant Indians did not even hesitate to accuse him of pusillanimity. To all of them, he replied in words at once earnest and gentle: 'What I have undertaken at no one's order, I can abandon at no one's order. I should be happy to have popular opinion on my side while performing my duty; should I not be so fortunate, however, that too may be just as well.' He believed that one's duty lay in working. I never observed that, while doing anything, he considered its effect on popular opinion from the point of view of his personal fortunes. If it ever became necessary to mount the gallows for the sake of the country, I believe he had the strength to do so fearlessly and with a smile on his face. I know that, often enough, mounting the gallows would have been a far easier thing for him than to be in the condition he had to pass through. He was in such painful situations more than once but he never gave way.

All these instances would seem to point to this lesson, that if we would learn anything from the life of this great patriot, it should be to emulate his religious attitude. All of us cannot go into the Central Legislative Assembly, nor do we always observe that doing so necessarily means serving the nation. We all cannot join the Public Service Commission and all those who do are not patriots. We may not, everyone of us, acquire his learning, nor do we see that every learned person is a servant of the country. All of us, however, can cultivate virtues like fearlessness, truthfulness, fortitude, justice, straightforwardness, firmness of purpose, and dedicate them to the service of the nation. This is the religious way. This is what the *mahavakya*, that political life should be spiritualized, means. He who follows this line will always know the path he should take. He will earn a share in the legacy left by the late Shri Gokhale. It is the divine assurance that anyone acting in this spirit will come by all the other gifts he needs. The life of the late Shri Gokhale is an irrefutable proof of this.

'Message of Gokhale's Life' (G.)
Mahatma Gandhini Vicharsrishti

[1] *Bhagavad Gita*, IX.27.

[2] 1842-1901; Indian judge, social reformer, author, and one of the founders of the Indian National Congress.

[3] The Royal Commission on Expenditure appointed in 1894 to consider the apportionment of military expenditure between England and India.

[4] (1859-1925); Viceroy of India, 1899-1905.

[5] In reply to addresses of welcome of the Mahajan Sabha and Madras Provincial Conference Committee at Madras on 24 Nov. 1913, Lord Hardinge said: 'Recently your compatriots in South Africa have taken matters into their own hands by organising what is called passive resistance to laws which they consider invidious and unjust—an opinion which we who watch their struggle from afar cannot but share. They have violated as they intend to violate those laws with full knowledge of the penalties involved and ready with all courage and patience to endure those penalties. In all this they have the sympathy of India deep and burning and not only of India but of all those who, like myself, without being Indians themselves, have feelings of sympathy for the people of this country.'

64. GOKHALE'S IDEAL OF SERVICE

[Before 19 February 1918]

As I was the first to come by the idea of bringing out a translation of the speeches of the late Mahatma Gokhale on his death anniversary, it is in a way appropriate that I myself write the foreword to the first volume. It is hoped that we will keep on celebrating the Gokhale anniversary. Every time to sing devotional songs, make speeches and then disperse is very much of a waste of time with no gain to anyone. In order that people may attach more importance to action than to speech-making and that they may derive some tangible benefit from the annual celebrations, the organizers of the anniversary resolved last year to publish, on the occasion, a useful book in the mother tongue. They decided, at the same time, what book was to be published and, naturally enough, the choice fell on the speeches of the late Mahatma.

It was everyone's wish that the translation should be an outstanding work in Gujarati literature and that every effort should be made to preserve in the translation the beauty of the holy word of the Mahatma as it stands in the original. This could not be secured with money but only through voluntary services. These we obtained, but, even so, the future alone can say whether the desired result has been achieved. The part to which this is a foreword has been translated by Shri Mahadev Haribhai Desai. This is no occasion to say anything of him by way of introduction. I shall only mention that he is a lover of Gujarati literature. He is no stranger to the subject; besides, he is one of the thousands of the late Mahatma's votaries. He has carried out his task with great enthusiasm and devotion, and one may justifiably hope, therefore, that this translation will earn a place in Gujarati literature.

During last year's anniversary celebrations, as soon as the Home Rule League of Bombay learnt that a decision to publish the volume was about to be announced, its secretaries wired an offer of generous help and later sanctioned a big amount, no less than three thousand rupees, for this project; and so the organizing committee had little worry left for collection of funds and its desire to ensure beauty of printing and the general get-up was satisfied even in these times of rising prices. The Home Rule League deserves congratulations on this large-hearted help. The foregoing paragraphs are but a foreword to the Foreword. In the Foreword itself, one must write something about the departed soul. What could a disciple, however, write about his master? How could he write it? It would be presumptuous for a disciple to do so.

The true disciple merges himself in the *guru* and so can never be a critic of the *guru*. *Bhakti* or devotion has no eye for shortcomings. There can be no cause for complaint if the public do not accept the eulogies of one who refuses to analyse the merits and shortcomings of his subject. The disciple's own actions are, in fact, his commentary on the master. I have often said that Gokhale was my political *guru*. That is why I consider myself incapable of writing about him. Whatever I write would seem imperfect in my eyes. I believe the relationship between the master and the disciple is purely spiritual. It is not based on arithmetical calculations. The relationship is formed on the instant, spontaneously, as it were, and never snaps once it is formed.

This relationship of ours was formed in the year 1896. I had no idea of its nature then; nor had he. About the same time, I had the good fortune to wait on the master's master [Justice Mahadev Govind Ranade[1]], Lokamanya Tilak, Sir Pherozeshah Mehta,[2] Justice Badruddin Tyabji,[3] Dr. Bhandarkar,[4] as also the leaders of Madras and Bengal. I was but a raw youth. Everyone of them showered his love on me. These were among the occasions which I can never forget while I live. But the peace of mind which my contacts with Gokhale gave me, those with others did not. I do not remember that any special affection was shown to me by Gokhale. If I were to measure and compare the love I experienced from them all, I have an impression that no one else showed such love to me as Dr. Bhandarkar did. He told me: 'I do not take any part in public affairs now. But, for your sake I will preside over the public meeting on the issue which you have at heart.' Still, it was only Gokhale who bound me to himself.

Our new relationship did not take shape immediately. But in 1902,[5] when I attended the Calcutta Congress, I became fully aware of my being in the position of a disciple. Now, again, I had the privilege of meeting almost all the leaders mentioned above. I saw that Gokhale had not only not forgotten me but had actually taken me under his charge. This had its tangible results. He dragged me to his quarters. During the Subjects Committee meeting, I felt helpless. While the various resolutions were under discussion, I could not, right till the end, gather enough courage to declare that I too had a resolution in my pocket on South Africa. It was not to be expected that the night would halt for my sake. The leaders were impatient to finish the business on hand. I was trembling with the fear that they would rise to leave any moment. I could not summon up courage to remind even Gokhale of my business. Just then he cried out, 'Gandhi has a resolution on South Africa; we must take it up.' My joy knew no bounds. This was my first experience of the Congress and I put great store by resolutions passed by it. There is no counting the occasions [of our meeting] that followed, and they are all sacred to me. For the present, however, I think I would do well to state what I have believed to be the guiding principle of his life and conclude this Foreword.

In these difficult and degenerate times, the pure spirit of religion is hardly in evidence anywhere. Men who go about the world calling themselves *rishis, munis* and *sadhus* rarely show this spirit in themselves. Obviously, they have no great treasure of the religious spirit to guard. In one beautiful phrase, Narasinha Mehta, best among the lovers of God, has shown in what that spirit consists:

> Vain, vain all spiritual effort
> Without meditation on the Self.

He said this out of his own vast experience. It tells us that religion does not necessarily dwell even in the man of great austerities or a great *yogi* who knows all the procedures of *yoga*. I have not the least doubt that Gokhale was wise in the truth of the Self. He never pretended to observe any religious practice but his life was full of the true spirit of religion.

Every age is known to have its predominant mode of spiritual effort best suited for the attainment of *moksha*. Whenever the religious spirit is on the decline, it is revived through such an effort in tune with the times. In this age, our degradation reveals itself through our political condition. Not taking a comprehensive view of things, we run away with the belief that, if but our political conditions improved, we would rise from this fallen state. This is only partially true. To be sure, we cannot rise again till our political condition changes for the better; but it is not true that we shall necessarily progress if our political condition undergoes a change, irrespective of the manner in which it is brought about. If the means employed are impure, the change will be not in the direction of progress but very likely the opposite. Only a change brought about in our political condition by pure means can lead to real progress. Gokhale not only perceived this right at the beginning of his public life but also followed the principle in action. Everyone had realized that popular awakening could be brought about only through political activity. If such activity was spiritualized, it could show the path to *moksha*. He placed this great ideal before his Servants of India Society and before the whole nation. He firmly declared that, unless our political movement was informed with the spirit of religion, it would be barren. The writer who took notice of his death in *The Times of India* drew particular attention to this aspect of Gokhale's mission and, doubting if his efforts to create political *sannyasis* would bear fruit, warned the Servants of India Society, which he left as his legacy, to be vigilant.

In this age, only political *sannyasis* can fulfil and adorn the ideal of *sannyasa*, others will more likely than not disgrace the *sannyasi*'s saffron garb. No Indian who aspires to follow the way of true religion can afford to remain aloof from politics. In other words, one who aspires to a truly religious life cannot fail to undertake public service as his mission, and we are today so much caught up in the political machine that service of the people is impossible without taking part in politics. In olden days, our peasants, though ignorant of who ruled them, led their simple lives free from fear; they can no longer afford to be so unconcerned. In the circumstances that obtain today, in following the path of religion they must take into account the political conditions. If our *sadhus*, *rishis*, *munis*, *maulvis* and priests realized the truth of this, we would have a Servants of India Society in every village, the spirit of religion would come to prevail all over India, the political system which has become odious would reform itself, India would regain the spiritual empire which, we know, it enjoyed in the days gone by, the bonds which hold India under subjection would be severed in an instant, and the ideal state which an ancient seer described in his immortal words would come into being: 'Iron would be used not for forging swords but for forging ploughshares, and the lion and the lamb would be friends and live together in love.' Gokhale's ideal in his life was to labour to bring about this state of affairs. That, indeed, is his message and I believe that whoever reads his writings with an open mind will recognize this message in every word of his.

Foreword to Volume of Gokhale's Speeches (G.)
Gopal Krishna Gokhalenan Vyakhyano, Vol. I

[1] 1842-1901; eminent judge, reformer, and a founder of the Indian National Congress.
[2] 1845-1915; prominent Indian leader, twice President of the Congress.
[3] 1844-1906; judge, legislator, President of the Congress.
[4] R. G. Bhandarkar (1837-1925); orientalist and reformer.
[5] The year was actually 1901.

§ 10. Rajchandra

65. THE CULTURE OF THE HEART

I was introduced to Raychandbhai[1] in July 1891, on the very day on which, returning from England, I landed in Bombay. At this time of the year the sea is stormy. The ship, therefore, had arrived late and it was already night. I stayed with Dr. Pranjivan Mehta, Barrister, now the well-known jeweller of Rangoon. Raychandbhai was his elder brother's son-in-law. The doctor himself introduced me to him. On the same day I was also introduced to Jhaveri Revashanker Jagjivandas, another elder brother of his. The doctor introduced Raychandbhai as 'a poet', and added, 'though a poet, he is in our business. He is a man of spiritual knowledge and a *shatavadhani*.' Someone suggested that I should utter a number of words in his presence, saying that no matter to what language they belonged he would repeat them in the same order in which I had uttered them. I could not believe this. I was a young man, had just returned from England, and was a little vain, too, of my knowledge of languages; in those days I was under the powerful spell of English. Having been to England made a man feel that he was heaven-born. I poured out all my store of knowledge, and first wrote out words from different languages—for how possibly could I afterwards remember them in their due order? I then read out the words. Raychandbhai repeated them slowly one after another and in the same order. I was pleased and astonished, and formed a high opinion about his memory. This was an excellent experience to break a little the binding spell of English on me.

The Poet did not know English at all. At the time I am speaking of, he was not more than twenty-five. His study in the Gujarati school was not much either. And even then he possessed such a powerful memory and such knowledge, and was respected by everyone round him! I was all admiration. The power of memory is not sold in schools. Knowledge, too, can be acquired without going to school if one wants it—is keen on it—and one need not go to England or elsewhere to command respect, for virtue is always respected. I learned these truths on the very day I landed in Bombay.

The acquaintance with the poet which began on this occasion grew over the years. Other persons possess a powerful memory, and one need not be dazzled by it. Knowledge of the Shastras, too, is found in plenty in many. But such persons, if they have no real culture, can give us nothing of value. A combination of powerful memory and knowledge of the Shastras will have real worth and will benefit the world only if they exist along with genuine culture of the heart.

Shrimad Rajchandra, Ch. 2

[1] Rajchandra Ravjibhai Mehta (1868-1901), poet, mystic, and connoisseur of pearls and diamonds.

66. THE PATH TO *MOKSHA*

November 5, 1926

When Shri Revashanker Jagjivan, whom I regard as an elder brother, asked me for a foreword to this edition of Shrimad

Rajchandra's letters and writings, I could not refuse his request. As I tried to think what I could say in such a foreword, I felt that it would serve two purposes if I gave the few chapters of my reminiscences of Rajchandra which I wrote in the Yeravda jail: one, that as my attempt, though incomplete, was undertaken purely in a spirit of religious devotion, it might help other *mumukshus* like me, and two, that those who did not know Shrimad Rajchandra in life might know a little about him and so find it easier to understand some of his writings.

The chapters which follow leave the story incomplete. I do not think that I can complete it, for, even if I get time, I do not feel inclined to go much further than the point where I left off. I, therefore, wish to complete the last chapter, which had remained unfinished, and include in it a few things.

In these chapters I have not touched upon one aspect of the subject which I think I should place before the readers. Some people assert that Shrimad was the twenty-fifth Tirthankar. Some others believe that he has attained *moksha*. Both these beliefs, I think, are improper. Either those who hold them do not know Shrimad or their definitions of Tirthankar or liberated soul are different from the commonly accepted ones. We may not lower the standard of truth even for the sake of those whom we love dearly. *Moksha* is a condition of supreme value. It is the highest state of the *atman*. It is so rare a condition that to attain it much more effort and patience are necessary than, say, for emptying the sea drop by drop with a blade of grass. A perfect description of that state is impossible. A Tirthankar will naturally command, without seeking them, the powers which belong to the state immediately preceding *moksha*. One who has attained freedom while still living in this body will suffer from no physical disease. In a body untroubled by desire there can be no disease. There can be no disease where there is no attachment. Where there is desire there is attachment and while there is attachment *moksha* is impossible. Shrimad had not attained the total freedom from attachment which should characterize a *mukta purusha* or the *vibhuti* which belongs to a Tirthankar. He had such freedom and such powers in a much larger measure than the ordinary man or woman, and so in common speech we may describe him as one who was free from attachment or who possessed superhuman powers. I am sure, however, that Shrimad had not attained to the perfect freedom from attachment which we attribute to a *mukta purusha* or acquired the *vibhuti* which we believe that a Tirthankar would manifest. I do not say this with the intention of pointing out any shortcoming in a great character worthy of our highest reverence; I say it in order to do justice both to him and to the cause of truth. We are all worldly creatures, whereas Shrimad was not. We shall have to wander from existence to existence, whereas Shrimad may have only one life more to live. We are perhaps running away from *moksha*, while Shrimad was flying towards it with the speed of wind.

This was no small achievement. Even so, I must say that he had not attained the supreme state so beautifully described by him. He himself said that he had come upon the Sahara in his journey and that he had failed to cross the desert. Shrimad Rajchandra, however, was a rare being. His writings are the quintessence of his experiences. Anyone who reads them, reflects over them and follows them in his life will find the path to *moksha* easier; his yearning for sense-pleasures will become progressively weaker, he will become disinterested in the affairs of this world, will cease to be attached to the life of the body and devote himself to the welfare of the *atman*.

The reader will see from this that Shrimad's writings are meant only for those who are qualified to study them. All readers will not find them interesting. Those who are inclined to be critical will get material for criticism. But those who have faith will find these writings of absorbing interest. I have always felt that Shrimad's writings breathe the spirit of truth. He did not write a single word in order to show off his knowledge. His aim in his writings was to share his inward bliss with his readers. I am sure that anyone who wishes to free himself from inner conflicts and is eager to know his duty in life will gain much from Shrimad's writings, whether such a reader is a Hindu or belongs to another faith.

And, hoping that the few reminiscences of Shrimad's life which I have written down will help the reader who is qualified to read his writings, I give them here as part of this Foreword.

Gandhi's Preface (G.)
Shrimad Rajchandra

67. CONTACT WITH RAJCHANDRA

Today is the birth anniversary of the late Shrimad Rajchandra, whose reminiscences I have started writing, that is, it is the *Kartiki Purnima*,[1] Samvat 1979. I am not attempting to write a biography of Shrimad; such an attempt is beyond my capacity. I do not have with me the material needed for a biography. If I wanted to write one, I would spend some time in his birth-place, the port of Vavania, examine the house in which he lived, see the places which were the scenes of his childhood play and wanderings, meet his childhood friends, visit the school which he attended, interview his friends, disciples and relatives and gather from them all information which was likely to be useful; only after I had done all this would I start writing the biography. But I have not visited these places or become acquainted with such persons.

And now I have begun to doubt even my capacity for writing these reminiscences. I remember to have remarked more than once that, if I had the time, I would write such reminiscences. One of his disciples, for whom I feel the highest respect, heard me say this, and I have undertaken this attempt mainly to satisfy him. I would be happy, in any case, to write these reminiscences of Shrimad Rajchandra, whom I used to call Raychandbhai or the Poet, out of my love and respect, and explain their significance to *mumukshus*. As it is, however, my attempt is merely intended to satisfy a friend. To be able to do justice to these reminiscences of his life, I should be well acquainted with the Jain way, which I must admit I am not. I will, therefore, write them from an extremely restricted point of view. I will content myself with a record of my memories of him and of the events in his life which had left an impression on me, and with a discussion of what I learned from those occasions. The benefit which I thus derived, or some benefit similar to that, will perhaps be derived by the reader who is a *mumukshu* from a perusal of these reminiscences.

I have used the word *mumukshu* advisedly. This attempt at writing my reminiscences is not intended for all classes of readers.

Three persons have influenced me deeply, Tolstoy, Ruskin

and Raychandbhai: Tolstoy through one of his books and through a little correspondence with him, Ruskin through one book of his, *Unto This Last*—which in Gujarati I have called *Sarvodaya,*—and Raychandbhai through intimate personal contact. When I began to feel doubts about Hinduism as a religion, it was Raychandbhai who helped me to resolve them. In the year 1893, I came into close contact with some Christian gentlemen in South Africa. Their lives were pure, and they were devoted to their religion. Their main work in life was to persuade followers of other faiths to embrace Christianity. Though I had come into contact with them in connection with practical affairs, they began to feel solicitude for my spiritual welfare. I realized that I had one duty: that until I had studied the teachings of Hinduism and found that they did not satisfy my soul, I should not renounce the faith in which I was born. I, therefore, started reading Hindu and other scriptures. I read books on Christianity and Islam. I carried on correspondence with some friends I had made in London. I placed my doubts before them.

I entered into correspondence with every person in India in whom I had some trust, Raychandbhai being the chief among them. I had already been introduced to him and a close bond had grown between us. I had respect for him, and so I decided to get from him everything he could give. The result was that I gained peace of mind. I felt reassured that Hinduism could give me what I needed. The reader will have some idea of how much my respect for Raychandbhai must have increased because of his being responsible for this result.

Nevertheless, I have not accepted him as my *guru*. I am still in search of one, and so far my feeling in regard to everyone whom I might think of as a *guru* has been 'No, not this'. One must have the requisite qualification to come upon a perfect *guru*, and I cannot claim to have it.

'Some Reminiscences of Rajchandbhai' (G.)
Shrimad Rajchandra, Ch. 1

[1] Full moon of the month of Kartika in the Hindu calendar.

68. FREEDOM FROM ATTACHMENT

When shall I know that state supreme,
When will the knots, outer and inner, snap?
When shall I, breaking the bonds that bind us fast,
Tread the path trodden by the wise and the great?

Withdrawing the mind from all interests,
Using this body solely for self-control,
He desires nothing to serve any ulterior end of his own,
Seeing nothing in the body to bring on a trace of the darkness of ignorance.

These are the first two verses of Raychandbhai's inspired utterance at the age of eighteen.

During the two years I remained in close contact with him, I felt in him every moment the spirit of *vairagya* which shines through these verses. One rare feature of his writings is that he always set down what he had felt in his own experience. There is in them no trace of unreality. I have never read any line by him which was written to produce an effect on others. He had always by his side a book on some religious subject and a note-book with blank pages. The latter he used for noting down any thoughts which occurred to him. Sometimes, it would be prose and sometimes poetry. The poem about the 'supreme state' must have been written in that manner.

Whatever he was doing at the moment, whether eating or resting or lying in bed, he was invariably disinterested towards things of the world. I never saw him being tempted by objects of pleasure or luxury in this world.

I watched his daily life respectfully, and at close quarters. He accepted whatever he was served at meals. His dress was simple, a *dhoti* and shirt, an *angarakhun* and a turban of mixed silk and cotton yarn. I do not remember that these garments used to be strikingly clean or carefully ironed. It was the same to him whether he squatted on the ground or had a chair to sit on. In the shop, he generally squatted on a *gadi*.

He used to walk slowly, and the passer-by could see that he was absorbed in thought even while walking. There was a strange power in his eyes; they were extremely bright, and free from any sign of impatience or anxiety. They bespoke single-minded attention. The face was round, the lips thin, the nose neither pointed nor flat and the body of light build and medium size. The skin was dark. He looked an embodiment of peace. There was such sweetness in his voice that one simply wanted to go on listening to him. The face was smiling and cheerful; it shone with the light of inner joy. He had such ready command of language that I do not remember his ever pausing for a word to express his thoughts. I rarely saw him changing a word while writing a letter. And yet the reader would never feel that any thought was imperfectly expressed, or the construction of a sentence was defective or the choice of a word faulty.

These qualities can exist only in a man of self-control. A man cannot become free from attachments by making a show of being so. That state is a state of grace for the *atman*. Anyone who strives for it will discover that it may be won only after a ceaseless effort through many lives. One will discover, if one struggles to get rid of attachments, how difficult it is to succeed in the attempt. The Poet made me feel that this state of freedom from attachment was spontaneous to him.

The first step towards *moksha* is freedom from attachment. Can we ever listen with pleasure to anyone talking about *moksha* so long as our mind is attached to a single object in this world? If at any time we seem to do so, it is only the ear which is pleased, in the same way, that is, as we may be pleased merely by the musical tune of a song without following its meaning. It will be a long time before such indulgence of the ear results in our adopting a way of life which could lead towards *moksha*. Without genuine *vairagya* in the mind, one cannot be possessed with a yearning for *moksha*. The poet was possessed by such yearning.

'*Vairagya*'
Shrimad Rajchandra, Ch. 3

69. VIGILANCE IN WORK

He is a true *Vanik* who never speaks an untruth,
He is a true *Vanik* who never gives short measure,

He is a true *Vanik* who honours his father's word,
He is a true *Vanik* who returns the principal with interest.
Good sense is the *Vanik*'s measure, and the king's measure his credit.
Should the *Vania* neglect business, suffering like a forest fire spreads far and wide.

SHAMAL BHATT

It is generally believed that the spheres of practical affairs or business and spiritual pursuits or *dharma* are distinct from and incompatible with each other, that it is madness to introduce *dharma* into business, for we should succeed in neither if we made any such attempt. If this belief is not false, there is no hope for us at all. There is not a single concern or sphere of practical affairs from where *dharma* can be kept out.

Raychandbhai showed through his life that, if a man is devoted to *dharma*, this devotion should be evident in every action of his. It is not true at all that *dharma* is something to be observed on the *Ekadashi* day or during the *Paryushan*, on the *Id* day or on a Sunday, in temples, churches or mosques, but not in the shop or the king's court; on the contrary, Raychandbhai used to say and hold, and demonstrated through his own conduct, that such a belief amounted to ignorance of the nature of *dharma*.

The business in which he was engaged was that of diamonds and pearls. He carried it on in partnership with Revashanker Jagjivan Jhaveri. He also ran a cloth shop side by side. I formed the impression that he was completely upright in his dealings. I was accidentally present sometimes when he negotiated a deal. His terms were always clear and firm. I never saw any 'cleverness' about them. If the other party tried it, he immediately saw through it, and would not tolerate it. On such occasions, he would even knit his brows in anger, and one could see a flash of redness in his eyes.

Raychandbhai disproved the prevalent idea that a man who is wise in the sphere of *dharma* will not be wise in the affairs of practical life. He displayed the utmost vigilance and intelligence in his business. He could judge the worth of diamonds and pearls with the utmost accuracy. Though he did not know English, he was quick in following the general substance of the letters and telegrams received from his agents in

Paris and took no time to see through their tricks. His guesses generally turned out to be correct.

Though he displayed such vigilance and intelligence in his business, he was never impatient or felt worried about his affairs. Even when he was attending the shop, some book on a religious subject would always be lying by his side and, as soon as he had finished dealing with a customer, he would open it, or would open the note-book in which he used to note down the thoughts which occurred to him. Every day he had men like me, in search of knowledge, coming to him. He would not hesitate to discuss religious matters with them. The Poet did not follow the general, and beautiful, rule of doing business and discussing *dharma* each at its proper time, of attending to one thing at a time. Being a *shatavadhani*, he could afford to violate it. Others who might seek to emulate him would fare as a man trying to ride two horses at the same time. Even for a man who is wholly devoted to *dharma* and is completely free from attachments, it would be best always to concentrate on what he is engaged in at the moment; in fact that would be the right thing for him to do. That would be a sign of his being a man of *yoga*. *Dharma* requires one to act in that manner. If any work, be it business or something else, is a worthy activity, it ought to be done with single-minded attention. For a *mumukshu*, inward meditation on the Self should be as spontaneous and continuous as breathing. He should not cease from it even for a moment. But even while meditating on the Self, he should be totally absorbed in the work he was doing.

I do not mean to say that the Poet did not live in this way. I have said above that he used to display the utmost vigilance in his business. I did, however, form the impression that the Poet extracted more work from his body than he should have done. Could it possibly mean imperfection in his *yoga*? It is a principle of *dharma* that one should even lay down one's life in discharging one's duty. But to undertake work beyond one's capacity and look upon it as one's duty is a form of attachment. I have always felt that the Poet did have this highly subtle attachment in him.

It often happens that a man accepts work beyond his capacity through spiritual motives and then finds it difficult to cope with it. We look upon this as virtue and admire it. But looked at from a spiritual point of view, that is, from the point of view of *dharma*, there is every possibility that the motive behind such work springs from a subtle form of ignorance.

If we are no more than instruments in this world, if it is true that we are given this body on hire and that our highest duty is to attain *moksha* through it as quickly as we may, then we must certainly give up everything which may serve as an obstacle in our path—that is the only true spiritual attitude.

Raychandbhai himself had explained to me, in a different form and in his own wonderful manner the argument I have advanced above. How, then, did he come to take upon himself certain tasks which worried him and brought on severe illness on him?

If I am right in believing that even Raychandbhai was temporarily overcome with spiritual ignorance in the form of a desire to do good, the truth of the line 'All creatures follow their nature, what then will constraint avail?'[1] is very well illustrated in his case, and this is all that it means. There are some who use these words of Krishna to justify self-indulgence; they altogether pervert their meaning. Raychandbhai's *prakriti* took him into deep waters despite himself. To undertake work in this manner may be an error, but it may be considered so only in the case of one who is nearing perfection. We, ordinary men and women, can do justice to a good cause only if we become mad after it. We shall close this argument here.

It is also sometimes believed that religious-minded men are so simple that everyone can deceive them, that they understand nothing about worldly affairs. If this belief is true, then the two *Avatars*, Krishnachandra and Ramachandra, should be looked upon not as incarnations but as mere ordinary men of the world. The Poet used to say that it should be impossible to deceive a person of perfect spiritual knowledge. A person may be religious-minded, that is, may be moral in his life, but may have no spiritual knowledge. What is required for *moksha*, however, is a happy combination of moral life and spiritual knowledge which is the result of one's own experience. In the presence of one who has acquired such knowledge, hypocrisy and fraud cannot keep their mask for long. Untruth cannot flourish in the presence of truth. In the presence of non-violence, violence ceases. Where the light of honesty shines, the darkness of deception vanishes. The moment a man of spiritual knowledge devoted to *dharma* sees a deceitful man, his heart melts with compassion. How can one who has seen the Self in him fail to understand another person? I cannot say that the Poet always demonstrated this truth in his life. People did occasionally cheat him in the name of religion. Such instances do not prove any flaw in the principles, but suggest how very difficult it is to acquire spiritual knowledge of absolute purity.

Despite these limitations, I have not observed in anyone else such a beautiful combination of practical ability and devotion to *dharma* as I did in the Poet.

'Business Life' (G.)
Shrimad Rajchandra, Ch. 4

[1] *Bhagavad Gita*, III. 33.

70. RAJCHANDRA ON RELIGION

Before we examine Raychandbhai's life of *dharma*, it is necessary to discuss the nature of *dharma* as explained by him.

Dharma does not mean any particular creed or dogma. Nor does it mean reading or learning by rote books known as *shastras* or even believing all that they say.

Dharma is a quality of the soul and is present, visibly or invisibly, in every human being. Through it we know our duty in human life and our true relation with other souls. It is evident that we cannot do so till we have known the Self in us. Hence *dharma* is the means by which we can know ourselves.

We may accept this means from wherever we get it, whether from India or Europe or Arabia. Anyone who has studied the scriptures of different faiths will say that the general nature of this means as expounded in them is the same. No *shastra* in any religion says that we may speak untruth or follow it in practice, nor that we may commit violence. Stating the quintessence of all *shastras*, Shankaracharya said: '*Brahma satyam jaganmithya.*'[1] The Koran-e-Sharif says the same thing in different words when it asserts that God is one and alone, and that nothing beside Him exists. The Bible says: 'I and my Father are one.' All these are different statements of the same truth. But imperfect human beings, expounding this one truth through their various understandings, have erected veritable prison-houses from which our minds have to escape. We, imperfect human beings, try to go forward with the help of others less imperfect than we, and imagine that beyond a certain stage there is no further way to go. In truth it is not so at all. After a certain stage is reached, the Shastras give no help; experience alone helps then. Hence sang Raychandbhai:

That state the Blessed one who has attained perfection of knowledge
  sees in his vision,
But cannot describe in words;
I have fixed my eyes on that supreme state as my goal,
But at present it is an aspiration beyond my power to realize.

Ultimately, therefore, it is the *atman* which wins *moksha* for itself.

Raychandbhai has expounded this essential truth in numerous ways in his writings. He had made a deep study of a number of books on *dharma*. He could follow Sanskrit and Magadhi languages without any difficulty. He had studied *Vedanta*, as also the *Bhagavata* and the *Gita*. As for books on Jain religion, he used to read every such book that he came across. His capacity for reading and absorbing was inexhaustible. He found one reading enough for grasping the substance of a book.

He had also read in translation the Koran and the Zend-Avesta.

He used to tell me that he was inclined towards Jain philosophy. He believed that the *Jinagamas* contained the perfection of spiritual knowledge. It is necessary that I should state this view of his. I look upon myself as altogether unqualified to express an opinion on it.

Raychandbhai did not, however, lack respect for other faiths. He even felt admiration for *Vedanta*. A *Vedantin* would naturally take the Poet to be a *Vedantin*. In all his discussions with me, he never told me that if I wished to attain *moksha* I should follow a particular *dharma* and no other. He advised me only to pay attention to my actions. When we discussed what books I should read, he took into consideration my personal inclination and the early family influences on me and advised me to continue the *Gita* which I was then reading. Other books which he suggested were *Panchikaran*, *Maniratnamala*, the chapter on *vairagya* in *Yogavasishtha*, *Kavyadohan*, Part I, and *Mokshamala* composed by himself.

Raychandbhai used to say that the different faiths were like so many walled enclosures in which men and women were confined. He whose one aim in life is to attain *moksha* need not give exclusive devotion to a particular faith.

>Live as you will,
>Attain to Hari anyhow.

This was Raychandbhai's principle too, as it was Akha's. He was always bored by religious controversy and rarely engaged himself in it. He would study and understand the excellence of each faith and explain it to the followers of that faith. Through my correspondence with him from South Africa, too, this is the lesson which I learned from him.

My own belief is that every religion is perfect from the point of view of its followers and imperfect from that of the followers of other faiths. Examined from an independent point of view, every religion is both perfect and imperfect. Beyond a certain stage, every *shastra* becomes a fetter hindering further progress; but, then, that is the stage reached by one who has transcended the *gunas*. If we follow Raychandbhai's point of view, no one need give up his faith and embrace another. Everyone may, following his own faith, win his freedom, that is, *moksha*, for to win *moksha* means to be perfectly free from attachments and aversions.

MOHANDAS KARAMCHAND GANDHI

'*Dharma*'
Shrimad Rajchandra, Ch. 5

[1] 'The *Brahman* alone is real, the world of appearance is false.'

71. RAJCHANDRA'S FAITH IN *AHIMSA*

Napa,
March 18, 1930

Kavi Rajchandra was born in a place called Vavania in Kathiawar. I came in touch with him in 1891, the day of my return from London, at Dr. P.J. Mehta's residence in Bombay. Kavi, as I used to call him, was nearly related to Dr. Mehta. He was introduced to me as a *shatavadhani*, i.e., one who can remember a hundred things at a time. Kavi was quite young at the time, not much older than I was then, i.e., 21 years. He had, however, given up all public exhibition of his powers and was given to purely religious pursuits. I was much struck by his simplicity and independence of judgment. He was free from all touch of blind orthodoxy. What struck me perhaps more was his combining business with religion in practice. A student of the philosophy of religion, he tried to practise what he believed. Himself a Jain, his toleration of the other creeds was remarkable. He had a chance of going to England for studies, but he would not go. He would not learn English. His schooling was quite elementary. But he was a genius. He knew Sanskrit, Magadhi and, I believe, Pali. He was a voracious reader of religious literature and had acquired through Gujarati sources a knowledge, enough for his purpose, of Islam, Christianity and Zoroastrianism. Such was the man who captivated my heart in religious matters as no other man has till now.

I have said elsewhere that in moulding my inner life Tolstoy and Ruskin vied with Kavi. But Kavi's influence was undoubtedly deeper if only because I had come in closest personal touch with him. His judgment appealed to my moral sense in the vast majority of cases. The bedrock of his faith was unquestionably *ahimsa*. His *ahimsa* was not of the crude type we witness today among its so-called votaries who confine their attention merely to the saving of aged cattle and insect life. His *ahimsa*, if it included the tiniest insect, also covered the whole of humanity.

Yet I never could regard Kavi as a perfect man. But of all the men I knew he appeared to me to be nearer perfection than the rest. Alas! he died all too young (thirty-three years) when he felt that he was surely going to see truth face to face. He has many worshippers but not as many followers. His writings, largely consisting of soulful letters to inquirers, have been collected and published. An attempt is being made to have them translated in Hindi. I know that they would bear an English translation. They are largely based on inward experience.

'A Great Seer'
Modern Review, June 1930

§ 11. Books Read

72. ESOTERIC CHRISTIANITY

Durban, Natal

The following books by the late Mrs. Anna Kingsford and Mr. Edward Maitland are offered for sale at their published prices. They are introduced in South Africa for the first time:

The Perfect Way, 7/6
Clothed With the Sun, 7/6
The Story of the New Gospel of Interpretation, 2/6
The New Gospel of Interpretation, 1/-
The Bible's Own Account of Itself, 1/-

The following are some of the opinions concerning the books:

'A fountain of light (*The Perfect Way*) interpretative and reconciliatory .... No student of divine things can dispense with it.'—*Light*, London

'Unequalled as a means of grace amongst all the English books of the century'.—*Occult World*

Some pamphlets bearing on the subject can be had free of charge at my office.

M. K. GANDHI
AGENT FOR THE ESOTERIC CHRISTIAN UNION
AND THE LONDON VEGETARIAN SOCIETY

'Books for Sale'
Natal Mercury, 28 Nov. 1894

73. SUFI MYSTICS

We have received for review two volumes of a series entitled *The Wisdom of the East* being published in Britain. The first is called *The Way of the Buddha* and the second, *Persian Mystics*, in which the author has assigned the first place to Jalaluddin Rumi.[1] An informative account of Sufis is followed by a narrative of the life of Jalaluddin and translations of some of his poems. In the author's view, a Sufi is a lover of God. Above everything else, the Sufis aspire after a pure heart and love of God. Jalaluddin was once observed dancing with joy at a funeral, and on being questioned what he meant thereby, the saint replied, 'When the human spirit, after years of imprisonment in the cage and dungeon of the body, is at length set free and wings its flight to the source whence it came, is it not an occasion for rejoicing?' We can see that, in the olden days, even women freely participated in such a way of life. Rabia Bibi was a Sufi herself. When asked if she hated the devil, she retorted that 'her love of God left her no time to hate anyone'. According to the Sufi point of view, no religion based on morality can be considered to be false. In reply to a question Jalaluddin said, 'The ways of God are as many as the number of souls of men.' Elsewhere he says, 'God's light is one but its rays are various in hue .... We can worship God along any path, provided it be with a true and sincere heart.'

Referring to the nature of true knowledge, Jalaluddin says that 'a blood-stain can be washed away with water, but the stain of ignorance can be washed clean only with the water of God's grace'. And then again, 'True knowledge is the knowledge of God.' When asked where one could find God, the poet replied, 'I saw the Cross and also Christians, but I did not find God on the Cross. I went to find Him in the temple, but in vain. I saw Him neither in Herat nor in Kandahar. He could be found neither on the hill nor in the cave. At last, I looked into my heart and found Him there, only there and nowhere else.' This is an excellent book to read. We can endlessly go on quoting passages such as those above. We should like to recommend the book to everyone. It will be of profit to all, Hindus and Muslims alike. It costs two shillings in Britain. The same firm has also published *Gulistan* of Sheikh

Saadi, priced at 1s. Then there is a book entitled *The Essence of the Koran*, costing 1s. *The Way of the Buddha* is priced at 2s and *The Way of Zoroaster* also at 2s. Other books are to follow. If any of our readers wants one or more of these books, he should send us the amount of the price quoted above, and an extra 6d for every book. We shall procure the book for him. The extra 6d is for postage.

'Jalaluddin Rumi' (G.)
Indian Opinion, 15 June 1907

[1] 1207-73; Sufi poet of Persia.

74. GUJARATI AND HINDI BOOKS

Tolstoy Farm,
[27 May 1911]

Chi. Harilal,

I have the letter you wrote before leaving Delagoa Bay. It is desirable that Rami[1] grows up under the influence of strictly Indian ideas. Accordingly, I think you have done right in deciding not to send her chocolates. However, I should caution you against taking the line that 'this must be done because Bapu wants it so'. From among the ideas I suggest, you should put only those into practice that appeal to you. I should like you to grow up in freedom. I know your motives are good. Whenever your ideas are mistaken, they will therefore get corrected automatically.

The prisoners have not been released so far but they will soon be.

It appears that the cable I sent you about your applying for registration has not reached you. I sent it care of Nanji Dulabhdas.

Keep reading *Indian Opinion* carefully while you are there.

The following Gujarati books are eminently worth reading: *Kavyadohan, Panchikaran, Maniratnamala, Dasbodh*, Chapter VI of *Yogavasishtha*—Hindi translation available—Poet Narmadashanker's[2] *Dharma vishe Vichar*, two volumes of Raychandbhai's writings.

And, of course, there are *Karanghelo* and other books. *Karanghelo* evidences the maturity of the Gujarati language. Taylor's grammar and his introduction to it are both very good. I have forgotten whether it is a preface or a separate essay on the Gujarati language.

I should advise you to make it a regular practice to read Tulsidasa's *Ramayana*. Most of the books I have listed at the end of *Indian Home Rule* are worth going through. In order to learn Sanskrit well, I would advise you always to start your reading with it. Only then will you remember and understand it all. You won't find it difficult after the first book is properly done. Do not take up the second book before you have mastered the first. Whenever you come across a Sanskrit verse, you should immediately try to understand its meaning in Gujarati.

Please write to me in detail and regularly.

Blessings from
MOHANDAS

Letter to Harilal Gandhi (G.)
SN 9532

[1] Harilal's daughter.
[2] Narmadashanker Lalshanker Dave (1833-89); poet and a pioneer of modern Gujarati literature.

75. BOOKS READ IN GAOL (1922)

APRIL 21, FRIDAY

Up to today, I have read the following books:[1]
1. *Master and His Teaching*
2. *Arm of God*
3. *Christianity in Practice*
4. *By an Unknown Disciple*
5. *Satyagraha aur Asahayoga*
6. The Koran
7. *The Way to Begin Life*
8. *Trips to the Moon*
9. *Indian Administration* (Thakore)

10. *The Ramayana*—Tulsidas

I have started baking *chapatis* since yesterday.

APRIL 22, SATURDAY

I finished reading *Natural History of Birds*.

Today the Superintendent called all political prisoners to meet him.

I had a talk with Deshpande.

APRIL 23, SUNDAY

Finished reading *The Young Crusader*.

Gave up lemons and sugar from today.

APRIL 26, WEDNESDAY

Yesterday I finished reading *A History of Scotland*—Book I.

The Rev. Lawrence has sent me *Bible View of the World*.

APRIL 29, SATURDAY

Finished reading the book sent by the Rev. Lawrence.

Glanced through a book on martyrs.

MAY 1, MONDAY

Finished reading *A History of Scotland*—Book II. Today they passed on to me ten lbs. of flour in one lot.

MAY 5, FRIDAY

Finished reading Farrar's *Seekers after God*. Stopped taking oranges since yesterday.

MAY 6, SATURDAY

Finished reading *A History of Scotland*. Today received a letter from the Government saying that my letter to Hakimji could not be forwarded to him.

Finished reading *Misar Kumari*.

MAY 12, FRIDAY

Finished reading *Stories from the History of Rome*. Today the Superintendent declined to give me a copy of the Government's order requiring him to keep back my letter to Hakimji. Consequently, I wrote one letter to the Government and one to Hakimji. I wrote to Hakimji only to inform him that, since the Government have refused to forward to him my letter uncensored, I have given up the intention of writing a quarterly letter.

MAY 15, MONDAY
Banker was today transferred to this ward. Wrote to the Superintendent, not officially but in a personal way, that I did not like his increasing again the supply of oranges to me. He should stop the supply of oranges and *chapatis* and the additional supply of milk to me.

MAY 16, TUESDAY
Mr. Jacob, head clerk to Mr. Griffiths, came to see me and talked to me on his behalf. The Superintendent declined to reduce the supply of oranges and told me, on the contrary, that he had orders to supply nine oranges to me.

> They are slaves who will not choose
> Hatred, scoffing and abuse,
> Rather than in silence shrink
> From the Truth they needs must think:
> They are slaves who dare not be
> In the right with two or three.
>   LOWELL
> (From *Tom Brown's School Days*)

MAY 17, WEDNESDAY
Finished reading *Tom Brown's Schooldays*. Some portions of it are beautiful.

> The Holy Supper is kept indeed
> In whatso we share with another's need—
> Not that which we give, but what we share,
> For the gift without the giver is bare;
> Who bestows himself with his alms feeds three,
> Himself, his hungering neighbour, and Me.
>   LOWELL

—from the same book as above.

MAY 20, SATURDAY
Finished reading Bacon's *The Wisdom of the Ancients*. Have given up *chapatis* since Wednesday. I am living, as an experiment, on four seers of milk, two ounces of raisins, four oranges and two lemons. Haji was taken to a dark cell yesterday.

MAY 28, SUNDAY
Read the history of India up to the Moghul dynasty. Went through Morris's grammar.

MAY 29, MONDAY
Finished reading *Chandrakant*, Part II, as also Patanjali's *Yogadarshan*.
   Nearly four weeks have passed.
   Started reading the Gujarati translation of Valmiki's *Ramayana*.

MAY 31, WEDNESDAY
Finished reading Kipling's *The Five Nations*.

JUNE 4, SUNDAY
Finished reading Edward Bellamy's *Equality*.

JUNE 6, TUESDAY
The Superintendent called and informed me that the Government has refused permission for the printing of the *Balpothi*.[2] It has permitted me to send for books mentioned in the list.

JUNE 7, WEDNESDAY
Finished reading *St. Paul in Greece* (by Davis).

JUNE 9, FRIDAY
Finished reading *Dr. Jekyll and Mr. Hyde*.

JUNE 14, WEDNESDAY
Finished reading *Pitt* by Lord Rosebery.

Truth is	Untruth is
gold	brass
silver	tin
light	darkness
heaven	hell
sky	the nether world
day	night
diamond	a pebble
a virtuous wife	a prostitute
celibacy	adultery
God	Satan
Ormuzd	Ahriman
Brahman	a soul in delusion

living	lifeless
virility	impotence
valour	cowardice
Rama	Ravana
deliverance	bondage
ambrosia	poison
life	death
goodness	evil
existence	non-existence
Truth is one	Untruth has many forms
Truth is a straight line	Untruth is a curved line
a right angle	...
an ocean	the Sahara Desert
restraint	self-indulgence
love	hatred

JUNE 17, SATURDAY
Finished reading Kipling's *Second Jungle Book*.

JUNE 21, WEDNESDAY
Finished reading *Faust*.

JUNE 24, SATURDAY
Finished reading John Howard's life.
   Received yesterday a parcel of five lbs. of raisins.

JUNE 25, SUNDAY
Finished reading Valmiki's *Ramayana*. Started reading *Shantiparva*,[3] Part I.

JUNE 28, WEDNESDAY
Finished reading Jules Verne's *Dropped from the Clouds*.

JULY 1, SATURDAY
Finished reading the life of Columbus by Irving. Anasuyabehn, Kanji and Dhirajlal came to see Shankerlal. Ba, Harilal, Ramdas, Maganlal, Mathuradas, and Manu came to see me.

JULY 5, WEDNESDAY
Warner came yesterday and gave me a box and some books. Commenced reading Girdhar's *Ramayana* and *The Crusades*. Finished reading Wilberforce's *Five Empires*.

JULY 10, MONDAY
Finished reading *Lays of Ancient Rome.*

JULY 12, WEDNESDAY
Received another parcel of five and a half seers of raisins.

JULY 13, THURSDAY
Finished reading *The Crusades*. Started reading Gibbon's *Rome*.

JULY 16, SUNDAY
Finished reading *Shantiparva*, Part I. Started reading Part II.

JULY 18, TUESDAY
Finished reading First Urdu Book.

JULY 22, SATURDAY
Finished reading Girdhar's *Ramayana*. Started reading *Shrimad Bhagavat*.

JULY 23, SUNDAY
Started reading *Krishnacharitra* by Jhaveri.

JULY 29, SATURDAY
Finished reading Krishnalal Jhaveri's *Krishnacharitra*.

AUGUST 4, FRIDAY
Finished reading Vaidya's *Krishnacharitra*.

AUGUST 7, MONDAY
Finished reading Gibbon, Vol. I. Started reading Vol. II.

AUGUST 10, THURSDAY
Finished reading the *Gita* by Tilak, *Shantiparva*—Part II, and *Bhagavat*—Part I. Started reading *Bhagavat*—Part II.

AUGUST 22, TUESDAY
Political prisoners were removed yesterday to the European ward. Today they were brought back to their original ward.

AUGUST 24, THURSDAY
Finished reading *Adiparva*.

AUGUST 27, SUNDAY
Finished reading *Bhagavat*—Part II. Started reading *Sabhaparva* on Friday. Started reading *Sarasvatichandra*.

AUGUST 28, MONDAY
Finished reading *Manusmriti*. Started reading *Ishopanishad*.

AUGUST 30, WEDNESDAY
Finished reading *Sabhaparva*. Started reading *Vanaparva*.

SEPTEMBER 1, FRIDAY
Finished reading Gibbon, Vol. II. Finished reading *Ishopanishad*.

SEPTEMBER 2, SATURDAY
Started reading Gibbon, Vol. III.

SEPTEMBER 3, SUNDAY
Finished reading *Sarasvatichandra*—Part I. Started reading Part II.

SEPTEMBER 6, WEDNESDAY
Finished reading *Sarasvatichandra*—Part II. Started reading Part III.

SEPTEMBER 9, SATURDAY
Finished reading *Sarasvatichandra*—Part III. Started reading Part IV.

SEPTEMBER 13, WEDNESDAY
I have decided to observe silence from 3 p.m. today up to 3 p.m. on Tuesday, with Major Jones's consent. The following exceptions will be made:
1. When others or I suffer.
2. When friends from outside come to see me.
3. If, in the meanwhile, I am removed to the ward of my Dharwar friends.
4. If an official like Mr. Hayward happens to visit us.
5. If Major Jones wishes to have a talk with me.

Bedsteads were received today for ...

SEPTEMBER 20, WEDNESDAY
Observance of silence ended yesterday. Experienced supreme joy during the period of silence. Finished reading *Sarasvatichandra*—Part IV today. Finished reading Kabir's poems. Started reading Jacob Boehmen. Wrote a letter of apology to Shankerlal. Have again started observing silence. It will end at 3 p.m. on Tuesday.

SEPTEMBER 23, SATURDAY
Finished reading Boehmen's *Supersensual Life*.

It is naught indeed but thine own hearing and willing that do hinder thee so that thou dost not see and hear God. (p. 14)

If thou rulest over the creatures externally only and not from the right internal ground of thy inward nature, then thy will and ruling is in a bestial kind or matter. (p. 18)

Thou art like all things and nothing is unlike thee. (p. 19)

If thou wilt be like all things, thou must forsake all things. (p. 20)

Let the hands or the head be at labour, thy heart ought nevertheless to rest in God. (p. 65)

Heaven is the turning in of the will to the love of God. (p. 83)

Hell is the turning in of the will into the wrath of God. (p. 83)

BOEHMEN, *Supersensual Life*

Started reading *Pro Christo et Ecclesia*.

SEPTEMBER 24, SUNDAY
Finished reading *Kathavalli Upanishad*.

SEPTEMBER 25, MONDAY
Finished reading *Pro Christo et Ecclesia*. Started reading *Satyartha Prakasha*. Finished reading *Vanaparva*.

SEPTEMBER 26, TUESDAY
Started reading *Virataparva* and *Galilean*.

SEPTEMBER 27, WEDNESDAY
Started reading *Jnaneshwari*.

SEPTEMBER 30, SATURDAY
Finished reading *Virataparva* and Gibbon, Vol. III.

OCTOBER 1, SUNDAY
Started Gibbon, Vol. IV, and *Udyogaparva*.

OCTOBER 3, TUESDAY
Finished reading *Galilean*.

OCTOBER 6, FRIDAY
Ba, Jamnalalji, Ramdas, Punjabhai and Kishorelal came to see me on Wednesday. Wrote a letter to Jamnalalji yesterday

about Ramdas. Wrote a letter to the Superintendent today about Gani and newspapers. Started reading *Philo Christus* and Fourth Urdu Book.

OCTOBER 15, SUNDAY
Finished reading *Udyogaparva*.

OCTOBER 16, MONDAY
Started reading *Bhishmaparva*.

OCTOBER 18, WEDNESDAY
Finished reading *Satyartha Prakasha*.

OCTOBER 22, SUNDAY
Finished reading *Bhishmaparva* and *Philo Christus*.

OCTOBER 23, MONDAY
Finished reading Gibbon. Started reading *Dronaparva* and *Prem Mitra*. Finished reading *Jnaneshwari*.

OCTOBER 24, TUESDAY
Finished reading *Prem Mitra*.

OCTOBER 25, WEDNESDAY
Started reading *Shad-darshan-samuchchaya* and *The Gospel and the Plough*. Started reading Nathuram Sharma's commentary on the *Gita*.

OCTOBER 28, SATURDAY
Finished reading *The Gospel and the Plough*.

NOVEMBER 6, MONDAY
Finished reading *Dronaparva*.

NOVEMBER 7, TUESDAY
Started reading *Karnaparva*. Shankerlal was taken ill yesterday; he vomited, etc.

NOVEMBER 11, SATURDAY
Finished reading *Karnaparva*.

NOVEMBER 12, SUNDAY
Started reading *Shalyaparva*.

NOVEMBER 17, FRIDAY
Finished reading *Shalyaparva*. Left off oranges from today as an experiment. Started reading *Anushasanparva*.

NOVEMBER 22, WEDNESDAY
Finished reading *Shad-darshan-samuchchaya*.

NOVEMBER 27, MONDAY
Finished reading *Urdu Reader* No. 3. Started reading *Urdu Reader* No. 4.

NOVEMBER 28, TUESDAY
Finished reading *Anushasanparva*. Started reading *Ashvamedhikaparva*.

DECEMBER 2, SATURDAY
Finished reading *Ashvamedhikaparva*. Started reading *Ashramvasik*.

DECEMBER 4, MONDAY
Finished reading the *Mahabharata*. Started reading the writings of poet Rajchandra. Had started reading the *Mahabharata* on June 25.

DECEMBER 5, TUESDAY
Had severe stomach-ache yesterday, so took castor oil today and started taking oranges. Also started taking raisins after an interval of nearly one month.

DECEMBER 6, WEDNESDAY
Commenced J. Brierly's *Ourselves and the Universe*.

DECEMBER 9, SATURDAY
To wish ill, to do ill, to speak ill or to think ill of anyone, we are equally forbidden without exception.

TERTULLIAN
(J. BRIERLY in *Ourselves and the Universe*)
Gave up raisins and oranges from Friday.

DECEMBER 15, FRIDAY
Completed J. Brierly's *Ourselves and the Universe*.

DECEMBER 16, SATURDAY
Started reading *What Christianity Means to Me* by Lyman Abbott. Ba was to come today, but did not.

DECEMBER 21
Wrote a letter yesterday to the Major about refusal of permission to Maganlal and others. Handed it today to Warner.

DECEMBER 25
Finished reading *What Christianity Means to Me*. Took raisins and figs sent by Anasuyabehn.

'Jail Diary, 1922' (G.)
SN 8039M

[1] Gandhi was taken to the Yeravda Central Jail on 21 Mar. 1922. During the period of his imprisonment there he read, in all, about 150 books on religion, literature, social and natural sciences. In the series *My Jail Experiences*, which appeared in *Young India* from April 1924 to October 1924, he offered detailed comments on some of these books.
[2] Gujarati Primer.
[3] One of the eighteen books into which the *Mahabharata* is divided.

76. BOOKS READ IN GAOL (1923)

JANUARY 3, WEDNESDAY
Finished reading *Steps to Christianity* yesterday. Started reading Trine's *My Philosophy and Religion*. Today the Major gave me a copy of the notification that the Inner Temple had removed my name from its Roll.

JANUARY 7, SUNDAY
Finished reading *My Philosophy and Religion* yesterday. Started reading Rabindranath's *Sadhana* and *Upanishad* [*Prakash*].

JANUARY 14, SUNDAY
Finished reading *Sadhana* yesterday.

FEBRUARY 4, SUNDAY
Finished reading Rajchandra's writings and *Ishopanishad* with a commentary. Reading *Kena*.[1] Completed the second reading of Urdu Book III. Finished reading *Auto-suggestion*. Ba came and saw me on January 27. Released Shankerlal from his vow on the 28th.

FEBRUARY 5, MONDAY
Finished reading *Helps to Bible Study*. Started reading Max Muller's translation of the *Upanishads* as also Wells's History.

FEBRUARY 22, THURSDAY
Finished reading Max Muller's translation of the *Upanishads*

as also *Upanishad Prakash*, Part III. Reading Part IV and Wells's History.

FEBRUARY 25, SUNDAY
Finished reading the *Upanishad* [*Prakash*], Part IV. Started reading Part V, *Kathavalli Upanishad*.

MARCH 2, FRIDAY
Finished reading Wells's History, Part II, on February 28. Started reading the Bible yesterday. Finished reading the leaflet on the worship of Vishnu. Started reading Wells's History, Part I.

MARCH 11, SUNDAY
Applied, on Wednesday, caustic soda to the eye for conjunctivitis.
  Finished reading the *Upanishad* [*Prakash*], Part V, on Thursday. Started reading Part VI. I could not spin on that day. Completed Urdu Book IV. Started reading Book V.

MARCH 16, FRIDAY
Finished reading Wells's History, Part I, yesterday. Today glanced through *Science of Peace* by Bhagwandas.

MARCH 19, MONDAY
Finished reading Kipling's *Barrack-room Ballads*. Reading Geddes's *Evolution of Cities*. Finished reading the pamphlet on Vedic religion.

MARCH 21, WEDNESDAY
Yesterday finished reading Geddes's *Evolution of Cities*. Today started reading a biography of Ramanuja. Received ten seers of raisins.

MARCH 22, THURSDAY
Finished reading the biography of Ramanujacharya. Started reading Sikh history.

MARCH 26, MONDAY
Started reading Mirza's *Ethics of Islam* yesterday.

MARCH 31, SATURDAY
Finished reading Sikh history and Mirza's *Ethics of Islam* yes-

terday and started reading Benjamin Kidd's *Social Evolution*.
Started reading Buhler's translation of *Manusmriti* today.

APRIL 4, WEDNESDAY
Finished reading Kidd's *Social Evolution* yesterday. Finished reading Buhler's preface to *Manusmriti* today. Started reading *Rise of the Sikh Power* by Gokulchand.

APRIL 9, MONDAY
Yesterday finished reading Gokulchand's *Rise of the Sikh Power* as also *Kabir's Songs* by the Poet. Started reading *Our Hellenic Heritage* by James today. Started reading Dadachandji's *Avesta* and Purani's translation of Aurobindo's *Gitanishkarsha*.

APRIL 17, TUESDAY
Finished reading James's *Our Hellenic Heritage*. Devdas came and saw me yesterday. Shankerlal was released today.

APRIL 19, THURSDAY
Sufishah Mullah Shah, when he was advised to flee from the wrath of Shah Jehan, is reported to have said:

I am not an impostor that I should seek safety in flight. I am an utterer of truth. Death and life are to me alike. Let my blood in another life also redden the impaling stake. I am living and eternal; death recoils from me, for my knowledge has vanquished death. The sphere where all colours are effaced has become my abode.

Mansuri Hallaj said:

To cut off the hands of a fettered man is easy, but to sever the links that bind me to the Divinity would be a task indeed.

CLAUDE FIELD in *Mystics and Saints of Islam*

Received today five seers of raisins.

APRIL 26, THURSDAY
Finished reading *Upanishad Prakash*, Parts VII-X (*Kathopanishad*). Today started reading Part XI commencing with *Prashnopanishad*. Completed on Saturday the second reading of Urdu Reader No. I. Severe pain in stomach on Saturday. Subsided on Monday. The Major looked after me very well. I suffered very much. On Saturday, could continue work and studies according to schedule, despite the pain. They remained sus-

pended from Sunday to Tuesday. Did not observe silence on account of pain. I believe the pain was due to my taking milk and bread as usual at 7 a.m. before the castor oil taken by me early on Saturday morning could act. Once before I had done precisely this. It had done no harm then, but this time it did. I draw two conclusions from this. First, the disease must be digging itself in slowly. Second, this body of mine will not stand the experiment of taking food before the purgative has had its effect. This result is both welcome and painful. God has been testing me on all sides—He does not permit me to see what He has been recording in His book. His wisdom is boundless.

APRIL 28, SATURDAY
Yesterday I finished reading Dadachandji's *Avesta* and started reading Spencer's *Elements of Sociology*. Today I started reading *History of Sikhism* by Macauliff.

MAY 9, WEDNESDAY
Col. Maddock[2] examined me last Saturday and informed me that most probably I was suffering from incipient dysentery. The Major started giving me injection of emetine since Sunday. It is about a week since Manzar Ali arrived. News was received today that Indulal also would be coming here. The Major delivered Andrews's letter to me today. Finished reading *Gitanishkarsha* yesterday.

MAY 16, WEDNESDAY
Indulal came yesterday. Col. Maddock examined me once again. Finished reading Herbert Spencer's *Elements of Sociology* today. Also glanced through Shivram Pherwani's *Social Efficiency*.

MAY 19, SATURDAY
I was taken to the European ward yesterday. Ba, Radha, Mani, Laxmi (Junior) and Jamnadas saw me yesterday. Yesterday I finished reading Wadia's *Message of Mahomed* and started reading *Message of Christ*. Finished reading *Prashnopanishad*.

MAY 20, SUNDAY
Started reading *Mandukopanishad*

Books Read 171

MAY 21, MONDAY
Finished reading Hasan's *Saints of Islam*. Started reading Moulton's *Early Zoroastrianism*.

MAY 27, SUNDAY
Finished reading Kaka's *Himalayno Pravas* and *History of Sikhism*, Part III. Commenced reading Part IV and also Chandrashankar's *Sitaharan*. Read Rolf Evelyn's *Bars and Shadows*.

MAY 31, THURSDAY
On Tuesday, took up again the spinning-wheel which had been abandoned for thirteen days. Finished reading Chandrashanker's *Sitaharan* yesterday. Today finished reading Moulton's *Early Zoroastrianism*.

JUNE 1, FRIDAY
Finished reading Kishorelal's book, *Buddha and Mahavira*, as also *History of Sikhism*, Part V.

JUNE 3, SUNDAY
Finished reading Kishorelal's *Rama and Krishna*, and also *History of Sikhism*, Part VI.

JUNE 6, WEDNESDAY
Finished reading the story of Aurobindo's imprisonment and *Mandukopanishad*.

JUNE 16, SATURDAY
Finished reading *Man and Superman* yesterday. Finished reading *Bhagyano Varas* today. Started reading English translation of *Markandeya Purana*.

JUNE 30, SATURDAY
At the beginning of this week, finished reading *Poorva Rang* by Kaka and Narahari and started reading the lectures given in the Puratatvamandir. Yesterday finished reading a book on an episode in the life of the Prophet in Urdu and started reading the account of the companions of the Prophet [*Usva-e-Sahaba*].

There was a discussion yesterday with Dalziel and the Major about the flogging of the Mulshi Peta prisoners.

JULY 2, MONDAY
Yesterday finished reading *Markandeya Purana* and started

reading Chapters XV and XVI of *Mandukyopanishad* and Chapter XVII of Gaudapadacharya's *Karika*.

Started reading today Buckle's *History of Civilization*, Part I.

JULY 7, SATURDAY
Finished reading the lecture series given at the Puratatvamandir. Started reading *Jaya-jayant*. Suffered great pain on Monday night. The fault was entirely mine. I ate more than I should have of the figs sent by Anasuyabehn. Boundless indeed is God's kindness. What else can be more conducive to welfare than immediate punishment for a sin?

JULY 10, TUESDAY
Yesterday finished reading the lecture series given at the Puratatvamandir and started reading Rabindranath's book on ancient literature.

Wrote a letter yesterday to the Superintendent about my commencing a fast from today. He, therefore, came and appealed to me to postpone the fast. He again called on me this morning and asked me to postpone it for 48 hours for his sake. I have agreed to do so. Mr. Griffiths came today at 2 p.m. and left after talking to me for two hours.

JULY 12, THURSDAY
Yesterday again Griffiths came with a message from the Governor. Finished reading the book on ancient literature yesterday. Started reading *Yugadharma*. I met Dastane and Dev in the presence of the Superintendent and Mr. Griffiths. After discussing the moral issues involved, they announced their decision to give up their fast.

JULY 13, FRIDAY
Chhaganlal, Kashi and others were to come to see me but did not.

JULY 22, SUNDAY
Ba, Chhaganlal, Amina, Ramdas and Manu saw me last Monday. Finished reading, during the week, the autobiography of Countess Tolstoy and Buckle's *History*, Part I. Reading Part II and *Kalapani-ni-katha*. Wrote a letter to Mr. Griffiths about Dastane and others on Tuesday.

JULY 30, MONDAY
Finished reading *Kalapani-ni-katha* last week. Finished reading *Sampattishastra*, Part I. Reading Part II. Finished reading *Juno Karar*[3] yesterday. Started reading *Navo Karar*[4] today.

AUGUST 8, WEDNESDAY
Finished reading Buckle's *History*, Part II, and *Gitagovind*.

AUGUST 12, SUNDAY
Finished reading the last part of the *Upanishad* [*Prakash*], covering *Aitareya Brahmana* and *Taittiriya Brahmana*. Started reading *Chhandogya Upanishad*. On Thursday, started reading Prof. James's *Varieties of Religious Experience*. Finished reading *Sampattishastra*.

AUGUST 15, WEDNESDAY
The Governor paid a visit on Monday. Wrote a letter today about the special division. Finished reading *Sahaba* today. Reading *Stories from the History of Rome*.

AUGUST 19, SUNDAY
Finished reading Buckle's *History*, Part III. Started reading Hopkins's *Origin and Evolution of Religion*.

AUGUST 23, THURSDAY
Finished reading Hopkins's book. Started reading Lecky's *European Morals*.

AUGUST 26, SUNDAY
Finished reading James's *Varieties of Religious Experience*. Four days ago finished reading Vinoba's *Maharashtra-Dharma*, Part I. Part II is about to be finished.

The Superintendent said yesterday that those who took raw milk did not need fruits and hence he refused to give fruits to Manzar Ali. Even for me, he said, they were not really necessary; so I stopped asking for oranges, lemons, etc. Today ate bananas from Manzar Ali's ration. Took milk raw.

AUGUST 28, TUESDAY
Today finished writing *Gitakosh*.[5] Started reading Holmes's *Freedom and Growth* yesterday.

Started living exclusively on raw milk from today. May God help me!

AUGUST 31, FRIDAY
Finished reading Holmes's *Freedom and Growth*. Started reading Haeckel's *Evolution of Man*.

Today the Major applied caustic soda to the eye for conjunctivitis.

SEPTEMBER 2, SUNDAY
Finished reading the Bible yesterday. Started reading an illustrated account of Jesus today.

Lost in weight by three lbs. during the last week.

SEPTEMBER 9, SUNDAY
Finished reading the illustrated account of Jesus and also Kavi's *Muktadhara* and *Dubtoon Vahan*.[6] Weight went up by one lb.; it is now 101 lbs.

SEPTEMBER 16, SUNDAY
Devdas, Narandas, Keshu and Kacho saw me on Monday. Finished reading the first part of Maulana Shibli's life of the Prophet, and also the preface to the Koran by Dr. Mahomed Ali.

SEPTEMBER 28, FRIDAY
This week finished reading Vivekanand's *Rajayoga* and Champakrai Jain's *Dharmani Ekata*. Finished reading life of the Prophet (by Maulana Shibli) today.

SEPTEMBER 30, SUNDAY
Started reading Nicholson's *Mystics of Islam* yesterday and finished it today. Started making the fair copy of *Gitakosh* today. Yesterday started reading *Sahaba Ekram*, Part II, and the unread portion of Urdu Reader No. V. Started reading Paul Carus's *Gospel of Buddha*. Today Major Jones came to bid me farewell.

OCTOBER 7, SUNDAY
Finished reading Paul Carus's *Gospel of Buddha* during the week. Reading Rhys Davids's Hibbert Lectures on Buddhism. Started reading Ameer Ali's *Spirit of Islam* today. Work on the fair copy of *Gitakosh* continues. Received a basket of fruits from Jamnalalji today.

Finished reading *Chhandogyopanishad* today and started reading *Brihadaranyak*.

Books Read

OCTOBER 14, SUNDAY
Ba, Avantikabai, Jamnalalji and Savatibai came and saw me on Wednesday.
    Finished reading Davids's *Hibbert Lectures on Buddhism*. Reading Sir Oliver Lodge's *Modern Problems*.

OCTOBER 21, SUNDAY
Finished reading Sir Oliver Lodge's *Modern Problems* and started the current issue of *Puratatva*.

OCTOBER 25, THURSDAY
Manzar Ali was taken to Prayag today.
    Finished reading Ameer Ali's book on Tuesday.
    Started reading Washington Irving's *Mahomed* yesterday. Started reading *Syadvada Manjari* today.

OCTOBER 26, FRIDAY
Abdul Gani was brought to this ward today.

NOVEMBER 4, SUNDAY
Abdul Gani started spinning on Wednesday.
    Finished reading Irving's *Mahomed*.
    Started reading *History of the Saracens* by Ameer Ali.

NOVEMBER 11, SUNDAY
Finished reading *Brihadaranyak Upanishad* on Tuesday. Started reading *History of Civilization in Europe* by Guizot on Thursday. Finished reading *Sahaba*, Part II, today. Will commence tomorrow Maulana Shibli's biography of Hasrat Omar.

NOVEMBER 12, MONDAY
Wrote a letter today to the Superintendent saying that I would have to give up oranges and raisins from Wednesday because he could not provide Abdul Gani the diet of the latter's choice.

NOVEMBER 18, SUNDAY
Have given up oranges and raisins since Wednesday last. Found today that I have lost three pounds in weight, but my physical strength remains unaffected.

NOVEMBER 24, SATURDAY
Today finished reading Ameer Ali's *History of the Saracens* and also making the fair copy of the *Gitakosh*. Yesterday finished

reading Guizot's *History of Civilization in Europe*. Today started reading Guizot's *History of Civilization in France*, Part II.

NOVEMBER 26, MONDAY

Yesterday started reading Motley's *Rise of the Dutch Republic*. Today commenced writing the history of *satyagraha* in South Africa. Finished reading Reese's autobiography and started reading Rajam Iyer's *Vedantabhraman*.

DECEMBER 9, SUNDAY

Today finished reading the first part of Motley's book and started the second part. Finished reading Rajam Iyer's *Vedantabhraman*.

On Wednesday finished reading Guizot's *History of Civilization in France*, Part II, and started Part III.

Finished reading *Syadvada Manjari* today. Started reading *Uttaradhyayan Sutra*. The experiment of doing without fruits is going on. Have been taking some bread with milk since Tuesday. Found that my weight had increased by two pounds; now it is 99.

DECEMBER 15, SATURDAY

Finished reading Guizot and started *Rosicrucian Mysteries*.

DECEMBER 16, SUNDAY

Finished reading the second part of Motley's book and started the third.

DECEMBER 23, SUNDAY

Ba, Mathuradas and Ramdas came to see me on Tuesday.

Ramabai Ranade came on Wednesday. At the instance of the Superintendent, wrote a letter to Harilal asking him to come and see me.

Have started taking fruits again since Tuesday evening. Last Sunday my weight stood as low as 96 and even the Superintendent got alarmed. Since Thursday started taking honey and increased the intake of bread to eight ounces.

I weighed 99 lbs. today. On Wednesday, finished reading *Rosicrucian Mysteries* and started Plato's *Dialogues*. Today finished reading the biography of Hasrat Omar and started reading Maulana Shibli's *Al Kalam* and also Woodroffe's *Shakta and Shakti*. Finished reading Motley's book.

Books Read

DECEMBER 30, SUNDAY
Finished reading *Uttaradhyayan Sutra* and started *Bhagavati Sutra*. Finished reading Woodroffe's *Shakta and Shakti*. On Thursday, finished reading the first part of Plato's *Dialogues* and started the second.

APPENDIX: LIST OF BOOKS

*157[7] *Natural History*
*158 *The Wisdom of the Ancients*
*159 *Natural Features of India*
*178 *Stories from the History of Rome*
See Diary of 23-4-1922
*205 *The Young Crusader*
*212 *Lives of Fathers and Martyrs*
*215 *Dropped from the Clouds*
 264 *Ivanhoe*
 282 *The Old Curiosity Shop*
*295 *The Five Empires*
 305 *Westward Ho*
*356 *Seekers after God*
    Equality: Bellamy
* 41 *The Five Nations:* Kipling
* 49 *Dr. Jekyll and Mr. Hyde*
  10 *The Second Jungle Book*
*107 *J. Howard*
 109 *Satires and Epistles of Horace*
*111 Goethe's *Faust*
 116 *Tropical Agriculture*
*125 *Lays of Ancient Rome*
 129 Primer of Marathi Language
*132 *Natural History of Birds*
 144 *Enoch Arden*
 148 *Historical English Grammar*
 149-50 Scott's Poetical Works
*152 *Life and Voyages of Columbus*
    Muktivivek—Translation of Vidyaranyaswami's book
    Kanta—Do
    Malati Madhava
    Siddhantasara
    Panchasati

Gulabsinh
Shrivritiprabhakar
Chatuh Sutri
Bhojprabandh
Vikramcharitra
Anubhavapradipika
Vastupalcharitra
Yogabindu
Kumarpalcharitra
Vivadtandav

'Jail Diary, 1923' (G.)
SN 8039

[1] *Kena Upanishad.*
[2] Surgeon-General at Sassoon Hospital, Poona.
[3] Gujarati translation of the Old Testament.
[4] Gujarati translation of the New Testament.
[5] Glossary to the *Gita*.
[6] Plays by Rabindranath Tagore.
[7] The significance of the asterisks is not known, but the numerals stand for catalogue numbers.

77. STUDIES IN GAOL

As a boy I had not much taste for reading anything outside my school books. They alone gave me enough food for thought; for it was natural for me to reduce to practice what I learnt at school. For home reading I had an intense dislike. I used to labour through home lessons because I had to. During my student days in England too the same habit persisted of not reading outside the books for examinations. When however I began life, I felt I ought to read for the sake of gaining general knowledge. But at the earliest period of my life it became one of storm and stress. It commenced with a fight with the then political agent of Kathiawar. I had therefore not much time for literary pursuits. In South Africa for one year I had fair leisure in spite of the battle for freedom that faced me. The year 1893 I devoted to religious striving. The reading was therefore wholly religious. After 1894 all the time for sustained reading I got was in the jails of South Africa. I had developed not only a taste for reading but for completing my knowledge of Sanskrit and studying Tamil, Hindi and Urdu. Tamil because I was in touch with so many Tamilians in South Africa and Urdu because I had dealings with so many Mussalmans. The South African jails had whetted my appetite and I was grieved when during my last incarceration in South Africa I was prematurely discharged.

When therefore the opportunity came to me in India, I hailed it with joy. I mapped out a rigid programme of studies at Yeravda to finish which six years were not enough. During the first three months I had a vague hope that India would rise to the occasion, complete the boycott of foreign cloth and unlock the prison gates. But I soon learnt that such was not to be the case. I saw at once that it meant laborious quiet organizing which could not take the nation anything less than five years. I had no desire whatsoever for being discharged before my time except by the peaceful constructive act of the nation even if it was not actually *swaraj*. I therefore settled down to studies with the zest of a youth of twenty-four instead of an old man of fifty-four with a broken constitution. I accounted for every minute of my time and would have been discharged a fair Urdu and Tamil scholar and well versed in Sanskrit. I would have satisfied my desire for reading original Sanskrit texts. But such was not to be the case. My studies were rudely interrupted by my unfortunate illness and consequent discharge. However the following list gives the reader an idea of my studies:

The Cambridge History of Scotland; The Master and His Teaching; Arm of God; Christianity in Practice; Tulsidas's *Ramayana* (Hindi); *Satyagraha and Asahayoga* (Hindi); the Koran; *The Way to Begin Life; Trips to the Moon* (Lucian); *Indian Administration* (Thakore); *Natural History of Birds; The Young Crusader; Bible View of the World Martyrs;* Farrar's *Seekers after God; Misra Kumari* (Gujarati); *Stories from the History of Rome; Tom Brown's School Days; Wisdom of the Ancients* (Bacon); *History of India* (Gujarati)—Chandrakant; Patanjali's *Yogadarshana* (Kania's translation); Valmiki's *Ramayana* (Gujarati translation); *Five Nations* (Kipling); *Equality* (Edward Bellamy); *St. Paul in Greece; The Strange Case of Dr. Jekyll and Mr. Hyde;* Rosebery's *Pitt; Jungle Book* (Kipling); *Faust; Life of John Howard;* Mahab-

harata—all the books (Gujarati translation); *Dropped from the Clouds* (Jules Verne); Irving's *Life of Columbus; Ramayana* by Girdhar (Gujarati); *Five Empires* (Wilberforce); *Lays of Ancient Rome*; *The Crusades*; Gibbon's *Rome*; *Urdu Readers*; *Bhagavata* (Gujarati translation); Bankim's *Krishnacharitra* (Jhaveri's translation); Vaidya's *Krishna* (Gujarati translation); Tilak's *Gita* (Gujarati translation); *Saraswatichandra* (Gujarati); *Manusmriti* (Gujarati translation); *Ishopanishad* (Aurobindo's commentary); Kabir's Songs; Jacob Boehmen's *Supersensual Life*; *Pro Christo et Ecclesia*; *Kathavali Upanishad* (Hindi commentary); *Galilean*; *Jnaneshwari* (Gujarati translation); *Philo Christus*; *Satyartha Prakasha* (Hindi); *Prem Mitra* (English); *The Six Systems* (Gujarati translations); *The Gospel and the Plough*; Nathuram's *Commentary on the Gita*; Shankara's *Commentary on the Gita*; Rajchandra's Letters and Writings; *Ourselves and the Universe* (J. Brierly); *What Christianity Means to Me* (Abbott); *Steps to Christianity*; *My Philosophy and Religion* (Trine); *Sadhana* (Rabindranath); Bhanu's *Commentaries on Upanishads*; Max Muller's *Upanishads;* Wells's *History*; The Bible; *Science of Peace* (Bhagwandas); *Barrack-room Ballads* (Kipling); *Evolution of Cities* (Geddes); *Life of Ramanuja*; Cunningham's *Sikhs*; Gokulchand's *Sikhs*; Macauliff's *Sikhs*; *Ethics of Islam*; *Social Evolution* (Kidd); *Manusmriti* (Buhler); *Our Hellenic Heritage* (James); *Avesta* (Dadachandji); *Gita* (Aurobindo); *Elements of Sociology* (Spencer); *Social Efficiency* (Pherwani); *Message of Mahomed* (Wadia); *Message of Christ* (Wadia); *Saints of Islam* (Hassan); *Early Zoroastrianism* (Moulton); *Travels in the Himalayas* (Gujarati); *Sita-haran* (Gujarati); *Buddha and Mahavira* (Gujarati); *Rama and Krishna* (Gujarati); *Man and Superman*; *Markandeya Purana* (Gujarati); *Poorva Rang* (Gujarati); *Life of Hasrat Umar* (Urdu); *Confessions of the Prophet* (Urdu); *History of Civilization* (Buckle); *Jaya and Jayant* (Gujarati); Rabindranath's Essays (Gujarati); Countess Tolstoy's *Defence*; *Kalapani-nikatha* (Gujarati); *Economics* (Gujarati); *Gita Govinda*; *Varieties of Religious Experience* (James); *Origin and Evolution of Religion* (Hopkins); Lecky's *European Morals*; *Maharashtra-dharma* (Marathi); *Freedom and Growth* (Holmes); *Evolution of Man* (Haeckel); *Muktadhara* (Gujarati)—Rabindranath; *Sinking Ship* (Gujarati)—Rabindranath; *Life of the Prophet* (Urdu)—Maulana Shibli; Dr. Mahomed Ali's Koran; *Rajayoga* (Vivekananda); *Confluence of*

Religions (Champakrai Jain); *Mystics of Islam* (Nicholson); *Gospel of Buddha* (Paul Carus); Rhys Davids's *Lectures on Buddhism*; *Spirit of Islam* (Ameer Ali); *Modern Problems* (Lodge); *Mahomed* (Washington Irving); *Syadvada Manjari*; *History of the Saracens* (Ameer Ali); *European Civilization* (Guizot); *Al Faruq* (Shibli); *Rise of the Dutch Republic* (Motley); *Musings of Saint Theresa*; *Vedanta* (Rajam Iyer); *Uttaradhyayan Sutra*; *Rosicrucian Mysteries*; *Dialogues of Plato*; *Al Kalam* (Urdu) Shibli; Woodroffe's *Shakta and Shakti*; *Bhagavati Sutra* (incomplete).

Let the reader however not imagine that I read all these books by choice. Some of them were useless and outside the jail would not have read them. Some of them were sent by friends known and unknown and I felt I was bound for their sakes at least to go through them. The Yeravda Jail has what may be called not a bad collection of English books. Some of them were really good books, such for instance as Farrar's *Seekers after God*, Lucian's *Trips to the Moon* or Jules Verne's *Dropped from the Clouds*, all of them excellent in their own way. Farrar's is an inspiring book giving the best side of the lives of Marcus Aurelius, Seneca and Epictetus. Lucian's book is a fine and instructive satire. Jules Verne teaches science in the guise of a story. His method is inimitable.

Many Christian friends were most attentive to me. I received books from them from America, England and India. I must confess that whilst I recognized their kind motive, I could not appreciate the majority of the books they sent. I wish I could say something of their gifts that would please them. But that would not be fair or truthful if I could not mean it. The orthodox books on Christianity do not give me any satisfaction. My regard for the life of Jesus is indeed very great. His ethical teaching, his common sense, his sacrifice command my reverence. But I do not accept the orthodox teaching that Jesus was or is God incarnate in the accepted sense or that he was or is the only son of God. I do not believe in the doctrine of appropriation of another's merit. His sacrifice is a type and an example for us. Every one of us has to be 'crucified' for salvation. I do not take the words 'Son' 'Father' and 'the Holy Ghost' literally. They are all figurative expressions. Nor do I accept the limitations that are sought to be put upon the teaching of *The Sermon on the Mount*. I can discover no justification in the New Testament for wars. I regard Jesus as one among the most illustrious teachers and prophets the world has seen. Needless to say I do not regard the Bible as an infallible record of the life and teachings of Jesus. Nor do I consider every word in the New Testament as God's own word. Between the Old and the New there is a fundamental difference. Whilst the Old contains some very deep truths, I am unable to pay it the same honours I pay the New Testament. I regard the latter as an extension of the teaching of the Old and in some matters rejection of the Old. Nor do I regard the New as the last word of God. Religious ideas like everything else are subject to the same law of evolution that governs everything else in this universe. Only God is changeless and as His message is received through the imperfect human medium, it is always liable to suffer distortion in proportion as the medium is pure or otherwise. I would therefore respectfully urge my Christian friends and well-wishers to take me as I am. I respect and appreciate their wish that I should think and be as they are even as I respect and appreciate a similar wish on the part of my Mussalman friends. I regard both the religions as equally true with my own. But my own gives me full satisfaction. It contains all that I need for my growth. It teaches me to pray not that others may believe as I believe but that they may grow to their full height in their own religion. My constant prayer therefore is for a Christian or a Mussalman to be a better Christian and a better Mahomedan. I am convinced, I know, that God will ask, asks us now, not what we label ourselves but what we are, i.e., what we do. With Him *deed* is everything, *belief* without deed is nothing. With Him doing is believing. The reader will pardon me for this digression. But it was necessary for me to deliver my soul over the Christian literature with which the Christian friends flooded me in the jail, if only to show my appreciation of their interest in my spiritual welfare.

That which I would not have missed was the *Mahabharata* and the Upanishads, the *Ramayana* and the *Bhagavata*. The Upanishads whetted my appetite for exploring the Vedic religion at its source. Its bold speculations afforded the keenest delight. And their spirituality satisfied the soul. At the same time I must confess that there was much in some of them that

I was unable to understand or appreciate in spite of the help of the copious notes of Professor Bhanu who has incorporated in them the whole of Shankara's commentaries and the substance of the others. The *Mahabharata* I had never read before except in scraps. I was even prejudiced against it, believing (falsely as it has now turned out) that it was nothing but a record of bloodshed and impossible long descriptions which would send me to sleep. I dreaded to approach the bulky volumes covering over closely printed six thousand pages. But having once commenced the reading, I was impatient to finish it, so entrancing it proved to be except in parts. I compared it, as I finished it in four months, not to a treasure chest in which you find nothing but polished gems limited as to quantity and quality but to an inexhaustible mine which the deeper one digs the more precious are the finds.

The *Mahabharata* is not to me a historical record. It is hopeless as a history. But it deals with eternal verities in an allegorical fashion. It takes up historical personages and events and transforms them into angels or devils as it suits the purpose of the poet whose theme is the eternal duel between good and evil, spirit and matter, God and Satan. It is like a mighty river which in its progress absorbs many streams, some even muddy. It is the conception of one brain. But it has undergone ravages and received accretions in process of time till it has become difficult always to say which is the original and which is apocryphal. The ending of it is magnificent. It demonstrates the utter nothingness of earthly power. The great sacrifice at the end is proved inefficacious in comparison with the sacrifice of the heart by a *Brahmin* who gave his little all, the last morsel, to a needy beggar. What is left to the virtuous Pandavas is poignant grief. The mighty Krishna dies helplessly. The numerous and powerful Yadavas because of their corruption die an inglorious death fighting amongst one another. Arjuna the unconquerable is conquered by a band of robbers, his Gandiv notwithstanding. The Pandavas retire leaving the throne to an infant. All but one die on the journey to heaven. And even Yudhishthira, the very embodiment of *dharma*, has to taste the foetid smell of hell for the lie he permitted himself to utter under stress. The inexorable law of cause and effect is allowed without exception to run its even course. The claim put forth in its behalf that it omits nothing that is useful or interesting and that is to be found in any other book is well sustained by this marvellous poem.

'My Jail Experiences—XI'
Young India, 4 Sept. 1924

78. ISLAM, ENGLISH HISTORY, AND THE *MAHABHARATA*

My Urdu studies proved as absorbing as the reading of *Mahabharata*. They grew on me as I proceeded. I approached this study with a light heart foolishly imagining that in two or three months I should be quite an adept in Urdu. But to my sorrow I discovered that it had been made into a language distinct from Hindi and that the tendency was growing in that direction. But that discovery only made me more determined than ever to be able to read and understand Urdu literature. I therefore gave nearly three hours per day to Urdu reading. The Urdu writers have purposely gone out of their way to use Arabic or Persian words even to the rejection of words current among Hindus and Mussalmans. They have rejected even the common grammar and imported Arabic or Persian grammar. The result is that the poor nationalist if he will keep in touch with the Mussalman thought, must study Urdu as a separate, new language. The Hindi writers I know have done no better or no less. Only I thought that the evil had not gone very deep and the separatist tendency was a mere passing phase. Now I see that if we are to have a common national language being a mixture of Hindi and Urdu, special and prolonged effort will have to be made to effect a juncture between the two streams which seem at present to be diverging more and more one from the other. In spite however of the difficulty I retain the opinion that it is necessary for a Hindu to complete his education to know literary Urdu as it is for a Mussalman to know literary Hindi. It is easy enough if begun early. This study may have no pecuniary value, it may not open up the treasures of Western knowledge. But its national value is beyond compare. I am the richer for my close study of Urdu. I wish I could even now complete it.

I know the Mussalman mind much better than I did two years ago. I was interested in the religious side of Urdu literature and therefore plunged, as soon as I was able, into Urdu religious books. Fates have always favoured me. Maulana Hasrat Mohani[1] had sent to Mr. Mansar Ali *Leaves from the Lives of the Companions of the Prophet*. As he was teaching me Urdu he passed the volumes on to me. And I went through them with the greatest diligence. The volumes, though they contain repetitions and would gain in lucidity for compression, were to me deeply interesting for the insight they gave me into the doings of the Prophet's many companions. How their lives were transformed as if by magic, what devotion they showed to the Prophet, how utterly unmindful they became of worldly wealth, how they used power itself for showing the utter simplicity of their lives, how they were untouched by the lust for gold, how reckless they were of their own lives in a cause they held sacred, is all told with a wealth of detail that carries conviction with it. When one notes their lives and then the lives of the present-day representatives of Islam in India, one is inclined to shed a tear of bitter grief.

I passed from the companions to the Prophet himself. The two bulky volumes written by Maulana Shibli are a creditable performance. But I have the same complaints about them that I have lodged about the diffuseness of the volumes devoted to the companions. But the diffuseness did not interfere with my interest to know how a Mussalman had treated the incidents of the life of one who has been almost uniformly maligned and abused in the West. When I closed the second volume, I was sorry there was not more for me to read of that great life. There are incidents in it which I do not understand, there are some I cannot explain. But I did not approach the study as a critic or a scoffer. I wanted to know the best of the life of one who holds today undisputed sway over the hearts of millions of mankind. And I found enough in the volumes to account for it. I became more than ever convinced that it was not the sword that won a place for Islam in those days in the scheme of life. It was the rigid simplicity, the utter self-effacement of the Prophet, the scrupulous regard for pledges, his intense devotion to his friends and followers, his intrepidity, his fearlessness, his absolute trust in God and his own mission. These and not the sword carried everything before them and surmounted every obstacle. As I do not regard any human being absolutely perfect, be he a prophet or an *Avatar*, it is unnecessary for me to be able to explain to the censor's satisfaction every detail of the Prophet's life. It is enough for me to know that he was a man among millions who tried to walk in the fear of God, died a poor man, wanted no grand mausoleum for his mortal remains and who did not forget even on his death-bed the least of his creditors. The teaching of the Prophet is no more responsible for the degrading intolerance or questionable proselytizing methods that one sees around himself than Hinduism is responsible for the degradation and intolerance of present-day Hindus.

From the Prophet I passed to the two volumes devoted to the life of Umar the unconquerable. As I pictured him before my mental eye walking to Jerusalem upbraiding some of his followers for aping the pomp of their neighbours, refusing to pray in a Christian church lest succeeding generations might claim to convert it into a mosque, granting the most liberal terms to the conquered Christians, and as I picture him declaring that the word of a follower of Islam, though pledged by one not authorized thereto, was as good as the written decree of the great Caliph himself, he commands my humble respect. His was an iron will. He weighed out the same justice to his daughter that he would weigh to an utter stranger. I fancy I understand the breaking of idols and wanton desecration of temples, the thoughtless intolerance of Hindu music now going on in our midst. These acts seem to me to be due to an utter misreading of the events in the life of the greatest of the Caliphs. I fear that the acts of this great and just man are being presented to the Mussalman masses in a most distorted fashion. I know that if he rose from his grave, he would disown the many acts of the so-called followers of Islam which are a crude caricature of those of the great Umar himself.

From this entrancing study, I went to the philosophical volumes called *Al Kalam*. These are difficult to understand. The language is highly technical. Mr. Abdul Gani however made my study fairly easy. I was only sorry my illness interrupted my study when I had only half finished the volumes.

Of the English books Gibbon[2] takes easily the first place. It was recommended to me years ago by so many English friends. I was determined to read Gibbon in the jail this time. I was glad of it. For me even history has a spiritual significance. As the author proceeds to trace the events in the life of the citizens of a single city who built up a world-empire, one traces the history of the soul. For Gibbon does not deal with trifles, he deals with vast masses of facts and arrays them before you in his own inimitable way. He deals with three civilizations, Pagan, Christian and Islamic, in sufficient detail to enable you to frame your own conclusions. His own compel attention. But he is as a historian jealous of his calling, faithful enough to give you all his data so as to enable you to judge for yourself.

Motley is another type. Gibbon traces the decay of a mighty empire. Motley extracts from a little republic the life of his hero. Gibbon's heroes are subservient to the story of a mighty empire. Motley's story of a State is subservient to that of one single life. The republic merges in William the Silent.

Add to these two Lord Rosebery's *Life of Pitt*. And you are perhaps then prepared to draw with me the conclusion that the dividing line between fact and fiction is very thin indeed and that even facts have at least two sides or as lawyers say facts are after all opinions. However I have no desire to engage the reader's attention upon my speculations on the value of history considered as an aid to the evolution of our race. I believe in the saying that a nation is happy that has no history. It is my pet theory that our Hindu ancestors solved the question for us by ignoring history as it is understood today and by building on slight events their philosophical structure. Such is the *Mahabharata*. And I look upon Gibbon and Motley as inferior editions of the *Mahabharata*. The immortal but unknown author of the *Mahabharata* weaves into his story sufficient of the supernatural to warn you against taking him literally. Gibbon and Motley are unnecessarily at pains to tell you they are giving you facts and nothing but facts. Lord Rosebery comes to the rescue and tells you that even the last words said 'to have been uttered by Pitt are disputed by his butler. The substance of all these stories is: Names and forms matter little, they come and go. That which is permanent and therefore necessary eludes the historian of events. Truth transcends history.

'My Jail Experiences—XI'
Young India, 11 Sept. 1924

> [1] 1875-1951; nationalist Muslim leader.
> [2] Edward Gibbon, *History of the Decline and Fall of the Roman Empire*.

79. MYSTICS, SIKHS, AND THE *GITA*

I must not omit to mention one little but precious book a dear friend sent. It was *Supersensual Life* by Jacob Boehmen. I notice it to enable me to share with the reader some of the striking passages I copied from it. Here they are:

It is naught indeed but thine own hearing and willing that do hinder them, so that thou dost not see and hear God.

If thou rulest over the creatures externally only and not from the right internal ground of thy inward nature, then thy will and ruling is in a bestial kind or matter.

Thou art like all things and nothing unlike thee.
If thou wilt be like all things, thou must forsake all things.

Let the hands and the head be at labour. Thy heart ought nevertheless to rest in God.

Heaven is the turning in of the will to the love of God.
Hell is the turning in of the will into the wrath of God.

Whilst I am turning over my scrappy notebook, I come across certain other passages collected in the course of my reading from other books.

Here is one for *satyagrahis*:

> They are slaves who will not choose
> Hatred, scoffing and abuse,
> Rather than in silence shrink
> From the truth they needs must think.
> They are slaves who dare not be
> In the right with two or three.
>
> LOWELL

(Copied from *Tom Brown's School Days*)

Another bearing on the same subject is from Claude Field's *Mystics and Saints of Islam*.

Sufi Shah Mullah Shah, when he was advised to flee from the wrath of Shah Jehan, is reported to have said, 'I am not an impostor that I should seek safety in flight. I am an utterer of truth. Death and life are to me alike. Let my blood in another life also redden the impaling stake. I am living and eternal. Death recoils from me, for my knowledge has vanquished death. The sphere where all colours are effaced has become my abode.' Mansuri Hallaj said, 'To cut off the hands of a fettered man is easy, but to sever the link that binds me to the Divinity would be a task indeed.'

Here is another from Lowell. It helps those who would give to the Malabar sufferers to do so in the right spirit and to share their very best.

> The Holy Supper is kept indeed
> In what we share with another's need.
> Not that which we give, but what we share.
> For the gift without the giver is bare.
> Who bestows himself with his alms feeds three,
> Himself, his hungering neighbour and Me.

The following will strengthen those who believe in the gospel of non-violence:

To wish ill, to do ill, to speak ill or to think ill of anyone, we are equally forbidden without exception.

TERTULLIAN

(Copied from J. Brierley's *Ourselves and the Universe*)

The last set of books I should like to mention are the histories of the Sikhs by Cunningham, Macauliff and Gokulchand Narang. All those books are good in their own way. It is impossible to appreciate the present Sikh struggle without understanding their previous history and the life of the Gurus. Cunningham's is a sympathetic record of events leading to the Sikh wars. Macauliff's is a life-story of the Gurus giving copious extracts from their compositions. It is a sumptuously printed publication. It loses its value because of its fulsome praise of the English rule and the author's emphasis on Sikhism as a separate religion having nothing in common with Hinduism. Gokulchand Narang's is a monograph supplying information not available in the two works mentioned.

Before concluding this review of my studies in the jail, let me draw the student-readers' attention to the value of doing things regularly and the way of making dry things interesting. I had a mind, for my own instruction and guidance, to prepare a concordance of the *Gita*. It is not a particularly interesting task to note down words and references and index them twice. I thought I should do it during the incarceration. At the same time I grudged giving much time to the task. My time-table was packed. I determined therefore to do what I could in 20 minutes daily. Doing it for such a short time freed the task from becoming a drudgery. On the contrary, I daily looked forward to it. When it came to re-indexing, it became absorbing. The curious may solve the knotty problem for themselves. The first indexing gave me the alphabetical order of the first letters of the words to be indexed. But how to rearrange the words under each letter in their alphabetical order was a problem to solve. I had never written a dictionary. I had therefore to discover my own method. I was glad when I made the discovery. It was so good that it became deeply interesting. It was neat, quick and infallible. The whole work took me nearly eighteen months to finish. I am now by referring to the concordance able to know where and how often a particular word occurs in the *Gita*. It has also a meaning attached to it. If I ever succeed in reducing to writing my thoughts on the *Gita*, I propose to share the concordance and the thoughts with the public.

'My Jail Experiences—XI'
Young India, 25 Sept. 1924

80. BOOKS READ IN GAOL (1932)

February 3-8, 1932

Chi. Narandas,

I got your packet just this evening.

Lady Vithaldas, Bhai Trivedi, Vidya Hingorani and Damodar came in the afternoon and saw me. I got the leather for the sandals. I have heard that Ramdas, Chhaganlal, Surendra, Somabhai and others, in all 190 persons, have arrived here. Most probably some of them will see us.

For your scabies, you should eat green leaves and tomatoes. Baths with potassium permanganate will certainly help.

At the bottom of the letter from C. P. Scott's son you write that you have sent by book post a biography of the late Scott. The book does not seem to have been received here. If you remember to have sent it, write to me again so that I may inquire further.

You have stated in the 'Ashram Samachar' that I spin 500 rounds in two days. I spin, not 500 rounds, but 500 yards, in two days. I spin 375 rounds in two days and it is my ambition to spin as many in one day. But I do not think I shall be able to fulfil it as quickly as I had hoped.

Night, February 6, 1932

I had a letter from Tilakam. The food does not seem to have agreed with him. Read my letter to him, and if he agrees, carry out my suggestions. Pulses do not agree with people who are used to meat. If they start with milk, they experience no difficulty. They ordinarily eat *rotli*, etc., even with meat. As a substitute for meat, they generally take pulses, but a stomach used to meat cannot digest pulses. In any case such persons must not eat any kind of nuts such as ground-nuts, etc. Milk and curds are but pure forms of non-vegetarian food.

February 8, 1932

You will find with this an article on the subject of the existence of God. It is a translation of a portion of an article published in *Young India*. I had read that portion for a gramophone record. The record is on sale now. Anand heard it. He took down the extract and sent a copy to me for translation. The Hindi article mentioned above is a translation of that extract. The translation should not be published now. Anand wants the Hindi translation to help him to make a Sindhi translation. Send it to him and ask him not to publish it just now. The inmates of the Ashram can read it. You may even publish it in the *Ashram Patrika*. If you do, publish it as an original article. Let Parasram correct its Hindi. I leave it to you to decide whether or not you should publish it in the *Patrika*. If your publishing it is likely to be misconstrued or if someone is likely to reproduce the article elsewhere, drop the idea. It will be enough if the inmates of the Ashram who are there read it, think over it and understand its meaning.

I read in the papers news about a few things having been seized from the Ashram. I expect I shall get more details from you.

I have some free time today and so I wish to describe a few details of the routine here.

Both of us get up at 3.40. After brushing, we pray. After that, we take warm water with honey and lemon-juice and then read till the stroke of five. From five to six, we walk. At six, if I feel the call of nature I answer it, and then sleep for about 20 minutes. I get up at 6.45, when the bell for opening the cells is given, and read up to seven. The *Sardar*, after answering the call of nature, walks about and sits down to breakfast when milk is brought. As he eats the breakfast, he reads from the newspaper, which has arrived by then. During the day, I read, write and spin. In between I take a nap twice. The *Sardar* walks about much longer than I do. He generally reads from the newspapers which are supplied to us. I have two meals a day. The *Sardar* occasionally eats salad or something else like that at twelve. I again take warm water and honey at that time. When the honey is exhausted, I add jaggery and lemon-juice. I once happened to remark that Mirabehn or Pyarelal used to prepare fruit for me to eat. Since then the *Sardar* has monopolized that duty. It is he who prepares dates and tomatoes for me and does it with great love. I have accepted this loving service without the slightest hesitation. Having mentioned the service rendered by Mirabehn and Pyarelal, it would be a useless attempt on my part to decline the *Sardar*'s help. He takes milk and bread in the morning. At four in the afternoon, he takes bread, curds and some vegetables and, generally salad. I used to take milk in the morning, with dates, tomatoes and, from the fruit sent by Lady Thackersey or Prof. Trivedi, some oranges, *chikus*, etc. At present, I eat dates and tomatoes and, in the morning, take half a pound of milk, and in the evening the same quantity of curds. But I see that I shall have to reduce the quantity of milk still further. My health is good, of course. Do you know that my weight has increased? I see that now I do not need very nourishing food, especially when I enjoy solitude and peace of mind. This was my experience last year. While in jail, I could live without milk and maintain my weight. After my release, I lost weight in a short time for want of milk and, therefore, resumed it. In England, it was only with that that I could keep up my strength. Here I may not perhaps need milk. On the contrary milk may even do me harm. Of course I will do no violence to my body. My aversion to milk remains. But I will give the body what it requires. Let no one, therefore, worry after reading this.

I will also mention some other work which the *Sardar* does. It is he who trims the envelope which you send and prepares it [for use again]. There are many such happy details which I can mention. But are not these enough? Just as we walk about between five and six in the morning, so do we again in the evening. Between six and seven, I read. Meanwhile, the *Sardar* prepares dates, babul sticks for brushing teeth, etc., for me and then joins me in the cell. At seven, the prayers begin. After prayers, reading and writing again, till 8.30. By nine, we are in bed. Both of us sleep in the open. From among newspapers, we get *The Times of India*, *The Bombay Chronicle*, *The Tribune*, *The Leader*, and *The Hindu*. From among weeklies, we get *Social Reformer*, and from among monthlies, *The Modern Review*. As for books, I get some from outside and I have brought some with me. They supply enough fare to both of us.

I have read the following books so far: Durant's *The Case for India*, Crozier's *A Word to Gandhi*, Brailsford's *Rebel India*, Al Haj Salmin's *Imam Hussain* and *Khalifa Ally*, Samuel Hoare's *Fourth Seal*, R. MacDonald's travelogue, *Survey of Matar Taluka*, Ramanathan's *Speech on Khadi*, Will Hayes's *Essence of Hinduism*, Ruskin's *St. George's Guild*, Shah's *Federal Finance*, Rothenstein's *Ruin of Egypt*, Hayes's *The Book of the Cow*, A. E.'s *Candle of Vision*, Kinley's *Money*, and *Shankh ane Kodi* (Gujarati). I am now reading the biography of Munshi Zaka Ullah written by Andrews and Shah's book, *Sixty Years of*

Economic Administration of India. The *Sardar* has read Hoare's and MacDonald's books and is now reading the book on Egypt. He gives plenty of time to newspapers and, in addition to the two hours which he gives to walking, as I have mentioned, he must certainly be spending two hours more in walking at other times.

Besides this, there are 36 letters.

Blessings from
BAPU

Letter to Narandas Gandhi (G.)
CW 8207

81. 'THE HOUND OF HEAVEN'

March 9, 1945

Chi. Munnalal,

You seem to have written the above in haste. Try and see if you can steady your mind. Read *The Hound of Heaven*, think over it and understand its meaning. You will not be happy anywhere if you turn your back upon the 'Hound'.

Blessings from
BAPU

Letter to Munnalal G. Shah (G.)
CW 5845

§12. Reading and Reflection

82. PEACE OF MIND

Yeravda Mandir,
May 26, 1930

Chi. Narandas,

Gujarati letters were received here last week, but they have not yet been given to me. I got Mirabehn's and Mathew's letters because they were in English. That is how it is! But this will not go on for many days—if God wills.

My weight is practically the same; perhaps it has gone up by half a pound.

The diet, too, continues to be the same. Now I get perfect curds from unboiled milk. The milk forms into curd in 24 hours. I had mixed a fairly good quantity of curd with the milk.

I have finished reading two books: *The Light of Asia* and *The Saints of Islam*. I am now reading a book on prisons written by an Inspector-General of Prisons of the Punjab, which was sent to me by a friend. I simply get no time for reading. Seven hours are taken up by the spinning-wheel, the *takli* and carding. On some days, when the wheel requires no repair, it is less than seven hours, and on some other days it is more. I am unhappy about this. I love work. My skill in the work increases because I have to do everything myself, and I discover very small faults. The yarn which I spin on the *takli* has improved very much in quality. The speed, too, has gone up.

I think very little about things outside. I am so busy with work that I get no time to think about them. I fix my eyes on the central teaching of the *Gita*, and so enjoy peace of mind. If I did not do that, though I am supplied newspapers, reading about all that happens would have made it difficult for me to experience peace. The prayers twice a day and the daily reading of the *Gita* have proved a great support to me.

Is there any news about Krishna Nair, Surajbhan and Jayanti Prakash? How is Satis Babu? Write to all to whom you send letters and tell them that they are not to be published. Friends may certainly read them.

How is Jamna's health?

Blessings from
BAPU

Letter to Narandas Gandhi (G.)
CW 8112

83. SELF-RESTRAINT IN READING

June 18, 1932

You will be surprised, but I would advise you just now to stop reading even the works of Raychandbhai and the *Gita*. Reflect on whatever you understand from the *Gita* verses and the *bhajans* which are recited or sung during prayers. This self-restraint will be difficult to exercise, but you will see its miraculous effect. At present, your reading has become your work. Whenever you are free, take up some useful work to do. Give up trying to understand things with your reason. This is the meaning of 'one step enough for me'. One should give up any aid which becomes a bond.

BAPU

[P. S.] There is no harm in your reading newspapers.

Letter to Bhagwanji P. Pandya (G.)
CW 348

84. ASSIMILATION OF READING

[Before 3 July 1932]

All the food that is digested is not transformed into blood, etc., but what is assimilated is transformed into the various elements which sustain or build the body. In the same way, what we read, should be assimilated by us, just as manure is assimilated by the tree and as a result it bears fruit.

Letter to Chhaganlal Joshi (G.)
Mahadevbhaini Diary, Vol. I, p. 271

85. READING AND THINKING

July 10, 1932

There is a class of people to which many of us belong. They read and read and read until they almost lose their power of thinking. To such people I suggest that they should stop reading and think over what they have previously read.

A Letter (G.)
Mahadevbhaini Diary, Vol. I, p. 284

86. READING WITHOUT REFLECTION

August 14, 1932

We are taught at school: 'Studying without thinking is useless.' This is literally true. Love of reading is good. Those who are prevented by lethargy from reading and studying have certainly dull minds. But those also who merely read and never reflect on what they read remain more or less dull. Some of them lose their eyesight in the bargain. Reading without reflecting is only a kind of mental disease.

Many of us read in that manner. They read but never reflect on what they read, much less put it in practice. We should, therefore, read less, but think over what we read and put it in practice. We may reject what we discover from experience to be unsuitable, and go on with the rest. Persons who follow this method will satisfy their needs with less reading, save much time and become fit for creative and responsible work.

Another benefit which accrues to persons who learn to think is worth noting. One cannot always get books to read or get time to read them. Persons to whom reading has become a habit are known to have gone crazy when they could not read. If, however, one forms the habit of thinking, the book of thoughts is always available to one, and there will then be no danger of one becoming crazy when not able to read.

I have deliberately used the phrase to 'learn to think'. That way lots of people think, carelessly and unprofitably. This is only a form of madness. Some get caught in futile speculation and, overcome with despair, even kill themselves. This is not the kind of thinking I am advising. What I am advising here is only that people should reflect over what they read. Suppose we hear or read a *bhajan* today. We should then reflect over it. We should try to discover its deeper meaning and think what we should accept from it and what we should reject. We should see if there is any error in its idea. If we have not understood its meaning, we should make an effort to understand it. This is called systematic thinking. I have given the simplest illustration. Everybody should apply it to his own case and draw necessary conclusions for himself, and may thus progress. One who follows this method will experience great inner joy and benefit from all that he reads.

BAPU

'Reading and Reflection—I' (G.)
MMU/II (Microfilm)

87. DISCOVERY OF MEANING

August 21, 1932

> Awake and arise, O traveller, it is morning now;
> It is no longer night that you still slumber.

If anybody understands these lines to mean simply: 'O traveller, arise, it is morning now. Is it night that you are still sleeping?'—then he has read the lines but not reflected over their meaning. For such a reader will rise early at dawn and feel satisfied. But a reader who wishes to think will ask: Who is this traveller? What is meant by 'It is morning'? What does the poet mean when he says that it is no longer night? What does sleeping mean? He will then discover daily new meanings in every line, and understand that the traveller represents every human being. For one who has faith in God, it is always morning. Night may also signify ignorance. This line applies to any person who is negligent, be it in ever so small a degree. Anybody who tells a lie also slumbers. The line is a call to such a person to wake up. We may thus read a wide meaning in the line and learn peace of mind through it. In other words, meditation on this single line can supply enough provision to a man for his spiritual journey, whereas a person who has memorized the four Vedas and also studied their meaning may find them a useless burden. I have given here but one illustration which occurred to me. If all of us decide in what direction we wish to progress and start thinking, we would discover new meanings in life and daily experience a new joy.

'Reading and Reflection—II' (G.)
MMU/II (Microfilm)

IV

Hind Swaraj, Modern Civilization, and Moral Progress

§1. *Hind Swaraj*

88. HIND SWARAJ[1]

I have written some chapters on the subject of Indian Home Rule which I venture to place before the readers of *Indian Opinion*. I have written because I could not restrain myself. I have read much, I have pondered much, during the stay, for four months in London, of the Transvaal Indian deputation. I discussed things with as many of my countrymen as I could. I met, too, as many Englishmen as it was possible for me to meet. I consider it my duty now to place before the readers of *Indian Opinion* the conclusions, which appear to me to be final. The Gujarati subscribers of *Indian Opinion* number about 800. I am aware that, for every subscriber, there are at least ten persons who read the paper with zest. Those who cannot read Gujarati have the paper read out to them. Such persons have often questioned me about the condition of India. Similar questions were addressed to me in London. I felt, therefore, that it might not be improper for me to ventilate publicly the views expressed by me in private.

These views are mine, and yet not mine. They are mine because I hope to act according to them. They are almost a part of my being. But, yet, they are not mine, because I lay no claim to originality. They have been formed after reading several books. That which I dimly felt received support from these books.

The views I venture to place before the reader are, needless to say, held by many Indians not touched by what is known as civilization, but I ask the reader to believe me when I tell him that they are also held by thousands of Europeans. Those who wish to dive deep and have time, may read certain books themselves. If time permit me, I hope to translate portions of such books for the benefit of the readers of *Indian Opinion*.

If the readers of *Indian Opinion* and others who may see the following chapters will pass their criticism on to me, I shall feel obliged to them.

The only motive is to serve my country, to find out the Truth and to follow it. If, therefore, my views are proved to be wrong, I shall have no hesitation in rejecting them. If they are proved to be right, I would naturally wish, for the sake of the motherland, that others should adopt them.

To make it easy reading, the chapters are written in the form of a dialogue between the reader and the editor.

MOHANDAS KARAMCHAND GANDHI

Kildonan Castle (G.)
22 Nov. 1909

CHAPTER I: THE CONGRESS AND ITS OFFICIALS

READER: Just at present there is a Home Rule wave passing over India. All our countrymen appear to be pining for National Independence. A similar spirit pervades them even in South Africa. Indians seem to be eager to acquire rights. Will you explain your views in this matter?

EDITOR: You have put the question well, but the answer is not easy. One of the objects of a newspaper is to understand popular feeling and to give expression to it; another is to arouse among the people certain desirable sentiments; and the third is fearlessly to expose popular defects. The exercise of all these three functions is involved in answering your question. To a certain extent the people's will has to be expressed; certain sentiments will need to be fostered, and defects will have to be brought to light. But, as you have asked the question, it is my duty to answer it.

READER: Do you then consider that a desire for Home Rule has been created among us?

EDITOR: That desire gave rise to the National Congress. The choice of the word 'National' implies it.

READER: That, surely, is not the case. Young India seems to

ignore the Congress. It is considered to be an instrument for perpetuating British Rule.

EDITOR: That opinion is not justified. Had not the Grand Old Man[2] of India prepared the soil, our young men could not have even spoken about Home Rule. How can we forget what Mr. Hume[3] has written, how he has lashed us into action, and with what effort he has awakened us, in order to achieve the objects of the Congress? Sir William Wedderburn[4] has given his body, mind and money to the same cause. His writings are worthy of perusal to this day. Professor Gokhale,[5] in order to prepare the nation, embraced poverty and gave twenty years of his life. Even now, he is living in poverty. The late Justice Budruddin Tyebj[6] was also one of those who, through the Congress, sowed the seed of Home Rule. Similarly, in Bengal, Madras, the Punjab and other places, there have been lovers of India and members of the Congress, both Indian and English.

READER: Stay, stay; you are going too far, you are straying away from my question. I have asked you about Home- or Self-Rule; you are discussing foreign rule. I do not desire to hear English names, and you are giving me such names. In these circumstances, I do not think we can ever meet. I shall be pleased if you will confine yourself to Home Rule. All other talk will not satisfy me.

EDITOR: You are impatient. I cannot afford to be likewise. If you will bear with me for a while, I think you will find that you will obtain what you want. Remember the old proverb that the tree does not grow in one day. The fact that you have checked me and that you do not want to hear about the well-wishers of India shows that, for you at any rate, Home Rule is yet far away. If we had many like you, we would never make any advance. This thought is worthy of your attention.

READER: It seems to me that you simply want to put me off by talking round and round. Those whom you consider to be well-wishers of India are not such in my estimation. Why, then, should I listen to your discourse on such people? What has he whom you consider to be the Father of the Nation done for it? He says that the English Governors will do justice and that we should co-operate with them.

EDITOR: I must tell you, with all gentleness, that it must be a matter of shame for us that you should speak about that great man in terms of disrespect. Just look at his work. He has dedicated his life to the service of India. We have learned what we know from him. It was the respected Dadabhai who taught us that the English had sucked our life-blood. What does it matter that, today, his trust is still in the English nation? Is Dadabhai less to be honoured because, in the exuberance of youth, we are prepared to go a step further? Are we, on that account, wiser than he? It is a mark of wisdom not to kick away the very step from which we have risen higher. The removal of a step from a staircase brings down the whole of it. When, out of infancy, we grow into youth, we do not despise infancy, but, on the contrary, we recall with affection the days of our childhood. If, after many years of study, a teacher were to teach me something, and if I were to build a little more on the foundation laid by that teacher, I would not, on that account, be considered wiser than the teacher. He would always command my respect. Such is the case with the Grand Old Man of India. We must admit that he is the author of nationalism.

READER: You have spoken well. I can now understand that we must look upon Mr. Dadabhai with respect. Without him and men like him, we should probably not have the spirit that fires us. How can the same be said of Professor Gokhale? He has constituted himself a great friend of the English; he says that we have to learn a great deal from them, that we have to learn their political wisdom, before we can talk of Home Rule. I am tired of reading his speeches.

EDITOR: If you are tired, it only betrays your impatience. We believe that those, who are discontented with the slowness of their parents and are angry because the parents would not run with their children, are considered disrespectful to their parents. Professor Gokhale occupies the place of a parent. What does it matter if he cannot run with us? A nation that is desirous of securing Home Rule cannot afford to despise its ancestors. We shall become useless, if we lack respect for our elders. Only men with mature thoughts are capable of ruling themselves and not the hasty-tempered. Moreover, how many Indians were there like Professor Gokhale, when he gave himself to Indian education? I verily believe that whatever Professor Gokhale does, he does with pure motives and with a view to serving India. His devotion to the Motherland is so great that he would give his life for it, if necessary. Whatever he says is said not to flatter anyone but because he believes it to be true. We are bound, therefore, to entertain the highest regard for him.

READER: Are we, then, to follow him in every respect?

EDITOR: I never said any such thing. If we conscientiously differed from him, the learned Professor himself would advise us to follow the dictates of our conscience rather than him. Our chief purpose is not to decry his work, but to believe that he is infinitely greater than we are, and to feel assured that compared with his work for India, ours is infinitesimal. Several newspapers write disrespectfully of him. It is our duty to protest against such writings. We should consider men like Professor Gokhale to be the pillars of Home Rule. It is a bad habit to say that another man's thoughts are bad and ours only are good and that those holding different views from ours are the enemies of the country.

READER: I now begin to understand somewhat your meaning. I shall have to think the matter over. But what you say about Mr. Hume and Sir William Wedderburn is beyond my comprehension.

EDITOR: The same rule holds good for the English as for the Indians. I can never subscribe to the statement that all Englishmen are bad. Many Englishmen desire Home Rule for India. That the English people are somewhat more selfish than others is true, but that does not prove that every Englishman is bad. We who seek justice will have to do justice to others. Sir William does not wish ill to India—that should be enough for us. As we proceed, you will see that, if we act justly, India will be sooner free. You will see, too, that if we shun every Englishman as an enemy, Home Rule will be delayed. But if we are just to them, we shall receive their support in our progress towards the goal.

READER: All this seems to me at present to be simply nonsensical. English support and the obtaining of Home Rule are two contradictory things. How can the English people tolerate Home Rule for us? But I do not want you to decide this question for me just yet. To spend time over it is useless. When you have shown how we can have Home Rule, perhaps I shall understand your views. You have prejudiced me against you by discoursing on English help. I would, therefore, beseech you not to continue this subject.

EDITOR: I have no desire to do so. That you are prejudiced against me is not a matter for much anxiety. It is well that I should say unpleasant things at the commencement. It is my duty patiently to try to remove your prejudice.

READER: I like that last statement. It emboldens me to say what I like. One thing still puzzles me. I do not understand how the Congress laid the foundation of Home Rule.

EDITOR: Let us see. The Congress brought together Indians from different parts of India, and enthused us with the idea of nationality. The Government used to look upon it with disfavour. The Congress has always insisted that the Nation should control revenue and expenditure. It has always desired self-government after the Canadian model. Whether we can get it or not, whether we desire it or not, and whether there is not something more desirable, are different questions. All I have to show is that the Congress gave us a foretaste of Home Rule. To deprive it of the honour is not proper, and for us to do so would not only be ungrateful, but retard the fulfilment of our object. To treat the Congress as an institution inimical to our growth as a nation would disable us from using that body.

CHAPTER II: THE PARTITION OF BENGAL

READER: Considering the matter as you put it, it seems proper to say that the foundation of Home Rule was laid by the Congress. But you will admit that this cannot be considered a real awakening. When and how did the real awakening take place?

EDITOR: The seed is never seen. It works underneath the ground, is itself destroyed, and the tree which rises above the ground is alone seen. Such is the case with the Congress. Yet, what you call the real awakening took place after the Partition of Bengal. For this we have to be thankful to Lord Curzon.[7] At the time of the Partition, the people of Bengal reasoned

Hind Swaraj

with Lord Curzon, but in the pride of power he disregarded all their prayers. He took it for granted that Indians could only prattle, that they could never take any effective steps. He used insulting language, and in the teeth of all opposition partitioned Bengal. That day may be considered to be the day of the partition of the British Empire. The shock the British power received through the Partition has never been equalled by any other act. This does not mean that the other injustices done to India are less glaring than that done by the Partition. The salt-tax is not a small injustice. We shall see many such things later on. But the people were ready to resist the Partition. At that time feeling ran high. Many leading Bengalis were ready to lose their all. They knew their power; hence the conflagration. It is now well-nigh unquenchable; it is not necessary to quench it either. The Partition will go, Bengal will be reunited, but the rift in the English barque will remain; it must daily widen. India awakened is not likely to fall asleep. The demand for the abrogation of the Partition is tantamount to a demand for Home Rule. Leaders in Bengal know this. British officials realize it. That is why the Partition still remains. As time passes, the Nation is being forged. Nations are not formed in a day; the formation requires years.

READER: What, in your opinion, are the results of the Partition?

EDITOR: Hitherto we have considered that for redress of grievances we must approach the throne, and if we get no redress we must sit still, except that we may still petition. After the Partition, people saw that petitions must be backed up by force, and that they must be capable of suffering. This new spirit must be considered to be the chief result of the Partition. That spirit was seen in the outspoken writings in the Press. That which the people said tremblingly and in secret began to be said and to be written publicly. The *swadeshi* movement was inaugurated. People, young and old, used to run away at the sight of an English face; it now no longer awes them. They do not fear even a row, or being imprisoned. Some of the best sons of India are at present in banishment.[8] This is something different from mere petitioning. Thus are the people moved. The spirit generated in Bengal has spread in the north to the Punjab, and in the south to Cape Comorin.

READER: Do you suggest any other striking result?

EDITOR: The Partition has not only made a rift in the English ship but has made it in ours also. Great events always produce great results. Our leaders are divided into two parties: the Moderates and the Extremists. These may be considered as the slow party and the impatient party. Some call the Moderates the timid party, and the Extremists the bold party. All interpret the two words according to their preconceptions. This much is certain—that there has arisen an enmity between the two. The one distrusts the other and imputes motives. At the time of the Surat Congress[9] there was almost a fight. I think that this division is not a good thing for the country, but I think also that such divisions will not last long. It all depends upon the leaders how long they will last.

CHAPTER III: DISCONTENT AND UNREST

READER: Then you consider the Partition to be a cause of the awakening? Do you welcome the unrest which has resulted from it?

EDITOR: When a man rises from sleep, he twists his limbs and is restless. It takes some time before he is entirely awakened. Similarly, although the Partition has caused an awakening, the comatose condition has not yet disappeared. We are still twisting our limbs and are still restless, and just as the state between sleep and awakening must be considered to be necessary, so may the present unrest in India be considered a necessary and, therefore, a proper state. The knowledge that there is unrest will, it is highly probable, enable us to outgrow it. Rising from sleep, we do not continue in a comatose state, but according to our ability, sooner or later, we are completely restored to our senses. So shall we be free from the present unrest which no one likes.

READER: What is the other form of unrest?

EDITOR: Unrest is, in reality, discontent. The latter is only now described as unrest. During the Congress period, it was labelled discontent. Mr. Hume always said that the spread of discontent in India was necessary. This discontent is a very useful thing. As long as a man is contented with his present lot, so long is it difficult to persuade him to come out of it. Therefore it is that every reform must be preceded by discontent. We throw away things we have, only when we cease to like them. Such discontent has been produced among us after reading the great works of Indians and Englishmen. Discontent has led to unrest, and the latter has brought about many deaths, many imprisonments, many banishments. Such a state of things will still continue. It must be so. All these may be considered good signs but they may also lead to bad results.

CHAPTER IV: WHAT IS *SWARAJ*?

READER: I have now learnt what the Congress has done to make India one nation, how the Partition has caused an awakening, and how discontent and unrest have spread through the land. I would now like to know your views on *swaraj*. I fear that our interpretation is not the same as yours.

EDITOR: It is quite possible that we do not attach the same meaning to the term. You and I and all Indians are impatient to obtain *swaraj*, but we are certainly not decided as to what it is. To drive the English out of India is a thought heard from many mouths, but it does not seem that many have properly considered why it should be so. I must ask you a question. Do you think that it is necessary to drive away the English, if we get all we want?

READER: I should ask of them only one thing, that is: 'Please leave our country.' If, after they have complied with this request, their withdrawal from India means that they are still in India, I should have no objection. Then we would understand that, in their language, the word 'gone' is equivalent to 'remained'.

EDITOR: Well then, let us suppose that the English have retired. What will you do then?

READER: That question cannot be answered at this stage. The state after withdrawal will depend largely upon the manner of it. If, as you assume, they retire, it seems to me we shall still keep their constitution and shall carry on the Government. If they simply retire for the asking, we should have an army, etc., ready at hand. We should, therefore, have no difficulty in carrying on the Government.

EDITOR: You may think so; I do not. But I will not discuss the matter just now. I have to answer your question, and that I can do well by asking you several questions. Why do you want to drive away the English?

READER: Because India has become impoverished by their Government. They take away our money from year to year. The most important posts are reserved for themselves. We are kept in a state of slavery. They behave insolently towards us and disregard our feelings.

EDITOR: If they do not take our money away, become gentle, and give us responsible posts, would you still consider their presence to be harmful?

READER: That question is useless. It is similar to the question whether there is any harm in associating with a tiger if he changes his nature. Such a question is sheer waste of time. When a tiger changes his nature, Englishmen will change theirs. This is not possible, and to believe it to be possible is contrary to human experience.

EDITOR: Supposing we get Self-Government similar to what the Canadians and the South Africans have, will it be good enough?

READER: That question also is useless. We may get it when we have the same powers; we shall then hoist our own flag. As is Japan, so must India be. We must own our navy, our army, and we must have our own splendour, and then will India's voice ring through the world.

EDITOR: You have drawn the picture well. In effect it means this: that we want English rule without the Englishman. You want the tiger's nature, but not the tiger; that is to say, you would make India English. And when it becomes English, it will be called not Hindustan but *Englistan*. This is not the *swaraj* that I want.

READER: I have placed before you my idea of *swaraj* as I think it should be. If the education we have received be of any use, if the works of Spencer, Mill and others be of any importance, and if the English Parliament be the Mother of Parliaments, I certainly think that we should copy the English people, and this to such an extent that, just as they do not allow others to obtain a footing in their country, so we should not allow them or others to obtain it in ours. What they have done in their own country has not been done in any other country. It is, therefore, proper for us to import their institutions. But now I want to know your views.

EDITOR: There is need for patience. My views will develop of themselves in the course of this discourse. It is as difficult for me to understand the true nature of *swaraj* as it seems to you to be easy. I shall therefore, for the time being, content myself with endeavouring to show that what you call *swaraj* is not truly *swaraj*.

CHAPTER V: THE CONDITION OF ENGLAND

READER: Then from your statement I deduce that the Government of England is not desirable and not worth copying by us.

EDITOR: Your deduction is justified. The condition of England at present is pitiable. I pray to God that India may never be in that plight. That which you consider to be the Mother of Parliaments is like a sterile woman and a prostitute. Both these are harsh terms, but exactly fit the case. That Parliament has not yet, of its own accord, done a single good thing. Hence I have compared it to a sterile woman. The natural condition of that Parliament is such that, without outside pressure, it can do nothing. It is like a prostitute because it is under the control of ministers who change from time to time. Today it is under Mr. Asquith,[10] tomorrow it may be under Mr. Balfour.[11]

READER: You have said this sarcastically. The term 'sterile woman' is not applicable. The Parliament, being elected by the people, must work under public pressure. This is its quality.

EDITOR: You are mistaken. Let us examine it a little more closely. The best men are supposed to be elected by the people. The members serve without pay[12] and therefore, it must be assumed, only for the public weal. The electors are considered to be educated and therefore we should assume that they would not generally make mistakes in their choice. Such a Parliament should not need the spur of petitions or any other pressure. Its work should be so smooth that its effects would be more apparent day by day. But, as a matter of fact, it is generally acknowledged that the members are hypocritical and selfish. Each thinks of his own little interest. It is fear that is the guiding motive. What is done today may be undone tomorrow. *It is not possible to recall a single instance in which finality can be predicted for its work.* When the greatest questions are debated, its members have been seen to stretch themselves and to doze. Sometimes the members talk away until the listeners are disgusted. Carlyle has called it the 'talking shop of the world'. Members vote for their party without a thought. Their so-called discipline binds them to it. If any member, by way of exception, gives an independent vote, he is considered a renegade. If the money and the time wasted by Parliament were entrusted to a few good men, the English nation would be occupying today a much higher platform. Parliament is simply a costly toy of the nation. These views are by no means peculiar to me. Some great English thinkers have expressed them. One of the members of that Parliament recently said that a true Christian could not become a member of it. Another said that it was a baby. And if it has remained a baby after an existence of seven hundred years, when will it outgrow its babyhood?

READER: You have set me thinking; you do not expect me to accept at once all you say. You give me entirely novel views. I shall have to digest them. Will you now explain the epithet 'prostitute'?

EDITOR: That you cannot accept my views at once is only right. If you will read the literature on this subject, you will have some idea of it. Parliament is without a real master. Under the Prime Minister, its movement is not steady but it is buffeted about like a prostitute. The Prime Minister is more concerned about his power than about the welfare of Parliament. His energy is concentrated upon securing the success of his party. His care is not always that Parliament shall do right. Prime Ministers are known to have made Parliament do things merely for party advantage. All this is worth thinking over.

READER: Then you are really attacking the very men whom we have hitherto considered to be patriotic and honest?

EDITOR: Yes, that is true; I can have nothing against Prime Ministers, but what I have seen leads me to think that they cannot be considered really patriotic. If they are to be considered honest because they do not take what are generally known as bribes, let them be so considered, but they are open to subtler influences. In order to gain their ends, they certainly bribe people with honours. I do not hesitate to say that they have neither real honesty nor a living conscience.

READER: As you express these views about Parliament, I would like to hear you on the English people, so that I may have your view of their Government.

EDITOR: To the English voters their newspaper is their Bible. They take their cue from their newspapers which are often dishonest. The same fact is differently interpreted by different newspapers, according to the party in whose interests they are edited. One newspaper would consider a great Englishman to be a paragon of honesty, another would consider him dishonest. What must be the condition of the people whose newspapers are of this type?

READER: You shall describe it.

EDITOR: These people change their views frequently. It is said that they change them every seven years. These views swing like the pendulum of a clock and are never steadfast. The people would follow a powerful orator or a man who gives them parties, receptions, etc. As are the people, so is their Parliament. They have certainly one quality very strongly developed. They will never allow their country to be lost. If any person were to cast an evil eye on it, they would pluck out his eyes. But that does not mean that the nation possesses every other virtue or that it should be imitated. If India copies England, it is my firm conviction that she will be ruined.

READER: To what do you ascribe this state of England?

EDITOR: It is not due to any peculiar fault of the English people, but the condition is due to modern civilization. It is a civilization only in name. Under it the nations of Europe are becoming degraded and ruined day by day.

CHAPTER VI: CIVILIZATION

READER: Now you will have to explain what you mean by civilization.

EDITOR: It is not a question of what I mean. Several English writers refuse to call that civilization which passes under that name. Many books have been written upon that subject. Societies have been formed to cure the nation of the evils of civilization. A great English writer[13] has written a work called *Civilization: Its Cause and Cure*. Therein he has called it a disease.

READER: Why do we not know this generally?

EDITOR: The answer is very simple. We rarely find people arguing against themselves. Those who are intoxicated by modern civilization are not likely to write against it. Their care will be to find out facts and arguments in support of it, and this they do unconsciously, believing it to be true. A man, whilst he is dreaming, believes in his dream; he is undeceived only when he is awakened from his sleep. A man labouring under the bane of civilization is like a dreaming man. What we usually read are the works of defenders of modern civilization, which undoubtedly claims among its votaries very brilliant and even some very good men. Their writings hypnotize us. And so, one by one, we are drawn into the vortex.

READER: This seems to be very plausible. Now will you tell me something of what you have read and thought of this civilization?

EDITOR: Let us first consider what state of things is described by the word 'civilization'. Its true test lies in the fact that people living in it make bodily welfare the object of life. We will take some examples. The people of Europe today live in better-built houses than they did a hundred years ago. This is considered an emblem of civilization, and this is also a matter to promote bodily happiness. Formerly, they wore skins, and used spears as their weapons. Now, they wear long trousers, and, for embellishing their bodies, they wear a variety of clothing, and, instead of spears, they carry with them revolvers containing five or more chambers. If people of a certain country, who have hitherto not been in the habit of wearing much clothing, boots, etc., adopt European clothing, they are supposed to have become civilized out of savagery. Formerly, in Europe, people ploughed their lands mainly by manual labour. Now, one man can plough a vast tract by means of steam engines and can thus amass great wealth. This is called a sign of civilization. Formerly, only a few men wrote valuable books. Now, anybody writes and prints anything he likes and poisons people's minds. Formerly, men travelled in waggons. Now, they fly through the air in trains at the rate of four hundred and more miles per day. This is considered the height of civilization. It has been stated that, as men progress, they shall be able to travel in airships and reach any part of the world in a few hours. Men will not need the use of their hands and feet. They will press a button, and they will have their clothing by their side. They will press another button, and they will have their newspaper. A third, and a motor-car will be in waiting for them. They will have a variety of delicately dished up food. Everything will be done by machinery. Formerly, when people wanted to fight with one another, they measured between them their bodily strength; now it is possible to take away thousands of lives by one man working behind a gun from a hill. This is civilization. Formerly, men worked in the open air only as much as they liked. Now thousands of workmen meet together and for the sake of maintenance work in factories or mines. Their condition is worse than that of beasts. They are obliged to work, at the risk of their lives, at most dangerous occupations, for the sake of millionaires.

Formerly, men were made slaves under physical compulsion. Now they are enslaved by temptation of money and of the luxuries that money can buy. There are now diseases of which people never dreamt before, and an army of doctors is engaged in finding out their cures, and so hospitals have increased. This is a test of civilization. Formerly, special messengers were required and much expense was incurred in order to send letters; today, anyone can abuse his fellow by means of a letter for one penny. True, at the same cost, one can send one's thanks also. Formerly, people had two or three meals consisting of home-made bread and vegetables; now, they require something to eat every two hours so that they have hardly leisure for anything else. What more need I say? All this you can ascertain from several authoritative books. These are all true tests of civilization. And if anyone speaks to the contrary, know that he is ignorant. This civilization takes note neither of morality nor of religion. Its votaries calmly state that their business is not to teach religion. Some even consider it to be a superstitious growth. Others put on the cloak of religion, and prate about morality. But, after twenty years' experience, I have come to the conclusion that immorality is often taught in the name of morality. Even a child can understand that in all I have described above there can be no inducement to morality. Civilization seeks to increase bodily comforts, and it fails miserably even in doing so.

This civilization is irreligion, and it has taken such a hold on the people in Europe that those who are in it appear to be half mad. They lack real physical strength or courage. They keep up their energy by intoxication. They can hardly be happy in solitude. Women, who should be the queens of households, wander in the streets or they slave away in factories. For the sake of a pittance, half a million women in England alone are labouring under trying circumstances in factories or similar institutions. This awful fact is one of the causes of the daily growing suffragette movement.

This civilization is such that one has only to be patient and it will be self-destroyed. According to the teaching of Mahomed this would be considered a Satanic Civilization. Hinduism calls it the Black Age. I cannot give you an adequate conception of it. It is eating into the vitals of the English nation. It must be shunned. Parliaments are really emblems of slavery. If you will sufficiently think over this, you will entertain the same opinion and cease to blame the English. They rather deserve our sympathy. They are a shrewd nation and I therefore believe that they will cast off the evil. They are enterprising and industrious, and their mode of thought is not inherently immoral. Neither are they bad at heart. I therefore respect them. Civilization is not an incurable disease, but it should never be forgotten that the English people are at present afflicted by it.

CHAPTER VII: WHY WAS INDIA LOST?

READER: You have said much about civilization—enough to make me ponder over it. I do not now know what I should adopt and what I should avoid from the nations of Europe, but one question comes to my lips immediately. If civilization is a disease and if it has attacked England, why has she been able to take India, and why is she able to retain it?

EDITOR: Your question is not very difficult to answer, and we shall presently be able to examine the true nature of *swaraj*; for I am aware that I have still to answer that question. I will, however, take up your previous question. The English have not taken India; we have given it to them. They are not in India because of their strength, but because we keep them. Let us now see whether these propositions can be sustained. They came to our country originally for purposes of trade. Recall the Company[14] Bahadur.[15] Who made it Bahadur? They had not the slightest intention at the time of establishing a kingdom. Who assisted the Company's officers? Who was tempted at the sight of their silver? Who bought their goods? History testifies that we did all this. In order to become rich all at once we welcomed the Company's officers with open arms. We assisted them. If I am in the habit of drinking *bhang* and a seller thereof sells it to me, am I to blame him or myself? By blaming the seller, shall I be able to avoid the habit? And, if a particular retailer is driven away, will not another take his place? A true servant of India will have to go to the root of the matter. If an excess of food has caused me indigestion, I shall certainly not avoid it by blaming water. He is a true physician who probes the cause of disease, and if you pose as a physician for the disease of India, you will have to find out its true cause.

READER: You are right. Now I think you will not have to argue much with me to drive your conclusions home. I am impatient to know your further views. We are now on a most interesting topic. I shall, therefore, endeavour to follow your thought, and stop you when I am in doubt.

EDITOR: I am afraid that, in spite of your enthusiasm, as we proceed further, we shall have differences of opinion. Nevertheless, I shall argue only when you stop me. We have already seen that the English merchants were able to get a footing in India because we encouraged them. When our Princes fought among themselves, they sought the assistance of Company Bahadur. That corporation was versed alike in commerce and war. It was unhampered by questions of morality. Its object was to increase its commerce and to make money. It accepted our assistance, and increased the number of its warehouses. To protect the latter it employed an army which was utilized by us also. Is it not then useless to blame the English for what we did at that time? The Hindus and the Mahomedans were at daggers drawn. This, too, gave the Company its opportunity and thus we created the circumstances that gave the Company its control over India. Hence it is truer to say that we gave India to the English than that India was lost.

READER: Will you now tell me how they are able to retain India?

EDITOR: The causes that gave them India enable them to retain it. Some Englishmen state that they took and they hold India by the sword. Both these statements are wrong. The sword is entirely useless for holding India. We alone keep them. Napoleon is said to have described the English as a nation of shop-keepers. It is a fitting description. They hold whatever dominions they have for the sake of their commerce. Their army and their navy are intended to protect it. When the Transvaal offered no such attractions, the late Mr. Gladstone[16] discovered that it was not right for the English to hold it. When it became a paying proposition, resistance led to war. Mr. Chamberlain[17] soon discovered that England enjoyed a suzerainty over the Transvaal. It is related that someone asked the late President Kruger[18] whether there was gold in the moon. He replied that it was highly unlikely because, if there were, the English would have annexed it. Many problems can be solved by remembering that money is their God. Then it follows that we keep the English in India for our base self-interest. We like their commerce; they please us by their subtle methods and get what they want from us. To blame them for this is to perpetuate their power. We further strengthen their hold by quarrelling amongst ourselves. If you accept the above statements, it is proved that the English entered India for the purposes of trade. They remain in it for the same purpose and we help them to do so. Their arms and ammunition are perfectly useless. In this connection I remind you that it is the British flag which is waving in Japan and not the Japanese. The English have a treaty with Japan for the sake of their commerce, and you will see that if they can manage it, their commerce will greatly expand in that country. They wish to convert the whole world into a vast market for their goods. That they cannot do so is true, but the blame will not be theirs. They will leave no stone unturned to reach the goal.

CHAPTER VIII: THE CONDITION OF INDIA

READER: I now understand why the English hold India. I should like to know your views about the condition of our country.

EDITOR: It is a sad condition. In thinking of it my eyes water and my throat gets parched. I have grave doubts whether I shall be able sufficiently to explain what is in my heart. It is my deliberate opinion that India is being ground down, not under the English heel, but under that of modern civilization. It is groaning under the monster's terrible weight. There is yet time to escape it, but every day makes it more and more difficult. Religion is dear to me and my first complaint is that India is becoming irreligious. Here I am not thinking of the Hindu or the Mahomedan or the Zoroastrian religion but of that religion which underlies all religions. We are turning away from God.

READER: How so?

EDITOR: There is a charge laid against us that we are a lazy people and that Europeans are industrious and enterprising. We have accepted the charge and we therefore wish to change our condition. Hinduism, Islam, Zoroastrianism, Christianity and all other religions teach that we should remain passive about worldly pursuits and active about godly pursuits, that we should set a limit to our worldly ambition and that our religious ambition should be illimitable. Our activity should be directed into the latter channel.

READER: You seem to be encouraging religious charlatanism. Many a cheat has, by talking in a similar strain, led the people astray.

EDITOR: You are bringing an unlawful charge against religion. Humbug there undoubtedly is about all religions. Where there is light, there is also shadow. I am prepared to maintain that humbugs in worldly matters are far worse than the humbugs in religion. The humbug of civilization that I am endeavouring to show to you is not to be found in religion.

READER: How can you say that? In the name of religion Hindus and Mahomedans fought against one another. For the same cause Christians fought Christians. Thousands of innocent men have been murdered, thousands have been burned and tortured in its name. Surely, this is much worse than any civilization.

EDITOR: I certainly submit that the above hardships are far more bearable than those of civilization. Everybody understands that the cruelties you have named are not part of religion although they have been practised in its name; therefore there is no aftermath to these cruelties. They will always happen so long as there are to be found ignorant and credulous people. But there is no end to the victims destroyed in the fire of civilization. Its deadly effect is that people come under its scorching flames believing it to be all good. They become utterly irreligious and, in reality, derive little advantage from the world. Civilization is like a mouse gnawing while it is soothing us. When its full effect is realized, we shall see that religious superstition is harmless compared to that of modern civilization. I am not pleading for a continuance of religious superstitions. We shall certainly fight them tooth and nail, but we can never do so by disregarding religion. We can only do so by appreciating and conserving the latter.

READER: Then you will contend that the Pax Britannica is a useless encumbrance?

EDITOR: You may see peace if you like; I see none.

READER: You make light of the terror that the Thugs,[19] the Pindaris[20] and the Bhils[21] were to the country.

EDITOR: If you give the matter some thought, you will see that the terror was by no means such a mighty thing. If it had been a very substantial thing, the other people would have died away before the English advent. Moreover, the present peace is only nominal, for by it we have become emasculated and cowardly. We are not to assume that the English have changed the nature of the Pindaris and the Bhils. It is, therefore, better to suffer the Pindari peril than that someone else should protect us from it and thus render us effeminate. I should prefer to be killed by the arrow of a Bhil than to seek unmanly protection. India without such protection was an India full of valour. Macaulay betrayed gross ignorance when he libelled Indians as being practically cowards. They never merited the charge. Cowards living in a country inhabited by hardy mountaineers and infested by wolves and tigers must surely find an early grave. Have you ever visited our fields? I assure you that our agriculturists sleep fearlessly on their farms even today; but the English and you and I would hesitate to sleep where they sleep. Strength lies in absence of fear, not in the quantity of flesh and muscle we may have on our bodies. Moreover, I must remind you who desire Home Rule that, after all, the Bhils, the Pindaris, and the Thugs are our own countrymen. To conquer them is your and my work. So long as we fear our own brethren, we are unfit to reach the goal.

CHAPTER IX: THE CONDITION OF INDIA (CONTINUED): RAILWAYS

READER: You have deprived me of the consolation I used to have regarding peace in India.

EDITOR: I have merely given you my opinion on the religious aspect, but when I give you my views as to the poverty of India, you will perhaps begin to dislike me because what you and I have hitherto considered beneficial for India no longer appears to me to be so.

READER: What may that be?

EDITOR: Railways, lawyers and doctors have impoverished the country so much so that, if we do not wake up in time, we shall be ruined.

READER: I do now, indeed, fear that we are not likely to agree at all. You are attacking the very institutions which we have hitherto considered to be good.

EDITOR: It is necessary to exercise patience. The true inwardness of the evils of civilization you will understand with difficulty. Doctors assure us that a consumptive clings to life even when he is about to die. Consumption does not produce apparent hurt—it even produces a seductive colour about a patient's face so as to induce the belief that all is well. Civilization is such a disease and we have to be very wary.

READER: Very well, then. I shall hear you on the railways.

EDITOR: It must be manifest to you that, but for the railways, the English could not have such a hold on India as they have. The railways, too, have spread the bubonic plague. Without them, the masses could not move from place to place. They are the carriers of plague germs. Formerly, we had natural segregation. Railways have also increased the frequency of famines because, owing to facility of means of locomotion, people sell out their grain and it is sent to the dearest markets. People become careless and so the pressure of famine increases. Railways accentuate the evil nature of man. Bad men fulfil their evil designs with greater rapidity. The holy places of India have become unholy. Formerly, people went to these places with very great difficulty. Generally, therefore, only the real devotees visited such places. Nowadays rogues visit them in order to practise their roguery.

READER: You have given a one-sided account. Good men can visit these places as well as bad men. Why do they not take the fullest advantage of the railways?

EDITOR: Good travels at a snail's pace—it can, therefore, have little to do with the railways. Those who want to do good are not selfish, they are not in a hurry, they know that to impregnate people with good requires a long time. But evil has wings. To build a house takes time. Its destruction takes none. So the railways can become a distributing agency for the evil one only. It may be a debatable matter whether railways spread famines, but it is beyond dispute that they propagate evil.

READER: Be that as it may, all the disadvantages of railways are more than counterbalanced by the fact that it is due to them that we see in India the new spirit of nationalism.

EDITOR: I hold this to be a mistake. The English have taught us that we were not one nation before and that it will require centuries before we become one nation. This is without foundation. We were one nation before they came to India. One thought inspired us. Our mode of life was the same. It was because we were one nation that they were able to establish one kingdom. Subsequently they divided us.

READER: This requires an explanation.

EDITOR: I do not wish to suggest that because we were one nation we had no differences, but it is submitted that our leading men travelled throughout India either on foot or in bullock-carts. They learned one another's languages and there was no aloofness between them. What do you think could have been the intention of those farseeing ancestors of ours who established Setubandha (Rameshwar) in the South, Jagannath in the East and Hardwar in the North as places of pilgrimage? You will admit they were no fools. They knew that worship of God could have been performed just as well at home. They taught us that those whose hearts were aglow with righteousness had the Ganges in their own homes. But they saw that India was one undivided land so made by nature. They, therefore, argued that it must be one nation. Arguing thus, they established holy places in various parts of India, and fired the people with an idea of nationality in a manner unknown in other parts of the world. And we Indians are one as no two Englishmen are. Only you and I and others who consider ourselves civilized and superior persons imagine that we are many nations. It was after the advent of railways that we began to believe in distinctions, and you are at liberty now to say that it is through the railways that we are beginning to abolish those distinctions. An opium-eater may argue the advantage of opium-eating from the fact that he began to understand the evil of the opium habit after having eaten it. I would ask you to consider well what I had said on the railways.

READER: I will gladly do so, but one question occurs to me even now. You have described to me the India of the pre-Mahomedan period, but now we have Mahomedans, Parsis and Christians. How can they be one nation? Hindus and Mahomedans are old enemies. Our very proverbs prove it. Mahomedans turn to the West for worship, whilst Hindus turn to the East. The former look down on the Hindus as idolaters. The Hindus worship the cow, the Mahomedans kill her. The Hindus believe in the doctrine of non-killing, the

Mahomedans do not. We thus meet with differences at every step. How can India be one nation?

CHAPTER X: THE CONDITION OF INDIA (CONTINUED): THE HINDUS AND THE MAHOMEDANS

EDITOR: Your last question is a serious one and yet, on careful consideration, it will be found to be easy of solution. The question arises because of the presence of the railways, of the lawyers and of the doctors. We shall presently examine the last two. We have already considered the railways. I should, however, like to add that man is so made by nature as to require him to restrict his movements as far as his hands and feet will take him. If we did not rush about from place to place by means of railways and such other maddening conveniences, much of the confusion that arises would be obviated. Our difficulties are of our own creation. God set a limit to man's locomotive ambition in the construction of his body. Man immediately proceeded to discover means of overriding the limit. God gifted man with intellect that he might know his Maker. Man abused it so that he might forget his Maker. I am so constructed that I can only serve my immediate neighbours, but in my conceit I pretend to have discovered that I must with my body serve every individual in the Universe. In thus attempting the impossible, man comes in contact with different natures, different religions, and is utterly confounded. According to this reasoning, it must be apparent to you that railways are a most dangerous institution. Owing to them, man has gone further away from his Maker.

READER: But I am impatient to hear your answer to my question. Has the introduction of Mahomedanism not unmade the nation?

EDITOR: India cannot cease to be one nation because people belonging to different religions live in it. The introduction of foreigners does not necessarily destroy the nation; they merge in it. A country is one nation only when such a condition obtains in it. That country must have a faculty for assimilation. India has ever been such a country. In reality, there are as many religions as there are individuals; but those who are conscious of the spirit of nationality do not interfere with one another's religion. If they do, they are not fit to be considered a nation. If the Hindus believe that India should be peopled only by Hindus, they are living in dreamland. The Hindus, the Mahomedans, the Parsis and the Christians who have made India their country are fellow countrymen, and they will have to live in unity, if only for their interest. In no part of the world are one nationality and one religion synonymous terms; nor has it ever been so in India.

READER: But what about the inborn enmity between Hindus and Mahomedans?

EDITOR: That phrase has been invented by our mutual enemy. When the Hindus and Mahomedans fought against one another, they certainly spoke in that strain. They have long since ceased to fight. How, then, can there be any inborn enmity? Pray remember this too, that we did not cease to fight only after British occupation. The Hindus flourished under Moslem sovereigns and Moslems under the Hindu. Each party recognized that mutual fighting was suicidal, and that neither party would abandon its religion by force of arms. Both parties, therefore, decided to live in peace. With the English advent quarrels re-commenced.

The proverbs you have quoted were coined when both were fighting; to quote them now is obviously harmful. Should we not remember that many Hindus and Mahomedans own the same ancestors and the same blood runs through their veins? Do people become enemies because they change their religion? Is the God of the Mahomedan different from the God of the Hindu? Religions are different roads converging to the same point. What does it matter that we take different roads so long as we reach the same goal? Wherein is the cause for quarrelling?

Moreover, there are deadly proverbs as between the followers of Shiva and those of Vishnu, yet nobody suggests that these two do not belong to the same nation. It is said that the Vedic religion is different from Jainism, but the followers of the respective faiths are not different nations. The fact is that we have become enslaved and, therefore, quarrel and like to have our quarrels decided by a third party. There are Hindu iconoclasts as there are Mahomedan. The more we advance in true knowledge, the better we shall understand that we need not be at war with those whose religion we may not follow.

READER: Now I would like to know your views about cow-protection.

EDITOR: I myself respect the cow, that is, I look upon her with affectionate reverence. The cow is the protector of India because, being an agricultural country, she is dependent on the cow. The cow is a most useful animal in hundreds of ways. Our Mahomedan brethren will admit this.

But, just as I respect the cow, so do I respect my fellow-men. A man is just as useful as a cow no matter whether he be a Mahomedan or a Hindu. Am I, then, to fight with or kill a Mahomedan in order to save a cow? In doing so, I would become an enemy of the Mahomedan as well as of the cow. Therefore, the only method I know of protecting the cow is that I should approach my Mahomedan brother and urge him for the sake of the country to join me in protecting her. If he would not listen to me I should let the cow go for the simple reason that the matter is beyond my ability. If I were overfull of pity for the cow, I should sacrifice my life to save her but not take my brother's. This, I hold, is the law of our religion.

When men become obstinate, it is a difficult thing. If I pull one way, my Moslem brother will pull another. If I put on superior airs, he will return the compliment. If I bow to him gently, he will do it much more so; and if he does not, I shall not be considered to have done wrong in having bowed. When the Hindus became insistent, the killing of cows increased. In my opinion, cow-protection societies may be considered cow-killing societies. It is a disgrace to us that we should need such societies. When we forgot how to protect cows, I suppose we needed such societies.

What am I to do when a blood-brother is on the point of killing a cow? Am I to kill him, or to fall down at his feet and implore him? If you admit that I should adopt the latter course, I must do the same to my Moslem brother.

Who protects the cow from destruction by Hindus when they cruelly ill-treat her? Whoever reasons with the Hindus when they mercilessly belabour the progeny of the cow with their sticks? But this has not prevented us from remaining one nation.

Lastly, if it be true that the Hindus believe in the doctrine of non-killing and the Mahomedans do not, what, pray, is the duty of the former? It is not written that a follower of the religion of *ahimsa* (non-killing) may kill a fellow-man. For him the way is straight. In order to save one being, he may not kill another. He can only plead—therein lies his sole duty.

But does every Hindu believe in *ahimsa*? Going to the root of the matter, not one man really practises such a religion because we do destroy life. We are said to follow that religion because we want to obtain freedom from liability to kill any kind of life. Generally speaking, we may observe that many Hindus partake of meat and are not, therefore, followers of *ahimsa*. It is, therefore, preposterous to suggest that the two cannot live together amicably because the Hindus believe in *ahimsa* and the Mahomedans do not.

These thoughts are put into our minds by selfish and false religious teachers. The English put the finishing touch. They have a habit of writing history; they pretend to study the manners and customs of all peoples. God has given us a limited mental capacity, but they usurp the function of the Godhead and indulge in novel experiments. They write about their own researches in most laudatory terms and hypnotize us into believing them. We in our ignorance then fall at their feet.

Those who do not wish to misunderstand things may read up the Koran, and they will find therein hundreds of passages acceptable to the Hindus; and the *Bhagavad-gita* contains passages to which not a Mahomedan can take exception. Am I to dislike a Mahomedan because there are passages in the Koran I do not understand or like? It takes two to make a quarrel. If I do not want to quarrel with a Mahomedan, the latter will be powerless to foist a quarrel on me; and, similarly, I should be powerless if a Mahomedan refuses his assistance to quarrel with me. An arm striking the air will become disjointed. If everyone will try to understand the core of his own religion and adhere to it, and will not allow false teachers to dictate to him, there will be no room left for quarrelling.

READER: But will the English ever allow the two bodies to join hands?

EDITOR: This question arises out of your timidity. It betrays our shallowness. If two brothers want to live in peace, is it possible for a third party to separate them? If they were to listen to evil counsels we would consider them to be foolish. Similarly, we Hindus and Mahomedans would have to blame our folly rather than the English, if we allowed them to put us asunder. A clay pot would break through impact, if not with one stone, then with another. The way to save the pot is not to keep it away from the danger point but to bake it so that no stone would break it. We have then to make our hearts of perfectly baked clay. Then we shall be steeled against all danger. This can be easily done by the Hindus. They are superior in numbers; they pretend that they are more educated; they are, therefore, better able to shield themselves from attack on their amicable relations with the Mahomedans.

There is mutual distrust between the two communities. The Mahomedans, therefore, ask for certain concessions from Lord Morley. Why should the Hindus oppose this? If the Hindus desisted, the English would notice it, the Mahomedans would gradually begin to trust the Hindus, and brotherliness would be the outcome. We should be ashamed to take our quarrels to the English. Everyone can find out for himself that the Hindus can lose nothing by desisting. That man who has inspired confidence in another has never lost anything in this world.

I do not suggest that the Hindus and the Mahomedans will never fight. Two brothers living together often do so. We shall sometimes have our heads broken. Such a thing ought not to be necessary, but all men are not equitable. When people are in a rage, they do many foolish things. These we have to put up with. But when we do quarrel, we certainly do not want to engage counsel and resort to English or any law-courts. Two men fight; both have their heads broken, or one only. How shall a third party distribute justice amongst them? Those who fight may expect to be injured.

CHAPTER XI: THE CONDITION OF INDIA (CONTINUED): LAWYERS

READER: You tell me that when two men quarrel they should not go to a law-court. This is astonishing.

EDITOR: Whether you call it astonishing or not, it is the truth. And your question introduces us to the lawyers and the doctors. My firm opinion is that the lawyers have enslaved India, have accentuated Hindu-Mahomedan dissensions and have confirmed English authority.

READER: It is easy enough to bring these charges, but it will be difficult for you to prove them. But for the lawyers, who would have shown us the road to independence? Who would have protected the poor? Who would have secured justice? For instance, the late Manomohan Ghose[22] defended many a poor man free of charge. The Congress, which you have praised so much, is dependent for its existence and activity upon the work of the lawyers. To denounce such an estimable class of men is to spell injustice, and you are abusing the liberty of the Press by decrying lawyers.

EDITOR: At one time I used to think exactly like you. I have no desire to convince you that they have never done a single good thing. I honour Mr. Ghose's memory. It is quite true that he helped the poor. That the Congress owes the lawyers something is believable. Lawyers are also men, and there is something good in every man. Whenever instances of lawyers having done good can be brought forward, it will be found that the good is due to them as men rather than as lawyers. All I am concerned with is to show you that the profession teaches immorality; it is exposed to temptation from which few are saved.

The Hindus and the Mahomedans have quarrelled. An ordinary man will ask them to forget all about it; he will tell them that both must be more or less at fault, and will advise them no longer to quarrel. But they go to lawyers. The latter's duty is to side with their clients and to find out ways and arguments in favour of the clients, to which they (the clients) are often strangers. If they do not do so, they will be considered to have degraded their profession. The lawyers, therefore, will, as a rule, advance quarrels instead of repressing them. Moreover, men take up that profession, not in order to help others out of their miseries, but to enrich themselves. It is one of the avenues of becoming wealthy and their interest exists in multiplying disputes. It is within my knowledge that they are glad when men have disputes. Petty pleaders actually manufacture them. Their touts, like so many leeches, suck the blood of the poor people. Lawyers are men who have little to do. Lazy people, in order to indulge in luxuries, take up such professions. This is a true statement. Any other argument is a mere pretension. It is the lawyers who have discovered that theirs is an honourable profession. They frame laws as they frame their own praises. They decide what fees they will charge and they put on so much side that poor people almost consider them to be heaven-born.

Why do they want more fees than common labourers? Why are their requirements greater? In what way are they more profitable to the country than the labourers? Are those who do good entitled to greater payment? And, if they have done anything for the country for the sake of money, how shall it be counted as good?

Those who know anything of the Hindu–Mahomedan quarrels know that they have been often due to the intervention of lawyers. Some families have been ruined through them; they have made brothers enemies. Principalities, having come under the lawyers' power, have become loaded with debt. Many have been robbed of their all. Such instances can be multiplied.

But the greatest injury they have done to the country is that they have tightened the English grip. Do you think that it would be possible for the English to carry on their Government without law courts? It is wrong to consider that courts are established for the benefit of the people. Those who want to perpetuate their power do so through the courts. If people were to settle their own quarrels, a third party would not be able to exercise any authority over them. Truly, men were less unmanly when they settled their disputes either by fighting or by asking their relatives to decide for them. They became more unmanly and cowardly when they resorted to the courts of law. It was certainly a sign of savagery when they settled their disputes by fighting. Is it any the less so, if I ask a third party to decide between you and me? Surely, the decision of a third party is not always right. The parties alone know who is right. We, in our simplicity and ignorance, imagine that a stranger, by taking our money, gives us justice.

The chief thing, however, to be remembered is that without lawyers courts could not have been established or conducted and without the latter the English could not rule. Supposing that there were only English judges, English pleaders and English police, they could only rule over the English. The English could not do without Indian judges and Indian pleaders. How the pleaders were made in the first instance and how they were favoured you should understand well. Then you will have the same abhorrence for the profession that I have. If pleaders were to abandon their profession, and consider it just as degrading as prostitution, English rule would break up in a day. They have been instrumental in having the charge laid against us that we love quarrels and courts as fish love water. What I have said with reference to the pleaders necessarily applies to the judges; they are first cousins; and the one gives strength to the other.

CHAPTER XII: THE CONDITION OF INDIA (CONTINUED): DOCTORS

READER: I now understand the lawyers; the good they may have done is accidental. I feel that profession is certainly hateful. You, however, drag in the doctors also, how is that?

EDITOR: The views I submit to you are those I have adopted. They are not original. Western writers have used stronger terms regarding both lawyers and doctors. One writer has likened the whole modern system to the Upas tree. Its branches are represented by parasitical professions, including those of law and medicine, and over the trunk has been raised the axe of true religion. Immorality is the root of the tree. So you will see that the views do not come right out of my mind but represent the combined experiences of many. I was at one time a great lover of the medical profession. It was my intention to become a doctor for the sake of the country. I no longer hold that opinion. I now understand why the medicine men (the *vaids*) among us have not occupied a very honourable status.

The English have certainly effectively used the medical profession for holding us. English physicians are known to have used their profession with several Asiatic potentates for political gain.

Doctors have almost unhinged us. Sometimes I think that quacks are better than highly qualified doctors. Let us consider: the business of a doctor is to take care of the body, or, properly speaking, not even that. Their business is really to rid the body of diseases that may afflict it. How do these diseases arise? Surely by our negligence or indulgence. I overeat, I have indigestion, I go to a doctor, he gives me medicine, I am cured. I overeat again, I take his pills again. Had I not taken the pills in the first instance, I would have suffered the punishment deserved by me and I would not have overeaten again. The doctor intervened and helped me to indulge myself. My body thereby certainly felt more at ease; but my mind became weakened. A continuance of a course of medicine must, therefore, result in loss of control over the mind.

I have indulged in vice, I contract a disease, a doctor cures me, the odds are that I shall repeat the vice. Had the doctor not intervened, nature would have done its work, and I would have acquired mastery over myself, would have been freed from vice and would have become happy.

Hospitals are institutions for propagating sin. Men take less care of their bodies and immorality increases. European doctors are the worst of all. For the sake of a mistaken care of the human body, they kill annually thousands of animals. They practise vivisection. No religion sanctions this. All say that it is not necessary to take so many lives for the sake of our bodies.

These doctors violate our religious instinct. Most of their medical preparations contain either animal fat or spirituous liquors; both of these are tabooed by Hindus and Mahomedans. We may pretend to be civilized, call religious prohibitions a superstition and want only to indulge in what we like. The fact remains that the doctors induce us to indulge, and the result is that we have become deprived of self-control and have become effeminate. In these circumstances, we are unfit to serve the country. To study European medicine is to deepen our slavery.

It is worth considering why we take up the profession of medicine. It is certainly not taken up for the purpose of serving humanity. We become doctors so that we may obtain honours and riches. I have endeavoured to show that there is no real service of humanity in the profession, and that it is injurious to mankind. Doctors make a show of their knowledge, and charge exorbitant fees. Their preparations, which are intrinsically worth a few pence, cost shillings. The populace, in its credulity and in the hope of ridding itself of some disease, allows itself to be cheated. Are not quacks then, whom we know, better than the doctors who put on an air of humaneness?

CHAPTER XIII: WHAT IS TRUE CIVILIZATION?

READER: You have denounced railways, lawyers and doctors. I can see that you will discard all machinery. What, then, is civilization?

EDITOR: The answer to that question is not difficult. I believe that the civilization India has evolved is not to be beaten in the world. Nothing can equal the seeds sown by our ancestors. Rome went, Greece shared the same fate; the might of the Pharaohs was broken; Japan has become westernized; of China nothing can be said; but India is still, somehow or other, sound at the foundation. The people of Europe learn their lessons from the writings of the men of Greece or Rome, which exist no longer in their former glory. In trying to learn from them, the Europeans imagine that they will avoid the mistakes of Greece and Rome. Such is their pitiable condition. In the midst of all this India remains immovable and that is her glory. It is a charge against India that her people are so uncivilized, ignorant and stolid, that it is not possible to induce them to adopt any changes. It is a charge really against our merit. What we have tested and found true on the anvil of experience, we dare not change. Many thrust their advice upon India, and she remains steady. This is her beauty: it is the sheet-anchor of our hope.

Civilization is that mode of conduct which points out to man the path of duty. Performance of duty and observance of morality are convertible terms. To observe morality is to attain mastery over our mind and our passions. So doing, we know ourselves. The Gujarati equivalent for civilization means 'good conduct'.

If this definition be correct, then India, as so many writers have shown, has nothing to learn from anybody else, and this is as it should be. We notice that the mind is a restless bird; the more it gets the more it wants, and still remains unsatisfied. The more we indulge our passions, the more unbridled they become. Our ancestors, therefore, set a limit to our indulgences. They saw that happiness was largely a mental condition. A man is not necessarily happy because he is rich, or unhappy because he is poor. The rich are often seen to be unhappy, the poor to be happy. Millions will always remain poor. Observing all this, our ancestors dissuaded us from luxuries and pleasures. We have managed with the same kind of plough as existed thousands of years ago. We have retained the same kind of cottages that we had in former times and our indigenous education remains the same as before. We have had no system of life-corroding competition. Each followed his own occupation or trade and charged a regulation wage.

It was not that we did not know how to invent machinery, but our forefathers knew that, if we set our hearts after such things, we would become slaves and lose our moral fibre. They, therefore, after due deliberation decided that we should only do what we could with our hands and feet. They saw that our real happiness and health consisted in a proper use of our hands and feet. They further reasoned that large cities were a snare and a useless encumbrance and that people would not be happy in them, that there would be gangs of thieves and robbers, prostitution and vice flourishing in them and that poor men would be robbed by rich men. They were, therefore, satisfied with small villages. They saw that kings and their swords were inferior to the sword of ethics, and they, therefore, held the sovereigns of the earth to be inferior to the *rishis* and the *fakirs*. A nation with a constitution like this is fitter to teach others than to learn from others. This nation had courts, lawyers and doctors, but they were all within bounds. Everybody knew that these professions were not particularly superior; moreover, these *vakils* and *vaids* did not rob people; they were considered people's dependants, not their masters. Justice was tolerably fair. The ordinary rule was to avoid courts. There were no touts to lure people into them. This evil, too, was noticeable only in and around capitals. The common people lived independently and followed their agricultural occupation. They enjoyed true Home Rule.

And where this cursed modern civilization has not reached, India remains as it was before. The inhabitants of that part of India will very properly laugh at your new-fangled notions. The English do not rule over them, nor will you ever rule over them. Those in whose name we speak we do not know, nor do they know us. I would certainly advise you and those like you who love the motherland to go into the interior that has yet been not polluted by the railways and to live there for six months; you might then be patriotic and speak of Home Rule.

Now you see what I consider to be real civilization. Those who want to change conditions such as I have described are enemies of the country and are sinners.

READER: It would be all right if India were exactly as you have described it, but it is also India where there are hundreds of child widows, where two-year-old babies are married, where twelve-year-old girls are mothers and housewives, where women practise polyandry, where the practice of *niyoga*[23] obtains, where, in the name of religion, girls dedicate themselves to prostitution, and in the name of religion sheep and goats are killed. Do you consider these also symbols of the civilization that you have described?

EDITOR: You make a mistake. The defects that you have shown are defects. Nobody mistakes them for ancient civilization. They remain in spite of it. Attempts have always been made and will be made to remove them. We may utilize the new spirit that is born in us for purging ourselves of these evils. But what I have described to you as emblems of modern civilization are accepted as such by its votaries. The Indian civilization, as described by me, has been so described by its votaries. In no part of the world, and under no civilization, have all men attained perfection. The tendency of the Indian civilization is to elevate the moral being, that of the Western civilization is to propagate immorality. The latter is godless, the former is based on a belief in God. So understanding and so believing, it behoves every lover of India to cling to the old Indian civilization even as a child clings to the mother's breast.

CHAPTER XIV: HOW CAN INDIA BECOME FREE?

READER: I appreciate your views about civilization. I will have to think over them. I cannot take them in all at once. What, then, holding the views you do, would you suggest for freeing India?

EDITOR: I do not expect my views to be accepted all of a sudden. My duty is to place them before readers like yourself. Time can be trusted to do the rest. We have already examined the conditions for freeing India, but we have done so indirectly; we will now do so directly. It is a world-known maxim that the removal of the cause of a disease results in the removal of the disease itself. Similarly if the cause of India's slavery be removed, India can become free.

READER: If Indian civilization is, as you say, the best of all, how do you account for India's slavery?

EDITOR: This civilization is unquestionably the best, but it is to be observed that all civilizations have been on their trial. That civilization which is permanent outlives it. Because the sons of India were found wanting, its civilization has been placed in jeopardy. But its strength is to be seen in its ability to survive the shock. Moreover, the whole of India is not touched. Those alone who have been affected by Western civilization have become enslaved. We measure the universe by our own miserable foot-rule. When we are slaves, we think that the whole universe is enslaved. Because we are in an abject condition, we think that the whole of India is in that condition. As a matter of fact, it is not so, yet it is as well to impute our slavery to the whole of India. But if we bear in mind the above fact, we can see that if we become free, India is free. And in this thought you have a definition of *swaraj*.

It is *swaraj* when we learn to rule ourselves. It is, therefore, in the palm of our hands. Do not consider this *swaraj* to be like a dream. There is no idea of sitting still. The *swaraj* that

I wish to picture is such that, after we have once realized it, we shall endeavour to the end of our life-time to persuade others to do likewise. But such *swaraj* has to be experienced, by each one for himself. One drowning man will never save another. Slaves ourselves, it would be a mere pretension to think of freeing others. Now you will have seen that it is not necessary for us to have as our goal the expulsion of the English. If the English become Indianized, we can accommodate them. If they wish to remain in India along with their civilization, there is no room for them. It lies with us to bring about such a state of things.

READER: It is impossible that Englishmen should ever become Indianized.

EDITOR: To say that is equivalent to saying that the English have no humanity in them. And it is really beside the point whether they become so or not. If we keep our own house in order, only those who are fit to live in it will remain. Others will leave of their own accord. Such things occur within the experience of all of us.

READER: But it has not occurred in history.

EDITOR: To believe that what has not occurred in history will not occur at all is to argue disbelief in the dignity of man. At any rate, it behoves us to try what appeals to our reason. All countries are not similarly conditioned. The condition of India is unique. Its strength is immeasurable. We need not, therefore, refer to the history of other countries. I have drawn attention to the fact that, when other civilizations have succumbed, the Indian has survived many a shock.

READER: I cannot follow this. There seems little doubt that we shall have to expel the English by force of arms. So long as they are in the country we cannot rest. One of our poets says that slaves cannot even dream of happiness. We are day by day becoming weakened owing to the presence of the English. Our greatness is gone; our people look like terrified men. The English are in the country like a blight which we must remove by every means.

EDITOR: In your excitement, you have forgotten all we have been considering. We brought the English, and we keep them. Why do you forget that our adoption of their civilization makes their presence in India at all possible? Your hatred against them ought to be transferred to their civilization. But let us assume that we have to drive away the English by fighting, how is that to be done?

READER: In the same way as Italy did it. What was possible for Mazzini[24] and Garibaldi[25] is possible for us. You cannot deny that they were very great men.

CHAPTER XV: ITALY AND INDIA

EDITOR: It is well that you have instanced Italy. Mazzini was a great and good man; Garibaldi was a great warrior. Both are adorable; from their lives we can learn much. But the condition of Italy was different from that of India. In the first instance, the difference between Mazzini and Garibaldi is worth noting. Mazzini's ambition was not and has not yet been realized regarding Italy. Mazzini has shown in his writings on the duty of man that every man must learn how to rule himself. This has not happened in Italy. Garibaldi did not hold this view of Mazzini's. Garibaldi gave and every Italian took arms. Italy and Austria had the same civilization; they were cousins in this respect. It was a matter of tit for tat. Garibaldi simply wanted Italy to be free from the Austrian yoke. The machinations of Minister Cavour[26] disgrace that portion of the history of Italy. And what has been the result? If you believe that because Italians rule Italy the Italian nation is happy, you are groping in darkness. Mazzini has shown conclusively that Italy did not become free. Victor Emmanuel [II] gave one meaning to the expression; Mazzini gave another. According to Emmanuel, Cavour and even Garibaldi, Italy meant the King of Italy and his henchmen. According to Mazzini, it meant the whole of the Italian people, that is, its agriculturists. Emmanuel was only its servant. The Italy of Mazzini still remains in a state of slavery. At the time of the so-called national war, it was a game of chess between two rival kings with the people of Italy as pawns. The working classes in that land are still unhappy. They, therefore, indulge in assassination, rise in revolt, and rebellion on their part is always expected. What substantial gain did Italy obtain after the withdrawal of the Austrian troops? The gain was only nominal. The reforms for the sake of which the war was supposed to have been undertaken have not yet been granted. The condition of the people in general still remains the same.

I am sure you do not wish to reproduce such a condition in India. I believe that you want the millions of India to be happy, not that you want the reins of government in your hands. If that be so, we have to consider only one thing: how can the millions obtain self-rule? You will admit that people under several Indian princes are being ground down. The latter mercilessly crush them. Their tyranny is greater than that of the English, and if you want such tyranny in India, then we shall never agree. My patriotism does not teach me that I am to allow people to be crushed under the heel of Indian princes if only the English retire. If I have the power, I should resist the tyranny of Indian princes just as much as that of the English. By patriotism I mean the welfare of the whole people, and if I could secure it at the hands of the English, I should bow down my head to them. If any Englishman dedicated his life to securing the freedom of India, resisting tyranny and serving the land, I should welcome that Englishman as an Indian.

Again, India can fight like Italy only when she has arms. You have not considered this problem at all. The English are splendidly armed; that does not frighten me, but it is clear that, to pit ourselves against them in arms, thousands of Indians must be armed. If such a thing be possible, how many years will it take? Moreover, to arm India on a large scale is to Europeanize it. Then her condition will be just as pitiable as that of Europe. This means, in short, that India must accept European civilization, and if that is what we want, the best thing is that we have among us those who are so well trained in that civilization. We will then fight for a few rights, will get what we can and so pass our days. But the fact is that the Indian nation will not adopt arms, and it is well that it does not.

READER: You are over-stating the facts. All need not be armed. At first, we shall assassinate a few Englishmen and strike terror; then, a few men who will have been armed will fight openly. We may have to lose a quarter of a million men, more or less, but we shall regain our land. We shall undertake guerilla warfare, and defeat the English.

EDITOR: That is to say, you want to make the holy land of India unholy. Do you not tremble to think of freeing India by assassination? What we need to do is to sacrifice ourselves. It is a cowardly thought, that of killing others. Whom do you suppose to free by assassination? The millions of India do not desire it. Those who are intoxicated by the wretched modern civilization think these things. Those who will rise to power by murder will certainly not make the nation happy. Those who believe that India has gained by Dhingra's act and other similar acts in India make a serious mistake. Dhingra was a patriot, but his love was blind. He gave his body in a wrong way; its ultimate result can only be mischievous.

READER: But you will admit that the English have been frightened by these murders, and that Lord Morley's[27] reforms are due to fear.

EDITOR: The English are both a timid and a brave nation. England is, I believe, easily influenced by the use of gunpowder. It is possible that Lord Morley has granted the reforms through fear, but what is granted under fear can be retained only so long as the fear lasts.

CHAPTER XVI: BRUTE FORCE

READER: This is a new doctrine, that what is gained through fear is retained only while the fear lasts. Surely, what is given will not be withdrawn?

EDITOR: Not so. The Proclamation of 1857[28] was given at the end of a revolt, and for the purpose of preserving peace. When peace was secured and people became simple-minded, its full effect was toned down. If I cease stealing for fear of punishment, I would recommence the operation as soon as the fear is withdrawn from me. This is almost a universal experience. We have assumed that we can get men to do things by force and, therefore, we use force.

READER: Will you not admit that you are arguing against yourself? You know that what the English obtained in their own country they obtained by using brute force. I know you have argued that what they have obtained is useless, but that does not affect my argument. They wanted useless things and they got them. My point is that their desire was fulfilled. What does it matter what means they adopted? Why should we not obtain our goal, which is good, by any means whatsoever, even by using violence? Shall I think of the means when I have to deal with a thief in the house? My duty is to drive him out anyhow. You seem to admit that we have received nothing, and that we shall receive nothing, by petitioning. Why, then, may we not do so by using brute force? And, to retain what we may receive, we shall keep up the fear by using the same force to the extent that it may be necessary. You will not find fault with a continuance of force to prevent a child from thrusting its foot into fire? Somehow or other we have to gain our end.

EDITOR: Your reasoning is plausible. It has deluded many. I have used similar arguments before now. But I think I know better now, and I shall endeavour to undeceive you. Let us first take the argument that we are justified in gaining our end by using brute force because the English gained theirs by using similar means. It is perfectly true that they used brute force and that it is possible for us to do likewise, but by using similar means we can get only the same thing that they got. You will admit that we do not want that. Your belief that there is no connection between the means and the end is a great mistake. Through that mistake even men who have been considered religious have committed grievous crimes. Your reasoning is the same as saying that we can get a rose through planting a noxious weed. If I want to cross the ocean, I can do so only by means of a vessel; if I were to use a cart for that purpose, both the cart and I would soon find the bottom. 'As is the God, so is the votary', is a maxim worth considering. Its meaning has been distorted and men have gone astray.

The means may be likened to a seed, the end to a tree; and there is just the same inviolable connection between the means and the end as there is between the seed and the tree. I am not likely to obtain the result flowing from the worship of God by laying myself prostrate before Satan. If, therefore, anyone were to say: 'I want to worship God; it does not matter that I do so by means of Satan', it would be set down as ignorant folly. We reap exactly as we sow. The English in 1833 obtained greater voting power by violence. Did they by using brute force better appreciate their duty? They wanted the right of voting, which they obtained by using physical force. But real rights are a result of performance of duty; these rights they have not obtained. We, therefore, have before us in England the farce of everybody wanting and insisting on his rights, nobody thinking of his duty. And, where everybody wants rights, who shall give them to whom? I do not wish to imply that they do no duties. They don't perform the duties corresponding to those rights; and as they do not perform that particular duty, namely, acquire fitness, their rights have proved a burden to them. In other words, what they have obtained is an exact result of the means they adopted. They used the means corresponding to the end. If I want to deprive you of your watch, I shall certainly have to fight for it; if I want to buy your watch, I shall have to pay you for it; and if I want a gift I shall have to plead for it; and, according to the means I employ, the watch is stolen property, my own property, or a donation. Thus we see three different results from three different means. Will you still say that means do not matter?

Now we shall take the example given by you of the thief to be driven out. I do not agree with you that the thief may be driven out by any means. If it is my father who has come to steal I shall use one kind of means. If it is an acquaintance I shall use another; and in the case of a perfect stranger I shall use a third. If it is a white man, you will perhaps say you will use means different from those you will adopt with an Indian thief. If it is a weakling, the means will be different from those to be adopted for dealing with an equal in physical strength; and if the thief is armed from top to toe, I shall simply remain quiet. Thus we have a variety of means between the father and the armed man. Again, I fancy that I should pretend to be sleeping whether the thief was my father or that strong armed man. The reason for this is that my father would also be armed and I should succumb to the strength possessed by either and allow my things to be stolen. The strength of my father would make me weep with pity; the strength of the armed man would rouse in me anger and we should become enemies. Such is the curious situation. From these examples we may not be able to agree as to the means to be adopted in each case. I myself seem clearly to see what should be done in all these cases, but the remedy may frighten you. I therefore hesitate to place it before you. For the time being I will leave you to guess it, and if you cannot, it is clear you will have to adopt different means in each case. You will also have seen that any means will not avail to drive away the thief. You will have to adopt means to fit each case. Hence it follows that your duty is *not* to drive away the thief by any means you like.

Let us proceed a little further. That well-armed man has stolen your property; you have harboured the thought of his act; you are filled with anger; you argue that you want to punish that rogue, not for your own sake, but for the good of your neighbours; you have collected a number of armed men, you want to take his house by assault; he is duly informed of it, he runs away; he too is incensed. He collects his brother-robbers, and sends you a defiant message that he will commit robbery in broad daylight. You are strong, you do not fear him, you are prepared to receive him. Meanwhile, the robber pesters your neighbours. They complain before you. You reply that you are doing all for their sake, you do not mind that your own goods have been stolen. Your neighbours reply that the robber never pestered them before, and that he commenced his depredations only after you declared hostilities against him. You are between Scylla and Charybdis. You are full of pity for the poor men. What they say is true. What are you to do? You will be disgraced if you now leave the robber alone. You, therefore, tell the poor men: 'Never mind. Come, my wealth is yours, I will give you arms, I will teach you how to use them; you should belabour the rogue; don't you leave him alone.' And so the battle grows; the robbers increase in numbers; your neighbours have deliberately put themselves to inconvenience. Thus the result of wanting to take revenge upon the robber is that you have disturbed your own peace; you are in perpetual fear of being robbed and assaulted; your courage has given place to cowardice.

If you will patiently examine the argument, you will see that I have not overdrawn the picture. This is one of the means. Now let us examine the other. You set this armed robber down as an ignorant brother; you intend to reason with him at a suitable opportunity: you argue that he is, after all, a fellow man; you do not know what prompted him to steal. You, therefore, decide that, when you can, you will destroy the man's motive for stealing. Whilst you are thus reasoning with yourself, the man comes again to steal. Instead of being angry with him, you take pity on him. You think that this stealing habit must be a disease with him. Henceforth, you, therefore, keep your doors and windows open, you change your sleeping-place, and you keep your things in a manner most accessible to him. The robber comes again and is confused as all this is new to him; nevertheless, he takes away your things. But his mind is agitated. He inquires about you in the village, he comes to learn about your broad and loving heart, he repents, he begs your pardon, returns you your things, and leaves off the stealing habit. He becomes your servant, and you find for him honourable employment. This is the second method. Thus, you see, different means have brought about totally different results. I do not wish to deduce from this that robbers will act in the above manner or that all will have the same pity and love like you, but I only wish to show that fair means alone can produce fair results, and that, at least in the majority of cases, if not indeed in all, the force of love and pity is infinitely greater than the force of arms. There is harm in the exercise of brute force, never in that of pity.

Now we will take the question of petitioning. It is a fact beyond dispute that a petition, without the backing of force, is useless. However, the late Justice Ranade[29] used to say that petitions served a useful purpose because they were a means of educating people. They give the latter an idea of their condition and warn the rulers. From this point of view, they are not altogether useless. A petition of an equal is a sign of courtesy; a petition from a slave is a symbol of his slavery. A petition backed by force is a petition from an equal and, when he transmits his demand in the form of a petition, it testifies to his nobility. Two kinds of force can back petitions. 'We shall hurt you if you do not give this', is one kind of force; it is the force of arms, whose evil results we have already examined. The second kind of force can thus be stated: 'If you do not concede our demand, we shall be no longer your petitioners. You can govern us only so long as we remain the governed; we shall no longer have any dealings with you.' The force implied in this may be described as love-force, soul-force, or, more popularly but less accurately, passive resistance. This force is indestructible. He who uses it perfectly understands his position. We have an ancient proverb which literally means: 'One negative cures thirty-six diseases.' The force of arms is powerless when matched against the force of love or the soul.

Now we shall take your last illustration, that of the child thrusting its foot into fire. It will not avail you. What do you really do to the child? Supposing that it can exert so much physical force that it renders you powerless and rushes into fire, then you cannot prevent it. There are only two remedies open to you—either you must kill it in order to prevent it from perishing in the flames, or you must give your own life because you do not wish to see it perish before your very eyes. You will not kill it. If your heart is not quite full of pity, it is possible that you will not surrender yourself by preceding the child and going into the fire yourself. You, therefore, helplessly allow it to go into the flames. Thus, at any rate, you are not using physical force. I hope you will not consider that it is still physical force, though of a low order, when you would forcibly prevent the child from rushing towards the fire if you could. That force is of a different order and we have to understand what it is.

Remember that, in thus preventing the child, you are minding entirely its own interest, you are exercising authority for its sole benefit. Your example does not apply to the English. In using brute force against the English you consult entirely your own, that is the national, interest. There is no question here either of pity or of love. If you say that the actions of the English, being evil, represent fire, and that they proceed to their actions through ignorance, and that therefore they occupy the position of a child and that you want to protect such a child, then you will have to overtake every evil action of that kind by whomsoever committed and, as in the case of the evil child, you will have to sacrifice yourself. If you are capable of such immeasurable pity, I wish you well in its exercise.

CHAPTER XVII: *SATYAGRAHA*—SOUL-FORCE

READER: Is there any historical evidence as to the success of what you have called soul-force or truth-force? No instance seems to have happened of any nation having risen through soul-force. I still think that evil-doers will not cease doing evil without physical punishment.

EDITOR: The poet Tulsidas has said: 'Of religion, pity, or love, is the root, as egotism of the body. Therefore, we should not abandon pity so long as we are alive.' This appears to me to be a scientific truth. I believe in it as much as I believe in two and two being four. The force of love is the same as the force of the soul or truth. We have evidence of its working at every step. The universe would disappear without the existence of that force. But you ask for historical evidence. It is, therefore, necessary to know what history means. The Gujarati equivalent means: 'It so happened.'[30] If that is the meaning of history, it is possible to give copious evidence. But, if it means the doings of kings and emperors, there can be no evidence of soul-force or passive resistance in such history. You cannot expect silver ore in a tin mine.

History, as we know it, is a record of the wars of the world, and so there is a proverb among Englishmen that a nation which has no history, that is, no wars, is a happy nation. How kings played, how they became enemies of one another, how they murdered one another, is found accurately recorded in history, and if this were all that had happened in the world, it would have been ended long ago. If the story of the universe had commenced with wars, not a man would have been found alive today. Those people who have been warred against have disappeared as, for instance, the natives of Australia of whom hardly a man was left alive by the intruders. Mark, please, that these natives did not use soul-force in self-defence, and it does not require much foresight to know that the Australians will share the same fate as their victims. 'Those that take the sword shall perish by the sword.' With us the proverb is that professional swimmers will find a watery grave.

The fact that there are so many men still alive in the world shows that it is based not on the force of arms but on the force of truth or love. Therefore, the greatest and most unimpeachable evidence of the success of this force is to be found in the fact that, in spite of the wars of the world, it still lives on.

Thousands, indeed tens of thousands, depend for their existence on a very active working of this force. Little quarrels of millions of families in their daily lives disappear before the exercise of this force. Hundreds of nations live in peace. History does not and cannot take note of this fact. History is really a record of every interruption of the even working of the force of love or of the soul. Two brothers quarrel; one of them repents and re-awakens the love that was lying dormant in him; the two again begin to live in peace; nobody takes note of this. But if the two brothers, through the intervention of solicitors or some other reason take up arms or go to law—which is another form of the exhibition of brute force,—their doings would be immediately noticed in the Press, they would be the talk of their neighbours and would probably go down to history. And what is true of families and communities is true of nations. There is no reason to believe that there is one law for families and another for nations. History, then, is a record of an interruption of the course of nature. Soul-force, being natural, is not noted in history.

READER: According to what you say, it is plain that instances of this kind of passive resistance are not to be found in history. It is necessary to understand this passive resistance more fully. It will be better, therefore, if you enlarge upon it.

EDITOR: *Satyagraha* is referred to in English as passive resistance. Passive resistance is a method of securing rights by personal suffering; it is the reverse of resistance by arms. When I refuse to do a thing that is repugnant to my conscience, I use soul-force. For instance, the Government of the day has passed a law which is applicable to me. I do not like it. If by using violence I force the Government to repeal the law, I am employing what may be termed body-force. If I do not obey the law and accept the penalty for its breach, I use soul-force. It involves sacrifice of self.

Everybody admits that sacrifice of self is infinitely superior to sacrifice of others. Moreover, if this kind of force is used in a cause that is unjust, only the person using it suffers. He does not make others suffer for his mistakes. Men have before now done many things which were subsequently found to have been wrong. No man can claim that he is absolutely in the right or that a particular thing is wrong because he thinks so, but it is wrong for him so long as that is his deliberate judgment. It is therefore meet that he should not do that which he knows to be wrong, and suffer the consequence whatever it may be. This is the key to the use of soul-force.

READER: You would then disregard laws—this is rank disloyalty. We have always been considered a law-abiding nation. You seem to be going even beyond the extremists. They say that we must obey the laws that have been passed, but that if the laws be bad, we must drive out the law-givers even by force.

EDITOR: Whether I go beyond them or whether I do not is a matter of no consequence to either of us. We simply want to find out what is right and to act accordingly. The real meaning of the statement that we are a law-abiding nation is that we are passive resisters. When we do not like certain laws, we do not break the heads of law-givers but we suffer and do not submit to the laws. That we should obey laws whether good or bad is a new-fangled notion. There was no such thing in former days. The people disregarded those laws they did not like and suffered the penalties for their breach. It is contrary to our manhood if we obey laws repugnant to our conscience. Such teaching is opposed to religion and means slavery. If the Government were to ask us to go about without any clothing, should we do so? If I were a passive resister, I would say to them that I would have nothing to do with their law. But we have so forgotten ourselves and become so compliant that we do not mind any degrading law.

A man who has realized his manhood, who fears only God, will fear no one else. Man-made laws are not necessarily binding on him. Even the Government does not expect any such thing from us. They do not say: 'You must do such and such a thing', but they say: 'If you do not do it, we will punish you.' We are sunk so low that we fancy that it is our duty and our religion to do what the law lays down. If man will only realize that it is unmanly to obey laws that are unjust, no man's tyranny will enslave him. This is the key to self-rule or home rule.

It is a superstition and ungodly thing to believe that an act of a majority binds a minority. Many examples can be given in which acts of majorities will be found to have been wrong and those of minorities to have been right. All reforms owe their origin to the initiation of minorities in opposition to majorities. If among a band of robbers a knowledge of robbing is obligatory, is a pious man to accept the obligation? So long as the superstition that men should obey unjust laws exists, so long will their slavery exist. And a passive resister alone can remove such a superstition.

To use brute-force, to use gunpowder, is contrary to passive resistance, for it means that we want our opponent to do by force that which we desire but he does not. And if such a use of force is justifiable, surely he is entitled to do likewise by us. And so we should never come to an agreement. We may simply fancy, like the blind horse moving in a circle round a mill, that we are making progress. Those who believe that they are not bound to obey laws which are repugnant to their conscience have only the remedy of passive resistance open to them. Any other must lead to disaster.

READER: From what you say I deduce that passive resistance is a splendid weapon of the weak, but that when they are strong they may take up arms.

EDITOR: This is gross ignorance. Passive resistance, that is, soul-force, is matchless. It is superior to the force of arms. How, then, can it be considered only a weapon of the weak? Physical-force men are strangers to the courage that is requisite in a passive resister. Do you believe that a coward can ever disobey a law that he dislikes? Extremists are considered to be advocates of brute force. Why do they, then, talk about obeying laws? I do not blame them. They can say nothing else. When they succeed in driving out the English and they themselves become governors, they will want you and me to obey their laws. And that is a fitting thing for their constitution. But a passive resister will say he will not obey a law that is against his conscience, even though he may be blown to pieces at the mouth of a cannon.

What do you think? Wherein is courage required—in blowing others to pieces from behind a cannon, or with a smiling face to approach a cannon and be blown to pieces? Who is the true warrior—he who keeps death always as a bosom-friend, or he who controls the death of others? Believe me that a man devoid of courage and manhood can never be a passive resister.

This, however, I will admit: that even a man weak in body is capable of offering this resistance. One man can offer it just as well as millions. Both men and women can indulge in it. It does not require the training of an army; it needs no jiu-jitsu. Control over the mind is alone necessary, and when that is attained, man is free like the king of the forest and his very glance withers the enemy.

Passive resistance is an all-sided sword, it can be used anyhow; it blesses him who uses it and him against whom it is used. Without drawing a drop of blood it produces far-reaching results. It never rusts and cannot be stolen. Competition between passive resisters does not exhaust. The sword of passive resistance does not require a scabbard. It is strange indeed that you should consider such a weapon to be a weapon merely of the weak.

READER: You have said that passive resistance is a speciality of India. Have cannons never been used in India?

EDITOR: Evidently, in your opinion, India means its few princes. To me it means its teeming millions on whom depends the existence of its princes and our own.

Kings will always use their kingly weapons. To use force is bred in them. They want to command, but those who have to obey commands do not want guns: and these are in a majority throughout the world. They have to learn either body-force or soul-force. Where they learn the former, both the rulers and the ruled become like so many madmen; but where they learn soul-force, the commands of the rulers do not go beyond the point of their swords, for true men disregard unjust commands. Peasants have never been subdued by the sword, and never will be. They do not know the use of the sword, and they are not frightened by the use of it by others. That nation is great which rests its head upon death as its pillow. Those who defy death are free from all fear. For those who are labouring under the delusive charms of brute-force, this picture is not overdrawn. The fact is that, in India, the nation at large has generally used passive resistance in all departments of life. We cease to co-operate with our rulers when they displease us. This is passive resistance.

I remember an instance when, in a small principality, the villagers were offended by some command issued by the prince. The former immediately began vacating the village. The prince became nervous, apologized to his subjects and withdrew his command. Many such instances can be found in India. Real Home Rule is possible only where passive resistance is the guiding force of the people. Any other rule is foreign rule.

READER: Then you will say that it is not at all necessary for us to train the body?

EDITOR: I will certainly not say any such thing. It is difficult to become a passive resister unless the body is trained. As a rule, the mind, residing in a body that has become weakened by pampering, is also weak, and where there is no strength of mind there can be no strength of soul. We shall have to improve our physique by getting rid of infant marriages and luxurious living. If I were to ask a man with a shattered body to face a cannon's mouth, I should make a laughing-stock of myself.

READER: From what you say, then, it would appear that it is not a small thing to become a passive resister, and, if that is so, I should like you to explain how a man may become one.

EDITOR: To become a passive resister is easy enough but it is also equally difficult. I have known a lad of fourteen years become a passive resister; I have known also sick people do likewise; and I have also known physically strong and otherwise happy people unable to take up passive resistance. After a great deal of experience it seems to me that those who want to become passive resisters for the service of the country have to observe perfect chastity, adopt poverty, follow truth, and cultivate fearlessness.

Chastity is one of the greatest disciplines without which the mind cannot attain requisite firmness. A man who is unchaste loses stamina, becomes emasculated and cowardly. He whose mind is given over to animal passions is not capable of any great effort. This can be proved by innumerable instances. What, then, is a married person to do is the question that arises naturally; and yet it need not. When a husband and wife gratify the passions, it is no less an animal indulgence on that account. Such an indulgence, except for perpetuating the race, is strictly prohibited. But a passive resister has to avoid even that very limited indulgence because he can have no desire for progeny. A married man, therefore, can observe perfect chastity. This subject is not capable of being treated at greater length. Several questions arise: How is one to carry one's wife with one, what are her rights, and other similar questions. Yet those who wish to take part in a great work are bound to solve these puzzles.

Just as there is necessity for chastity, so is there for poverty. Pecuniary ambition and passive resistance cannot well go together. Those who have money are not expected to throw it away, but they *are* expected to be indifferent about it. They must be prepared to lose every penny rather than give up passive resistance.

Passive resistance has been described in the course of our discussion as truth-force. Truth, therefore, has necessarily to be followed and that at any cost. In this connection, academic questions such as whether a man may not lie in order to save a life, etc., arise, but these questions occur only to those who wish to justify lying. Those who want to follow truth every time are not placed in such a quandary; and if they are, they are still saved from a false position.

Passive resistance cannot proceed a step without fearlessness. Those alone can follow the path of passive resistance who are free from fear, whether as to their possessions, false honour, their relatives, the government, bodily injuries or death.

These observances are not to be abandoned in the belief that they are difficult. Nature has implanted in the human breast ability to cope with any difficulty or suffering that may come to man unprovoked. These qualities are worth having, even for those who do not wish to serve the country. Let there be no mistake, as those who want to train themselves in the use of arms are also obliged to have these qualities more or less. Everybody does not become a warrior for the wish. A would-be warrior will have to observe chastity and to be satisfied with poverty as his lot. A warrior without fearlessness cannot be conceived of. It may be thought that he would not need to be exactly truthful, but that quality follows real fearlessness. When a man abandons truth, he does so owing to fear in some shape or form. The above four attributes, then, need not frighten anyone. It may be as well here to note that a physical-force man has to have many other useless qualities which a passive resister never needs. And you will find that whatever extra effort a swordsman needs is due to lack of fearlessness. If he is an embodiment of the latter, the sword will drop from his hand that very moment. He does not need its support. One who is free from hatred requires no sword. A man with a stick suddenly came face to face with a lion and instinctively raised his weapon in self-defence. The man saw that he had only prated about fearlessness when there was none in him. That moment he dropped the stick and found himself free from all fear.

CHAPTER XVIII: EDUCATION

READER: In the whole of our discussion, you have not demonstrated the necessity for education; we always complain of its absence among us. We notice a movement for compulsory education in our country. The Maharaja Gaekwar has introduced it in his territories. Every eye is directed towards them. We bless the Maharaja for it. Is all this effort then of no use?

EDITOR: If we consider our civilization to be the highest, I have regretfully to say that much of the effort you have described is of no use. The motive of the Maharaja and other great leaders who have been working in this direction is perfectly pure. They, therefore, undoubtedly deserve great praise. But we cannot conceal from ourselves the result that is likely to flow from their effort.

What is the meaning of education? It simply means a knowledge of letters. It is merely an instrument, and an instrument may be well used or abused. The same instrument that may be used to cure a patient may be used to take his life, and so may a knowledge of letters. We daily observe that many men abuse it and very few make good use of it; and if this is a correct statement, we have proved that more harm has been done by it than good.

The ordinary meaning of education is a knowledge of letters. To teach boys reading, writing and arithmetic is called primary education. A peasant earns his bread honestly. He has ordinary knowledge of the world. He knows fairly well how he should behave towards his parents, his wife, his children and his fellow-villagers. He understands and observes the rules of morality. But he cannot write his own name. What do you propose to do by giving him a knowledge of letters? Will you add an inch to his happiness? Do you wish to make him discontented with his cottage or his lot? And even if you want to do that, he will not need such an education. Carried away by the flood of western thought we came to the conclusion, without weighing pros and cons, that we should give this kind of education to the people.

Now let us take higher education. I have learned Geography, Astronomy, Algebra, Geometry, etc. What of that? In what way have I benefited myself or those around me? Why have I learned these things? Professor Huxley has thus defined education:

That man I think has had a liberal education who has been so trained in youth that his body is the ready servant of his will and does with ease and pleasure all the work that as a mechanism it is capable of; whose intellect is a clear, cold logic[al] engine ... whose mind is stored with a knowledge of the fundamental truths of nature ... whose passions are trained to come to heel by a vigorous will, the servant of a tender conscience ... who has learnt to hate all vileness and to respect others as himself. Such a one and no other, I conceive, has had a liberal education, for he is in harmony with nature. He will make the best of her and she of him.

If this is true education, I must emphatically say that the sciences I have enumerated above I have never been able to use for controlling my senses. Therefore, whether you take elementary education or higher education, it is not required for the main thing. It does not make men of us. It does not enable us to do our duty.

READER: If that is so, I shall have to ask you another question. What enables you to tell all these things to me? If you had not received higher education, how would you have been able to explain to me the things that you have?

EDITOR: You have spoken well. But my answer is simple: I do not for one moment believe that my life would have been wasted, had I not received higher or lower education. Nor do I consider that I necessarily serve because I speak. But I do desire to serve and in endeavouring to fulfil that desire, I make use of the education I have received. And, if I am making good use of it, even then it is not for the millions, but I can use it only for such as you, and this supports my contention. Both you and I have come under the bane of what is mainly false education. I claim to have become free from its ill effect, and I am trying to give you the benefit of my experience and in doing so, I am demonstrating the rottenness of this education.

Moreover, I have not run down a knowledge of letters in all circumstances. All I have now shown is that we must not make of it a fetish. It is not our Kamadhuk.[31] In its place it can be of use and it has its place when we have brought our senses under subjection and put our ethics on a firm foundation. And then, if we feel inclined to receive that education, we may make good use of it. As an ornament it is likely to sit well on us. It now follows that it is not necessary to make this education compulsory. Our ancient school system is enough. Character-building has the first place in it and that is primary education. A building erected on that foundation will last.

READER: Do I then understand that you do not consider English education necessary for obtaining Home Rule?

EDITOR: My answer is yes and no. To give millions a knowledge of English is to enslave them. The foundation that Macaulay laid of education has enslaved us. I do not suggest that he had any such intention, but that has been the result. Is it not a sad commentary that we should have to speak of Home Rule in a foreign tongue?

And it is worthy of note that the systems which the Europeans have discarded are the systems in vogue among us.

Their learned men continually make changes. We ignorantly adhere to their cast-off systems. They are trying each division to improve its own status. Wales is a small portion of England. Great efforts are being made to revive a knowledge of Welsh among Welshmen. The English Chancellor, Mr. Lloyd George, is taking a leading part in the movement to make Welsh children speak Welsh. And what is our condition? We write to each other in faulty English, and from this even our M. A.'s are not free; our best thoughts are expressed in English; the proceedings of our Congress are conducted in English; our best newspapers are printed in English. If this state of things continues for a long time, posterity will—it is my firm opinion—condemn and curse us.

It is worth noting that, by receiving English education, we have enslaved the nation. Hypocrisy, tyranny, etc., have increased; English-knowing Indians have not hesitated to cheat and strike terror into the people. Now, if we are doing anything for the people at all, we are paying only a portion of the debt due to them.

Is it not a painful thing that, if I want to go to a court of justice, I must employ the English language as a medium, that when I become a barrister, I may not speak my mother-tongue and that someone else should have to translate to me from my own language? Is not this absolutely absurd? Is it not a sign of slavery? Am I to blame the English for it or myself? It is we, the English-knowing Indians, that have enslaved India. The curse of the nation will rest not upon the English but upon us.

I have told you that my answer to your last question is both yes and no. I have explained to you why it is yes. I shall now explain why it is no.

We are so much beset by the disease of civilization, that we cannot altogether do without English education. Those who have already received it may make good use of it wherever necessary. In our dealings with the English people, in our dealings with our own people, when we can only correspond with them through that language, and for the purpose of knowing how disgusted they have themselves become with their civilization, we may use or learn English, as the case may be. Those who have studied English will have to teach morality to their progeny through their mother-tongue and to teach them another Indian language; but when they have grown up, they may learn English, the ultimate aim being that we should not need it. The object of making money thereby should be eschewed. Even in learning English to such a limited extent we shall have to consider what we should learn through it and what we should not. It will be necessary to know what sciences we should learn. A little thought should show you that immediately we cease to care for English degrees, the rulers will prick up their ears.

READER: Then what education shall we give?

EDITOR: This has been somewhat considered above, but we will consider it a little more. I think that we have to improve all our languages. What subjects we should learn through them need not be elaborated here. Those English books which are valuable, we should translate into the various Indian languages. We should abandon the pretension of learning many sciences. Religious, that is ethical, education will occupy the first place. Every cultured Indian will know in addition to his own provincial language, if a Hindu, Sanskrit; if a Mahomedan, Arabic; if a Parsee, Persian; and all, Hindi. Some Hindus should know Arabic and Persian; some Mahomedans and Parsees, Sanskrit. Several Northerners and Westerners should learn Tamil. A universal language for India should be Hindi, with the option of writing it in Persian or Nagari characters. In order that the Hindus and the Mahomedans may have closer relations, it is necessary to know both the characters. And, if we can do this, we can drive the English language out of the field in a short time. All this is necessary for us, slaves. Through our slavery the nation has been enslaved, and it will be free with our freedom.

READER: The question of religious education is very difficult.

EDITOR: Yet we cannot do without it. India will never be godless. Rank atheism cannot flourish in this land. The task is indeed difficult. My head begins to turn as I think of religious education. Our religious teachers are hypocritical and selfish; they will have to be approached. The *Mullas*, the *Dasturs* and the *Brahmins* hold the key in their hands, but if they will not have the good sense, the energy that we have derived from English education will have to be devoted to religious education. This is not very difficult. Only the fringe of the ocean has been polluted and it is those who are within the fringe who alone need cleansing. We who come under this category can even cleanse ourselves because my remarks do not apply to the millions. In order to restore India to its pristine condition, we have to return to it. In our own civilization there will naturally be progress, retrogression, reforms, and reactions; but one effort is required, and that is to drive out Western civilization. All else will follow.

CHAPTER XIX: MACHINERY

READER: When you speak of driving out Western civilization, I suppose you will also say that we want no machinery.

EDITOR: By raising this question, you have opened the wound I have received. When I read Mr. Dutt's *Economic History of India*, I wept; and as I think of it again my heart sickens. It is machinery that has impoverished India. It is difficult to measure the harm that Manchester has done to us. It is due to Manchester that Indian handicraft has all but disappeared.

But I make a mistake. How can Manchester be blamed? We wore Manchester cloth and this is why Manchester wove it. I was delighted when I read about the bravery of Bengal. There were no cloth-mills in that Presidency. They were, therefore, able to restore the original hand-weaving occupation. It is true Bengal encourages the mill-industry of Bombay. If Bengal had proclaimed a boycott of *all* machine-made goods, it would have been much better.

Machinery has begun to desolate Europe. Ruination is now knocking at the English gates. Machinery is the chief symbol of modern civilization; it represents a great sin.

The workers in the mills of Bombay have become slaves. The condition of the women working in the mills is shocking. When there were no mills, these women were not starving. If the machinery craze grows in our country, it will become an unhappy land. It may be considered a heresy, but I am bound to say that it were better for us to send money to Manchester and to use flimsy Manchester cloth than to multiply mills in

India. By using Manchester cloth we only waste our money; but by reproducing Manchester in India, we shall keep our money at the price of our blood, because our very moral being will be sapped, and I call in support of my statement the very mill-hands as witnesses. And those who have amassed wealth out of factories are not likely to be better than other rich men. It would be folly to assume that an Indian Rockefeller would be better than the American Rockefeller. Impoverished India can become free, but it will be hard for any India made rich through immorality to regain its freedom. I fear we shall have to admit that moneyed men support British rule; their interest is bound up with its stability. Money renders a man helpless. The other thing which is equally harmful is sexual vice. Both are poison. A snake-bite is a lesser poison than these two, because the former merely destroys the body but the latter destroy body, mind and soul. We need not, therefore, be pleased with the prospect of the growth of the mill-industry.

READER: Are the mills, then, to be closed down?

EDITOR: That is difficult. It is no easy task to do away with a thing that is established. We, therefore, say that the non-beginning of a thing is supreme wisdom. We cannot condemn millowners; we can but pity them. It would be too much to expect them to give up their mills, but we may implore them not to increase them. If they would be good they would gradually contract their business. They can establish in thousands of households the ancient and sacred handlooms and they can buy out the cloth that may be thus woven. Whether the millowners do this or not, people can cease to use machine-made goods.

READER: You have so far spoken about machine-made cloth, but there are innumerable machine-made things. We have either to import them or to introduce machinery into our country.

EDITOR: Indeed, our goods even are made in Germany. What need, then, to speak of matches, pins and glassware? My answer can be only one. What did India do before these articles were introduced? Precisely the same should be done today. As long as we cannot make pins without machinery, so long will we do without them. The tinsel splendour of glassware we will have nothing to do with, and we will make wicks, as of old, with home-grown cotton and use hand-made earthen saucers for lamps. So doing, we shall save our eyes and money and support *swadeshi* and so shall we attain Home Rule.

It is not to be conceived that all men will do all these things at one time or that some men will give up all machine-made things at once. But, if the thought is sound, we shall always find out what we can give up and gradually cease to use it. What a few may do, others will copy; and the movement will grow like the cocoanut of the mathematical problem. What the leaders do, the populace will gladly do in turn. The matter is neither complicated nor difficult. You and I need not wait until we can carry others with us. Those will be the losers who will not do it, and those who will not do it, although they appreciate the truth, will deserve to be called cowards.

READER: What, then, of the tram-cars and electricity?

EDITOR: This question is now too late. It signifies nothing. If we are to do without the railways we shall have to do without the tram-cars. Machinery is like a snake-hole which may contain from one to a hundred snakes. Where there is machinery there are large cities; and where there are large cities, there are tram-cars and railways; and there only does one see electric light. English villages do not boast of any of these things. Honest physicians will tell you that where means of artificial locomotion have increased, the health of the people has suffered. I remember that when in a European town there was a scarcity of money, the receipts of the tramway company, of the lawyers and of the doctors went down and people were less unhealthy. I cannot recall a single good point in connection with machinery. Books can be written to demonstrate its evils.

READER: Is it a good point or a bad one that all you are saying will be printed through machinery?

EDITOR: This is one of those instances which demonstrate that sometimes poison is used to kill poison. This, then, will not be a good point regarding machinery. As it expires, the machinery, as it were, says to us: 'Beware and avoid me. You will derive no benefits from me and the benefit that may accrue from printing will avail only those who are infected with the machinery-craze.'

Do not, therefore, forget the main thing. It is necessary to realize that machinery is bad. We shall then be able gradually to do away with it. Nature has not provided any way whereby we may reach a desired goal all of a sudden. If, instead of welcoming machinery as a boon, we should look upon it as an evil, it would ultimately go.

CHAPTER XX: CONCLUSION

READER: From your views I gather that you would form a third party. You are neither an extremist nor a moderate.

EDITOR: That is a mistake. I do not think of a third party at all. We do not all think alike. We cannot say that all the moderates hold identical views. And how can those who want only to serve have a party? I would serve both the moderates and the extremists. Where I differ from them, I would respectfully place my position before them and continue my service.

READER: What, then, would you say to both the parties?

EDITOR: I would say to the extremists: 'I know that you want Home Rule for India; it is not to be had for your asking. Everyone will have to take it for himself. What others get for me is not Home Rule but foreign rule; therefore, it would not be proper for you to say that you have obtained Home Rule if you have merely expelled the English. I have already described the true nature of Home Rule. This you would never obtain by force of arms. Brute force is not natural to Indian soil. You will have, therefore, to rely wholly on soul-force. You must not consider that violence is necessary at any stage for reaching our goal.'

I would say to the moderates: 'Mere petitioning is derogatory; we thereby confess inferiority. To say that British rule is indispensable is almost a denial of the Godhead. We cannot say that anybody or anything is indispensable except God. Moreover, commonsense should tell us that to state that, for the time being, the presence of the English in India is a necessity, is to make them conceited.

'If the English vacated India, bag and baggage, it must not be supposed that she would be widowed. It is possible that those who are forced to observe peace under their pressure would fight after their withdrawal. There can be no advantage in suppressing an eruption; it must have its vent. If, therefore, before we can remain at peace, we must fight amongst ourselves, it is better that we do so. There is no occasion for a third party to protect the weak. It is this so-called protection which has unnerved us. Such protection can only make the weak weaker. Unless we realize this, we cannot have Home Rule. I would paraphrase the thought of an English divine and say that anarchy under Home Rule were better than orderly foreign rule. Only, the meaning that the learned divine attached to Home Rule is different from Indian Home Rule according to my conception. We have to learn, and to teach others, that we do not want the tyranny of either English rule or Indian rule.'

If this idea were carried out, both the extremists and the moderates could join hands. There is no occasion to fear or distrust one another.

READER: What, then, would you say to the English?

EDITOR: To them I would respectfully say: 'I admit you are my rulers. It is not necessary to debate the question whether you hold India by the sword or by my consent. I have no objection to your remaining in my country, but although you are the rulers, you will have to remain as servants of the people. It is not we who have to do as you wish, but it is you who have to do as we wish. You may keep the riches that you have drained away from this land, but you may not drain riches henceforth. Your function will be, if you so wish, to police India; you must abandon the idea of deriving any commercial benefit from us. We hold the civilization that you support to be the reverse of civilization. We consider our civilization to be far superior to yours. If you realize this truth, it will be to your advantage and, if you do not, according to your own proverb, you should only live in our country in the same manner as we do. You must not do anything that is contrary to our religions. It is your duty as rulers that for the sake of the Hindus you should eschew beef, and for the sake of Mahomedans you should avoid bacon and ham. We have hitherto said nothing because we have been cowed down, but you need not consider that you have not hurt our feelings by your conduct. We are not expressing our sentiments either through base selfishness or fear, but because it is our duty now to speak out boldly. We consider your schools and law courts to be useless. We want our own ancient schools and courts to be restored. The common language of India is not English but Hindi. You should, therefore, learn it. We can hold communication with you only in our national language.

'We cannot tolerate the idea of your spending money on railways and the military. We see no occasion for either. You may fear Russia; we do not. When she comes we shall look after her. If you are with us, we may then receive her jointly. We do not need any European cloth. We shall manage with articles produced and manufactured at home. You may not keep one eye on Manchester and the other on India. We can work together only if our interests are identical.

'This has not been said to you in arrogance. You have great military resources. Your naval power is matchless. If we wanted to fight with you on your own ground, we should be unable to do so, but if the above submissions be not acceptable to you, we cease to play the part of the ruled. You may, if you like, cut us to pieces. You may shatter us at the cannon's mouth. If you act contrary to our will, we shall not help you; and without our help, we know that you cannot move one step forward.

'It is likely that you will laugh at all this in the intoxication of your power. We may not be able to disillusion you at once; but if there be any manliness in us, you will see shortly that your intoxication is suicidal and that your laugh at our expense is an aberration of intellect. We believe that at heart you belong to a religious nation. We are living in a land which is the source of religions. How we came together need not be considered, but we can make mutual good use of our relations.

'You, English, who have come to India are not good specimens of the English nation, nor can we, almost half-Anglicized Indians, be considered good specimens of the real Indian nation. If the English nation were to know all you have done, it would oppose many of your actions. The mass of the Indians have had few dealings with you. If you will abandon your so-called civilization and search into your own scriptures, you will find that our demands are just. Only on condition of our demands being fully satisfied may you remain in India; and if you remain under those conditions, we shall learn several things from you and you will learn many from us. So doing we shall benefit each other and the world. But that will happen only when the root of our relationship is sunk in a religious soil.'

READER: What will you say to the nation?

EDITOR: Who is the nation?

READER: For our purposes it is the nation that you and I have been thinking of, that is, those of us who are affected by European civilization, and who are eager to have Home Rule.

EDITOR: To these I would say: 'It is only those Indians who are imbued with real love who will be able to speak to the English in the above strain without being frightened, and only those can be said to be so imbued who conscientiously believe that Indian civilization is the best and that the European is a nine days' wonder. Such ephemeral civilizations have often come and gone and will continue to do so. Those only can be considered to be so imbued who, having experienced the force of the soul within themselves, will not cower before bruteforce, and will not, on any account, desire to use brute-force. Those only can be considered to have been so imbued who are intensely dissatisfied with the present pitiable condition, having already drunk the cup of poison.

'If there be only one such Indian, he will speak as above to the English and the English will have to listen to him.

'These are not demands, but they show our mental state. We shall get nothing by asking; we shall have to take what we want, and we need the requisite strength for the effort and that strength will be available to him only who will act thus:

1. He will only on rare occasions make use of the English language.

2. If a lawyer, he will give up his profession and take up a hand-loom.

3. If a lawyer, he will devote his knowledge to enlightening both his people and the English.

4. If a lawyer, he will not meddle with the quarrels between parties but will give up the courts, and from his experience induce the people to do likewise.

5. If a lawyer, he will refuse to be a judge, as he will give up his profession.

6. If a doctor, he will give up medicine, and understand that rather than mending bodies, he should mend souls.

7. If a doctor, he will understand that no matter to what religion he belongs, it is better that bodies remain diseased rather than that they are cured through the instrumentality of the diabolical vivisection that is practised in European schools of medicine.

8. Although a doctor, he will take up a hand-loom, and if any patients come to him, will tell them the cause of their diseases, and will advise them to remove the cause rather than pamper them by giving useless drugs; he will understand that if by not taking drugs, perchance the patient dies, the world will not come to grief and that he will have been really merciful to him.

9. Although a wealthy man, yet regardless of his wealth, he will speak out his mind and fear no one.

10. If a wealthy man, he will devote his money to establishing hand-looms, and encourage others to use hand-made goods by wearing them himself.

11. Like every other Indian, he will know that this is a time for repentance, expiation and mourning.

12. Like every other Indian, he will know that to blame the English is useless, that they came because of us, and remain also for the same reason, and that they will either go or change their nature only when we reform ourselves.

13. Like others, he will understand that at a time of mourning, there can be no indulgence, and that, whilst we are in a fallen state, to be in gaol or in banishment is much the best.

14. Like others, he will know that it is superstition to imagine it necessary that we should guard against being imprisoned in order that we may deal with the people.

15. Like others, he will know that action is much better than speech; that it is our duty to say exactly what we think and face the consequences and that it will be only then that we shall be able to impress anybody with our speech.

16. Like others, he will understand that we shall become free only through suffering.

17. Like others, he will understand that deportation for life to the Andamans is not enough expiation for the sin of encouraging European civilization.

18. Like others, he will know that no nation has risen without suffering; that, even in physical warfare, the true test is suffering and not the killing of others, much more so in the warfare of passive resistance.

19. Like others, he will know that it is an idle excuse to say that we shall do a thing when the others also do it; that we should do what we know to be right, and that others will do it when they see the way; that when I fancy a particular delicacy, I do not wait till others taste it; that to make a national effort and to suffer are in the nature of delicacies; and that to suffer under pressure is no suffering.'

READER: This is a large order. When will all carry it out?

EDITOR: You make a mistake. You and I have nothing to do with the others. Let each do his duty. If I do my duty, that is, serve myself, I shall be able to serve others. Before I leave you, I will take the liberty of repeating:

1. Real home-rule is self-rule or self-control.

2. The way to it is passive resistance: that is soul-force or love-force.

3. In order to exert this force, *swadeshi* in every sense is necessary.

4. What we want to do should be done, not because we object to the English or because we want to retaliate but because it is our duty to do so. Thus, supposing that the English remove the salt-tax, restore our money, give the highest posts to Indians, withdraw the English troops, we shall certainly not use their machine-made goods, nor use the English language, nor many of their industries. It is worth noting that these things are, in their nature, harmful; hence we do not want them. I bear no enmity towards the English but I do towards their civilization.

In my opinion, we have used the term '*swaraj*' without understanding its real significance. I have endeavoured to explain it as I understand it, and my conscience testifies that my life henceforth is dedicated to its attainment.

APPENDICES

Some Authorities and Testimonies by Eminent Men

I. SOME AUTHORITIES

The following books are recommended for perusal to follow up the study of the foregoing:

The Kingdom of God Is within You (Tolstoy)
What Is Art? (Tolstoy)
The Slavery of Our Times (Tolstoy)
The First Step (Tolstoy)
How Shall We Escape? (Tolstoy)
Letter to a Hindoo (Tolstoy)
The White Slaves of England (Sherard)
Civilization, Its Cause and Cure (Carpenter)
The Fallacy of Speed (Taylor)
A New Crusade (Blount)
On the Duty of Civil Disobedience (Thoreau)
Life without Principle (Thoreau)
Unto This Last (Ruskin)
A Joy for Ever (Ruskin)
Duties of Man (Mazzini)
Defence and Death of Socrates (from Plato)
Paradoxes of Civilization (Max Nordau)
Poverty and Un-British Rule in India (Naoroji)
Economic History of India (Dutt)
Village Communities (Maine)

II. TESTIMONIES BY EMINENT MEN

The following extracts from Mr. Alfred Webb's valuable collection show that the ancient Indian civilization has little to learn from the modern:

J. SEYMOUR KEAY, M. P.
BANKER IN INDIA AND INDIA AGENT (WRITING IN 1883)

It cannot be too well understood that our position in India has never been in any degree that of civilians bringing civilization to savage races. When we landed in India we found there a hoary civilization, which, during the progress of thousands of years, had fitted into the character and adjusted itself to the wants of highly intellectual races. The civilization was not perfunctory, but universal and all-pervading—furnishing the country not only with political systems, but with social and domestic institutions of the most ramified description. The beneficent nature of these institutions as a whole may be judged from their effects on the character of the Hindu race. Perhaps there are no other people in the world who show so much in their character the advantageous effects of their own civilization. They are shrewd in business, acute in reasoning, thrifty, religious, sober, charitable, obedient to parents, reverential to old age, amiable, law-abiding, compassionate towards the helpless and patient under suffering.

VICTOR COUSIN (1792–1867)
FOUNDER OF SYSTEMATIC ECLECTICISM IN PHILOSOPHY

On the other hand when we read with attention the poetical and philosophical movements of the East, above all, those of India, which are beginning to spread in Europe, we discover there so many truths, and truths, so profound, and which make such a contrast with the meanness of the results at which the European genius has sometimes stopped, that we are constrained to bend the knee before that of the East, and do see in this cradle of the human race the native land of the highest philosophy.

FRIEDRICH MAX MÜLLER

If I were to ask myself from what literature we here in Europe, we who have been nurtured almost exclusively on the thoughts of Greeks and Romans, and of one Semitic race, the Jewish, may draw that corrective which is most wanted in order to make our inner life more perfect, more comprehensive, more universal, in fact more truly human, a life, not for this life only, but a transfigured and eternal life—again I should point to India.

FREDERICK VON SCHLEGEL

It cannot be denied that the early Indians possessed a knowledge of the true God; all their writings are replete with sentiments and expressions, noble, clear, and severely grand, as deeply conceived and reverently expressed as in any human language in which men have spoken of their God.... Among nations possessing indigenous philosophy and metaphysics together with an innate relish for these pursuits, such as at present characterizes Germany, and, in olden times was the proud distinction of Greece, Hindustan holds the first rank in point of time.

ABBÉ J. A. DUBOIS
MISSIONARY IN MYSORE

[Extracts from a letter dated December 15, 1820, Seringapatam]

December, 1820, Seringapatam:

The authority of married women within their houses is chiefly exerted in preserving good order and peace among the persons who compose their families; and a great many among them discharge this important duty with a prudence and a discretion which have scarcely a parallel in Europe. I have known families composed of between thirty and forty persons, or more, consisting of grown-up sons and daughters, all married and all having children, living together under the superintendence of an old matron—their mother or mother-in-law. The latter, by good management, and by accommodating herself to the temper of the daughters-in-law, by using, according to circumstances, firmness or forbearance, succeeded in preserving peace and harmony during many years amongst so many females, who had all jarring tempers. I ask you whether it would be possible to attain the same end, in the same circumstances, in our countries, where it is scarcely possible to make two women living under the same roof to agree together.

In fact, there is perhaps no kind of honest employment in a civilized country in which the Hindu females have not a due share. Besides the management of the household, and the care of the family, which as already noticed is under their control, the wives and daughters of husbandmen attend and assist their husbands and fathers in the labours of agriculture. Those of tradesmen assist theirs in carrying on their trade. Merchants are attended and assisted by theirs in their shops. Many females are shopkeepers on their own account; and *without a knowledge of the alphabet* or of the decimal scale, they keep by other means their accounts in excellent order, and are considered as still shrewder than the males themselves in their commercial dealings.

J. YOUNG

SECRETARY, SAVON MECHANICS INSTITUTES WITHIN RECENT YEARS

Those races (the Indian viewed from a moral aspect) are perhaps the most remarkable people in the world. They breathe in an atmosphere of moral purity, which cannot but excite admiration, and this is especially the case with the poorer classes, who, notwithstanding the privations of their humble lot, appear to be happy and contented. True children of nature, they live on from day to day, taking no thought for the morrow and thankful for the simple fare which Providence has provided for them. It is curious to witness the spectacle of coolies of both sexes returning home at nightfall after a hard day's work often lasting from sunrise to sunset. In spite of fatigue from the effects of the unremitting toil, they are, for the most part, gay and animated, conversing cheerfully together and occasionally breaking into snatches of light-hearted song. Yet what awaits them on their return to the hovels which they call home? A dish of rice for food, and the floor for a bed. Domestic felicity appears to be the rule among the Natives, and this is the more strange when the customs of marriage are taken into account, parents arranging all such matters. Many Indian households afford examples of the married state in its highest degree of perfection. This may be due to the teachings of the Shastras, and to the strict injunctions which they inculcate with regard to marital obligation; but it is no exaggeration to say that husbands are generally devotedly attached to their wives, and in many instances the latter have the most exalted conception of their duties towards their husbands.

COLONEL THOMAS MUNRO
THIRTY-TWO YEARS' SERVICE IN INDIA

If a good system of agriculture, unrivalled manufacturing skill, a capacity to produce whatever can contribute to convenience or luxury; schools established in every village for teaching reading, writing and arithmetic; the general practice of hospitality and charity among each other; and, above all, a treatment of the female sex, full of confidence, respect and delicacy, are among the signs which denote a civilized people, then the Hindus are not inferior to the nations of Europe; and if civilization is to become an article of trade between the two countries, I am convinced that this country [England] will gain by the import cargo.

SIR WILLIAM WEDDERBURN, BART.

The Indian village has thus for centuries remained a bulwark against political disorder, and the home of the simple domestic and social virtues. No wonder, therefore, that philosophers and historians have always dwelt lovingly on this ancient institution which is the natural social unit and the best type of rural life: self-contained, industrious, peace-loving, conservative in the best sense of the word.... I think you will agree with me that there is much that is both picturesque and attractive in this glimpse of social and domestic life in an Indian village. It is a harmless and happy form of human existence. Moreover, it is not without good practical outcome.

[1] Originally written in Gujarati during Gandhi's return journey from England on the *Kildonan Castle* and published in *Indian Opinion*, 11 and 18 Dec. 1909. Issued as a booklet in January 1910, it was proscribed in India by the Government of Bombay on 24 Mar. 1910. This hastened Gandhi's decision to publish the English translation, issued by the International Printing Press, Phoenix. Ganesh & Co., Madras, brought out the first Indian edition in 1919 with Gandhi's foreword dated 28 May 1919. The sixth edition was issued in 1924. In the same year appeared an American edition by H. T. Mazumdar, with the title *Sermon on the Sea*. The text adopted here is that of the Revised New Edition published in 1939 by the Navajivan Press, Ahmedabad.

[2] Dadabhai Naoroji (1825–1917).
[3] A. O. Hume, one of the founders of the Indian National Congress.
[4] President, Indian National Congress at Bombay (1889) and at Allahabad (1910).
[5] Distinguished Indian leader and statesman, educationist and reformer.

⁶ Judge of the Bombay High Court and President of the Indian National Congress at Madras (1887).
⁷ Viceroy of India, 1899-1905.
⁸ Bal Gangadhar Tilak was in Mandalay prison at this time.
⁹ In 1907.
¹⁰ Herbert Henry Asquith (1852-1928), Prime Minister of Great Britain, 1908-16.
¹¹ Arthur James Balfour, Prime Minister of Great Britain, 1902-5.
¹² Payment to members began in 1911.
¹³ Edward Carpenter.
¹⁴ East India Company.
¹⁵ Literally, 'brave', here 'powerful', 'sovereign'.
¹⁶ William Ewart Gladstone (1809-98), Prime Minister of Great Britain, 1868-74, 1880-5, 1886, and 1892-4.
¹⁷ Joseph Chamberlain (1836-1914), Secretary of State for the Colonies, 1895.
¹⁸ Stephanus Johannes Paulus Kruger (1825-1904), Boer leader and State President of the South African Republic.
¹⁹ Predatory groups that used to loot, rob, and kill people.
²⁰ Mounted marauders during the seventeenth and eighteenth centuries.
²¹ Tribe in Central India and Gujarat.
²² (1844-96), lawyer and Congressman; first Indian barrister; founder and editor of the *Indian Mirror*.
²³ Insemination by a person other than one's husband.
²⁴ Giuseppe Mazzini (1805-72).
²⁵ Giuseppe Garibaldi (1807-82), Italian soldier and patriot, one of the leaders in the struggle for unification of Italy.
²⁶ Count Camillo Benso Cavour (1810-61), distinguished Italian statesman, who, as Premier to Victor Emmanuel, King of Sardinia (proclaimed King of Italy in 1861), did much for the unification of Italy which was achieved in 1870.
²⁷ Morley was Secretary of State for India. The Morley-Minto Reforms came into force on 15 Nov. 1909.
²⁸ Queen Victoria's Proclamation of 1858.
²⁹ Mahadev Govind Ranade (1842-1901), distinguished Indian judge, social reformer, author, and one of the founders of the Indian National Congress.
³⁰ Literally, '*Itihas* [history] means, "it so happened".'
³¹ Mythical cow, yielding whatever is wished for.

§ 2. On *Hind Swaraj*

89. IMPERFECTIONS IN *HIND SWARAJ*

Johannesburg,
March 20, 1910

It is not without hesitation that the translation of *Hind Swaraj* is submitted to the public. A European friend,[1] with whom I discussed the contents, wanted to see a translation of it and, during our spare moments, I hurriedly dictated and he took it down. It is not a literal translation but it is a faithful rendering of the original. Several English friends have read it, and whilst opinions were being invited as to the advisability of publishing the work, news was received that the original was seized in India. This information hastened the decision to publish the translation without a moment's delay. My fellow-workers at the International Printing Press shared my view and, by working overtime—a labour of love,—they have enabled me to place the translation before the public in an unexpectedly short time. The work is being given to the public at what is practically cost price. But, without the financial assistance of the many Indians who promised to buy copies for themselves and for distribution, it might never have seen the light of day.

I am quite aware of the many imperfections in the original. The English rendering, besides sharing these, must naturally exaggerate them, owing to my inability to convey the exact meaning of the original. Some of the friends who have read the translation have objected that the subject-matter has been dealt with in the form of a dialogue. I have no answer to offer to this objection except that the Gujarati language readily lends itself to such treatment and that it is considered the best method of treating difficult subjects. Had I written for English readers in the first instance, the subject would have been handled in a different manner. Moreover, the dialogue, as it has been given, actually took place between several friends, mostly readers of *Indian Opinion*, and myself.

Whilst the views expressed in *Hind Swaraj* are held by me, I have but endeavoured humbly to follow Tolstoy, Ruskin,

Thoreau, Emerson and other writers, besides the masters of Indian philosophy. Tolstoy has been one of my teachers for a number of years. Those who want to see a corroboration of the views submitted in the following chapters will find it in the words of the above-named masters. For ready reference, some of the books are mentioned in the Appendices.

I do not know why *Hind Swaraj* has been seized in India. To me, the seizure constitutes further condemnation of the civilization represented by the British Government. There is in the book not a trace of approval of violence in any shape or form. The methods of the British Government are, undoubtedly, severely condemned. To do otherwise would be for me to be a traitor to Truth, to India, and to the Empire to which I own allegiance. My notion of loyalty does not involve acceptance of current rule or government, irrespective of its righteousness or otherwise. Such notion is based upon the belief—not in its present justice or morality but—in a future acceptance by Government of that standard of morality in practice which it at present vaguely and hypocritically believes in, in theory. But I must frankly confess that I am not so much concerned about the stability of the Empire as I am about that of the ancient civilization of India which, in my opinion, represents the best that the world has ever seen. The British Government in India constitutes a struggle between the Modern Civilization, which is the Kingdom of Satan, and the Ancient Civilization, which is the Kingdom of God. The one is the God of War, the other is the God of Love. My countrymen impute the evils of modern civilization to the English people and, therefore, believe that the English people are bad, and not the civilization they represent. My countrymen, therefore, believe that they should adopt modern civilization and modern methods of violence to drive out the English. *Hind Swaraj* has been written in order to show that they are following a suicidal policy, and that, if they would but revert to their own glorious civilization, either the English would adopt the latter and become Indianised or find their occupation in India gone.

It was at first intended to publish the translation as a part of *Indian Opinion*, but the seizure of the original rendered such a course inadvisable. *Indian Opinion* represents the Transvaal

Passive Resistance struggle and ventilates the grievances of British Indians in South Africa generally. It was, therefore, thought desirable not to publish through a representative organ views which are held by me personally and which may even be considered dangerous or disloyal. I am naturally anxious not to compromise a great struggle by any action of mine which has no connection with it. Had I not known that there was a danger of methods of violence becoming popular, even in South Africa, had I not been called upon by hundreds of my countrymen, and not a few English friends, to express my opinion on the Nationalist movement in India, I would even have refrained, for the sake of the struggle, from reducing my views to writing. But, occupying the position I do, it would have been cowardice on my part to postpone publication under the circumstances just referred to.

Preface to 'Indian Home Rule'
Indian Opinion, 2 Apr. 1910

[1] Herman Kallenbach.

90. KNOWING THE FUNDAMENTALS

Johannesburg,
[29 March 1910]

Chi. Narandas,

I have received your letter.

I can appreciate your inability to come here without respected Khushalbhai's[1] permission. It is your duty to act according to his wishes.

You can assist the objectives of our struggle here even if you remain there. Now that *Hind Swaraj* has been proscribed, I see that a strenuous fight will have to be put up there too. To do that you must build your character. Do you know the fundamental principles of our religion? You will, perhaps, say that you are able to recite the whole of the *Gita* and also know its meaning and wonder why I am asking you about fundamental principles. Knowing the fundamentals, as I interpret it, means putting them into practice. The first attribute of the divine heritage is 'fearlessness'. I hope you remember that verse.[2] Have you attained to the state of 'fearlessness' to any extent? Will you do what is right fearlessly, even at the cost of your life? Practise fearlessness and try to attain that state till you succeed. You will be able to do a lot if you achieve that. In this context you should remember the lives of Prahlad, Sudhanva, and others. Please do not think that all these are legends. There have been many Indians in the past who have done such deeds and that is why we memorize the stories of their lives. We should not think that Prahlad and Sudhanva, Harishchandra and Shravana do not exist in India even today. We shall meet them when we deserve. They are not to be found in the *chawls* of Bombay. You cannot expect a wheat crop from rocky soil. I shall not write more. Do ponder over the attributes of the divine heritage again. Read this letter bearing in mind those attributes and then try to act accordingly. Do read afresh the chapters on *satyagraha* in Hind Swaraj and ponder over them. Do ask me any questions when you feel like doing so. You may live in Bombay but be quite sure that Bombay is a veritable hell, absolutely useless.

Blessings from
MOHANDAS

Letter to Narandas Gandhi (G.)
CW 4925

[1] Gandhi's cousin, the addressee's father.
[2] *Bhagavad Gita*, XVI. 1-3.

91. REQUEST TO TOLSTOY

Johannesburg,
Transvaal,
South Africa,
April 4, 1910

Dear Sir,

You will recollect my having carried on correspondence with you whilst I was temporarily in London. As a humble follower of yours, I send you herewith a booklet which I have written. It is my own translation of a Gujarati writing. Curiously enough, the original writing has been confiscated by the Government of India. I, therefore, hastened the above publication of the translation. I am most anxious not to worry you, but, if your health permits it and if you can find the time to go through the booklet, needless to say I shall value very highly your criticism of the writing. I am sending also a few copies of your *Letter to a Hindoo*, which you authorised me to publish. It has been translated in one of the Indian languages also.

I am,
Your obedient servant
M. K. GANDHI

Count Leo Tolstoy
Yasnaya Polyana
Russia

Letter to Leo Tolstoy
D. G. Tendulkar, *Mahatma*, Vol. I

92. PANIC OVER *HIND SWARAJ*

The Bombay *Government Gazette* of the 24th March last notifies that *Hind Swarajya, Universal Dawn, Mustafa Kamel Pasha's Speech*, and *Defence of Socrates or The Story of a True Warrior*—all publications of the International Printing Press—have been forfeited to His Majesty for the reason that they 'contain matter declared to be seditious'.

Hind Swarajya, in the shape of *Indian Home Rule*, is before our readers. *Universal Dawn* is a Gujarati rendering of Ruskin's *Unto This Last*. *Mustafa Kamel Pasha's Speech* is a Gujarati translation of the Egyptian patriot's speech delivered just before his death before a vast audience at Cairo. *Defence of Socrates or The Story of a True Warrior* is a Gujarati rendering of Plato's immortal work printed in order to illustrate the virtue and the true nature of passive resistance. All these publications, except *Hind Swarajya*, have been before the public for a considerable time. They are intended to impart a lofty, moral tone to the reader and are, in our opinion, works capable of being put into children's hands without any danger whatsoever.

But we have no right to complain. We consider this activity on the part of the Government of India a passing phase. They are in a state of panic and, wishing to do something, they intend to stop the circulation of literature that shows the slightest independence of spirit. This over-zeal is bound to kill itself. The really dangerous publications will seek all kinds of dubious and devious methods of circulation, and we fear that they will, on that account, be read by the very class whom the Government intend that they do not reach.

In these circumstances, we, who are uncompromising advocates of passive resistance, have only one course left open to us. The repression cannot affect us. Our views can only remain the same, and they must find expression on every due occasion, regardless of personal consequences.

We sympathize with the Government of India in their desire to stop the spread of methods of violence. We would do and give much to stop it. But the only way we know to eradicate the disease is to popularize passive resistance of the right stamp. Any other way, especially repression, must inevitably fail in the long run.

'Our Publications'
Indian Opinion, 7 May 1910

93. AIM OF *HIND SWARAJ*[1]

I wrote *Hind Swaraj* in 1909 on board the ship during the return voyage from England. Copies of the book having been confiscated in the Bombay Presidency, I published a translation in 1910. It is now five years since the ideas in it were expressed in public. And during this period, many persons have had discussions with me on those ideas. Englishmen and Indians, both have written to me about them, several expressing dissent. At the end of it all, however, I find that the convictions I stated in the book have grown stronger. If only I had the time, I could set forth the same ideas more elaborately, with additional arguments and illustrations. I see no reason at all for revising them.

Numerous requests having been received for a second edition of *Hind Swaraj*, the inmates of Phœnix and the pupils there have found time to print it as a labour of love.

I should like to comment on one thing only. I have gathered an impression that, though *Hind Swaraj* does not advocate the use of physical force at any time and in any circumstances, and advocates always the use of soul-force to gain the desired end, the result of its teaching has been to create hatred for the British and to suggest that they should be expelled through armed fighting or use of violence otherwise. I was unhappy to know this. Such was by no means my object in writing *Hind Swaraj*, and I can only say that those who have drawn from it the foregoing conclusion have totally failed to understand the book.

I, for one, bear no ill-will against the British or against any people or individuals. All living creatures are of the same substance as all drops of water in the ocean are the same in substance. I believe that all of us, individual souls, living in this ocean of spirit, are the same with one another with the closest bond among ourselves. A drop that separates soon dries up and any soul that believes itself separate from others is likewise destroyed. For myself, I am an uncompromising enemy of the present-day civilization of Europe. I tried to elaborate my view in *Hind Swaraj* and show that it is not the British that are responsible for the misfortunes of India but we who have succumbed to modern civilization. India can be free this very moment if we turn our back on this modern civilization and go back to our ancient way of life, which embodied the right ethical principles.

The key to an understanding of *Hind Swaraj* lies in the idea that worldly pursuits should give way to ethical living. This way of life has no room for violence in any form against any human being, black or white.

Preface to *Hind Swaraj* (G.)
Indian Opinion, 29 Apr. 1914

[1] Preface to the second Gujarati edition of *Hind Swaraj* which was published in May 1914.

94. MEANS TO REGENERATION

Bombay,
May 28, 1919

I have re-read this booklet more than once. The value at the present moment lies in re-printing it as it is. But if I had to revise it, there is only one word I would alter in accordance with a promise made to an English friend. She took exception to my use of the word 'prostitute' in speaking of the Parliament. Her fine taste recoiled from the indelicacy of the expression. I remind the reader that the booklet purports to be a free translation of the original which is in Gujarati.

After years of endeavour to put into practice the views expressed in the following pages, I feel that the way shown therein is the only true way to *swaraj*. *Satyagraha*—the law of love—is the law of life. Departure from it leads to disintegration. A firm adherence to it leads to regeneration.

M. K. GANDHI

Foreword
Indian Home Rule (Ganesh & Co., 4th edn.)

95. HIGHER SIMPLICITY AND RENUNCIATION

It is certainly my good fortune that this booklet of mine is receiving wide attention. The original is in Gujarati. It had a chequered career. It was first published in the columns of the *Indian Opinion* of South Africa. It was written in 1908[1] during my return voyage from London to South Africa in answer to the Indian school of violence, and its prototype in South Africa. I came in contact with every known Indian anarchist in London. Their bravery impressed me, but I feel that their zeal was misguided. I felt that violence was no remedy for India's ills, and that her civilization required the use of a different and higher weapon for self-protection. The *satyagraha* of South Africa was still an infant hardly two years old. But it had developed sufficiently to permit me to write of it with some degree of confidence. It was so much appreciated that it was published as a booklet. It attracted some attention in

India. The Bombay Government prohibited its circulation. I replied by publishing its translation. I thought that it was due to my English friends that they should know its contents. In my opinion it is a book which can be put into the hands of a child. It teaches the gospel of love in the place of that of hate. It replaces violence with self-sacrifice. It pits soul-force against brute force. It has gone through several editions and I commend it to those who would care to read it. I withdraw nothing except one word of it, and that in deference to a lady friend. I have given the reason for the alteration in the preface to the Indian edition.

The booklet is a severe condemnation of 'modern civilization'. It was written in 1908. My conviction is deeper today than ever. I feel that if India would discard 'modern civilization', she can only gain by doing so.

But I would warn the reader against thinking that I am today aiming at the *swaraj* described therein. I know that India is not ripe for it. It may seem an impertinence to say so. But such is my conviction. I am individually working for the self-rule pictured therein. But today my corporate activity is undoubtedly devoted to the attainment of parliamentary *swaraj* in accordance with the wishes of the people of India. I am not aiming at destroying railways or hospitals, though I would certainly welcome their natural destruction. Neither railways nor hospitals are a test of a high and pure civilization. At best they are a necessary evil. Neither adds one inch to the moral stature of a nation. Nor am I aiming at a permanent destruction of law courts, much as I regard it as a 'consummation devoutly to be wished for'. Still less am I trying to destroy all machinery and mills. It requires a higher simplicity and renunciation than the people are today prepared for.

The only part of the programme which is now being carried out in its entirety is that of non-violence. But I regret to have to confess that even that is not being carried out in the spirit of the book. If it were, India would establish *swaraj* in a day. If India adopted the doctrine of love as an active part of her religion and introduced it in her politics, *swaraj* would descend upon India from heaven. But I am painfully aware that that event is far off as yet.

I offer these comments because I observe that much is being

quoted from the booklet to discredit the present movement. I have even seen writings suggesting that I am playing a deep game, that I am using the present turmoil to foist my fads on India, and am making religious experiments at India's expense. I can only answer that *satyagraha* is made of sterner stuff. There is nothing reserved and nothing secret in it. A portion of the whole theory of life described in *Hind Swaraj* is undoubtedly being carried into practice. There is no danger attendant upon the whole of it being practised. But it is not right to scare away people by reproducing from my writings passages that are irrelevant to the issues before the country.

'*Hind Swaraj* or the "Indian Home Rule"'
Young India, 26 Jan. 1921

[1] An error for 1909, repeated in the next paragraph.

96. RETURN TO COMMON SENSE

GANDHIJI: I know, I would be overwhelmed with affection if ever I went to America. But as I have already explained to other friends I cannot as yet think of going there, without having finished my work here. I must work away amongst my own people, and not swerve from my path. Dr. Ward writing to me the other day said he was entirely at one with me in thinking that my visit would not be of much use in the present circumstances. And don't you think he is right? I know crowds would gather around me to hear me, I would get receptions everywhere but beyond that my visit would have no other result.

KELLY: Don't you think, Mr. Gandhi, we are ready to receive your message? Look at the gathering under the auspices of the Fellowship of Faiths. No less than ten faiths were represented there, and when a lecture about you was broadcast millions listened to it with intense interest. Mr. John Haines Holmes also earnestly desires you to pay a visit. We are growing, and we would like to accelerate the growth.

I know you are growing. But a gentle, steady growth would be more enduring than growth induced by lecturing campaigns and fireworks display. You must, at present, study my message through my writings and try to live up to it if it is acceptable to you. I could not hope to make you live up to it unless I have succeeded in making my own people do it. Every moment of my time is therefore usefully employed here and I would be doing violence to my inner being if I left work and proceeded to America.

Mrs. Kelly and Mrs. Langeloth seemed to be convinced and they now turned to putting a question or two before they left. 'Mr. Gandhi, is it true that you are a reactionary? I have heard some of your own people say so.'

What do they mean by 'reactionary'? If they mean that I am a civil resister and law-breaker I have been that all these years. If they mean that I have discarded all other methods and adopted non-violence, symbolized by the spinning-wheel, they are right.

Mrs. Kelly could not say, but I could well guess what was at the back of her mind from the questions that followed. In his remarkable autobiography, Henry Ford refers to a species of reformers whom he calls 'reactionary' meaning thereby those who want to go back to an old order of things. And Mrs. Kelly's next question was, 'Is it true that you object to railways, steamships and other means of speedy locomotion?'

It is and it is not! You should really get the book in which I have expounded my views in this connection—*Indian Home Rule*. It is true in the sense that under ideal conditions we should not need these things. It is not true in the sense that in these days it is not easy to sever ourselves from those things. But is the world any the better for those quick instruments of locomotion? How do these instruments advance man's spiritual progress? Do they not in the last resort hamper it? And is there any limit to man's ambition? Once we were satisfied with travelling a few miles an hour, today we want to negotiate hundreds of miles in an hour, one day we might desire to fly through space. What will be the result? Chaos—we would be tumbling upon one another, we would be simply smothered.

But do the masses desire these things?

They do. I have seen mobs getting almost mad on Sundays and holidays. In London a long unbreakable train of motor cars at every corner is quite a usual phenomenon. And what is all this worry and fateful hurry for? To what end? I tell you if by some sudden catastrophe all these instruments were to be destroyed I would not shed a single tear. I would say, it is a proper storm and a proper cleansing.

But supposing you need to go to Calcutta, how would you go unless by train?

Certainly by train. But why should I need to go to Calcutta? Under ideal conditions, as I have said, I need not traverse those long distances, not at any rate in the shortest possible time. I shall explain myself. Today two good people come from America with a kind and loving message. But along with the two come two hundred with all sorts of motives. For aught we know a large number may be coming just in search of further avenues of exploitation. Is that the benefit of quick locomotion to India?

I see, but how can we get back to the ideal condition of things?

Not easily. It is an express moving at a terrific speed that we are in. We cannot all of a sudden jump out of it. We cannot go back to the ideal state all at a jump. We can look forward to reaching it some day.

In short, the reactionary turn, if at all it was, meant a return to common sense, meant a restoration of what appears to common sense to be a natural order as distinguished from the present unnatural order, in a word not everything overturned or everything petrified but everything restored to its proper place.
But I do not think the friends quite saw the drift of the argument. For they too were hurrying through space. They had to catch a train, and were afraid to get to the station too late!

Interview with Mrs Langeloth and Mrs Kelly
Young India, 21 Jan. 1926

97. RELEVANCE OF *HIND SWARAJ*

Segaon,
[14 July 1938]

I welcome your advertising the principles in defence of which *Hind Swaraj* was written.[1] The English edition is a translation of the original which was in Gujarati. I might change the language here and there if I had to rewrite the booklet. But after the stormy thirty years through which I have since passed, I have seen nothing to make me alter the views expounded in it. Let the reader bear in mind that it is a faithful record of conversations I had with workers, one of whom was an avowed anarchist. He should also know that it stopped the rot that was about to set in among some Indians in South Africa. The reader may balance against this the opinion of a dear friend, who, alas, is no more, that it was the production of a fool.

M. K. GANDHI

'A Message to *Aryan Path*'
Aryan Path, Sept. 1938

[1] A special issue of the *Aryan Path* exclusively devoted to *Hind Swaraj* came out in September. The contributors included such eminent Western thinkers as Frederick Soddy, G. D. H Cole, C. Delisle Burns, John Middleton Murry, Hugh I'Anson Fausset, Gerald Heard, and Irene Rathbone. Though none of them was wholly in agreement with the thesis advanced in *Hind Swaraj*, they all nevertheless saw the importance of the work. Frederick Soddy, for instance, though disagreeing with Gandhi, suggested that 'anyone who wishes to change the world would do well to study' the book. To G. D. H. Cole Gandhi's case against the West looked 'infinitely stronger' than it did in 1908, when 'the ephemeral civilization of the West looked stable under Western eyes'. C. Delisle Burns saw the highest merit of *Hind Swaraj* 'in its emphasis upon moral issues and opposition to the pursuit of private wealth and power'. John Middleton Murry considered *Hind Swaraj* a 'great book', 'one of the spiritual classics of the world', and compared Gandhi's vision of real *swaraj* to the Christian vision of the Kingdom of Heaven. Gerald Heard thought *Hind Swaraj* superior to Rousseau's *Social Contract* and Karl Marx's *Das Kapital*, for it did not mark the end of an age, but the beginning of a new order. Irene Rathbone found the book 'enormously powerful': its language forced her 'by its tremendous honesty to search my own honesty'. She thought the work 'simple and logical, economical, condensed, poetic'. Hugh Fausset summed up *Hind Swaraj* as a book whose purpose was 'to save India, not from Englishmen, but from the modern civilization which is eating into the vitals of the West'. He had enough trust in the indestructible vitality of the human spirit to be sure that it would not allow itself to be lastingly mechanized. The machine was

exposing ruthlessly the consequences that must happen to men and nations who had lost their integrity. *Hind Swaraj* was an attempt at restoring man to his true estate, and was, for this reason, 'one of the best modern handbooks of that real revolution which must happen in us all, if we are to fulfil the creative purpose of life'.

98. DRAWING THE LINE

Panchgani,
June 14, 1945

Chi. Krishnachandra,

I have your long letter from Nagpur. As for Balkoba, all that can be said is that there has been no steady gain from *parpati*. Now we have to see how he can regain his strength. It will be good if he does not find it too hot to be taken back into the room.

I still abide by whatever I wrote about railways, etc., in *Hind Swaraj*. But that applies to an ideal state. It is possible that we may never reach that state. Let us not worry about it. It is for this reason that I have said that if we do not have railways and other such facilities, we should not feel unhappy. We should never make it our duty to multiply such facilities. At the same time we should also not make a duty of giving up these things. We should have a free and easy attitude in such matters. We must use these facilities as little as possible. There will be all types of people in our society. There certainly are today. We have to live with them. Non-attachment is the only proper *dharma* under these circumstances. The only thing we must be careful about is that we do not deceive ourselves. Your statement that trains, etc., should be shunned even as theft, adultery and falsehood is not correct. The important reason for this is that even society considers theft, etc., to be immoral. Trains, etc., have not been, nor need they be so considered. All that we may say is that we should not consider trains, etc., as means of enjoyment. I have repeatedly pointed out in my articles where to draw the line. Read them and if you give a little thought you will easily be able to draw the line.

Some small books have been written as an aid to the study of *Das Kapital*. Reading them would help.

Do not worry about the discharge. You can stop worrying by intelligently observing the rules [of *brahmacharya*].

Blessings from
BAPU

Letter to Krishnachandra (H.)
GN 4515

99. EMPHASIS ON ESSENTIALS

October 5, 1945

Chi. Jawaharlal,

I have long been intending to write to you but can do so only today. I have also been wondering whether I should write in English or Hindustani. In the end I have decided to write in Hindustani.

I take first the sharp difference of opinion that has arisen between us. If such a difference really exists people should also know about it, for the work of *swaraj* will suffer if they are kept in the dark. I have said that I fully stand by the kind of governance which I have described in *Hind Swaraj*. It is not just a way of speaking. My experience has confirmed the truth of what I wrote in 1909. If I were the only one left who believed in it, I would not be sorry. For I can only testify to the truth as I see it. I have not *Hind Swaraj* in front of me. It is better that I redraw the picture today in my own language. Then it would not matter to me whether or no the picture tallies with that of 1909, nor should it to you. I do not have to establish what I had said before. What is worth knowing is only what I have to say today.

I believe that if India, and through India the world, is to achieve real freedom, then sooner or later we shall have to go and live in the villages—in huts, not in palaces. Millions of people can never live in cities and palaces in comfort and peace. Nor can they do so by killing one another, that is, by resorting to violence and untruth. I have not the slightest doubt that, but for the pair, truth and non-violence, mankind will be doomed. We can have the vision of that truth and non-violence only in the simplicity of the villages. That simplicity resides in the spinning-wheel and what is implied by the spinning-wheel. It does not frighten me at all that the world seems to be going in the opposite direction. For the matter of that, when the moth approaches its doom it whirls round faster and faster till it is burnt up. It is possible that India will not be able to escape this moth-like circling. It is my duty to try, till my last breath, to save India and through it the world from such a fate. The sum and substance of what I want to say is that the individual person should have control over the things that are necessary for the sustenance of life. If he cannot have such control the individual cannot survive. Ultimately, the world is made up only of individuals. If there were no drops there would be no ocean. This is only a rough and ready statement. There is nothing new in this.

But even in *Hind Swaraj* I have not said all this. While I appreciate modern thought, I find that an ancient thing, considered in the light of this thought, looks so sweet. You will not be able to understand me if you think that I am talking about the villages of today. My ideal village still exists only in my imagination. After all every human being lives in the world of his own imagination. In this village of my dreams the villager will not be dull—he will be all awareness. He will not live like an animal in filth and darkness. Men and women will live in freedom, prepared to face the whole world. There will be no plague, no cholera and no smallpox. Nobody will be allowed to be idle or to wallow in luxury. Everyone will have to do body labour. Granting all this, I can still envisage a number of things that will have to be organized on a large scale. Perhaps there will even be railways and also post and telegraph offices. I do not know what things there will be or will not be. Nor am I bothered about it. If I can make sure of the essential thing, other things will follow in due course. But if I give up the essential thing, I give up everything.

The other day, at the final day's meeting of the Working Committee, we had taken a decision to the effect that the

Working Committee would meet for two or three days to work out this very thing. I shall be happy if it meets. But even if it does not meet, I want that we two should understand each other fully. And this for two reasons. Our bond is not merely political. It is much deeper. I have no measure to fathom that depth. This bond can never be broken. I therefore want that we should understand each other thoroughly in politics as well. The second reason is that neither of us considers himself as worthless. We both live only for India's freedom, and will be happy to die too for that freedom. We do not care for praise from any quarter. Praise or abuse are the same to us. They have no place in the mission of service. Though I aspire to live up to 125 years rendering service, I am nevertheless an old man, while you are comparatively young. That is why I have said that you are my heir. It is only proper that I should at least understand my heir and my heir in turn should understand me. I shall then be at peace.

One thing more. I had written to you about Kasturba Trust and Hindustani. You had said you would write after thinking things over. I find that your name is already figuring in the Hindustani Sabha. Nanavati reminded me that he had approached you and Maulana Saheb and that you had appended your signature. That was in 1942. That was long ago. You know where Hindustani stands today. If you still stand by that signature, I wish to get some work out of you in this regard. It will not involve much running about, but some work will be called for.

The work of Kasturba Memorial Trust is rather complicated. I realize that if what I have said above is going to irk you or is irking you, you will not feel comfortable even in the Kasturba Trust.

The last point concerns the sparks that are flying about in the conflict with Sarat Babu. I have been pained by the episode. I have been unable to trace it to its roots. If what you have told me is all there is to it and nothing more remains to be said, then I do not have to inquire further. But if an explanation seems necessary, I very much want to hear it.

If we have to meet to thrash out all these matters, then we should find time for a meeting.

You are working very hard. I trust you are in good health and Indu is well.

Blessings from
BAPU

Letter to Jawaharlal Nehru (H.)
Gandhi–Nehru Papers

§ 3. Civilization

100. DESTRUCTIVE MATERIALISM

M. K. Gandhi
Agent for The Esoteric Christian Union
& The London Vegetarian Society

Durban,
January 21, 1895

TO
The Editor
The Natal Advertiser

Sir,

You will oblige me by letting me draw the attention of your readers to the notice that appears in your advertisement columns about the Esoteric Christian Union and the London Vegetarian Society.

The system represented by the Union establishes the unity and common source of all the great religions of the world, and points out, as the books advertised will amply show, the utter inadequacy of materialism which boasts of having given the world a civilization which was never witnessed before, and which is alleged to have done the greatest good to humanity, all the while conveniently forgetting that its greatest achievements are the invention of the most terrible weapons of destruction, the awful growth of anarchism, the frightful disputes between capital and labour and the wanton and diabolical cruelty inflicted on innocent dumb, living animals in the name of science, 'falsely so called'.

There seem to be, however, signs of reaction setting in—the almost phenomenal success of the Theosophical Society, the gradual acceptance by the clergy of the doctrine of holiness, and what is more, the acceptance by Professor Max Muller of the doctrine of reincarnation so conclusively demonstrated in *The Perfect Way*, his statement that it was gaining ground among the thinking minds in England and elsewhere, and the publication of *The Unknown Life of Jesus Christ*. It is not possible to secure these works in S. Africa. My knowledge of them is, therefore, confined to their reviews. All these and many such facts are, I submit, unmistakable signs of a return from the materialistic tendencies, which have made us so cruelly selfish, to the unadulterated esoteric teachings of not only Jesus Christ, but also of Buddha, Zoroaster and Mahomed, who are no longer so generally denounced by the civilized world as false prophets, but whose and Jesus's teachings are beginning to be acknowledged to be complementary of one another.

I regret that I am unable yet to advertise books on vegetarianism, as they have, by mistake, been forwarded to India, and will, therefore, take some time before they arrive in Durban. I may, however, state one valuable fact with regard to the efficacy of vegetarianism. There is no more potent instrument of evil than drunkenness, and I may be allowed to say that all those who suffer from the craving for drink, but would like really to be free from the curse, have only to give a trial for at least one month to a diet chiefly consisting of brown bread and oranges or grapes, to secure an entire freedom from the craving. I have myself carried on a series of experiments, and can testify that on a vegetarian diet, without any condiments and consisting of a liberal supply of juicy fresh fruits, I have lived comfortably, without tea, coffee, or cocoa, and even water, for days together. Hundreds in England have become vegetarian for this reason, and having once been inveterate tipplers, have now reached a stage when the very smell of grog or whisky is an offence to their tastes. Dr. B.W. Richardson, in his *Food for Man*, recommends pure vegetarianism as a cure for drunkenness. In a comparatively hot country like Natal, where there is a plentiful supply of fruits and vegetables, a bloodless diet should prove very beneficial in every way, apart from its immeasurable superiority to flesh foods, on grounds scientific, sanitary, economical, ethical, and spiritual.

It is, perhaps, needless to mention that the sale of E.C.U. books is not at all a money-making concern. In certain cases the books have even been given away. They will be gladly lent in some cases. I shall be very happy to correspond with any of your readers who may want any further information, either about the E.C.U. or the L.V.S., or to have a quiet chat on these (to me at any rate) momentous questions.

I would conclude with what Rev. John Pulsford, D.D., has to say with regard to the teaching of the E.C.U.:

It is impossible for a spiritually intelligent reader to doubt that these teachings were received from within the astral veil. They are full of the concentrated and compact wisdom of the Holy Heavens, and of God. If the Christians knew their own religion, they would find in these priceless records Lord Christ and His vital process abundantly illustrated and confirmed. That such communications are possible, and are permitted to be given to the world, is a sign, and a most promising sign, of our age.

I am, etc.
M. K. GANDHI

'Inadequacy of Materialism'
Natal Advertiser, 1 Feb. 1895

101. INSTABILITY OF CIVILIZATION

The catastrophe at Paris[1] must have filled all the portions of the globe where the news reached with gloom. We can well imagine the feelings of the victims and the survivors. To us, these untoward happenings are not merely accidents but we look upon them as divine visitations from which we, if we chose, may learn rich lessons. To us, they show a grim tragedy behind all the tinsel splendour of the modern civilization. The ceaseless rush in which we are living does not leave any time for contemplating the full results of events such as have placed

Paris in mourning for the time being. The dead will be soon forgotten, and in a very short time, Paris will again resume its usual gaiety as if nothing whatsoever had happened. Those, however, who will give the accident, if so it may be called, more than a passing thought, cannot fail to realize that behind all the splendour and behind all the glittering appearances there is something very real which is missed altogether. To us, the meaning is quite clear, namely, that all of us have to live the present life merely as a preparation for a future, far more certain and far more real. Nothing that the modern civilization can offer in the way of stability can ever make any more certain that which is inherently uncertain; that, when we come to think of it, the boast about the wonderful discoveries and the marvellous inventions of science, good as they undoubtedly are in themselves, is, after all, an empty boast. They offer nothing substantial to the struggling humanity, and the only consolation that one can derive from such visitations has to come from a firm faith not in the theory, but in the fact, of the existence of a future life and real Godhead. And that alone is worth having or worth cultivating which would enable us to realise our Maker and to feel that, after all, on this earth we are merely sojourners.

'Accident?'
Indian Opinion, 20 Aug. 1903

[1] A disastrous fire in the underground Electric Railway on 10 Aug. 1903, in which eighty-four persons were killed and many injured.

102. MODERN CIVILIZATION AND ANCIENT WISDOM

[London]
October 14, 1909

My Dear Henry,

I have your cable from Madras. I am sorry Mr. Doke's book is not yet ready. I have just got two advance copies, but I suppose it is not necessary for me to send one out to you. As soon as the copies are ready, I shall ask Mr. Cooper to send 250 to Mr. Natesan.

I enclose herewith cutting from *The Times* reporting the Madras meeting. You will see also a cablegram from Pretoria. I do not know what it means. Surely, negotiations have gone on after Smuts's departure. However, we have to work as if they have fallen through. The report of the Indian Immigration Commission is good at the present juncture. I think that when you are in Calcutta, an attempt should be made to have an All-India deputation to wait on Lord Minto.[1] You may get Sir Charles Turner to join you, though I can well imagine your difficulty, but whether he does or not, there would be no difficulty in having the deputation, and a representative may travel from Madras, Bombay, Allahabad, Lahore, etc. I am writing with reference to your appointment as a delegate to the Congress and the Mahomedan Conference. They will be held, I suppose, about the same time, but if it is the same day, you will have to use your discretion whether you will go to the Mahomedan Conference or whether you will go to the Congress. Along passive resistance lines, it appears to me that the Mahomedan Conference will be the best. I take it, too, that you will go to Aligarh.

I am still creeping. I had thought that there would be a letter from Lord Crewe in reply to mine at once, but up to the time of dictating this (Thursday morning) there is none, and until his authority for publishing the net result of the negotiations is received, I feel that nothing can be done. It is now problematical whether, even if a reply is received this week from him, I can finish the work of education before the 30th inst. As you will be seeing practically the whole of India—a privilege I have myself not yet been able to enjoy— I think I should jot down the definite conclusions to which I have almost arrived after more matured observations made here.

The thing was brewing in my mind, but there was no certain clear light. The heart and brain became more active after I accepted the invitation of the Peace and Arbitration Society to speak to them on 'East and West'. It came off last night. I think this meeting was a splendid success; they were earnest folk, but some insolent questions were put on the South African situation. You will be surprised to learn that even in Hampstead there were men enough to stand up for

Civilization

the tragedy in South Africa, and to talk all the claptrap about the Indian trader being a canker, and what not. A dear old lady got up and said that I had uttered disloyal sentiments and, just as we have to deal with idolaters in South Africa who would think of and cling to form and superficiality as in the case of finger-impressions, so had I last night in the Friends' Meeting House. My main purpose was, in all the questions that were addressed to me, forgotten, and details were warmly taken up and discussed. The following are the conclusions:

(1) There is no impassable barrier between East and West.

(2) There is no such thing as Western or European civilization, but there is a modern civilization, which is purely material.

(3) The people of Europe, before they were touched by modern civilization, had much in common with the people of the East; anyhow, the people of India and, even today, Europeans who are not touched by modern civilization are far better able to mix with the Indians than the offspring of that civilization.

(4) It is not the British people who are ruling India, but it is modern civilization, through its railways, telegraphs, telephones, and almost every invention which has been claimed to be a triumph of civilization.

(5) Bombay, Calcutta, and the other chief cities of India are the real plague spots.

(6) If British rule was replaced tomorrow by Indian rule based on modern methods, India would be no better, except that she would be able then to retain some of the money that is drained away to England; but, then, Indians would only become a second or fifth edition of Europe or America.

(7) East and West can only and really meet when the West has thrown overboard modern civilization, almost in its entirety. They can also seemingly meet when East has also adopted modern civilization. But that meeting would be an armed truce, even as it is between, say, Germany and England, both of which nations are living in the Hall of Death in order to avoid being devoured, the one by the other.

(8) It is simply impertinence for any man or any body of men to begin or contemplate reform of the whole world. To

attempt to do so by means of highly artificial and speedy locomotion is to attempt the impossible.

(9) Increase of material comforts, it may be generally laid down, does not in any way whatsoever conduce to moral growth.

(10) Medical science is the concentrated essence of Black Magic. Quackery is infinitely preferable to what passes for high medical skill.

(11) Hospitals are the instruments that the Devil has been using for his own purpose, in order to keep his hold on his kingdom. They perpetuate vice, misery and degradation, and real slavery.

(12) I was entirely off the track when I considered that I should receive a medical training. It would be sinful for me in any way whatsoever to take part in the abominations that go on in the hospitals.

If there were no hospitals for venereal diseases, or even for consumptives, we should have less consumption, and less sexual vice amongst us.

(13) India's salvation consists in unlearning what she has learnt during the past fifty years.

The railways, telegraphs, hospitals, lawyers, doctors, and such like have all to go, and the so-called upper classes have to learn to live conscientiously and religiously and deliberately the simple peasant life, knowing it to be a life giving true happiness.

(14) Indians should wear no machine-made clothing, whether it comes out of European mills or Indian mills.

(15) England can help India to do this, and then she will have justified her hold of India. There seem to be many in England today who think likewise.

(16) There was true wisdom in the sages of old having so regulated society as to limit the material condition of the people: the rude plough of perhaps five thousand years ago is the plough of the husbandman today. Therein lies salvation. People live long, under such conditions, in comparative peace much greater than Europe has enjoyed after having taken up modern activity, and I feel that every enlightened man, certainly every Englishman, may, if he chooses, learn this truth and act according to it.

There is much more than I can write upon today, but the above is enough food for reflection. You will be able to check me when you find me to be wrong.

You will notice, too, that it is the true spirit of passive resistance that has brought me to the above almost definite conclusions. As a passive resister, I am unconcerned whether such a gigantic reformation, shall I call it, can be brought about among people who derive their satisfaction from the present mad rush. If I realize the truth of it, I should rejoice in following it, and, therefore, I could not wait until the whole body of people had commenced. All of us who think likewise have to take the necessary step; and the rest, if we are in the right, must follow. The theory is there: our practice will have to approach it as much as possible. Living in the midst of the rush, we may not be able to shake ourselves free from all taint. Every time I get into a railway car, use a motor-bus, I know that I am doing violence to my sense of what is right. I do not fear the logical result on that basis. The visiting of England is bad, and any communication between South Africa and India by means of Ocean's grey-hounds is also bad, and so on. You and I can, and may, outgrow those things in our present bodies, but the chief thing is to put our theory right. You will be seeing there all sorts and conditions of men. I, therefore, feel that I should no longer withhold from you what I call the progressive step I have taken mentally.

If you agree with me, it will be your duty to tell the revolutionaries and everybody else that the freedom they want, or they think they want, is not to be obtained by killing people or doing violence, but by setting themselves right, and by becoming and remaining truly Indian. Then the British rulers will be servants and not masters. They will be trustees and not tyrants, and they will live in perfect peace with the whole of the inhabitants of India. The future, therefore, lies not with the British race, but with the Indians themselves, and if they have sufficient self-abnegation and abstemiousness, they can make themselves free this very moment, and when we have arrived in India at the simplicity which is still ours largely and which was ours entirely until a few years ago, it will still be possible for the best Indians and the best Europeans to see one another throughout the length and breadth of India and act as the leaven. When there was no rapid locomotion, traders and preachers went on foot, from one end of the country to the other, braving all the dangers, not for pleasure, not for recreating their health, (though all that followed from their tramps), but for the sake of humanity. Then were Benares and other places of pilgrimage holy cities, whereas today they are an abomination.

You will recollect you used to rate me for talking to my children in Gujarati. I now feel more and more convinced that I was absolutely right in refusing to talk to them in English. Fancy a Gujarati writing to another Gujarati in English, which as you would properly say, he mispronounces and writes ungrammatically. I should certainly never commit the ludicrous blunders in writing in Gujarati that I do in writing or speaking in English. I think that, when I speak in English to an Indian or a foreigner, I in a measure unlearn the language. If I want to learn it well and if I want to attune my ear to it, I can only do so by talking to an Englishman, and by listening to an Englishman speaking.

Now I think I have given you a terrible dose, I hope you will be able to digest it. It is very likely that you with your great imagination and sound common sense have perhaps, in your varied experience there, probably come to the conclusions independently of me. After all, they are not new but they have only now assumed such a concrete form and taken a violent possession of me.

I have just received the following cablegram from Johannesburg:

Smuts tells newspapers he awaits Secretary State's answer regarding his proposals. London Committee continues for present.

This cablegram means that the question is being somewhat agitated in Johannesburg and that Smuts is no longer sanguine about smashing passive resistance. It shows, too, if Lord Crewe made a supreme effort, he could bring about a settlement. However, we can but fight on. So the London Committee continues. This does not alter the situation, and it eases Ritch's position.

Poor Mrs. Ritch will have to undergo another and further

operation. She may not even survive it. It will be a great relief to her if the living death turns into real death.

Later—Millie was here after the foregoing portion of this letter was finished. As I considered it to be so important, I read it to her. This was followed by a fruitful discussion, which you can picture to yourself.

Mr. Ali Imam is still here. He leaves, I believe, on Monday.

Yours sincerely,
M. K. GANDHI

Letter to H.S.L. Polak
SN 5127

[1] (1845-1914); Viceroy and Governor-General of India, 1905-10.

103. CIVILIZATION AND CONSCIENCE

May 10, 1910

Dear Mr. Wybergh,[1]

I am exceedingly obliged to you for your very full and valuable criticism of the little pamphlet on Indian Home Rule. I shall with very great pleasure send your letter to *Indian Opinion* for publication, and shall treat this reply likewise.

I entirely reciprocate the sentiments you express in the last paragraph of your letter. I am quite aware that my views will lead to many differences of opinion between my staunchest friends and those whom I have come to regard with respect and myself, but these differences, so far as I am concerned, can neither diminish respect nor affect friendly relations.

I am painfully conscious of the imperfections and defects you point out in your letter, and I know how unworthy I am to handle the very important problems dealt with in the booklet. But, having had the position of a publicist practically forced upon me by circumstances, I felt bound to write for those for whom *Indian Opinion* caters. The choice lay between allowing the readers of *Indian Opinion*, anxious though they were for guidance, to drift away in the matter of the insane violence that is now going on in India, or giving them, no matter how humble, a lead that they were asking for. The only way I saw of mitigating violence was the one sketched in the pamphlet.

I share your views that a superficial reader will consider the pamphlet to be a disloyal production, and I admit, too, that those who will not distinguish between men and measures, between modern civilization and its exponents, will come to that conclusion. And I accept your proposition that I discourage violence only because I think it to be both wrong and ineffective, and not because the object sought to be attained is wrong, that is to say, if it were ever possible, which I hold it is not, to detach the object from the means adopted to attain it. I hold that Home Rule obtained by violence would be totally different in kind from that obtained by the means suggested by me.

I have ventured utterly to condemn modern civilization because I hold that the spirit of it is evil. It is possible to show that some of its incidents are good, but I have examined its tendency in the scale of ethics. I distinguish between the ideals of individuals who have risen superior to their environment, as also between Christianity and modern civilization. Its activity is by no means confined to Europe. Its blasting influence is now being exhibited in full force in Japan. And it now threatens to overwhelm India. History teaches us that men who are in the whirlpool, except in the cases of individuals, will have to work out their destiny in it; but I do submit that those who are still outside its influence, and those who have a well-tried civilization to guide them, should be helped to remain where they are, if only as a measure of prudence. I claim to have tested the life which modern civilization has to give, as also that of the ancient civilization, and I cannot help most strongly contesting the idea that the Indian population requires to be roused by 'the lash of competition and the other material and sensuous, as well as intellectual, stimuli'; I cannot admit that these will add a single inch to its moral stature. Liberation in the sense in which I have used the term is undoubtedly the immediate aim of all humanity. It does not, therefore, follow that the whole of it can reach it in the same time. But if that liberation is the best thing attainable by mankind, then, I submit, it is wrong to lower the ideal for anyone. All the Indian Scriptures have certainly preached

Civilization

incessantly liberation as an immediate aim, but we know that this preaching has not resulted in 'activity in the lower worlds' being abandoned.

I admit that the term 'passive resistance' is a misnomer. I have used it because, generally speaking, we know what it means. Being a popular term, it easily appeals to the popular imagination. The underlying principle is totally opposed to that of violence. It cannot, therefore, be that 'the battle is transferred from the physical to the mental plane'. The function of violence is to obtain reform by external means; the function of passive resistance, that is, soul-force, is to obtain it by growth from within; which, in its turn, is obtained by self-suffering, self-purification. Violence ever fails; passive resistance is ever successful. The fight of a passive resister is none the less spiritual because he fights to win. Indeed, he is obliged to fight to win, that is, to obtain the mastery of self. Passive resistance is always moral, never cruel; and any activity, mental or otherwise, which fails in this test is undoubtedly not passive resistance.

Your argument tends to show that there must be complete divorce between politics and religion or spirituality. That is what we see in everyday life under modern conditions. Passive resistance seeks to rejoin politics and religion and to test every one of our actions in the light of ethical principles. That Jesus refused to use soul-force to turn stones into bread only supports my argument. Modern civilization is at present engaged in attempting that impossible feat. The use of soul-force for turning stones into bread would have been considered, as it is still considered, as black magic. Nor can I hold with you that motives alone can always decide the question of a particular act being right or wrong. An ignorant mother may, from the purest motives, administer a dose of opium to her child. Her motives will not cure her of her ignorance, nor, in the moral world purge her of the offence of killing her child. A passive resister, recognising this principle and knowing that, in spite of the purity of his motives, his action may be utterly wrong, leaves judgment to the Supreme Being, and, in attempting to resist what he holds to be wrong, suffers only in his own person.

Throughout the *Bhagavad Gita*, I can see no warrant for holding that a man who can only control 'the organs of action' but cannot help 'dwelling in his mind on the objects of the senses' had better use the organs of action until the mind, too, is under control. In ordinary practices, we call such use an indulgence, and we know, too, that, if we can control the flesh even while the spirit is weak, always wishing that the spirit were equally strong, we will certainly arrive at a right correspondence. I think the text you have quoted refers to a man who, for making a show, appears to be controlling the organs of action, whilst deliberately in his mind dwelling on the objects of the senses.

I agree with you entirely that a pure passive resister cannot allow himself to be regarded as a martyr nor can he complain of the hardships of prison or any other hardships, nor may he make political capital out of what may appear to be injustice or ill-treatment, much less may he allow any matter of passive resistance to be advertised. But all action unfortunately is mixed. Purest passive resistance can exist only in theory. The anomalies you point out only emphasize the fact that the Indian passive resisters of the Transvaal are, after all, very fallible human beings and yet very weak, but I can assure you that their object is to make their practice correspond with pure passive resistance as nearly as possible, and, as the struggle progresses, pure spirits are certainly rising in our midst.

I am free to admit also that all passive resisters are not fired with the spirit of love or of truth. Some of us are undoubtedly not free from vindictiveness and the spirit of hatred; but the desire in us all is to cure ourselves of hatred and enmity. I have noticed, too, that those who simply became passive resisters under the glamour of the newness of the movement or for selfish reasons have fallen away. Pretended self-suffering cannot last long. Such men never were passive resisters. It is necessary to discuss the subject of passive resistance somewhat impersonally. If you say that physical sufferings of soldiers have vastly exceeded those of the Transvaal passive resisters, I agree with you entirely; but the sufferings of world-known passive resisters who deliberately walked into funeral pyres or into boiling cauldrons were incomparably greater than those of any soldier it is possible to name.

I cannot pretend to speak for Tolstoy, but my reading of his works has never led me to consider that, in spite of his merciless analysis of institutions organised and based upon force, that is governments, he in any way anticipates or contemplates that the whole world will be able to live in a state of philosophical anarchy. What he has preached, as, in my opinion, have all world-teachers, is that every man has to obey the voice of his own conscience, and be his own master, and seek the Kingdom of God from within. For him there is no government that can control him without his sanction. Such a man is superior to all government. And can it be ever dangerous for a lion to tell a number of other lions who in their ignorance consider themselves to be merely lambs that they, too, are not lambs but lions? Some very ignorant lions will no doubt contest the knowing lion's proposition. There will, no doubt, on that account be confusion also, but, no matter how gross the ignorance may be, it will not be suggested that the lion who knows should sit still and not ask his fellow-lions to share his majesty and freedom.

It has indeed occurred to me that an anti-Asiatic league which from pure though entirely misguided motives wishes to deport Asiatics from the Transvaal, because it may consider them to be an evil, would be certainly justified, from its own view-point, in violently attaining its object. It is not open to passive resisters, if they are not weak, to complain of such, in their opinion, high-handed action, but for them deportation and worse must be a welcome relief from having to submit to a course of action which is repugnant to their conscience. I hope you will not fail to see the beauty of passive resistance in your own illustration. Supposing that these deportees were capable of offering physical violence against forcible deportation, and yet from pure choice elected to be deported rather than resist deportation, will it not show superior courage and superior moral fibre in them?

Yours sincerely,
M. K. GANDHI

Letter to W. J. Wybergh
Indian Opinion, 21 May 1910

[1] Member of the Legislative Assembly, Transvaal.

104. FAITH IN INDIAN CIVILIZATION

Indore,
March 30, 1918

We often think that changes of the kind that take place in Europe will also occur in India; that when some big transformation comes about, people who know beforehand how to prepare themselves for it win through and those who fail to take account of this are destroyed; that mere movement is progress and that our advancement lies in it. We think that we shall be able to progress through the great discoveries that have been made in the continent of Europe. But this is an illusion. We are inhabitants of a country which has so long survived with its own civilization. Many a civilization of Europe is destroyed, but India, our country, survives as a witness to its own civilization. All scholars agree in testifying that the civilization of India is the same today as it was thousands of years ago. But, now, there is reason to suspect that we no longer have faith in our civilization. Every morning we do our worship and prayer, recite the verses composed by our forbears, but we do not understand their significance. Our faith is turning in another direction.

So long as the world goes on, the war between the Pandavas and the Kauravas will also continue. The books of almost all the religions say that the war between the gods and Satan goes on for ever. The question is how we are to make our preparations. I have come here to tell you that you should have faith in your civilization and keep to it steadfastly. If you do this, India will one day hold sway over the entire world.

Our leaders say that, in order to fight the West, we have to adopt the ways of the West. But please rest assured that it will mean the end of Indian civilization. India's face is turned away from your modern trend; that India you do not know. I have travelled much and so come to know the mind of India and I have discovered that it has preserved its faith in its ancient civilization. The *swaraj* of which we hear will not be achieved the way we are working for it. The Congress League Scheme, or any other scheme which is even better, will not get us *swaraj*. We shall get *swaraj* through the way in which we live our lives. It cannot be had for the asking. We can never gain it through copying Europe.

That European civilization is Satanic we see for ourselves. An obvious proof of this is the fierce war that is going on at present. It is so terrible that the Mahabharata War was nothing in comparison. This should be a warning to us and we should remember that our sages have given us the immutable and inviolate principles that our conduct should be godly and that it should be rooted in *dharma*. We should follow these principles alone. So long as we do not follow *dharma*, our wish will not be fulfilled, notwithstanding all the grandiose schemes we may devise. Even if Mr. Montagu offers us *swaraj* today we can in no way benefit from that *swaraj*. We must make use of the legacy left us by our *rishis* and *munis*.

The whole world knows that the *tapasya* that was practised in ancient India is found nowhere else. Even if we want an empire for India, we can get it through no other method but that of self-discipline. We can be certain that once the spirit of discipline comes to pervade our lives, we shall be able to get anything we may want.

Truth and non-violence are our goal. Non-violence is the supreme *dharma*, there is no discovery of greater import than this. So long as we engage in mundane actions, so long as soul and body are together, some violence will continue to occur through our agency. But we must renounce at least the violence that it is possible for us to renounce. We should understand that the less violence a religion permits, the more is the truth contained in it. If we can ensure the deliverance of India, it is only through truth and non-violence. Lord Willingdon, the Governor of Bombay, has said that he feels greatly disappointed when he meets Indians for they do not express what is in their minds but only what would be agreeable to him, so that he never knows the real position. Many people have this habit of hiding their own sentiments when in the presence of an important person and suiting their talk to his pleasure. They do not realize how cruelly they deceive themselves and harm the truth. One must say what one feels. It is impertinence to go against one's reason. One must not hesitate the least to tell what one feels to anyone, be he a

Minister of the Government or even a more exalted person. Deal with all with truth and non-violence.

Love is a rare herb that makes a friend even of a sworn enemy and this herb grows out of non-violence. What in a dormant state is non-violence becomes love in the waking state. Love destroys ill will. We should love all—whether Englishmen or Muslims. No doubt, we should protect the cow. But we cannot do so by fighting with Muslims. We cannot save the cow by killing Muslims. We should act only through love; thus alone shall we succeed. So long as we do not have unshakeable faith in truth, love and non-violence, we can make no progress. If we give up these and imitate European civilization, we shall be doomed. I pray to Suryanarayan that India may not turn away from her civilization. Be fearless. So long as you live under various kinds of fears, you can never progress, you can never succeed. Please do not forget our ancient civilization. Never, never give up truth and love. Treat all enemies and friends with love. If you wish to make Hindi the national language, you can do so in a short time through the principles of truth and non-violence.

Speech on Indian Civilization (H.)
Mahatma Gandhi

105. ANCIENT GLORY AND PRESENT INERTIA

Editors often have something or other ready at hand to fill space in newspapers. In English this is called 'evergreen', that is, always fresh. You can publish the matter any time. I unexpectedly came across something like this in *The [Bombay] Chronicle*. It contains the following information.

The Hindus invented the decimal system. Geometry and Algebra were first developed in India, and so too Trigonometry. The first five hospitals to be built were in India. The physicians of ancient Europe used Indian drugs. Hindus investigated the anatomy of the human body in the sixth century B.C. and about the same time acquired the art of surgery. People in ancient India knew the art of casting iron pillars of the same kind which they make now. India specialized in carving caves. Alexander, when he invaded India, found republican states in the Punjab and in Sind. In ancient India, women enjoyed all those rights for which women of Europe are fighting hard at present. Municipalities were in existence in the times of Chandragupta. It was the Hindus who perfected the science of grammar. The *Ramayana* and the *Mahabharata* still remain unrivalled.

I do not know how far these statements are true, but this I know, that, if the late Justice Ranade were alive today and heard such talk of India's past glory, he would certainly have asked, 'So what?' He used to say that no people could progress by merely dwelling on its past glory. If at all we do so, it should be only in order that we may be able to add to it. Where is the man who can write the *Ramayana* today? Where are the morals of ancient times? Where is the ability of those days? And devotion to duty? We have added nothing to the drugs discovered thousands of years ago, nor do we even have adequate knowledge about those mentioned in the ancient books. We witness the same poverty in respect of all the other gifts mentioned above, borrowing as we do everything from Europe. I, at any rate, feel that so long as we have not revived in the present the glory which was ours in the past, wisdom lies in not speaking about it. That wealth which has no exchange value, which the world does not recognize as such, brings not credit, but only humiliation and is in the nature of a burden. If, as we believe, we had these gifts in ancient times, we ought to be able to give evidence of them again. We are indeed heirs of a brave people, but by confessing at the same time our inability to be worthy of that heritage we shall achieve nothing. We shall see hereafter how we can become so worthy.

'Living on the Past' (G.)
Navajivan, 20 June 1920

106. COWARDICE AND HYPOCRISY IN INDIA

As a son cannot live on his father's reputation for long, so the people of India cannot maintain their prosperity only on the strength of the glory of ancient India. We saw last week that at present there is poverty, not prosperity, in India.

We have to think over the causes and remedies for this state of affairs.

Akbar's successors lost the splendour of the Mogul Empire of his time, because they lost, one by one, Akbar's qualities of character. Jehangir lost one, Shahjehan one more, Aurangzeb more still and his successors lost almost all. The result was that they lost the Empire to the British. The Indian people in modern times have behaved like Akbar's successors.

Not wanting to admit this, we blame the British for everything; we fell because of their cunning, they robbed us of our wealth and left us beggars, we cannot even breathe without their permission; how are we to blame then?

Though there is much exaggeration in this charge, there is also some truth in it. What is the cause of the control the British have acquired over us? May it not be our own fault? Who were the people tempted by the rupees of the East India Company? If that Company did business according to its own fashion, does the fault lie with it? If a wine merchant sells wine, can the consumer throw the responsibility on him? If I pay to the usurer interest equal to the principal, how is that the usurer's fault? I, at any rate, cannot speak ill of him. A writer has said that as long as there are persons who can be deceived, so long will deceivers exist.

We shall not advance by finding fault with and hating the British. As long as we have not got rid of the shortcomings which enabled the British to get a hold here, we shall remain slaves.

And yet we are always pointing out their faults to the British and will continue to do so. This is what the Congress has mainly done. Speakers who do this are as numerous as leaves on a tree. I believe, therefore, that it will be more fruitful to try and see our own faults than to dwell on those of the British. The saying 'If we are good, the entire world is good' is not to be lightly brushed aside. There is much force in it. If we remain upright, nobody will be able to corrupt us. It is a principle of medical science that so long as one's blood is free from impurity, the poisonous air outside can have no effect on it. That is why, during an epidemic, some people are attacked while others are not. Likewise, had we been incorruptible, the East India Company could have done nothing and at the present time, too, officers like Michael O'Dwyer would have lost their jobs.

What are our failings, then, because of which we are helpless and cannot stop the profuse flow of wealth from our country, and in virtue of which our children get no milk, three crores of our people get only one meal a day, raids occur in broad daylight in Kheda district, and epidemics like the plague and cholera cannot be eradicated in our country while they can in others? How is it that the haughty Sir Michael O'Dwyer and the insolent General Dyer can crush us like so many bugs and the priest in Simla can write unworthy things about us; how is it that an intolerable injustice has been done to us in the Punjab? On the Khilafat issue, the British Prime Minister has gone back on his word. In both these matters, we seem to feel helpless.

The reason is our inveterate selfishness, our inability to make sacrifices for the country, our dishonesty, our timidity, our hypocrisy and our ignorance. Everybody is selfish, more or less, but we seem to be more selfish than others. We make some self-sacrifice in family matters, but very little of it for national work. Just look at our streets, our cities and our trains. In all these, we can see the condition of the country. How little attention is paid to the convenience of others in streets, in the town as a whole and in trains? We do not hesitate to throw refuse out of our courtyard on to the street; standing in the balcony, we throw out refuse or spit, without pausing to consider whether we are not inconveniencing the passers-by. When we are building a house, we take little thought of the inconvenience that may be caused to our neighbours. In cities, we keep the tap open and, thinking that it is not our water which flows away, we allow it to run waste. The same thing is seen in the trains. We secure a seat for ourselves by hook or by crook and, if possible, prevent others from getting in. No matter if others are inconvenienced, we start smoking. We do not hesitate to throw banana skins and sugar-cane peelings right in front of our neighbours. When we go to draw water from a tap, we take little thought for others. Many such instances of our selfishness can be listed.

Where so much selfishness exists, how can one expect self-sacrifice? Does the business man cleanse his business of dishonesty for the sake of his country? Does he forgo his profit? Does he stop speculation in cotton for his country's sake? Is any effort made to keep down milk prices by giving up the profit from its export? How many give up a job when necessary, for the sake of the country? Where are the men who will reduce their luxuries and adopt simplicity and use the money so saved for the country? If it is necessary for the country's sake to go to jail, how many will come forward?

Our dishonesty is there for all to see. We believe that business can never be carried on honestly. Those who have the chance never refuse a bribe. We have the worst experience of corruption in the railways. We can get our work done only if we bribe the railway police, the ticket master and the guard. Even for securing a railway ticket, we have to use dishonest means or shut one's eyes to them. The contents of railway parcels which can be opened ever so slightly, if not of those which are well-packed, are sure to be pilfered.

Our hypocrisy is only a little less than that of the British. We have experience of this every moment. In our meetings and in all other activities of ours, we try to show ourselves other than what we are.

We have made cowardice especially our own. Nobody wants bloodshed in connection with non-co-operation, and yet it is out of this fear of bloodshed that we do not want to do anything. We are so possessed by the fear of the Government's armed might that we dare not take any step. And so we submit to force in every matter and allow dacoits to plunder us in broad daylight.

What shall I say about our hypocrisy? It has increased in every field. Weakness is always accompanied by hypocrisy. Moreover, where the people want to be upright but cannot be so, hypocrisy will naturally increase; for, if we are not upright, we are anxious to seem so and thus we add another moral weakness to the one which we already possess. Hypocrisy has entered our religion as well, and that so fully that the marks which we put on our forehead, the rosary and things of that kind have ceased to be tokens of piety and become signs of impiety.

The origin of all is ignorance, to be sure. It is because we are ignorant of our strength that other weaknesses grow. We doubt the very existence of the *atman* in us, have no faith in its powers. This ignorance will not disappear merely with education. It can go only with a change in our ways of thinking. Literacy is necessary only to the extent that it develops our thinking power and teaches us to distinguish between good and evil.

Hence, so long as we have not given up our selfishness and learnt to be mindful of the interests of others, have not learnt self-sacrifice, have not taken refuge in truth, eschewed fear and become brave, shed hypocrisy and banished ignorance, the country will not prosper in any real sense.

'Living on the Past' (G.)
Navajivan, 27 June 1920

107. CIVILIZATION AND EDUCATION

[15 February 1925]

The Thakore Saheb expounded fine ideas about education, but was pessimistic about being able to put them into effect in a small State like this. But there is no justification for his pessimism. In fact a State enjoys many advantages in being small. The people of Rajkot are such that it should be easy to win their co-operation. The smaller countries of Europe, like Sweden, Norway and Switzerland—countries about which the world has not heard much because they were not involved in the last War—boast a civilization in no way inferior to that of the bigger countries; they have carried out many successful experiments in the field of education. The problems of big countries are also big. I can well understand what difficulties a person in the position of Lord Reading generally has to contend with. With so many parties and interests to consider, and the field of work so large, what effective work can be done? It is, therefore, only in small countries that promising plans can be easily implemented. The Gujarat Vidyapith has been, in a way, doing what the Thakore Saheb explained.

If we run one model institution with ideal students, many more institutions of the same kind will grow out of it. A zero can produce nothing, for it cannot be multiplied, whereas one can grow into many. There is, therefore, no cause for despair. The cause of despair generally lies in the man himself. The *atman* is its own friend and its own enemy.[1] We may not put any limits to what human effort can achieve. There can be any such limits only if we can see above us any obstruction to upward flight. For rising the sky is the only limit. There is a limit to how low we can fall. God Himself has created such limit in the form of land, rock, water, etc. Hence we need not despair. I advise the people to take the fullest advantage of the Ruler, and should like to say to the latter that he has done much, but much more needs to be done.

The Ruler and the subjects should develop the completest mutual understanding and confidence. If it is true to say 'As the Ruler, so the subjects', it is equally true to say 'As the subjects, so the Ruler.' If you yourselves do nothing, however much the Ruler may wish to help, he, too, can do nothing. If you let hypocrisy, flattery and wickedness govern your lives, the Ruler's life is bound to mirror those evils. I have to mention this because the saying, 'Salt is better even than honey', is still true.

Speech at Jain Students' Hostel, Rajkot (G.)
Navajivan, 1 Mar. 1925

[1] *Bhagavad Gita*, VI.5.

108. SCIENCE AND CIVILIZATION

[13 March 1925]

It is a common superstition in India, and more so outside India—because, that is what I find from my correspondence in Europe and America—that I am an opponent, a foe, of science. Nothing can be farther from truth than a charge of this character. It is perfectly true, however, that I am not an admirer of science unmixed with something that I am about to say to you. I think that we cannot live without science, if we keep it in its right place. But I have learnt so much during my wanderings in the world about the misuse of science that I have often remarked, or made such remarks, as would lead people to consider that I was really an opponent of science. In my humble opinion there are limitations even to scientific search, and the limitations that I place upon scientific search are the limitations that humanity imposes upon us.

I was only the other day discussing with a friend on the uses of science, and at that time I told him a story of my life which I propose to repeat to you. I told him that there was a time in my life when I very nearly went in for medicine and I told him also that had I gone in for it, probably, I would have become a celebrated physician or a celebrated surgeon or both; because really I am a lover of both these branches and I feel that I could have rendered a great deal of service in that department. But when I understood from a medical friend—and he was a distinguished doctor—that I would have to practise vivisection I recoiled with horror from it.

Probably, some of you will laugh at my horror, but I do not want you to laugh at it. I want you to consider carefully what I am really saying. I feel that we are placed on this earth to adore our Maker, to know ourselves, in other words, to realize ourselves and therefore to realize our destiny. Vivisection cannot add, in my opinion, an inch to our moral height. It may—though many medical men tell me that it is not an absolutely correct statement—it may, I say, bring in some relief to a man whose body is ailing. But I must honestly confess to you that I believe in placing limitations upon the remedies for keeping the body alive. After all it is a broken weed to rely upon. It may slip out of our hands at any moment. I recovered from the skilful handling of Col. Maddock[1] from the operation that was performed by him upon me. But there was no guarantee whatsoever that after my recovery I may not fall under a stroke of lightning or under some other accident. Such being the case, I feel that we have got to explore whether we should restrain ourselves or whether we might let ourselves go.

I have only given you one illustration of the limitation that I would place upon scientific research and upon the uses of science. Therefore, I would simply say—as I have said to so many of the students of India, and I have the good fortune to enjoy the confidence of the student world and the good fortune to come in contact with the thousands and thousands of students all over India, and therefore I would not hesitate to tell them—that they must make up their minds about one thing at least in life, viz., to understand what they are in this world for. I place the same view in all humility before professors and teachers and it is for that reason that I have so often written and spoken upon and against the materialistic tendency of modern civilization—I will not say Western civilization though as it so happens for the time being, the two have become convertible terms. But there is another aspect also which I would like to place before you. Many students go in for science not for the sake of knowledge but for the sake of livelihood that their scientific studies might give them. It is true not only for students belonging to colleges of science, but it is equally true of students belonging to any other college. But seeing that science is one of the few things in which you have to go in for accuracy of thought and accuracy of handling, the warning that I wish to utter to you will perhaps come home to you with greater force than to others.

I would like you to keep the two most brilliant examples we have in our own dear country, and those two are Drs. J. C. Bose[2] and P. C. Ray.[3] At least to the students of science, they must be household words, household names. I believe that they are household names to the whole of the educated India. They went in for science for the sake of science and we know what they have achieved. They never thought of what the profession of science would bring them in the shape of money or fame. They cultivated it for the sake of it and Sir J. C. Bose once told me that he had accepted the limitations for himself long before I had uttered a single word about how we should apply our minds to science and I speak upon his authority that all his researches have been devoted in order to enable us to come nearer our Maker.

But students in India labour under one very serious disability. Those who go in for this class of education or for higher education are drawn from the middle class. Unfortunately for us and unfortunately for our country, the middle classes have almost lost the use of their hands and I hold it to be utterly

Civilization

impossible for a boy to understand the secrets of science or the pleasures and the delights that scientific pursuits can give, if that boy is not prepared to use his hands, to tuck up his sleeves and labour like an ordinary labourer in the streets.

I well remember the classes that I used to attend on chemistry. It seemed to me then to be one of the dullest subjects. I know now what an interesting subject it is. Although I am an adorer of all my teachers, I must confess to you that the blame was not mine but that of my teacher. He asked me to learn by heart all those awful sounding names without knowing what they were. He never agreed to place even the different metals before me. I had simply to learn things by heart. He brought frightful notes carefully written by him, read those notes to us; we had to copy those notes and memorize them. I revolted and failed in that one subject, so much so that he might not have issued to me the certificate for going in for my Matriculation Examination. Fortunately for me, I was at that time ill; he took pity on me and issued the certificate. Had it been so, he would really have blamed me for not having passed in the Chemistry paper instead of blaming himself.

So, the professors and teachers—I except you, Sir, and your race from the category—the Indian teachers and professors and the Indian students, all sail in the same boat. Science is essentially one of those things in which theory alone is of no value whatsoever—unless you have practical knowledge and unless you conduct practical experiments. I wonder how far you go in for practical experiments and how far you take the keenest delight in it. If you go in for science in the right spirit then I know that there is nothing so great or so valuable for making us accurate in thought and accurate in action. Unless our hands go hand in hand with our heads we would be able to do nothing whatsoever.

Unfortunately we, who learn in colleges, forget that India lives in her villages and not in her towns.

India has 700,000 villages and you, who receive a liberal education, are expected to take that education or the fruits of that education to the villages. How will you infect the people of the villages with your scientific knowledge? Are you then learning science in terms of the villages and will you be so handy and so practical that the knowledge that you derive in a college so magnificently built—and I believe equally magnificently equipped—you will be able to use for the benefit of the villagers?

Lastly then, I place before you the instrument to which you may apply your scientific knowledge and that is the humble spinning-wheel. Seven *lakhs* villages in India are today pining for want of that simple instrument. It was in every home and every cottage of India only a century ago, and at that time, India was not a lazy country that it is today. Her agriculturists—and agriculturists form 85 per cent of the population—were not forced to be idle for at least four months in the year. That is not what I am telling you. This is not my testimony. This is the testimony of another scientist, that is Mr. Higginbottom. He has lately been giving evidence before the Taxation Committee and he said that India's poverty, which was growing, would not diminish but would grow unless the Indian millions had a supplementary occupation. Now apply your scientific means to finding out what such supplementary occupation can be which will serve the needs of 700,000 villages scattered over a surface 1,900 miles long and 1,500 miles broad, and I assure you, you will come to the same irresistible conclusion that I have, that nothing but the spinning-wheel can do it.

The spinning-wheel has gone out of use now. Wherever I go I ask for a spinning-wheel and instead of a spinning-wheel I get a toy. I cannot get good yarn, which will give you good *khaddar*, out of toys. It is for you to make the spinning-wheel hum. I present to you the noble example of Dr. P. C. Ray who is the maker of the Bengal Chemical Works. It is a growing concern which has furnished a calling for hundreds of students. But Dr. Ray is a scientist of scientists and he wanted to give the benefit of his scientific knowledge to the villagers of India. Because he was working at the time of the Khulna famine he saw the secret of the spinning-wheel and you know today that he is devoting his life only to the spinning-wheel propaganda and the noble band of workers under him, all scientists, are endeavouring to perfect the spinning-wheel, to perfect every accessory required for the spinning-wheel. It is a noble calling. It is worthy of scientists.

May it also find an abiding place in your hearts. I thank you for giving me this patient hearing.

Speech to Students, Trivandrum
The Hindu, 19 Mar. 1925

¹ Surgeon in the Sassoon Hospital, Poona, who operated upon Gandhi for appendicitis in January 1924.
² 1858-1937; botanist; Fellow of the Royal Society; founded the Bose Institute near Calcutta.
³ 1861-1944; chemist and patriot.

109. INDUSTRIALISM AND INDIA

The following extract from a letter of a passionate lover of *khaddar* will be read with interest:

I believe in *khaddar*. I see the mission of *khaddar* clear as crystal. It simplifies and hence purifies life. It binds us to the poor by the tie of service. It is the only insurance against poverty which is killing the body and the soul of the nation, for at least as far as the illiterate millions are concerned there is no question of the soul without the body. Realized Yoga and its votaries might talk of it, but for the millions soul is mockery without body. Last and not least, *charkha* is the only insurance against violent social outbreaks as are now flooding Europe with blood and passions. *Charkha* brings the masses and the classes together and as long as India accepts it Bolshevism and kindred violent eruptions would be impossible. These things convince me of the vital need of the *charkha*. But there is only one difficulty. Can it work? Can it succeed? Can we now plant again the *charkha* in its old place of sanctity in every home? Is it not too late? Before you went to prison I never would have questioned thus. There was room for hope. But now it is not all hope. And there is Bertrand Russell who says that industrialism is like a force of nature and India, too, will be submerged whether we want it or no. Only such people say we should find our own solution for industrialism. There is truth in what they say. Industrialism is flooding all the world and, after the flood, they are finding their own solutions. Take Europe. I do not believe that Europe will perish. I have too much faith in human nature and human nature will find the remedy sooner or later. Can India, even if she wants to, isolate herself and get out of the clutches of industrialism?

The argument to which this lover of *khaddar* has been involuntarily and irresistibly drawn is Satan's old device. He always goes with us half way, and then suddenly insinuates that it is no good going further and points to the seeming impossibility of further progress. He applauds virtue, but immediately says that it is not given to man to attain it.

Now, the difficulty that has occurred to the friend is a difficulty that faces a reformer at every step. Have not untruth and hypocrisy permeated society? Yet those who believe in the ultimate triumph of truth, persist in it in the absolute hope of success. A reformer never permits time to run against him, for he defies that ancient enemy. Of course, industrialism is like a force of Nature, but it is given to man to control Nature and to conquer her forces. His dignity demands from him resolution in the face of overwhelming odds. Our daily life is such a conquest. An agriculturist knows it only too well.

What is industrialism but a control of the majority by a small minority? There is nothing attractive about it, nor is there anything inevitable in it. If the majority simply wills to say 'no' to the blandishments of the minority, the latter is powerless for mischief.

It is good to have faith in human nature. I live because I have that faith. But that faith does not blind me to the fact of history that, whilst in the ultimate all is well, individuals and groups called nations have before now perished. Rome, Greece, Babylon, Egypt and many others are a standing testimony in proof of the fact that nations have perished before now because of their misdeeds. What may be hoped for is that Europe, on account of her fine and scientific intellect, will realize the obvious and retrace her steps, and from the demoralizing industrialism she will find a way out. It will not necessarily be a return to the old absolute simplicity. But it will have to be a reorganization in which village life will predominate, and in which brute and material force will be subordinated to the spiritual force.

Lastly, we must not be entrapped by false analogies. European writers are handicapped for want of experience and accurate information. They cannot guide us beyond a certain measure if they have to generalize from European examples which cannot be on all fours with Indian conditions, because in Europe they have nothing like the conditions of India, not even excluding Russia. What may be, therefore, true of Europe is not necessarily true of India. We know, too, that each nation has its own characteristics and individuality. India has her own; and if we are to find out a true solution for her many ills, we shall have to take all the idiosyncrasies of her constitution into account, and then prescribe a remedy. I claim that to industrialize India in the same sense as Europe is to attempt the impossible.

India has stood many a storm. Each has left its own indelible mark it is true, but she has hitherto dauntlessly maintained her individuality. India is one of the few nations of the earth which have witnessed the fall of many civilizations, herself remaining scatheless. India is one of the few nations on the earth which have retained some of their ancient institutions although they have been overlaid with superstition and error. But she has hitherto shown an inherent capacity for purging herself of error and superstition. My faith in her ability to solve the economic problem that faces her millions has never been so bright as it is today, especially after my study of the conditions in Bengal.

'Snares of Satan'
Young India, 6 Aug. 1925

110. THE SCIENTIFIC SPIRIT

A student who is carrying on post-graduate studies in America writes:

I am one of those who are extremely interested in the utilization of Indian resources as one of the means for remedying the poverty of India. This is my sixth year in this country. My special field is wood-chemistry. I would have entered executive services or taken up medical studies if I were not so profoundly convinced of the importance of the industrial development of India ... Would you approve of my going into industrial enterprise, say pulp and paper manufacture? What is your attitude in general on the question of adopting a sane, humanitarian industrial policy for India? Do you stand for the progress of science? I mean such progress which brings blessings to mankind, e.g., the work of Pasteur of France and that of Dr. Benting of Toronto.

I answer this question publicly as so many inquiries are received by me from students all over, and as so much misconception exists regarding my views on science. I should have no objection whatsoever to industrial enterprise such as the student has in view. Only I would not call it necessarily humanitarian. A humanitarian industrial policy for India means to me a glorified revival of hand-spinning, for through it alone can pauperism, which is blighting the lives of millions of human beings in their own cottages in this land, be immediately removed. Everything else may thereafter be added, so as to increase the productive capacity of this country. I would therefore have all young men with a scientific training to utilize their skill in making the spinning-wheel, if it is possible, a more efficient instrument of production in India's cottages.

I am not opposed to the progress of science as such. On the contrary the scientific spirit of the West commands my admiration and if that admiration is qualified, it is because the scientist of the West takes no note of God's lower creation. I abhor vivisection with my whole soul: I detest the unpardonable slaughter of innocent life in the name of science and humanity so-called, and all the scientific discoveries stained with innocent blood I count as of no consequence. If the circulation of blood theory could not have been discovered without vivisection, the human kind could well have done without it. And I see the day clearly dawning when the honest scientist of the West will put limitations upon the present methods of pursuing knowledge. Future measurements will take note not only of the human family but of all that lives, and even as we are slowly but surely discovering that it is an error to suppose that Hindus can thrive upon the degradation of a fifth of themselves or that peoples of the West can rise or live upon the exploitation and degradation of the Eastern and African nations, so shall we realize in the fulness of time, that our dominion over the lower order of creation is not for their slaughter, but for their benefit equally with ours. For I am as certain that they are endowed with a soul as that I am.

The same student asks:

I would like to know your very frank evaluation of the work of Christian missionaries in India. Do you believe that Christianity has some contribution to make to the life of our country? Can we do without Christianity?

In my opinion Christian missionaries have done good to us indirectly. Their direct contribution is probably more harmful than otherwise. I am against the modern method of proselytizing. Years' experience of proselytizing both in South Africa and India has convinced me that it has not raised the general moral tone of the converts who have imbibed the superficialities of European civilization, and have missed the teaching of Jesus. I must be understood to refer to the general tendency and not to brilliant exceptions. The indirect contribution, on the other hand, of Christian missionary effort is great. It has stimulated Hindu and Mussalman religious research. It has forced us to put our own houses in order. The great educational and curative institutions of Christian missions I also count, amongst indirect results, because they have been established, not for their own sakes, but as an aid to proselytizing.

The world, and therefore we, can no more do without the teaching of Jesus than we can without that of Mahomed or the Upanishads. I hold all these to be complementary to one another, in no case exclusive. Their true meaning, their interdependence and interrelation, have still to be revealed to us. We are but indifferent representatives of our respective faiths which we believe more often than not.

The third question put by the student is as follows:

In our united states of India, are we going to leave the present native States intact, or have a democracy there? In order to have political unity, what should be our common language? Why can't we make it English?

The Indian States are even now changing their character, be it ever so imperceptibly. They cannot be autocratic when the bulk of India becomes democratic. What, however, Indian democracy will be no one can tell. It is easy enough to foresee the future, if English were our common language. For it would be then the democracy of a mere handful. But if we desire to realize, as we must, the political unity of the vast mass of

Indian humanity, he must be a prophet who would foretell the future. And the common language of the vast mass can never be English. It is as a matter of course a resultant of Hindi and Urdu or Hindustani as I would call it. Our English speech has isolated us from the millions of our countrymen.

We have become foreigners in our own land. The manner in which English speech has permeated the political-minded men of India constitutes in my humble opinion a crime against the country, indeed humanity; because we are a stumbling-block in the progress of our own country, and the progress of what is after all a continent must mean the progress of humanity and *vice versa*. Every English-educated Indian who has penetrated the villages has realized this burning truth, even as I have. I have profound admiration for the English language and many noble qualities of the English people, but I have no manner of doubt in my mind that the English language and the English people occupy a place in our life which retards our progress and theirs as well.

'A Student's Questions'
Young India, 17 Dec. 1925

III. DEMOCRATIC *SWARAJ*

Right conduct is not like Euclid's right line. It is like a beautiful tree, not one of whose millions of leaves is like any other. Though, therefore, they are from one seed and belong to the same tree, there is none of the uniformity of a geometrical figure about any part of a tree. And yet we know that the seed, the branches and the leaves are one and the same. We know, too, that no geometrical figure can bear comparison with a full-blossomed tree in point of beauty and grandeur.

Therefore, where the correspondent sees inconsistency, I see neither contradiction nor insanity in my life. It is true that, as a man cannot see his back, so can he not see his errors or insanity. But the sages have often likened a man of religion to a lunatic. I, therefore, hug the belief that I may not be insane and may be truly religious. Which of the two I am in truth can only be decided after my death.

I never asked my audience to substitute the spinning-wheel for the rosary. I only suggested that they could go on spinning taking the name 'Narayana' simultaneously. And whilst today the whole country is on fire, I think it behoves us all to fill the buckets of the spinning-wheel with the water of yarn, and extinguish the fire with the name of 'Narayana' on our lips.

I want to see the spinning-wheel everywhere, because I see pauperism everywhere. Not until and unless we have fed and clothed the skeletons of India will religion have any meaning for them. They are living the cattle-life today and we are responsible for it. The spinning-wheel is, therefore, a penance for us. Religion is service of the helpless. God manifests Himself to us in the form of the helpless and the stricken. But we, in spite of our forehead marks, take no notice of them, i.e., of God. God is and is not in the Vedas. He who reads the spirit of the Vedas sees God therein. He who clings to the letter of the Vedas is a *vedia*—a literalist. Narasinha Mehta does indeed sing the praise of the rosary, and the praise is well-merited where it is given. But the same Narasinha has sung:

Of what avail is the *tilaka* and the *tulsi*, of what avail is the rosary and the muttering of the Name, what avail is the grammatical interpretation of the Veda, what avail is the mastery of the letters? All these are devices to fill the belly and nothing worth without their helping to a realization of the *Parabrahm*.

The Mussalman does count the beads of his *tasbih*, and the Christian of the rosary. But both would think themselves fallen from religion if their *tasbih* and rosary prevented them from running to the succour of one who, for instance, was lying stricken with a snake-bite. Mere knowledge of the Vedas cannot make our *Brahmins* spiritual preceptors. If it did, Max Muller would have become one. The *Brahmin* who has understood the religion of today will certainly give Vedic learning a secondary place and propagate the religion of the spinning-wheel, relieve the hunger of the millions of his starving countrymen and only then, and not until then, lose himself in Vedic studies.

I have certainly regarded spinning superior to the practice of denominational religions. But that does not mean that the latter should be given up. I only mean that a *dharma* which has to be observed by the followers of all religions transcends them, and hence I say that a *Brahmin* is a better *Brahmin*, a Mussalman a better Mussalman, a *Vaishnava* a better *Vaishnava*, if he turns the wheel in the spirit of service.

I certainly did not repeat the divine word 'Rama', nor count the beads on account of a feeling that my end was near. But I was too weak then to turn the wheel. I do count the rosary whenever it helps me in concentrating on Rama. When, however, I rise to a pitch of concentration where the rosary is more a hindrance than a help, I drop it. If it was possible for me to turn the wheel in my bed, and if I felt that it would help me in concentrating my mind on God, I would certainly leave the rosary aside and turn the wheel. If I am strong enough to turn the wheel, and I have to make a choice between counting beads or turning the wheel, I would certainly decide in favour of the wheel, making it my rosary, so long as I found poverty and starvation stalking the land. I do look forward to a time when even repeating the name of Rama will become a hindrance. When I have realized that Rama transcends even speech, I shall have no need to repeat the name. The spinning-wheel, the rosary and the *Ramanam* are all the same to me. They subserve the same end, they teach me the religion of service. I cannot practise *ahimsa* without practising the religion of service, and I cannot find the truth without practising the religion of *ahimsa*. And there is no religion other than Truth. Truth is Rama, Narayana, Ishwara, Khuda, Allah, God. As Narasinha says, 'The different shapes into which gold is beaten gives rise to different names and forms; but ultimately it is all gold.'

I have nothing to withdraw from what I have said about machines in *Indian Home Rule*, and a reference will show that I have included the printing press in the machines. It must be remembered that it is not Indian Home Rule depicted in that book that I am placing before India. I am placing before the nation parliamentary, i.e., democratic *swaraj*. I do not suggest today a destruction of all the machines, but I am making the spinning-wheel the master-machine. *Indian Home Rule* depicts an ideal State. The fact that I cannot come up to the ideal condition of things laid down therein is to be attributed to my weakness. I believe that there is no religion greater than *ahimsa,* and yet I cannot escape the *himsa* which is inevitably involved in the processes of eating and drinking. The ideal of *ahimsa* is, however, ever before me; therefore, even in these processes, I do endeavour to restrain myself. I am striving every moment to reduce even those functions to a minimum.

What I have said about hospitals is also true. And yet I suppose I shall resort to the few medicines I hold lawful, so long as I retain the least attachment for my body. I went to the hospital as a prisoner. I did not run away from it immediately on my release, because I thought it my duty to remain under the care of those who had treated me with courtesy and kindness.

I am, however, ashamed at the very fact of my illness, inasmuch as I believe that a man should never fall ill. It is humiliating for me to take any medicine, and the more so that it was at all necessary to take me to the hospital.

I have never preferred killing a dacoit to winning him over with love. But he who is not equal to that love, who cannot muster all the love that the act demands, has the right to protect his proteges and his property even by killing the dacoit.

It is a gross error to liken the Englishmen to dacoits. The dacoits loot you by sheer violence, the Englishmen do so chiefly by seducing us. There is thus a great difference of method in the two. A liquor-vendor also robs me of my soul by selling his liquor. Should I suggest killing him, or non-co-operating with him? But if an Englishman brutally assaults you, or a liquor-vendor forcibly tries to pour liquor down your throat, and if you will not win both over by love, then it is open to you to engage them in an armed combat. It would make no difference if the aggressors in the case were one or many, weak or strong.

I have ventured to give a reply to the above letter, but I have a doubt whether it was proper to do so. I have assumed the correspondent's object to be pure; hence only I took the trouble to reply to him. But I think it can be seen from my reply that usually a great deal of erroneous thinking is found in such correspondence.

It seems the life of many educated people has become devoid of thought. As long as one cannot deduce a corollary from a principle, one can be held to have no knowledge of the principle at all. If the correspondent had gone deep into the subject and thought profoundly over it, he himself could have deduced all the answers that I have given. Truly speaking, all these replies are already there in my earlier writings. But I find from the letters I receive that correspondents' lazy thinking is a common fault and hence I have to give this reply. But I advise every reader and correspondent that they should think deeply on every subject, because by doing so they will save themselves from many misconceptions.

'Reading without reflection is fruitless.'

'Rosary or the Spinning-Wheel' (G.)
Navajivan, 10 Aug. 1924
Young India, 14 Aug. 1924

§4. East and West

112. UNION OF EAST AND WEST

[LONDON,
13 October 1909]

Mr. Gandhi[1] said that the question of East and West presented a vast and complex problem. He had had 18 years' experience of contact between East and West and had endeavoured to study the question, and he felt that he might give an audience such as the present one the results of his observations. As he thought of the subject, his heart sank within him. He would have to say many things which would seem repugnant to his audience, and use hard words. He would also have to speak against a system under which he had been brought up. He hoped they would bear with him if he hurt their feelings. He would have to break many idols which he and his countrymen had worshipped, and which his audience may have worshipped. He then referred to the lines in Kipling's poem, that 'East is East and West is West and never the twain shall meet', and said he considered that doctrine to be a doctrine of despair, and inconsistent with the evolution of humanity.

He felt it utterly impossible to accept a doctrine of that nature. Another English poet, Tennyson, had in his 'Vision' clearly foretold the union between East and West and it was because he (the lecturer) believed in that vision that he had cast in his lot with the people of South Africa, who were living there in very great difficulties. It was because he thought it possible for the two peoples to live together in perfect equality that he found himself in South Africa. If he had believed in Kipling's doctrine, he would never have lived there. There had been individual instances of English and Indian people living together under the same rule without a jarring note, and what was true of individuals could be made true of nations. To a certain extent it was true that there was no meeting place between civilizations. The barriers between the Japanese and the Europeans were daily vanishing, because the Japanese assimilated Western civilization. It seemed to him that the chief characteristic of modern civilization [was that it] worshipped the body more than the spirit, and gave everything for the glorifying of the body. Their railways, telegraphs and telephones, did they tend to help them forward to a moral elevation? When he cast his eyes upon India, what was represented there today under British rule?

Modern civilization ruled India. What had it done? He hoped he would not shock his hearers when he said that civilization had done no good to India. There was there a network of railways and telegraphs and telephones; we had given them a Calcutta, a Madras, a Bombay, a Lahore and a Benares—these were symbols of slavery rather than of freedom. He noticed that these modern travelling facilities had reduced their holy places to unholy places. He could picture to himself Benares of old, before there was a mad rush of civilization, and he had seen the Benares of today with his own eyes, an unholy city. He saw the same thing here as in India. The mad activity had unhinged us and, although he was living under the system, it seemed to him desirable that he should speak to them in that strain. He knew it was impossible for the two peoples in India to live together until the British changed their ways. We had offended the religious susceptibilities of the Hindus by sport in their sacred places. Unless this mad rush was changed, a calamity must come. One way would be for them to adopt modern civilization; but far be it from him to say that they should ever do so. India would then be the football of the world, and the two nations would be flying at each other. India was not yet lost, but had been immersed in lethargy. There were many things which could not be understood, for which we must be patient; but one thing was certain, and that was that, so long as this mad rush lasted, with its glorification of the body, the soul within, which was imperishable, must languish.

Speech at Friends' Meeting House, Hampstead
India, 22 Oct. 1909

[1] Gandhi spoke on 'East and West' at a meeting held under the auspices of the Hampstead Peace and Arbitration Society at the Friends' Meeting House. C. E. Maurice presided.

113. TEST OF CIVILIZATION

[1 December 1931]

Q. How can Christian pacifists and internationalists help India?

A. First of all they can do so by a thorough scientific study of the question, so that events shall not nonplus them, and so that they shall not be subject to vacillation. There are people who sometimes hug me and sometimes revile me. They are subject to the passing moment. I want them to assimilate the truth about the movement in India so that they are not easily changed. If there are such people, then the movement is safe. Otherwise it has no roots. This study must also be followed by corporate action based on the truth they have assimilated.

Peace may arise out of strife, for all strife is not antipacific. To stand with folded hands is not to achieve reform.

I have been told that by suffering myself I hurt the feelings of those who are opposed to me. Yes, certainly I do. That is what I want to do. Surely you do not want your opponent to be so hard-hearted that he is indifferent to what others suffer. Of course, the sufferings must not be wanton and not merely for the sake of suffering. That would be terrible. I only suffer if I must suffer. When the suffering is there, the suffering must be borne; it is a necessity.

Is not this the process of conversion? Instead of overthrowing your adversary, and compelling him either to yield or perish, you permit yourself to be overthrown and to suffer. If it hurts him to see you suffer, that is what you want. The pacifists in this country do not believe in the fundamental law of peace. They must be prepared to suffer with those who suffer.

It has been said to me, 'Surely it is not necessary to impose this suffering upon ourselves? Why cannot the object be obtained by way of negotiation?' I reply, 'Argument has never convinced any man, but, on the contrary, conviction precedes argument.' If that were not so, all books would appeal to all men alike. I have been touched by books which made no appeal to millions, because I already had the conviction within me.

Take my vegetarianism. I was born a vegetarian. I was a vegetarian by the vow I made before my mother. Then I read Salt's *Plea for Vegetarianism* and I was convinced, but the conviction was already in me. Similarly with Ruskin's *Unto This Last*. I was trying to follow that life, but Ruskin made it real in my own life. He changed it, but the conviction was already there. To others, in whom the conviction was not already, the same book would make no appeal.

Q. How, Mr. Gandhi, can *satyagraha* be effective when followed merely as a method and not as a principle?

A. *Satyagraha* means utter insistence upon truth. When a man insists on truth, it gives him power. If a man without real perception uses it, he is taking its name in vain. I may refuse to acknowledge the rule of the road because of some principle involved. Another man may do so because he finds it inconvenient. We are both doing the same thing, but in the one case there is moral backing for the action, but not in the other. One of us is a civil resister, the other is a criminal resister. But the danger has its own corrective in that ultimately you have to suffer, and not many will invoke suffering from an impure motive.

The true conscientious objector is correct in his conduct, for he has a spiritual backing. But the act is correct whether there is spiritual backing or not. The difference is that the conduct in one case is correct throughout, and, in the other only up to a point.

Q. You have often said that Western civilization is Satanic. What are its Satanic elements, and are none of these elements present in Indian civilization?

A. Western civilization is material, frankly material. It measures progress by the progress of matter—railways, conquest of disease, conquest of the air. These are the triumphs of civilization according to Western measure. No one says, 'Now the people are more truthful or more humble.' I judge it by my own test and I use the word 'Satanic' in describing it. You set such store by the temporal, external things. The essential of Eastern civilization is that it is spiritual, immaterial. The fruits of Western civilization the East may approach with avidity but with a sense of guilt. Your idea is the more you want the better you are, and you don't fall far short in your belief. Your civilization has gone from one stage to another. There is no end to it. You are proud of your conquest over nature, but this makes no appeal to me. You might see me fly tomorrow, but I should be feeling guilty about it. Suppose all your London tubes and buses were taken away. I should say, 'Thank God I shall be able to walk to my quarters at Bow, even if it takes me three hours.'

A final question put to Mr. Gandhi was as to whether he found the spirit which he sought in any of the Western religious books. He replied at once:

Yes. For instance, some years ago my friend Henry Polak gave me Thomas à Kempis's *Imitation of Christ*. I read it through at a sitting and I thought I was reading an Eastern book.

Q. You mean a universal book?

A. Well, when I use the term 'Eastern' I mean 'Universal'. The term is one of my little 'foot rules'.

Interview with Journalists
The Friend, 11 Dec. 1931
Reconciliation, Jan. 1932

114. EVERYTHING ON ITS MERITS

March 8, 1945

Forgetting East and West we should consider everything on its own merits.

Note to Gope Gurbuxani (H.)
GN 1324

§ 5. Modern Civilization

115. ESOTERIC CHRISTIANITY AND MODERN CIVILIZATION

Durban,
November 26, 1894

TO
The Editor
The Natal Mercury
Sir,

You will greatly oblige me by allowing me to draw the attention of your readers to an advertisement that appears in your advertisement columns with regard to the Esoteric Christian Union. The system of thought expounded by the books advertised is not, by any means, a new system but a recovery of the old, presented in a form acceptable to the modern mind. It is, moreover, a system of religion which teaches universality, and is based on eternal verities and not on phenomena or historical facts merely. In that system, there is no reviling Mahomed or Buddha in order to prove the superiority of Jesus. On the other hand, it reconciles the other religions with Christianity which, in the opinion of the authors, is nothing but one mode (among many) of presentation of the same eternal truth. The many puzzles of the Old Testament find herein a solution at once complete and satisfactory.

If there is any one of your readers who has found the present-day materialism and all its splendour to be insufficient for the needs of his soul, if he has a craving for a better life, and if, under the dazzling and bright surface of modern civilization, he finds that there is much that is contrary to what one would expect under such a surface, and above all, if the modern luxuries and the ceaseless feverish activity afford no relief, to such a one I beg to recommend the books referred to. And I promise that, after a perusal, he will find himself a better man, even though he may not thoroughly identify himself with the teaching.

If there is anyone who would like to have a chat on the subject, it would afford me the greatest pleasure to have a quiet interchange of views. In such a case, I would thank any such gentleman to correspond with me personally. I need hardly mention that the sale of the books is not a pecuniary concern. Could Mr. Maitland, the President of the Union, or its agent here, afford to give them away, they would gladly do so. In many cases, the books have been sold at less than cost price. In a few, they have even been given away. A systematic distribution for nothing has been found impossible. The books will be gladly lent in some cases.

I would try to conclude with a quotation from a letter of the late Abbé Constant to the authors: '*Humanity has always and everywhere asked itself these three supreme questions: Whence come we? What are we? Whither go we?* Now these questions at length find an answer complete, satisfactory, and consolatory in *The Perfect Way*.'

I am, etc.
M. K. GANDHI

'Esoteric Christian Union'
Natal Mercury, 3 Dec. 1894

116. VIOLENCE AND CIVILIZATION

[London,
30 October 1909]

My Lord,

I have for some time past been wishing to place before Your Lordship the result of my observations made here during my brief stay on the nationalist movement among my countrymen.

If you will permit me to say so, I would like to say that I have been much struck by Your Lordship's candour, sincerity and honesty of which one notices nowadays such an absence among our great public men. I have noticed too that your imperialism does not blind you to matters of obvious justice and that your love of India is genuine and great. All this coupled with my desire to withhold nothing from Your Lordship regarding my own activity about Indian matters as they may have a direct or an indirect bearing on the struggle in the Transvaal, emboldens if it does not require me to inform you of what I have seen.

I have made it a point to see Indians here of every shade of opinion. Opposed as I am to violence in any shape or form, I have endeavoured specially to come into contact with the so-called extremists who may be better described as the party of violence. This I have done in order if possible to convince them of the error of their ways. I have noticed that some of the members of this party are earnest spirits, possessing a high degree of morality, great intellectual ability and lofty sacrifice. They wield an undoubted influence on the young Indians here. They are certainly unsparing in their efforts to impress upon the latter their convictions. One of them came to me with a view to convince me that I was wrong in my methods and that nothing but the use of violence, covert or open or both, was likely to bring about redress of the wrongs they consider they suffer.

An awakening of the national consciousness is unmistakable. But among the majority it is in a crude shape and there is not a corresponding spirit of self-sacrifice. Everywhere I have noticed impatience of British rule. In some cases the hatred of the whole race is virulent. In almost all cases distrust of British statesmen is writ large on their minds. They are supposed to do nothing unselfishly. Those who are against violence are so only for the time being. They do not disapprove of it. But they are too cowardly or too selfish to avow their opinions publicly. Some consider that the time for violence is not yet. I have practically met no one who believes that India can ever become free without resort to violence.

I believe that repression will be unavailing. At the same time, I feel that the British rulers will not give liberally and in time. The British people appear to me to be obsessed by commercial selfishness. The fault is not of men but of the system and the system is represented by the present civilization which has produced its blasting effect as well on the people here as on India. India suffers additionally only in so far as it is exploited in the interest of foreign capitalists. The true remedy lies, in my humble opinion, in England discarding modern civilization which is ensouled by this spirit of selfishness and materialism, is vain and purposeless and is a negation of the spirit of Christianity. But this is a large order. It may then be just possible that the British rulers in India may at least do as the Indians do and not impose upon them the modern civilization. Railways, machinery and corresponding increase of indulgent habits are the true badges of slavery of the Indian people as they are of Europeans. I, therefore, have no quarrel with the rulers. I have every quarrel with their methods.

I no longer believe as I used to in Lord Macaulay as a benefactor through his minute on education. And I do think that a great deal too much is being made of *pax Britannica*. To me the rise of the cities like Calcutta and Bombay is a matter for sorrow rather than congratulation. India has lost in having broken up a part of her village system. Holding these views, I share the national spirit but I totally dissent from the methods whether of the extremists or of the moderates. For either party relies ultimately on violence. Violent methods must mean acceptance of modern civilization and therefore of the same ruinous competition we notice here and consequent destruction of true morality. I should be uninterested in the fact as to who rules. I should expect rulers to rule according to my wish; otherwise I cease to help them to rule me. I become a passive resister against them. Passive resistance is soul-force exerted against physical force. In other words love conquering hatred.

I do not know how far I have made myself understood and I do not know how far I carry you with me in my reasoning. But I have put the case in the above manner before my countrymen. My purpose in writing to Your Lordship is twofold. The first is to tell Your Lordship that, whenever I can get the time, I would like to take my humble share in national regeneration and the second is either to secure Your Lordship's cooperation in the larger work if it ever comes to me or to invite your criticism.

The information I have given Your Lordship is quite confidential and not to be made use of prejudicially to my countrymen. I feel that no useful purpose will be served unless the truth is known and proclaimed.

If you will pursue the inquiry further, I shall be pleased to answer any questions you may wish to put. Mr. Ritch has full knowledge of the contents of this letter. If a discussion is considered necessary, I am at your service.

In conclusion, I hope I have not unduly or unwarrantably trespassed upon your courtesy and attention.

I remain, etc.,

Letter to Lord Ampthill
SN 5152

117. HIGH-HANDEDNESS AND HYPOCRISY

We saw in *Hind Swaraj* that it is not so much from British rule that we have to save ourselves as from Western civilization. Clearly, if Englishmen settle down in India as Indians, they will cease to be foreigners. If they cannot bring themselves to do so, it will be our duty to create conditions in which it will be impossible for them to stay on.

The writings of Englishmen themselves often tell us how wicked Western civilization is. There was a storm of protest in England against the alleged high-handedness of the Spanish authorities when Ferrer was put to death. The letter in the *Daily News* of October 22 which the famous author, Mr. G. K. Chesterton, wrote, pointing out that this was sheer hypocrisy on their part, will bear summarizing even today. Mr. Chesterton says:

We have been hysterically protesting against what Spain has done, but that is so much hypocrisy and nothing else. It is out of our pride that we take up such an attitude. In fact, we are just as bad as

Spain, in certain respects much worse. We have no political executions in England because we have no political rebellions in our country and not because we are a religious people. Wherever we do have rebellions, there we do have executions, much more mean, reckless and savage than the execution of Ferrer. The hanging of the Fenians at Manchester has been admitted by all lawyers to have been in contempt of logic and law. The killing of Scheepers in South Africa is a thing of which even the Imperialists are now ashamed. A few harmless peasants at Denshawai objected to the looting of their property; they were tortured and hanged. When our rulers react with such brutality and baseness to small and ineffectual local risings, how would they behave if confronted with a rising in London itself similar to the one in Spain? We are at peace, not because we do not exploit religion but because we have sunk silently under the domination of our rulers. If we have no rebellions, we are guilty of crimes worse than the death of Ferrer. A private soldier the other day committed suicide in order to avoid a flogging. This suicide is more hideous than the execution of Ferrer under the pressure of strong emotions in a time of excitement. Yet the incident attracted no attention in England, because we are the one people in Europe who are successfully oppressed.

In view of such shortcomings in the civilization of this people which dazzles us so much, we had better consider whether we should tolerate it in India or banish it while we have still time to do so. It is a civilization which grinds down the masses and in which a few men capture power in the name of the people and abuse it. The people are deceived because it is under cover of their name that these men act.

'Shortcomings of Western Civilization' (G.)
Indian Opinion, 22 Jan. 1910

118. NATURE AND CIVILIZATION

Nature works unceasingly according to her laws, but man violates them constantly. In different ways and at different times, Nature tells man that there is nothing in the world which is not subject to change. It is hardly necessary to give illustrations. As Mr. Malabari[1] says in a poem of his: 'They come but to leave.' We sing in a *gazal*: 'How many fairy-like creatures there have been, graced with youth, how great the men who left.' And yet every extraordinary occurrence startles us and sets us thinking. There has been one such in Paris. The river at Paris rose in such a heavy flood that huge buildings were washed off. A picture-gallery was in imminent danger. Strongly-built roads, on which millions of pounds had been spent, sagged at places. Men were drowned. Some who escaped drowning were buried alive. Rats, deprived of their food, attacked children. How did this happen? The people of Paris had built the city to last for ever. Nature has given a warning that even the whole of Paris may be destroyed. It certainly would have been, had the floods subsided a day later.

Of course, the people of Paris will not realize the futility of rebuilding the palatial structures. It will never occur to them that even these new buildings of theirs will come down again. Engineers, in their conceit, will have more grandiose plans now and pour out money like water, forgetting and making others forget the deluge; such is the obsession of present-day civilization.

Are we to behave in the same way? Shall we copy such wild, thoughtless people? Only those who forget God will engage in such ostentation. The question then arises why we should fight against the Transvaal legislation, why we do not advise everyone to take up the rosary. To anyone who may ask this question, we shall reply that that is the very advice we have given, and give again. What we do not advise is the mere ostentatious bead-telling in the manner of that pious fraud, the crane in the fable. We realize the meaning of the drama that Nature is enacting, and that is why we appeal to the Transvaal Indians and the Indians of South Africa, with all the strength at our command: 'Understand Nature's purposes and ponder over them; all your ostentatious ways will lead you nowhere. Telling beads on the rosary will be no answer to the Government's attack on your manliness and its attempt to enslave you. The servant of God will never consent to be the slave of any man. Do not be afraid of the despotic laws of the Government. You will have no reason for fear if you are not unduly attached to your wealth. If you cling to truth, it will always be with you, it will never forsake you; it cannot be submerged in floods. We advise you not to trust anything that the floods may wash away. We invite you to be firm in truth, which is the sole support for one to cling to. You may enjoy whatever you can, consistently with your loyalty to truth. You will then have no cause for regret. You will not then pursue enjoyments at any cost, for you will know that enjoyments are momentary but that truth is eternal and will abide with you for ever. To live thus is to follow the path of religion. Because the Government in its despotism opposes such an attempt, we call it irreligious. This is the essence of all religions and without it no religion will be true to itself.'

'Paris Havoc' (G.)
Indian Opinion, 5 Feb. 1910

[1] Behramji Malabari (1863-1912), a Parsee journalist, poet, and social reformer.

119. MESHES OF CIVILIZATION

[2 April 1910]

Chi. Maganlal,

Your letter to hand. I return it to you so that you can understand my reply to it.

I shall try to answer the questions you have raised. But even then you may not understand thoroughly. You will perhaps find the explanations you have sought from *Hind Swaraj* itself if you read it afresh once or twice.

There is no doubt that we shall have to go back to the extent to which we have imbibed modern civilization. This part of the task is the most difficult one, but it will have to be done. When we take a wrong path there is no alternative but to go back. We have got to free ourselves from attachment to the things we are enjoying. For this it is necessary that we begin to feel disgust for them. Whatever means and instruments appear to us to be beneficial are not going to be given up. Only he who realizes that there is more harm than the apparent benefit from a particular thing will give it up. I personally feel that no benefit has been derived from our being able to send letters quickly. When we give up railways and such other means we shall not bother ourselves about writing letters. A thing which is really free from fault may be used to a certain extent. We who are engulfed in this civilization may avail ourselves of postal and other facilities as long as we are so engulfed. If we make use of these things with knowledge and understanding we shall not go crazy over them, and instead of increasing our preoccupations we shall gradually reduce them. He who will understand this will not be tempted to take the post or the railway to the villages which do not have these. You and I should not remain passive and increase the use of steamers and other evil means for fear that these things cannot be abolished forthwith and that all the people will not give them up. Even if one man reduces or stops their use, others will learn to do so. He who believes that it is good to do so will go on doing so irrespective of others. This is the only way of spreading the truth; there is no other in the world.

It is very difficult to get rid of our fondness for Parliament. It was no doubt barbarous when people tore off the skin, burned persons alive and cut off their ears or nose; but the tyranny of Parliament is much greater than that of Chengiz Khan, Tamerlane and others. Hence it is that we are caught in its meshes. Modern tyranny is a trap of temptation and therefore does greater mischief. One can withstand the atrocities committed by one individual as such; but it is difficult to cope with the tyranny perpetrated upon a people in the name of the people. It seems to have happened in the past that some rulers were like King Foolishman while others turned out to be wise. Had Edward alone been our ruler it would not have been so objectionable; but every Englishman is ruling over you and me. Please ponder over the meaning of this statement. I do not refer here to people's fondness for this world. The common man in India at least believes that the Parliament is a hoax. Even an extraordinarily intelligent man, caught in the meshes of this civilization, loses his sanity in Parliament.

By saying that mercy cannot have any effect on the Pindaris you have denied the very existence of the soul or its essential attribute. Lord Patanjali[1] has emphasized the greatness of mercy, etc., in such a way that we feel delighted even while thinking of those virtues. The real fact is that fear has taken deep root in us and consequently truth, mercy and such other virtues do not develop. And then we think that mercy has no effect on cruel people. If we show mercy to the person who shows mercy to us it is no mercy; it is only the return for mercy.

We should be considered weak if someone protects us free of charge or even if we pay him for doing so. If we have to seek outside help to be free from the menace of the Pindaris, etc., we are unfit for *swaraj*. If we would subdue them with physical force, we shall have to develop that force in ourselves. We shall not then have to pay blackmail or tribute. A woman seeks her husband's protection as a matter of right; but she is considered an *abala* (weak) after all.

Swaraj is for those who understand it. You and I can enjoy it even today. All the others will have to learn to do likewise. What is secured for us by others is not *swaraj* but *pararaj*, i.e., foreign rule, whether they be Indians or Englishmen.

In calling the cow-protection societies cow-killing societies, I have but stated the truth; for their object is to rescue the cow or protect her by bringing pressure on Mussalmans.

To rescue the cow by paying money is no protection of the cow; it is a way to teach the butcher to be deceitful. If we try to coerce the Mussalmans they will slaughter more cows. But if we persuade them or offer *satyagraha* against them they will protect her. No cow-protection society is necessary for doing this. That body should be for teaching Hinduism to the Hindus. It is better to kill an ox by a single blow of the sword than to kill it by starving it, by pricking it, by over-working it and thus torturing it.

It would be very confusing to take the examples of Shri Ramachandra and others literally. I have never imagined the possibility of a Ravana in the physical form of a man with ten heads and twenty arms. But to imagine that he was a huge passionate senseless animal and that he was killed by Shri Ramachandra representing the divine essence may appeal to the intellect. Tulsidasji[2] has described Ramachandraji as the forces of the Sun who is the destroyer of pride, infatuation, and the darkness of the night of excessive attachment. Do you think we shall have the least desire left in us to destroy anybody when we are rid of all pride, infatuation and attachment? If you say 'no', how could Ramachandraji who was free from pride, infatuation and attachment and who was an ocean of mercy destroy Ravana? However, let us first attain his stage, like Lakshmana give up sleep and observe *brahmacharya* for fourteen years and then see where physical force could be used.

I want to say that everything is achieved by humility. The example you gave of the Transvaal is quite appropriate. It is not enough merely to profess orally to have the above sentiment; it should stand the test when the occasion comes. Think of the numberless adversities Harishchandra had to face before his devotion to truth was proved. Think of the suffering Sudhanva had to undergo before his *bhakti* was proved to be genuine. We may not consider these as mere legends. It may be that the names and forms were different; but they who have composed these stories have given their own experiences through them. Even in the Transvaal the babbling of persons like me are being put to the test. Also bear in mind that many who were regarded as *satyagrahis* have proved to be insincere demagogues. Who, then, should be regarded as true *satyagrahis*? Of course, they who possess virtues like compassion, etc. Nowhere has it been said that suffering may not have to be undergone. And what does suffering after all mean? It is the mind, says the *Gita*, which is the cause of our bondage as well as of our freedom. Sudhanva was thrown into boiling oil. The person who got him thrown into it thought that he was inflicting suffering on Sudhanva; but for the latter it was a grand opportunity to show the intensity of his devotion.

It will never happen that all are equally rich or equally poor at the same time. But if we consider the good and evil aspects [of the various professions] it seems that the world is sustained by farmers. Farmers are of course poor. If a lawyer would boast of his altruism or spirituality, let him earn his livelihood through physical labour and carry on his legal practice without charging anything for it. You will not easily realize that the lawyer is lazy. Just as a sensuous man, even when exhausted by indulging in passions, remains engrossed in sensual pleasures, so a lawyer, even when he is exhausted, goes on straining his nerves to the breaking point in his practice in the hope of getting wealth and attaining to greatness and later on passing a life of luxury and comfort. This is his objective.

I am conscious that there is a little exaggeration in this; but, what I have said above is true for the most part.

What service will an army of doctors render to the country? What great things are they going to achieve by dissecting dead bodies, by killing animals, and by cramming worthless dicta for five or seven years? What will the country gain by the ability to cure physical diseases? That will simply increase our attachment to the body. We can formulate a plan for preventing the growth of disease even without the knowledge of medical science. This does not mean that there should be no doctors or physicians at all. They will always be with us. The point is that many a young man who gives an undue importance to this profession and wastes hundreds of rupees and several years qualifying for it, ought not to do so. We must know that we are not, nor are we going to be, benefited in the least by allopathic doctors.

I hope I have replied to all your questions. Please do not carry unnecessarily on your head the burden of emancipating India. Emancipate your own self. Even that burden is very great. Apply everything to yourself. Nobility of soul consists in realizing that you are yourself India. In your emancipation is the emancipation of India. All else is make-believe. If you feel interested, do persevere. You and I need not worry about others. If we bother about others, we shall forget our own task and lose everything. Please ponder over this from the point of view of altruism, not of selfishness. If you want to ask anything more, please do.

Blessings from
MOHANDAS

Letter to Maganlal Gandhi (G.)
R. Patel, *Gandhiji ni Sadhana*

[1] The sage who systematized the *Yoga darshana* (philosophy).
[2] The author of *Ramacharitamanasa*, a Hindi version of the *Ramayana*.

120. HYPOCRISY AND CIVILIZATION

An English journal called *The New Age* has published a cartoon on this subject, which we reproduce in this issue. It shows an army on the march. Behind, there is a grotesque figure, that of a general. On the body of this terrible form are hanging a gun emitting smoke in every direction and swords dripping with blood, and on its head a cannon. There is the drawing of a skull on a badge hanging on one side. On the arm, moreover, there is a cross. In the mouth, held in the teeth, there is a dagger dripping with blood. On the shoulder is seen a belt studded with live cartridges. The drawing is entitled 'March of Civilization'. No one who reads this description of the cartoon can help becoming grave. On reflection, we cannot help feeling that Western civilization is as cruel as, perhaps more cruel than, the terrible expression on the face of the man in the cartoon. The sight which fills one with the utmost indignation is that of the cross in the midst of weapons dripping with blood. Here the hypocrisy of the new civilization reaches its climax. In former times, too, there used to be bloody wars, but they were free from the hypocrisy of modern civilization. While drawing our readers' attention to this cartoon, we want to give them at the same time a glimpse of the divine light of *satyagraha*. On the one side, look at the picture of civilization drawn above, a civilization grown as terrible as a wolf through its hunger for wealth and its greedy pursuit of worldly pleasures. On the other, look at the figure of a *satyagrahi* who, out of his loyalty to truth, to his nature as a spiritual being and out of a desire to obey God's command, submits to the suffering inflicted by wicked men, with fortitude in his breast, with a smile on his face and without a single tear in his eyes.

Of the two pictures, towards which will the reader feel attracted? We are sure it is the vision of the *satyagrahi* which will touch the heart of mankind, and that the effect will grow deeper as his sufferings increase. Is there anyone who, looking at this cartoon alone, does not feel in his heart that *satyagraha* is the only way in which mankind can attain freedom and strength? We admit, of course, that to be shot dead or hanged when trying to shoot another does test one's fortitude; but dying in the attempt to kill another does not require even a hundredth part of the fortitude and courage implicit in the suffering that a *satyagrahi* goes through, in the slow, prolonged torture that he calmly endures in facing a bullet without firing one in return. No one wields a sword strong enough to bear down the force of *satyagraha*; on the contrary, a man brandishing a sword of steel has to give ground when confronted by a sword sharper than his. That is the reason why the story of a *satyagrahi* is read with a feeling of reverence. One who is not strong enough to practise *satyagraha* is naturally tempted to resort to brute force, which is, in comparison, quite easy to employ. There are some desperate Indians who, in their mad obsession with *swaraj* for India, seem to imagine that *satyagraha* is bound to be followed by resort to brute force—that is, that *satyagraha* is but one step in the effort to key oneself up to the fanaticism of violence. It would not be wrong to compare persons holding such views to the frog in the well who sought to conceive the ocean [as a big, big well]. The truth of the matter is that the man who cannot cultivate to its utmost limit the capacity for endurance required in *satyagraha* turns in his impatience to brute force and, growing desperate, takes a blind leap in an effort to end his suffering quickly. Such a man has never been a *satyagrahi*. He does not want to understand what *satyagraha* means.

'Terrible Civilization of the West' (G.)
Indian Opinion, 2 Apr. 1910

121. GLAMOUR OF CIVILIZATION

[28 August 1925]

You have very adroitly got rid of the obligation of bestowing some little praise upon the speaker of the evening. I wish I could do likewise and avoid anything to speak at all. I had hoped that a little bit of praise from you would give me some encouragement, but that was not to be my lot. But I am convinced from the manner in which you have been conducting these proceedings[1] and in which you have garlanded this little girl[2] that you at least are not guilty of any race hatred.

But in India, at the present moment, the young generation is undoubtedly face to face with this very problem. Is it possible to love one's country and not to hate those who rule over one's country, whose domination we do not want, whose domination we dislike from the bottom of our hearts? The answer has been in the hearts of many young men that it is impossible to love one's country and not to hate those who rule over one's country. Some of them expressed their opinion in broad daylight, a few of them translate that opinion into action. Many, however, harbour this opinion in secret and feed upon that opinion.

I have been a student of this question, not since my return to India in 1915, but ever since I entered into public life and public service. That was in 1894. But I have come deliberately to the conclusion that love of one's country, namely nationalism, is perfectly consistent with the love of those whose rule, whose domination, whose methods we do not like. I was face to face with that problem in my dealings with the South African Government or, more accurately speaking, the then Natal Government; later on with the Transvaal Government and later still with the Union Government.

Most of you are aware of the disabilities—the glaring disabilities—under which our countrymen labour in that subcontinent—South Africa. It is enough; those disabilities are really enough to make one hate one's fellow beings, if one did not preserve one's sanity. You find there injustice rampant for no cause save that you do not have the same colour of skin. There shall be no equality between the white and the coloured races—so runs the constitution of the Union Government. It was at one time an article of the Transvaal Government constitution, but the constitution has today been adopted by the Union Government. When you come to India you find, though not the same thing, much the same thing and very often one finds it most difficult to reconcile the two things, love of one's country and love also of one whom you may consider to be the tiger. It is beside the point whether you are just and correct in your estimate or whether you are incorrect, but the impression left upon your mind is that you are labouring under the grossest form of tyranny, grossest form of injustice. How shall you then love the tiger?

Let me put it in another way—not necessarily that you should love the tiger, but love is an active force and the subject of this evening is—Is it necessary to hate the tiger? Is hatred essential for nationalism? You may not love, but must you also hate? The answer, as I have said before, in the minds of many people is undoubtedly that you must hate. Some, I know, consider it their duty to hate the tiger and they cite instances from modern constitutions, they cite the late disastrous War in Europe, they cite wars of which they have learnt in history; they cite also the law, and they say society hangs on the gallows those who are guilty of murder. Is not that a sign of hatred? There certainly is no love.

Would not one love one's father, would not one love one's dearest ones, even if they might err? Would one wish them to be hanged on the gallows? One would pray for their reformation but not for their punishment, and yet, it is said, perhaps with a great deal of justification, that society will break into pieces if under the law of sanction punishment was withdrawn, abolished or suspended. With those illustrations before them, the young men rush to the conclusion that those who consider that hatred is not essential for nationalism are in the wrong. I do not blame them. They have to be pitied; they command my sympathy, but I have not a shadow of doubt in my mind that they are labouring under the grossest delusion; and so long as they retain that attitude, so long as a large body of men and women retain that attitude the progress of this country, the progress of the world is retarded. It does not matter to me that all those illustrations that I have placed before you can be cited in order to justify their conduct.

The world is weary of it. We see the fatigue overcoming the Western nations. We see that this song of hate has not benefited humanity. Let it be the privilege of India to turn a new leaf and set a lesson to the world. Is it necessary that three hundred millions of people should hate one hundred thousand Englishmen? That is the concrete term to which I can reduce this evening's subject. In my humble opinion it is derogatory to the dignity of mankind, it is derogatory to the dignity of India to entertain for one single moment hatred towards Englishmen. That does not mean that you are to be blind to the excesses that English rulers have been found to commit in India. I have drawn this particular distinction between the evil and the evil-doer. Hate the evil but not the evil-doer. We ourselves, every one of us, are full of evil. And we want the world to be patient with us, to be forgiving, to be gentle with us. I would like the same thing to be meted out to the Englishmen. Heaven knows no one in India perhaps can claim better than myself to have spoken as fiercely and as fearlessly of the many misdeeds of English rulers and the corrupt nature of the system under which we are governed. My freedom from hatred—I would even go so far as to claim for myself individually—my love of those who consider themselves to be my enemies, does not make me blind to their faults. That is no love which is extended simply because of the possession of some virtues fancied or real in the beloved. If I am true to myself, if I am true to mankind, if I am true to humanity, I must understand all the faults that human flesh is heir to. I must understand the weaknesses of my opponents, the vices of my opponents and, yet, in spite of these vices, not hate but even love them. It is by itself a force. Brute force has been handed down to us from generation to generation. We have used it and we have found what it has done for Europe and what it has done for the world. The glamour of European civilization does not dazzle us. Scratch beneath the surface and you will find there very little to choose.

Do not for one moment consider that I condemn all that is Western. For the time being I am dealing with the predominant character of modern civilization, do not call it Western civilization, and the predominant character of modern civilization is the exploitation of the weaker races of the earth. The predominant character of modern civilization is to dethrone God and enthrone Materialism. I have not hesitated to use the word 'Satan'. I have not hesitated to call this system of Government under which we are labouring 'Satanic'. And I withdraw not one word from it. But, however, I shall not deal with it this evening. If I begin to devise means of punishing the evil-doer, my business is to love them and by patient and gentle handling to convert them. Non-co-operation or *satyagraha*, therefore, is not a hymn of hate. I know that many who call themselves *satyagrahis* or non-co-operators do not deserve to bear that name. They have done a violence to their own creed. They were not real representatives of this principle.

Real non-co-operation is non-co-operation with evil and not with the evil-doer. Sometimes it is difficult, I know, to distinguish between evil and evil-doer. But how are you to non co-operate with the evil and not with the evil-doer? I do not want to go into the whole of the intricate doctrine. I can simply comment on what is going on during these 5 or 6 years. If we understand the secret of this doctrine and the beautiful consistency between hating evil and not hating evil-doers, I have said that all that we need today is to extend the law that we apply in our domestic relations to the political field, and, therefore, to the relations between the rulers and the ruled and you will find the true solution. What does a father do to a son who is inclined to do evil and become corrupted? He does not punish him nor does he encourage him, but he tries to correct him.

Your non-co-operation is intended not to encourage evil. That is the meaning. One of the greatest writers has said that if the world ceases to encourage evil, evil will die of inanition. If we simply find out for ourselves to what extent we are responsible for the evil that exists in society today, we will soon see that evil will soon be gone from society. But we tolerate it under a false sense of love. I am not talking of the blind love that dotes on an erring son and pats him on the back while he errs, nor am I speaking of the son who, under a false sense of loyalty to his father, tolerates evil in his father. I am not talking of that. I am talking of the love that discriminates, that is intelligent, that is not blind to a single fault. That is the love of reform, and the moment we have seized the secret, that very moment the evil goes out of sight.

I talk of the relations between the two races. Think of the many evils from which we are suffering today in Hindu society. Let alone Mussalmans, let alone Christians, Parsis and others. The majority of us are Hindus. How should we deal with the evil that is rampant in Hinduism? Shall we hate those who consider untouchability part and parcel of Hindu religion and quote scriptures in favour of untouchability or shall we remove untouchability by our persistent conduct? The secret, then, is suffering, but not to subject the evil-doers to suffering, but to take the suffering upon our own shoulders. If we reform Hinduism of the many abuses that have crept into it, we shall only do so by taking the instance of Vaikom. It comes to me naturally because it is through praise that you see the finished example. I know every one of the brave young men. I think I know every one of them who are working in Vaikom under terrible difficulties. They have undergone suffering which I cannot possibly describe here in the few moments, but I dare give this testimony to you that these young men have not erred by a hair's breadth. I mean the young men of Vaikom. I don't say individuals have not erred, but they have kept their record absolutely clean. The result is that they have not yet got rid of the whole of abuses, but I have not a shadow of doubt in my mind that in Travancore today untouchability has lost its foothold, it is fast dying simply because of the determination of a handful of youth who plunged themselves down into the midst of Vaikom and invited suffering on their own shoulders. That is really the secret. In my humble opinion hatred is not essential for nationalism. Race hatred will kill the real national spirit.

Let us understand what nationalism is. We want freedom for our country. We do not want sufferings for other countries: we do not want the exploitation of other countries; we do not want the degradation of other countries. For my part I don't want the freedom of India if it means the disappearance of Englishmen, if it means the extinction of Englishmen. I want the freedom of my country so that other countries may learn something from this free country of mine. I want freedom of my country so that the resources of my country might be utilized for the benefit of mankind just as the cult of patriotism teaches us today that the individual has to die for the family, the family has to die for the village, the village for the district, the district for the Province and the Provinces for the whole nation, and when we indulge in provincialism I as a Gujarati say, Gujarat first, Bengal and the rest of Provinces next. There is no nationalism in it. On the contrary if I live in Gujarat and prepare Gujarat, I should prepare Gujarat so that the vast resources of Gujarat might be placed at the disposal of Bengal, nay, of the whole of India, that Gujarat may die for the whole of India. My love, therefore, of nationalism, or my idea of nationalism is that my country may become free—free that if need be the whole of the country may die—so that the human race may live. There is no room here for race hatred. Let that be our nationalism.

A question from Mr. Chapman, the Librarian of the Imperial Library, at the conclusion of the speech, drew a telling reply from Gandhiji. 'Was not the insistence on political freedom and political equality on the part of Indians conducive to race hatred, when Indians themselves were not capable of ruling themselves' was in effect Mr. Chapman's question.

If you have drawn the deduction from what I said that we should tolerate your rule so long as we are not capable of managing our affairs, you are mistaken. We can develop that capacity only by resistance to the system. And may I say that the questioner unconsciously betrayed his own race prejudice when he referred to Indians as incapable of ruling themselves. Underlying that prejudice is the idea of superiority and the conceit that the Englishmen are born to manage the affairs of the world. That is an idea to fight which my whole life has been dedicated. Unless the Englishmen are dislodged from that position there is no peace in India, nor any peace for the weaker races of the earth. It is the absolute right of India to misgovern herself. My heart rebels against any foreigner imposing on my country the peace which is here called 'Pax Britannica'.

Speech at Meccano Club, Calcutta (G.)
Forward, 29 Aug. 1925
Young India, 10 Sept. 1925

[1] The meeting to which admission was by tickets was held at the Overtoun Hall. The proceeds were sent to the All-Bengal Deshabandhu Memorial Fund. Revd. T. E. T. Shore presided.

[2] Here Gandhi pointed to a five-year-old girl.

122. APPEAL TO THE MASSES

A European friend thus writes:

What can be done, what would you suggest that could be tried in favour of the starving millions of the West? By starving millions I mean the masses of the European and American proletariat who are being driven to the abyss, who live a life not worth the name, full of the direst privations, who can nourish no dream of future relief by any form of *swaraj*, who are perhaps more hopeless than the millions of India because the faith in God, the consolation of religion, has left them to be replaced by nothing but hatred.

The iron hands which press down the Indian nation are at work there also. The devilish system is at work in each of these independent countries; politics do not count as there is a close solidarity of greed. Vice is devastating these masses who naturally try to escape the hell of their life at any cost, at the cost of making it a greater hell, and who have no longer the outlet of religious hopes, as Christianity by siding for centuries with the powerful and the greedy has lost all credit.

Of course, I expect Mahatmaji to answer that the only way to salvation for these masses, if there is any left, if the whole Western world is not already doomed, lies in the application of a disciplined non-violent resistance carried on on a large scale. But there are no traditions of *ahimsa* in the European soil and mind. Even the spreading of the doctrine would encounter huge difficulties, what about its right understanding and application!

The problem underlying the question so sincerely put by the friend lies outside my orbit. I, therefore, attempt an answer merely in courteous recognition of friendship between the questioner and myself. I confess that no value attaches to my answer, save what we attach to every considered argument. I know neither the diagnosis of the European disease nor the remedy in the same sense that I claim to know both in the case of India.

I, however, feel that fundamentally the disease is the same in Europe as it is in India, in spite of the fact that in the former country the people enjoy political self-government. No mere transference of political power in India will satisfy my ambition, even though I hold such transference to be a vital necessity of Indian national life. The peoples of Europe have no doubt political power but no *swaraj*. Asian and African races are exploited for their partial benefit, and they, on their part, are being exploited by the ruling class or caste under the sacred name of democracy. At the root, therefore, the disease appears to be the same as in India. The same remedy is, therefore, likely to be applicable. Shorn of all the camouflage, the exploitation of the masses of Europe is sustained by violence.

Violence on the part of the masses will never remove the

disease. Anyway, up to now experience shows that success of violence has been short-lived. It has led to greater violence. What has been tried hitherto has been a variety of violence and artificial checks dependent mainly upon the will of the violent. At the crucial moment these checks have naturally broken down. It seems to me, therefore, that sooner or later, the European masses will have to take to non-violence if they are to find their deliverance. That there is no hope of their taking to it in a body and at once does not baffle me. A few thousand years are but a speck in the vast time circle. Someone has to make a beginning with a faith that will not flinch. I doubt not that the masses, even of Europe, will respond, but what is more emergent in point of time is not so much a large experiment in non-violence as a precise grasp of the meaning of deliverance.

From what will the masses be delivered? It will not do to have a vague generalization and to answer 'from exploitation and degradation'. Is not the answer this that they want to occupy the status that capital does today? If so, it can be attained only by violence. But if they want to shun the evils of capital, in other words, if they would revise the viewpoint of capital, they would strive to attain a juster distribution of the products of labour. This immediately takes us to contentment and simplicity, voluntarily adopted. Under the new outlook multiplicity of material wants will not be the aim of life, the aim will be rather their restriction consistently with comfort. We shall cease to think of getting what we can, but we shall decline to receive what all cannot get.

It occurs to me that it ought not to be difficult to make a successful appeal to the masses of Europe in terms of economics, and a fairly successful working of such an experiment must lead to immense and unconscious spiritual results. I do not believe that the spiritual law works on a field of its own. On the contrary, it expresses itself only through the ordinary activities of life. It thus affects the economic, the social, and the political fields. If the masses of Europe can be persuaded to adopt the view I have suggested, it will be found that violence will be wholly unnecessary to attain the aim and they can easily come to their own by following out the obvious corollaries of non-violence. It may even be that what seems to me to be so natural and feasible for India, may take longer to permeate the inert Indian masses than the active European masses. But I must reiterate my confession that all my argument is based on suppositions and assumptions and must, therefore, be taken for what it is worth.

'What of the West'
Young India, 3 Sept. 1925

123. CONTROL OF EXCESSES

A Danish friend sends me translation of extracts from an article printed in *Gads Danske Magasin*. The heading he has given to the extracts is 'European Civilization and Gandhi'. In adopting his heading for *Young India* I have omitted my name as I have omitted references to my views in the extracts. My views are nothing new to the readers of *Young India*. Here is the translation received:

These extracts present a very lurid picture but probably they are true in substance. That the sum total of the activities of European nations is a denial of the teachings of the Sermon on the Mount will not, I think, be gainsaid. I have reproduced the extracts merely to emphasize the necessary caution against our being lifted off our feet by the dazzle and the glitter of European arms. If the foregoing picture were the whole of Europe it would be sad for Europe as for the world. Fortunately there is a considerable body of men and women of Europe who are devoting the whole of their energy to combat the war-fever and the breathless pursuit after material wealth and enjoyment. There are reasons for hoping that this body is daily gaining in numbers and in influence. May it be the privilege of India to take part in the new awakening and to advance it, instead of retarding it by succumbing to the European excesses which the best mind of Europe condemns in unmeasured terms and is manfully struggling to bring under effective control.

'European Civilization'
Young India, 15 Oct. 1925

124. PEACE FROM WITHIN

Bora Dada has received a letter from Germany from which I take the following:

Corruption cries to the sky. All bad men live in wealth but all good men have a hard struggle to fight out; the poorest of all are we town clerks, for our salary is very small, 35 dollars a month, and so is our life a perpetual starvation.

I often desire fervently to come and see India, to sit at the feet of Mr. Gandhi. I am quite alone. I have neither wife nor children. A poor sick niece who has none but me keeps my house. I should become a priest, if there was not my poor niece. I cannot leave her in misery. However I am an academician. I have studied classical and modern foreign languages. I have also studied Mysticism and Buddhism. I cannot find a better place nor a better salary. That is so in the Germany of today.

Before the terrible War 15 years ago I was an independent man, an investigator; now, after the terrible decline in value of our money-standard, I am a beggar like a thousand other learned men in Germany. Now I am 45 years old and you cannot think how desperate and hopeless I am, what a great disgust I feel in Europe. Here the men have no soul and are wild beasts who devour one another. Could I go to India? Could I become an Indian philosopher? I believe in India and I hope India will save us.

The opening lines of this letter might well have been written by any Indian clerk. His position is no better than the German clerk's. In India too 'bad men live in wealth and good men have a hard struggle to fight out'. It is therefore a case of distance lending enchantment to the view. Friends like this German writer must be warned against regarding India as better than Germany or any other country. Let him realize that riches are no test of goodness. Indeed poverty is the only test. A good man voluntarily embraces poverty. If the writer was at one time in affluent circumstances, Germany was at that time exploiting other countries. The remedy lies with every individual in every country. Each one has to find his peace from within. And peace to be real must be unaffected by outside circumstances. The writer says that, but for his poor niece, he might have become a priest. This seems to me to be a distorted view.

It would almost appear that the writer's present state is somewhat better than that of a priest of his imagination. For now he has at least one poor person to look after. Under the priestly licence, he would have none to look after! The fact, however, is that as a true priest he would have hundreds of nieces and even nephews to look after. As a priest the sphere of his responsibility would be as wide as that of the universe. Whereas now he slaves for himself and his niece, as a priest he would be expected to slave for the whole of distressed mankind. I would then venture to advise this friend and others like him without adopting the clerical robe to identify themselves with all in distress. They would then have all the advantages of the priestly calling without being exposed to its terrible temptations.

The German friend would like to become an Indian philosopher. I assure him that there are no territorial distinctions in philosophy. An Indian philosopher is as good or as bad as a European philosopher.

One thing the writer, in my opinion, has guessed somewhat correctly. Though India has her share of wild and soulless two-footed beasts, probably the tendency of the average Indian mind is to discard the wild beast in it. And it is my certain conviction that, if India retains the way she chose in 1921, Europe has reason to hope much from India. She chose then with the greatest deliberation the way of truth and peace and symbolized it in her acceptance of the *charkha* and non-co-operation with all that was evil. From all I know of her, she has not yet rejected it and is not likely to.

'A Cry from Germany'
Young India, 19 Nov. 1925

125. THE HUBBUB OF CITIES

An Englishman writing to his relatives in London thus gives vent to his feelings after having been in New York for 48 hours:

It is all quite true—skyscrapers, iced water, elevators express to the 25th floor, subways, Negroes; I never quite believed it before. But that's all I know. I have been here 48 hours—never such a 48 hours before—I can't last much longer. I've been walked about, talked at, dined, lunched, theatred; I'm so tired, I can hardly see. Incredible, inconceivable. My timetable is arranged to the minute—I am telephoned to wherever I am to see that I am moving on the next engagement. By a subterfuge I have escaped. I am to go out to dinner in an hour or so. You must not expect anything more than postcards. It's very cold out—freezing—while it boils within. My head goes into solution in these temperatures.

Englishmen will sympathize with me when I say that I felt about as uncomfortable reaching London for the first time as the writer of the foregoing did on reaching New York. And I know that a villager going to Bombay feels similarly bewildered and lost in finding himself in the midst of the hubbub and bustle of Bombay.

'What is it like?'
Young India, 15 Apr. 1926

126. SIMPLICITY AND ARTIFICIALITY

Satyagraha Ashram,
Sabarmati,
March 21, 1928

Dear Friend.

It was a pleasure to receive your letter after such a long time. I am sending you the two books you mention and I am adding a third—Hand-spinning Essay, the *Guide to Health* and *Takli Teacher*.

Now about the 2nd paragraph. I would just like to say that whilst I am a passionate devotee of simplicity in life, I have also discovered that it is worthless unless the echo of simplicity comes from within. The modern organized artificiality of so-called civilized life cannot have any accord with true simplicity of heart. Where the two do not correspond, there is always either gross self-deception or hypocrisy.

Yours sincerely,

T. de Manziarly

Letter to T. de Manziarly
SN 14267

§6. Progress Moral and Material

127. ECONOMIC AND MORAL PROGRESS

[22 December 1916]

Mr. M. K. Gandhi delivered an instructive lecture on 'Does economic progress clash with real progress?' at a meeting of the Muir Central College Economic Society held on Friday evening in the physical science theatre. The Hon. Pandit Madan Mohan Malaviya presided ... Mr. Gandhi delivered the following lecture:

When I accepted Mr. Kapildeva Malaviya's invitation to speak to you upon the subject of this evening, I was painfully conscious of my limitations. You are an economic society. You have chosen distinguished specialists for the subjects included in your syllabus for this year and the next. I seem to be the only speaker ill-fitted for the task set before him. Frankly and truly, I know very little of economics, as you naturally understand them. Only the other day, sitting at an evening meal, a civilian friend deluged me with a series of questions on my crankisms. As he proceeded in his cross-examination, I being a willing victim, he found no difficulty in discovering my gross ignorance of the matters. I appeared to him to be handling with a cocksureness worthy only of a man who knows not that he knows not. To his horror and even indignation, I suppose, he found that I had not even read books on economics by such well-known authorities as Mill, Marshall, Adam Smith and a host of such other authors. In despair, he ended by advising me to read these works before experimenting in matters economic at the expense of the public. He little knew that I was a sinner past redemption.

My experiments continue at the expense of trusting friends. For, there come to us moments in life when about some things we need no proof from without. A little voice within us tells us, 'You are on the right track, move neither to your left nor right, but keep to the straight and narrow way.' With such help we march forward slowly indeed, but surely and steadily. That is my position. It may be satisfactory enough for me, but it can in no way answer the requirements of a society such as yours. Still it was no use my struggling against Mr. Kapildeva

Malaviya. I knew that he was intent upon having me to engage your attention for one of your evenings. Perhaps you will treat my intrusion as a welcome diversion from the trodden path. An occasional fast after a series of sumptuous feasts is often a necessity. And as with the body, so, I imagine, is the case with the reason. And if your reason this evening is found fasting instead of feasting, I am sure it will enjoy with the greater avidity the feast that Rao Bahadur Pandit Chandrika Prasad has in store for you for the 12th of January.

Before I take you to the field of my experiences and experiments, it is perhaps best to have a mutual understanding about the title of this evening's address: *Does economic progress clash with real progress?* By economic progress, I take it, we mean material advancement without limit and by real progress we mean moral progress, which again is the same thing as progress of the permanent element in us. The subject may therefore be stated thus: 'Does not moral progress increase in the same proportion as material progress?' I know that this is a wider proposition than the one before us. But I venture to think that we always mean the larger one even when we lay down the smaller. For we know enough of science to realise that there is no such thing as perfect rest or repose in this visible universe of ours. If therefore material progress does not clash with moral progress, it must necessarily advance the latter. Nor can we be satisfied with the clumsy way in which sometimes those who cannot defend the larger proposition put their case. They seem to be obsessed with the concrete case of thirty millions of India stated by the late Sir William Wilson Hunter to be living on one meal a day. They say that before we can think or talk of their moral welfare, we must satisfy their daily wants. With these, they say, material progress spells moral progress. And then is taken a sudden jump: what is true of thirty millions is true of the universe. They forget that hard cases make bad law. I need hardly say to you how ludicrously absurd this deduction would be. No one has ever suggested that grinding pauperism can lead to anything else than moral degradation. Every human being has a right to live and therefore to find the wherewithal to feed himself and where necessary to clothe and house himself. But, for this very simple performance, we need no assistance from economists or their laws.

'Take no thought for the morrow'[1] is an injunction which finds an echo in almost all the religious scriptures of the world. In well-ordered society, the securing of one's livelihood should be and is found to be the easiest thing in the world. Indeed, the test of orderliness in a country is not the number of millionaires it owns, but the absence of starvation among its masses. The only statement that has to be examined is whether it can be laid down as a law of universal application that material advancement means moral progress.

Now let us take a few illustrations. Rome suffered a moral fall when it attained high material affluence. So did Egypt and so perhaps most countries of which we have any historic record. The descendants, kinsmen of the royal and divine Krishna, too, fell when they were rolling in riches. We do not deny to the Rockefellers and the Carnegies possession of an ordinary measure of morality but we gladly judge them indulgently. I mean that we do not even expect them to satisfy the highest standard of morality. With them material gain has not necessarily meant moral gain. In South Africa, where I had the privilege of associating with thousands of our countrymen on most intimate terms, I observed almost invariably that the greater the possession of riches, the greater was their moral turpitude. Our rich men, to say the least, did not advance the moral struggle of passive resistance as did the poor. The rich men's sense of self-respect was not so much injured as that of the poorest. If I were not afraid of treading on dangerous ground, I would even come nearer home and show you that possession of riches has been a hindrance to real growth. I venture to think that the scriptures of the world are far safer and sounder treatises on laws of economics than many of the modern text-books.

The question we are asking ourselves this evening is not a new one. It was addressed to Jesus two thousand years ago. St. Mark[2] has vividly described the scene. Jesus is in his solemn mood; he is earnest. He talks of eternity. He knows the world about him. He is himself the greatest economist of his time. He succeeded in economising time and space—he transcended them. It is to him at his best that one comes running, kneels down, and asks: 'Good Master, what shall I do that I may inherit eternal life?' And Jesus said unto him: 'Why callest thou me good? There is none good but one, that is God. Thou knowest the commandments. Do not commit adultery, Do not kill, Do not steal, Do not bear false witness, Defraud not, Honour thy father and mother.' And he answered and said unto him: 'Master, all these have I observed from my youth.' Then Jesus beholding him, loved him and said unto him: 'One thing thou lackest. Go thy way, sell whatever thou hast and give to the poor, and thou shalt have treasure in heaven—come take up the cross and follow me.' And he was sad at that saying and went away grieved—for he had great possessions. And Jesus looked round about and said unto his disciples: 'How hardly shall they that have riches enter into the kingdom of God.' And the disciples were astonished at his words. But Jesus answereth again and saith unto them: 'Children, how hard it is for them that trust in riches to enter into the kingdom of God. It is easier for a camel to go through the eye of a needle than for a rich man to enter into the kingdom of God!'

Here you have an eternal rule of life stated in the noblest words the English language is capable of producing. But the disciples nodded unbelief as we do even to this day. To him they said as we say today: 'But look how the law fails in practice. If we sell all and have nothing, we shall have nothing to eat. We must have money or we cannot even be reasonably moral.' So they state their case thus. 'And they were astonished out of measure saying among themselves: "Who then can be saved?"' And Jesus looking upon them saith: 'With men it is impossible but not with God, for with God all things are possible.' Then Peter began to say unto him: 'Lo, we have left all, and have followed thee.' And Jesus answered and said: 'Verily I say unto you there is no man that has left house or brethren or sisters, or father or mother, or wife or children or lands for my sake and the Gospels, but he shall receive one hundred fold, now in this time houses and brethren and sisters and mothers and children and lands with persecutions and in the world to come eternal life. But many that are first shall be last and the last first.' You have here the result or reward, if you prefer the term, of following the law.

I have not taken the trouble of copying similar passages from the other non-Hindu scriptures and I will not insult you by quoting in support of the law stated by Jesus passages from the writings and sayings of our own sages, passages even stronger if possible than the Biblical extracts I have drawn your attention to. Perhaps the strongest of all the testimonies in favour of the affirmative answer to the question before us are the lives of the greatest teachers of the world. Jesus, Mahomed, Buddha, Nanak, Kabir, Chaitanya, Shankara, Dayanand, Ramkrishna were men who exercised an immense influence over and moulded the character of thousands of men. The world is the richer for their having lived in it. And they were all men who deliberately embraced poverty as their lot.

I should not have laboured my point as I have done, if I did not believe that, in so far as we have made the modern materialistic craze our goal, in so far are we going downhill in the path of progress. I hold that economic progress in the sense I have put it is antagonistic to real progress. Hence the ancient ideal has been the limitation of activities promoting wealth. This does not put an end to all material ambition. We should still have, as we have always had, in our midst people who make the pursuit of wealth their aim in life. But we have always recognised that it is a fall from the ideal. It is a beautiful thing to know that the wealthiest among us have often felt that to have remained voluntarily poor would have been a higher state for them. That you cannot serve God and Mammon is an economic truth of the highest value. We have to make our choice. Western nations today are groaning under the heel of the monster-god of materialism. Their moral growth has become stunted. They measure their progress in £.s.d. American wealth has become standard. She is the envy of the other nations. I have heard many of our countrymen say that we will gain American wealth but avoid its methods. I venture to suggest that such an attempt if it were made is foredoomed to failure.

We cannot be 'wise, temperate and furious'[3] in a moment. I would have our leaders teach us to be morally supreme in the world. This land of ours was once, we are told, the abode of the gods. It is not possible to conceive gods inhabiting a land which is made hideous by the smoke and the din of mill chimneys and factories and whose roadways are traversed by rushing engines dragging numerous cars crowded with men mostly who know not what they are after, who are often absent-minded, and whose tempers do not improve by being uncomfortably packed like sardines in boxes and finding themselves in the midst of utter strangers who would oust them if they could and whom they would in their turn oust similarly. I refer to these things because they are held to be symbolical of material progress. But they add not an atom to our happiness. This is what Wallace, the great scientist, has said as his deliberate judgement.

In the earliest records which have come down to us from the past, we find ample indications that general ethical considerations and conceptions, the accepted standard of morality, and the conduct resulting from these were in no degree inferior to those which prevail to-day.

In a series of chapters, he then proceeds to examine the position of the English nation under the advance in wealth it has made. He says:

This rapid growth of wealth and increase of our power over nature put too great a strain upon our crude civilization, on our superficial Christianity, and it was accompanied by various forms of social immorality almost as amazing and unprecedented.

He then shows how factories have risen on the corpses of men, women and children, how as the country has rapidly advanced in riches, it has gone down in morality. He shows this by dealing with insanitation, life-destroying trades, adulteration, bribery and gambling. He shows how, with the advance of wealth, justice has become immoral, deaths from alcoholism and suicide have increased, the average of premature births and congenital defects has increased, and prostitution has become an institution. He concludes his examination by these pregnant remarks:

The proceedings of the divorce courts show other aspects of the result of wealth and leisure, while a friend who had been a good deal in London society assured me that both in country houses and in London various kinds of orgies were occasionally to be met with which would hardly have been surpassed in the period of the most dissolute emperors. Of war, too, I need say nothing. It has always been more or less chronic since the rise of the Roman Empire; but there is now undoubtedly a disinclination for war among all civilized peoples. Yet the vast burden of armaments, taken together with the most pious declarations in favour of peace, must be held to show an almost total absence of morality as a guiding principle among the governing classes.

Under the British aegis, we have learnt much, but it is my firm belief that there is little to gain from Britain in intrinsic morality, that if we are not careful, we shall introduce all the vices that she has been a prey to, owing to the disease of materialism. We can profit by that connection only if we keep our civilization, and our morals, straight, i.e., if instead of boasting of the glorious past, we express the ancient moral glory in our own lives and let our lives bear witness to our past. Then we shall benefit her and ourselves. If we copy her because she provides us with rulers, both they and we shall suffer degradation. We need not be afraid of ideals or of reducing them to practice even to the uttermost. Ours will only then be a truly spiritual nation when we shall show more truth than gold, greater fearlessness than pomp of power and wealth, greater charity than love of self. If we will but clean our houses, our palaces and temples of the attributes of wealth and show in them the attributes of morality, we can offer battle to any combinations of hostile forces without having to carry the burden of a heavy militia. Let us seek first the kingdom of God and His righteousness and the irrevocable promise is that everything will be added with us. These are real economics. May you and I treasure them and enforce them in our daily life.

An interesting discussion followed in the course of which several students put questions to the lecturer....
Prof. Jevons said ... It was necessary for economists to exist. It was not their business to lay down what the end should be. That was the business of philosophers....
Prof. Gidwani, president of the society, thanked the lecturer for his address....

Prof. Higginbottom said that there was no economic problem which could be separated from the moral problem....

Mr. Gandhi in the course of his remarks referred to Mr. Jevons's remark about the need for economists and said that it was said that dirt was matter misplaced. So also when an economist was misplaced, he was hurtful. He certainly thought that the economist had a place in the economy of nature when he occupied the humble sphere for which he was created. If an economist did not investigate the laws of God and show them how to distribute wealth so that there might not be poverty, he was a most unwelcome intrusion on the Indian soil. He would also suggest for the reflection of their economic students and professors that what might be good for England and America need not necessarily be good for India. He thought that most of the economic laws which were consistent with moral laws were of universal application, but there might be in their restricted application some distinction and difference. So he would utter the note of warning that Indian conditions being in some respects so essentially different from the English and American conditions, it was necessary to bring to bear on the matters that presented themselves to the economists a fresh mind. If they did so, both Indians and the economists would derive benefit. Mr. Higginbottom, he said, was studying the real economics that were so necessary for India and reducing his studies inch by inch to practice and that was the safest guide to follow, whether they were students or professors. Referring to a question by a student, he said that a man should not hoard money for selfish ends, but if he wished to hoard money as a trustee for the millions of India, he would say that he might have as much riches as he could. Ordinarily, economists prescribed laws for the rich people. It was against those economists that he would always cry out.

As regards another question, whether factories should not be replaced by cottage industries, Mr. Gandhi spoke approvingly of the suggestion but said that the economists should first of all examine with patience their indigenous institutions. If they were rotten, they must be wiped out and if there were remedies which could be suggested for their betterment, they should improve them.

As regards intercourse with other nations, he said that he did not think that they necessarily advanced one little bit in their moral growth by bringing their masses with others into physical contact and pointed to Indians in South Africa as an instance. The rapid locomotion such as steamers, trains and others dislocated so many of their ideals and created a great deal of mischief.

As regards the question what was the minimum and the maximum wealth a man should have—he would answer in the words of Jesus, Ramkrishna and others who said 'none'.

The Hon. Pandit Madan Mohan Malaviya in his concluding remarks offered a cordial vote of thanks to Mr. Gandhi for his excellent address. The ideals which Mr. Gandhi put before them, he said, were so high that he did not expect that all of them would be prepared to subscribe to all of them. But he was sure they would agree with the main object he put before them, namely, that they should go for the welfare of man as the test of all economic questions with which they dealt....

Speech at Muir College Economic Society, Allahabad
The Leader, 25 Dec. 1916

[1] Matthew 6: 34.
[2] Mark 10: 17-31.
[3] 'Who can be wise, amazed, temperate and furious,/Loyal and neutral, in a moment?/No man.' (*Macbeth*, II. iii)

128. MORAL GROWTH AND MATERIAL PROSPERITY

The only claim I have on your indulgence is that some months ago I attended with Mr. Ewbank a meeting of mill-hands to whom he wanted to explain the principles of co-operation. The *chawl* in which they were living was as filthy as it well could be. Recent rains had made matters worse. And I frankly confess that had not it been for Mr. Ewbank's great zeal for the cause he has made his own, I should have shirked the task. But there we were, seated on a fairly worn out *charpai*, surrounded by men, women and children. Mr. Ewbank opened fire on a man who had put himself forward and who wore not a particularly innocent countenance. After he had engaged him and the other people about him in Gujarati conversation, he wanted me to speak to the people. Owing to the suspicious looks of the man who was first spoken to, I naturally pressed home the moralities of co-operation. I fancy that Mr. Ewbank rather liked the manner in which I handled the subject. Hence, I believe, his kind invitation to me to tax your patience for a few moments upon a consideration of co-operation from a moral standpoint.

My knowledge of the technicality of co-operation is next to nothing. My brother Devdhar has made the subject his own. Whatever he does naturally attracts me and predisposes me to think that there must be something good in it and the handling of it must be fairly difficult. Mr. Ewbank very kindly placed at my disposal some literature too on the subject. And I have had a unique opportunity of watching the effect of some co-operative effort in Champaran. I have gone through Mr. Ewbank's ten main points which are like the commandments, and I have gone through the twelve points of Mr. Collins of Behar, which remind me of the law of the twelve tables. There are so-called agricultural banks in Champaran. They were to me disappointing efforts, if they were meant to be demonstrations of the success of co-operation. On the other hand, there is quiet work in the same direction being done by Mr. Hodge, a missionary whose efforts are leaving their impression on those who come in contact with him. Mr. Hodge is a co-operative enthusiast and probably considers that the results which he sees flowing from his efforts are due to the working of co-operation. I who was able to watch the two efforts had no hesitation in inferring that the personal equation counted for success in the one and failure in the other instance.

I am an enthusiast myself, but twenty-five years of experimenting and experience have made me a cautious and discriminating enthusiast. Workers in a cause necessarily, though quite unconsciously, exaggerate its merits and often succeed in turning its very defects into advantages. In spite of my caution I consider the little institution I am conducting in Ahmedabad as the finest thing in the world. It alone gives me sufficient inspiration. Critics tell me that it represents a soulless soul-force and that its severe discipline has made it merely mechanical. I suppose both—the critics and I—are wrong. It is, at best, a humble attempt to place at the disposal of the nation a home where men and women may have scope for free and unfettered development of character, in keeping with the national genius, and if its controllers do not take care, the discipline that is the foundation of character, may frustrate the very end in view. I would venture, therefore, to warn enthusiasts in co-operation against entertaining false hopes.

With Sir Daniel Hamilton, it has become a religion. On the 13th January last, he addressed the students of the Scottish Churches College, and in order to point a moral he instanced Scotland's poverty of two hundred years ago and showed how that great country was raised from a condition of poverty to plenty. He said:

There were two powers which raised her—the Scottish Church and the Scottish banks. The Church manufactured the men and the banks manufactured the money to give to the men a start in life.... The Church disciplined the nation in the fear of God which is the beginning of wisdom and in the parish schools of the Church, the children learned that the chief end of man's life was to glorify God and to enjoy Him for ever. Men were trained to believe in God and in themselves, and on the trustworthy character so created, the Scottish banking system was built.

Sir Daniel then shows that it was possible to build up the marvellous Scottish banking system only on the character so built. So far there can only be perfect agreement with Sir Daniel, for 'Without character there is no co-operation' is a sound maxim. But he would have us go much further. He thus waxes eloquent on co-operation:

Whatever may be your day-dreams of India's future, never forget this that it is to weld India into one, and so enable her to take her rightful place in the world, that the British Government is here; and the welding hammer in the hand of the Government is the co-operative movement.

In his opinion, it is the panacea of all the evils that afflict India at the present moment. In its extended sense it can justify the claim on one condition which need not be mentioned here; in the limited sense in which Sir Daniel has used it, I venture to think, it is an enthusiast's exaggeration. Mark his peroration:

Credit, which is only Trust and Faith, is becoming more and more the money power of the world, and in the parchment bullet into which is impressed the faith which removes mountains, India will find victory and peace.

Here there is evident confusion of thought. The credit which is becoming the money power of the world has little moral basis and is not a synonym for Trust or Faith, which are purely moral qualities.

After twenty years' experience of hundreds of men, who had dealings with banks in South Africa, the opinion I had so often heard expressed has become firmly rooted in me, that the greater the rascal, the greater the credit he enjoys with his banks. The banks do not pry into his moral character; they are satisfied that he meets his over-drafts and promissory notes punctually. The credit system has encircled this beautiful globe of ours like a serpent's coil, and if we do not mind, it bids fair to crush us out of breath. I have witnessed the ruin of many a home through the system, and it has made no difference whether the credit was labelled co-operative or otherwise. The deadly coil has made possible the devastating spectacle in Europe, which we are helplessly looking on. It was perhaps never so true as it is today that as in law so in war the longest purse finally wins. I have ventured to give prominence to the current belief about credit system in order to emphasise the point that the co-operative movement will be a blessing to India only to the extent that it is a moral movement strictly directed by men fired with religious fervour. It follows, therefore, that co-operation should be confined to men wishing to be morally right, but failing to do so, because of grinding poverty or of the grip of the *mahajan*. Facility for obtaining loans at fair rates will not make immoral or unmoral men moral. But the wisdom of the State or philanthropists demands that they should help, on the onward path, men struggling to be good.

Too often do we believe that material prosperity means moral growth. It is necessary that a movement which is fraught with so much good to India should not degenerate into one for merely advancing cheap loans. I was therefore delighted to read the recommendation in the *Report of the Committee on Co-operation in India*, that

> they wish clearly to express their opinion that it is to true co-operation alone, that is, to a co-operation which recognises the moral aspect of the question that Government must look for the amelioration of the masses and not to a pseudo-co-operative edifice, however imposing, which is built in ignorance of co-operative principles.

With this standard before us, we will not measure the success of the movement by the number of co-operative societies formed, but by the moral condition of the co-operators. The Registrars will in that event ensure the moral growth of existing societies before multiplying them. And the Government will make their promotion conditional, not upon the number of societies they have registered, but the moral success of the existing institutions. This will mean tracing the course of every pice lent to the members. Those responsible for the proper conduct of co-operative societies will see to it that the money advanced does not find its way into the toddy-sellers' till or into the pockets of the keepers of gambling dens. I would excuse the rapacity of the *mahajan* if it has succeeded in keeping the gambling die or toddy from the ryot's home.

A word perhaps about the *mahajan* will not be out of place. Co-operation is not a new device. The ryots co-operate to drum out monkeys or birds that destroy their crops. They co-operate to use a common thrashing floor. I have found them co-operate to protect their cattle to the extent of their devoting their best land for the grazing of their cattle. And they have been found co-operating against a particularly rapacious *mahajan*. Doubt has been expressed as to the success of co-operation because of the tightness of the *mahajan's* hold on the ryots. I do not share the fears. The mightiest *mahajan* must, if he represents an evil force, bend before co-operation, conceived as an essentially moral movement. But my limited experience of the *mahajan* of Champaran has made me revise the accepted opinion about his 'blighting influence'. I have found him to be not always relentless, not always exacting of the last pie. He sometimes serves his clients in many ways or even comes to their rescue in the hour of their distress. My observation is so limited that I dare not draw any conclusions from it, but I respectfully enquire whether it is not possible to make a serious effort to draw out the good in the *mahajan* and help him or induce him to throw out the evil in him. May he not be induced to join the army of co-operation, or has experience proved that he is past praying for?

I note that the movement takes note of all indigenous industries. I beg publicly to express my gratitude to Government for helping me in my humble effort to improve the lot of the weaver. The experiment I am conducting shows that there is a vast field for work in this direction. No well-wisher of India, no patriot dare look upon the impending destruction of the hand-loom weaver with equanimity. As Dr. Mann has stated, this industry used to supply the peasant with an additional source of livelihood and an insurance against famine. Every Registrar who will nurse back to life this important and graceful industry will earn the gratitude of India. My humble effort consists of, firstly, in making researches as to the possibilities of simple reforms in the orthodox hand-looms, secondly, in weaning the educated youth from the craving for Government or other service and the feeling that education renders him unfit for independent occupation and inducing him to take to weaving as a calling as honourable as that of a barrister or a doctor, and, thirdly, by helping those who have abandoned their occupation to revert to it. I will not weary the audience with any statement on the first two parts of the experiment. The third may be allowed a few sentences as it has a direct bearing upon the subject before us. I was able to enter upon it only six months ago. Five families that had left off the calling have reverted to it and they are doing a prosperous business. The Ashram supplies them at their door with the yarn they need; it volunteers to take delivery of the cloth woven, paying them cash at the market rate. The Ashram merely loses interest on the loan advanced for the yarn. It has as yet suffered no loss and is able to restrict its loss to a minimum by limiting the loan to a particular figure. All future transactions are strictly cash. We are able to command a ready sale for the cloth received. The loss of interest, therefore, on the transaction is negligible. I would like the audience to note its purely moral character from start to finish.

The Ashram depends for its existence on such help as *friends* render it. We, therefore, can have no warrant for charging interest. The weavers could not be saddled with it. Whole families that were breaking to pieces are put together again. The use of the loan is predetermined. And we the middlemen being volunteers obtain the privilege of entering into the lives of these families I hope for their and our betterment. We cannot lift them without being lifted ourselves. This last relationship has not yet been developed, but we hope at an early date to take in hand the education too of these families and not rest satisfied till we have touched them at every point. This is not too ambitious a dream. God willing, it will be a reality some day. I have ventured to dilate upon the small experiment to illustrate what I mean by co-operation to present it to others for imitation. Let us be sure of our ideal. We shall ever fail to realise it, but we should never cease to strive for it. Then there need be no fear of 'co-operation of scoundrels' that Ruskin so rightly dreaded.

'The Moral Basis of Co-operation'
Indian Review, October 1917

V

Politics and Religion

§ 1. Politics and Religion

129. EQUANIMITY IN POLITICS

Lord Selborne made a significant speech while laying the foundation-stone of the new Town Hall at Germiston. The speech consists of advice on morality and politics. Addressed as it is to the whites, it is perhaps of less direct interest to us from a political standpoint. But from the moral point of view, Lord Selborne's words deserve to be pondered over. Accordingly, we summarise the speech below:

'The Municipalities exercise a necessary influence on the lives of us whites, who are politically very active. They are the schools which prepare men for the administration of the State. It is there that the seeds of the freedom of a people are sown and nurtured. The British prefer a harsh but self-governing political system to a mild alien rule. At all times and everywhere, the Municipalities have been the chief forum of public opinion. The Municipality serves as a training-ground not only for its members but also for the electorate and for others who discuss public affairs. The electors should never forget how to offer legitimate criticism. This is a region in which engaging storms blow. Storms are of two kinds: actual storms and the metaphorical ones in politics. One who preserves one's equanimity during an actual storm can be said to have a cool and collected temperament. In the same way, one who remains unperturbed during a political storm is considered a man of firm convictions. I should consider a person reliable who sticks to his own way of life in fair weather and foul. A man should prove that he is absolutely firm in his principles regardless of whether or not people interpret his words or actions aright.'

'Lord Selborne' (G.)
Indian Opinion, 16 June 1906

130. SOUL-POWER AND PHYSICAL FORCE

[27 April 1915]

Mr. Chairman and dear friends, Madras has wellnigh exhausted the English vocabulary in using adjectives of virtue with reference to my wife and myself and, if I may be called upon to give an opinion as to where I have been smothered with kindness, love and attention, I would have to say: it is Madras. But, as I have said so often, I believe it of Madras. So it is no wonder to me that you are lavishing all these kindnesses with unparalleled generosity, and now the worthy President of the Servants of India Society[1]—under which Society I am going through a period of probation—has, if I may say so, capped it all. Am I worthy of these things? My answer from the innermost recesses of my heart is an emphatic 'No'. But I have come to India to become worthy of every adjective that you may use, and all my life will certainly be dedicated to prove worthy of them, if I am to be a worthy servant.

And so it is that you have sung that beautiful national song,[2] on hearing which all of us sprang to our feet. The poet[3] has lavished all the adjectives that he possibly could to describe Mother India. He describes Mother India as sweet-smelling, sweet-speaking, fragrant, all-powerful, all-good, truthful, a land flowing with milk and honey, and having ripe fields, fruits and grains, and inhabited by a race of men of whom we have only a picture in the great Golden Age. He pictures to us a land which shall embrace in its possession the whole of the world, the whole of humanity by the might or right not of physical power but of soul-power. Can we sing that hymn? I ask myself, 'Can I, by any right, spring to my feet when I listen to that song?' The poet no doubt gave us a picture for our realisation the words of which simply remain prophetic, and it is for you, the hope of India, to realise every word that the poet has said in describing this Motherland of ours. To-day I feel that these adjectives are very largely misplaced in his description of the Motherland, and it is for you and for me to make good the claim that the poet has advanced on behalf of his Motherland.

You—the students of Madras as well as students all over India—are you receiving an education which will make you worthy to realise that ideal and which will draw the best out of you, or is it an education which has become a factory for making Government employees or clerks in commercial offices? Is the goal of the education that you are receiving that of mere employment whether in the Government departments or other departments? If that be the goal of your education, if that is the goal that you have set before yourselves, I feel and I fear that the vision which the poet pictured for himself is far from being realised. As you have heard me say perhaps, or as you have read, I am and I have been a determined opponent of modern civilization. I want you to turn your eyes today upon what is going on in Europe and if you have come to the conclusion that Europe is today groaning under the heels of the modern civilization, then you and your elders will have to think twice before you can emulate that civilization in our Motherland. But I have been told: 'How can we help it, seeing that our rulers bring that culture to our Motherland?' Do not make any mistake about it at all. I do not for one moment believe that it is for any rulers to bring that culture to you unless you are prepared to accept it, and if it be that the rulers bring that culture before us, I think that we have forces within ourselves to enable us to reject that culture without having to reject the rulers themselves.

I have said on many a platform that the British race is with us. I decline to go into the reasons why that race is with us, but I do believe that it is possible for India if she would but live up to the tradition of the sages of whom you have heard from our worthy President, to transmit a message through this great race, a message not of physical might, but a message of love. And then, it will be your privilege to conquer the conquerors not by shedding blood but by sheer force of spiritual predominance. When I consider what is going on today in

India, I think it is necessary for us to say what our opinion is in connection with the political assassinations and political dacoities. I feel that these are purely a foreign importation which cannot take root in this land. But you the student world have to beware, lest mentally or morally you give one thought of approval to this kind of terrorism. I, as a passive resister, will give you another thing very substantial for it. Terrorise yourself; search within; by all means resist tyranny wherever you find it; by all means resist encroachment upon your liberty, but not by shedding the blood of the tyrant. That is not what is taught by our religion. Our religion is based upon *ahimsa*, which in its active form is nothing but love, love not only to your neighbours, not only to your friends but love even to those who may be your enemies.

One word more in connection with the same thing. I think that if we were to practise truth, to practise *ahimsa*, we must immediately see that we also practise fearlessness. If our rulers are doing what in our opinion is wrong, and if we feel it our duty to let them hear our advice even though it may be considered sedition, I urge you to speak sedition—but at your peril. You must be prepared to suffer the consequences. And when you are ready to suffer the consequences and not hit below the belt, then I think you will have made good your right to have your advice heard even by the Government.

I ally myself with the British Government, because I believe that it is possible for me to claim equal partnership with every subject of the British Empire. I today claim that equal partnership. I do not belong to a subject race. I do not call myself a member of a subject race. But there is this thing: it is not for the British governors to give you; it is for you to take the thing. I want and I can take the thing. That I want only by discharging my obligations. Max Muller[4] has told us—we need not go to Max Muller to interpret our own religion but he says, our religion consists of the four letters 'D-u-t-y' and not the five letters 'R-i-g-h-t'. And if you believe that all that we want can grow from better discharge of our duty, then think always of your duty and fighting along those lines, you will have no fear of any man, you will fear only God. That is the message that my master—if I may say so, your master, too—Mr. Gokhale has given to us. What is that message then?

It is in the constitution of the Servants of India Society and that is the message by which I wish to be guided in my life. The message is to spiritualize the political life and the political institutions of the country. We must immediately set about realising its practice. The students cannot be away from politics. Politics is as essential to them as religion.

Politics cannot be divorced from religion. Politics divorced from religion becomes debasing. Modern culture and modern civilization are such politics. My views may not be acceptable to you, I know. All the same I can only give you what is stirring me to my very depths. On the authority of my experiences in South Africa, I claim that your countrymen who had not that modern culture but who had that strength of the *Rishis* of old, who have inherited the *tapascharya* performed by the *Rishis*, without having known a single word of English literature and without knowing anything whatsoever of the present modern culture, they are able to rise to their full height. And what has been possible for the uneducated and illiterate countrymen of ours in South Africa is ten times possible for you and for me today in this sacred land of ours. May that be your privilege and may that be my privilege!

Speech at YMCA, Madras
The Hindu, 28 Apr. 1915
Speeches and Writings of Mahatma Gandhi

[1] V. S. Srinivasa Sastri.
[2] *Vande Mataram.*
[3] Bankim Chandra Chatterji.
[4] (1823-1900), German orientalist.

131. SPIRITUALIZING POLITICAL LIFE

[8 May 1915]

My dear Countrymen, before I perform this ceremony to which you have called me, I wish to say this to you that you have given me a great opportunity or rather a privilege on this great occasion. I saw in the recitation,[1] the beautiful recitation that was given to me, that God is with them whose garment was dusty and tattered. My thoughts immediately went to the end of my garment; I examined and found that it is not dusty and it is not tattered; it is fairly spotless and clean. God is not in me. There are other conditions attached; but in these conditions too I may fail; and you, my dear countrymen, may also fail; and if we do tend this well, we should not dishonour the memory of one whose portrait you have asked me to unveil this morning. I have declared myself his disciple in the political field and I have him as my *Rajya Guru*; and this I claim on behalf of the Indian people. It was in 1896 that I made this declaration, and I do not regret having made the choice.

Mr. Gokhale taught me that the dream of every Indian, who claims to love his country, should be to act in the political field, should be not to glorify in language, but to spiritualise the political life of the country, and the political institutions of the country. He inspired my life and is still inspiring [it]; and in that I wish to purify myself and spiritualise myself. I have dedicated myself to that ideal. I may fail, and to what extent I may fail, I call myself to that extent an unworthy disciple of my master.

What is the meaning of spiritualising the political life of the country? What is the meaning of spiritualising myself? That question has come before me often and often and to you it may seem one thing, to me it may seem another thing; it may mean different things to the different members of the Servants of India Society itself. It shows much difficulty and it shows the difficulties of all those who want to love their country, who want to serve their country and who want to honour their country. I think political life must be an echo of private life and that there cannot be any divorce between the two.

I was by the side of that saintly politician to the end of his life and I found no ego in him. I ask you, members of the Social Service League, if there is no ego in you. If he wanted to shine,—he wanted to shine in the political field of his country,—he did so not in order that he might gain public applause, but in order that his country might gain. He developed every particular faculty in him, not in order to win the praise of the world for himself, but in order that his country might gain. He did not seek public applause, but they were showered upon him, they were thrust upon him; he wanted that his country might gain and that was his great inspiration.

There are many things for which India is blamed, very rightly, and if you should add one more to our failure, the blame will descend not only on you but also on me for having participated in today's functions. But I have great faith in my countrymen.

You ask me to unveil this portrait[2] today, and I will do so in all sincerity and sincerity should be the end of your life.

Speech at Government High School, Bangalore
Indian Review, May 1915

[1] From Tagore's *Gitanjali*.
[2] Of G. K. Gokhale.

132. SHADOW AND SUBSTANCE IN POLITICS

Laburnum Road,
Bombay,
August 4, 1919

Dear Mr. Arundale,[1]

I have read and re-read your kind letter for which I thank you. I am publishing the letter in *Young India* together with this reply.

Much as I should like to follow your advice, I feel that I am incompetent for the task set forth by you in your letter. I am fully aware of my limitations. My bent is not political but religious and I take part in politics because I feel that there is no department of life which can be divorced from religion and because politics touch the vital being of India almost at every point. It is therefore absolutely necessary that the political relations between Englishmen and ourselves should be put on a sound basis. I am endeavouring to the best of my ability to assist in the process. I do not take much interest in the reforms because they are in safe hands and because reforms *cum* Rowlatt legislation mean to my mind a stalemate. Rowlatt legislation represents a poisonous spirit. After all, the English civilians can, unless Indian opinion produces a healthy reaction upon them, reduce the reforms practically to a nullity. They distrust us and we distrust them. Each considers the other as his natural enemy. Hence the Rowlatt legislation. The Civil Service has devised the legislation to keep us down. In my opinion, that legislation is like the coil of the snake round the Indian body. The obstinacy of the Government in clinging to the hateful legislation in spite of the clearest possible demonstration they have had of public opinion against it makes me suspect the worst. With the views enunciated above, you will not wonder at my inability to interest myself in the reforms. Rowlatt legislation blocks the way. And my life is dedicated among other things to removing the block.

Let there be no mistake. Civil resistance has come to stay. It is an eternal doctrine of life which we follow consciously or unconsciously in many walks of life. It is the new and extended application of it which has caused misgivings and excitement. Its suspension is designed to demonstrate its true nature, and to throw the responsibility for the removal of the Rowlatt legislation on the Government as also the leaders (you among them) who have advised me to suspend it. But if within a reasonable time the legislation is not removed, civil resistance will follow as surely as day follows night. No weapon in the Government armoury can either overcome or destroy that eternal force. Indeed a time must come when civil resistance will be recognized as the most efficacious, if also the most harmless, remedy for securing redress of grievances.

You suggest the desirability of unity. I think unity of goal we have. But parties we shall always have—and we may not find a common denominator for improvements. For some will want to go further than some others. I see no harm in a wholesome variety. What I would rid ourselves of is distrust of one another and imputation of motives. Our besetting sin is not our differences but our littleness. We wrangle over words, we fight often for shadow and lose the substance. As Mr. Gokhale used to say, our politics are a pastime of our leisure hours when they are not undertaken as a stepping-stone to a career in life.

I would invite you and every editor to insist on introducing charity, seriousness and selflessness in our politics. And our disunion will not jar as it does today. It is not our differences that really matter. It is the meanness behind that is undoubtedly ugly.

The Punjab sentences are inextricably mixed up with the Rowlatt agitation. It is therefore as imperatively necessary to have them revised as it is to have the Act removed. I agree with you that the Press Act requires overhauling. The Government are actually promoting sedition by high-handed executive action. And I was sorry to learn that Lord Willingdon[2] is reported to have taken—the sole responsibility for the—in my opinion unwarranted action[3] against *The Hindu* and the *Swadesha Mitran*. By it, they have not lost in prestige or popularity. They have gained in both. Surely there are judges enough in the land who would convict where a journalist has overstepped the bounds of legitimate criticism and uttered sedition. I am not enamoured of the Declaration of Rights business. When we have changed the spirit of the English civilian, we shall have made considerable headway with the Declaration of Rights. We must be honourable friends, or equally honourable enemies. We shall be neither, unless we are manly, fearless and independent. I would have us to treasure Lord Willingdon's advice and say 'no' when we mean 'no' without fear of consequences. This is unadulterated civil resistance. It is the way to friendliness and friendship. The other is the age-worn method of open violence on honourable lines in so far as violence can be allowed to be honourable. For me the roots of violence are in dishonour. I have therefore ventured to present to India the former, in its complete form called *satyagraha*, whose roots are always in honour.

Yours sincerely,
M. K. GANDHI

Letter to G. S. Arundale
Young India, 6 Aug. 1919

[1] In reply to his letter of 26 July, appealing to Gandhi that, since civil disobedience had been suspended, he should join in working the Montagu-Chelmsford Constitutional Reforms.

[2] 1866-1941; Governor of Bombay; later, Viceroy of India, 1931-6.

[3] The Government demanded a security of Rs 2,000 from each of these Madras newspapers and banned *The Hindu* in the Punjab and in Burma.

133. 'TRUTH EVEN UNTO THE WICKED'

[Delhi,
After 18 January 1920]

I naturally feel the greatest diffidence about joining issue with the Lokamanya[1] in matters involving questions of interpretation of religious works. But there are things in or about which instinct transcends even interpretation. For me there is no conflict between the two texts quoted by the Lokamanya. The Buddhist text lays down an eternal principle. The text from the *Bhagavad Gita* shows to me how the principle of conquering hate by love, untruth by truth, can and must be applied. If it be true that God metes out the same measure to us that we mete out to others, it follows that if we would escape condign punishment, we may not return anger but gentleness even against anger. And this is the law not for the unworldly but essentially for the worldly. With deference to the Lokamanya, I venture to say that it betrays mental laziness to think that the world is not for *sadhus*. The epitome of all religions is to promote *purushartha*, and *purushartha* is nothing but a desperate attempt to become *sadhu*, i.e., to become a gentleman in every sense of the term.

Finally, when I wrote the sentence about 'everything being fair in politics' according to the Lokamanya's creed, I had in mind his oft-repeated quotation.[2]

To me it enunciates bad law. And I shall not despair of the Lokamanya with all his acumen agreeably surprising India one day with a philosophical dissertation proving the falsity of the doctrine. In any case I pit the experience of a third of a century against the doctrine underlying 'wickedness unto the wicked'. The true law is 'truth even unto the wicked'.

M. K. GANDHI

'Note on Tilak's Letter'
Young India, 28 Jan. 1920

---

[1] An honorific, signifying 'revered by the people'. B. G. Tilak, the militant patriot, came to be widely known as 'the Lokamanya'.

[2] 'Wickedness unto the wicked.'

134. MEN AND SYSTEMS

A journalist has made these remarks ...[1]

Parliament is indeed barren. I do not imagine that its nature can change in India. I live, however, in the hope that our Parliament will only remain barren and not give birth to a wicked son. I cannot abandon practical considerations. The ideal is one only, namely, *Ramarajya*. But where can we find Rama? The journalist says, 'whom the people approve'. People means Parliament and, in our view, whomsoever the Parliament approves is a virtuous man or woman. I am suggesting many ways to ensure that the voice of Parliament is really the voice of the people and not that of hired voters. With this end in view I am searching for a device which will enable us to listen to the voice of the entire people. All systems are bound to be defective. We are looking for a system which will yield maximum benefit to India. Good men can transform a bad system into a good one—like the wise housewife who transforms dust into grains. Wicked men can misuse the best of systems and make it defective, like a foolish housewife who allows bright food grains to decay into dust. I am therefore on the look-out for good men in India and employing devices to sort out such men. But what can a man do? He can only make an honest effort. The fruit lies in the hands of God. The efforts of many, and not one, are required for securing the desired fruit. Many other factors determine the fruit. Therefore 'one step is enough' for us.

The same journalist says ...[2]

All this criticism is true; but these errors are unavoidable. Should we abandon truth because falsehood parades under the garb of truth? Man has to cultivate the inner sense. It does not belong to every man as a natural gift. Its cultivation needs spiritual surroundings and constant effort. It is a delicate plant. Children do not have anything like an inner voice. Those who are considered barbarous have no inner sense. The inner sense reflects the impact produced by a cultivated intellect on the heart. It would therefore be ridiculous if every man claimed to possess an inner voice.

Even then, there is not the least cause for apprehension if many people claim it. Injustice cannot be justified in the name of conscience. Moreover, those who work under the false pretext of an inner voice are not prepared to undergo suffering. Their activities will come to a halt after a little while. It would, therefore, not harm the world if any number of people made this claim. Those who have played pranks with this delicate device are likely to ruin themselves, not others. Newspapers provide many instances of this truth. Many newspapers are at present engaged in spreading poison in the name of public service. This business will not, however, last long. One day or other, people are bound to get sick of it. The Punjab is the greatest culprit in this respect. It is surprising that such dirty journals can manage to pay their way. Why do people encourage them? As long as moneyed men are there, thieves would not starve. Similarly, when a section of the people are prepared to read poisonous writings, such journals are bound to prosper. The only cure consists in the cultivation of a clean public opinion.

'My Notes' (G.)
Navajivan, 24 Aug. 1924

[1] Not reproduced here. They were critical of the British parliamentary system.
[2] Not reproduced here.

135. VOLUNTARY POVERTY

[London,
23 September 1931]

You will be astonished to hear from me that, although to all appearances my mission is political, I would ask you to accept my assurance that its roots are—if I may use that term—spiritual. It is commonly known, though perhaps not believed, that I claim that at least my politics are not divorced from morality, from spirituality, from religion. I have claimed—and the claim is based upon extensive experience—that a man who is trying to discover and follow the will of God cannot possibly leave a single field of life untouched. I came also, in the course of my service, to the conclusion that if there was any field of life where morality, where truth, where fear of God, were not essential, that field should be given up entirely.

But I found also that the politics of the day are no longer a concern of kings, but that they affect the lowest strata of society. And I found, through bitter experience that, if I wanted to do social service, I could not possibly leave politics alone.

Do not please consider that I want to speak to you tonight about politics and somehow or other connect voluntary poverty with politics. That is not my intention. I have simply given you an introduction how I came to believe in the necessity of voluntary poverty for any social worker or for any political worker who wanted to remain untouched by the hideous immorality and untruth that one smells today in ordinary politics. The stench that comes from that life has appeared to some to be so suffocating that they came to the conclusion that politics were not for a god-fearing man.

Had that been really so, I feel that it would have been a disaster for mankind. Find out for yourselves, in the light of what I am now saying, whether directly or indirectly every activity of yours today in this one of the greatest cities of the world is not touched by politics.

Well, then, when I found myself drawn into the political coil, I asked myself what was necessary for me in order to remain absolutely untouched by immorality, by untruth, by what is known as political gain.

In the course of my search, I made several discoveries which I must, for tonight, leave alone. But, if I am not mistaken, this necessity for poverty came to me first of all.

I do not propose to take you through all the details of that act or performance—interesting and, to me, sacred though they are—but I can only tell you that it was a difficult struggle in the beginning and it was a wrestle with my wife and—as I can vividly recall—with my children also.

Be that as it may, I came definitely to the conclusion that, if I had to serve the people in whose midst my life was cast and of whose difficulties I was witness from day to day, I must discard all wealth, all possessions.

I cannot tell you with truth that, when this belief came to me, I discarded everything immediately. I must confess to you that progress at first was slow. And now, as I recall those days of struggle, I remember that it was also painful in the beginning. But, as days went by, I saw that I had to throw overboard many other things which I used to consider as mine, and a time came when it became a matter of positive joy to give up those things. And one after another then, by almost geometric progression, the things slipped away from me. And, as I am describing my experiences, I can say a great burden fell off my shoulders, and I felt that I could now walk with ease and do my work also in the service of my fellowmen with great comfort and still greater joy. The possession of anything then became a troublesome thing and a burden.

Exploring the cause of that joy, I found that, if I kept anything as my own, I had to defend it against the whole world. I found also that there were many people who did not have the thing, although they wanted it; and I would have to seek police assistance also if hungry, famine-stricken people, finding me in a lonely place, wanted not merely to divide the thing with me but to dispossess me. And I said to myself: if they want it and would take it, they do so not from any malicious motive, but they would do it because theirs was a greater need than mine.

And then I said to myself: possession seems to me to be a crime. I can only possess certain things when I know that others, who also want to possess similar things, are able to do so. But we know—every one of us can speak from experience—that such a thing is an impossibility. Therefore, the only thing that can be possessed by all is non-possession, not to have anything whatsoever. In other words, a willing surrender.

You might then well say to me: but you are keeping many things on your body even as you are speaking about voluntary poverty and not possessing anything whatsoever! And your taunt would be right, if you only superficially understood the meaning of the thing that I am speaking about just now. It is really the spirit behind. Whilst you have the body, you will have to have something to clothe the body with also. But then you will take for the body not all that you can get, but the least possible, the least with which you can do. You will take for your house not many mansions, but the least cover that you can do with. And similarly with reference to your food and so on.

Now you see that there is here a daily conflict between what you and we understand today as civilization and the state which I am picturing to you as a state of bliss and a desirable state. On the one hand, the basis of culture or civilization is understood to be the multiplication of all your wants. If you have one room, you will desire to have two rooms, three rooms, the more the merrier. And similarly, you will want to have as much furniture as you can put in your house, and so on, endlessly. And the more you possess the better culture you represent, or some such thing. I am putting it, perhaps, not as nicely as the advocates of that civilization would put it, but I am putting it to you in the manner I understand it.

And, on the other hand, you find the less you possess the less you want, the better you are. And better for what? Not for enjoyment of this life, but for enjoyment of personal service to your fellow beings; service to which you dedicate yourselves, body, soul and mind.

Well, here you find there is ample room for hypocrisy and humbug, because a man or a woman may easily deceive himself or herself and deceive his or her neighbours also, by saying: 'In spirit I have given up all possessions, and yet externally I am possessing these things; you must not examine my deed, you must examine my intention; and of my intention only I must remain the sole witness.' That is a trap, and a death trap. How are you then to justify the possession even of a piece of cloth two or three or four yards, say, in length and a yard in width? How can you justify even the possession of that piece of cloth in order to cover your body somewhat, when you know that, if you left that piece of cloth alone, even that would be taken over by someone—not maliciously again—but because he would want it for he has not even so much as that piece of cloth? I am witness, eye-witness, of millions of human beings who have not even so much as that piece of cloth. How are you then to justify your act of possessing this thing with your intention not to possess anything at all?

Well, there is a remedy provided for this dilemma, this difficulty, this contradiction in life—that if you must possess these things, you must hold them at the disposal of those who want them. What happens is that, if somebody comes and wants your piece of cloth, you are not going to keep it from him, you are not going to shut any doors, you are certainly not going to the policeman to ask him to help you to keep these things.

And you have also got to be content with what the world will give you. The world may give you that piece of cloth or may not because, if you do not possess anything, naturally you do not possess the token coin with which you may buy clothing or food. You have got then to live purely on the charity of the world. And even when charitable people give you something, that something does not become your possession. You simply retain it with the fullest intention of that thing being surrendered to anybody who wishes to take it. If somebody comes and uses force against you to dispossess you, you may not go and report to the next policeman you meet and say you have been assaulted. You will not have been assaulted.

Well, that, to my mind, is the meaning of voluntary poverty. I have given you an ideal. Dr. Royden[1] has claimed that I am the greatest exponent of voluntary poverty in the world. I must, in all humility, disown any such claim whatsoever. And this I say to you not because of false modesty, but I say it to you sincerely, believing it to be true. I have given you but a little of my conception of voluntary poverty. And I must own to you that I am far from having realized that ideal in its fullness. In order to realize that ideal in its fullness, there must be a definite intention and conviction in my mind that I do not want to, I must not, possess anything on this earth as my property, not even this body, because this body also is a possession.

If you believe with me—as you must believe with me if you are church-goers, that is, if you believe in God—you believe that body and soul are not one and the same thing, but that the body is a house only, a temporary residence for a soul or a spirit within; and if you believe that, as you do believe, I take it—then it follows that even the body is not yours. It has been given to you as a temporary possession, and it can also be taken from you by Him who has given it to you.

Therefore, having that absolute conviction in me, such must be my constant desire, that this body also may be surrendered at the will of God, and while it is at my disposal, must be used not for dissipation, not for self-indulgence, not for pleasure, but merely for service and service the whole of our waking hours.

And if this is true with reference to the body, how much more with reference to clothing and many other things that we use?

Having got that conviction and held it for so many years, I am here to give you my evidence against myself, that I have not reached that perfect state of voluntary poverty. I am a poor man, in the sense you understand of struggling to reach that ideal, not poor in the sense in which we ordinarily use the word poor.

As a matter of fact, when I was once challenged by someone, I was able to claim that to my neighbours, and people in the world I seemed to be the richest man on earth, for the richest man is really one who, possessing nothing, has everything at his disposal.

And those who have actually followed out this vow of voluntary poverty to the fullest extent possible (to reach absolute perfection is an impossibility, but the fullest possible extent for a human being) those who have reached the ideal of that state, they testify that, when you dispossess yourself of everything you have, you really possess all the treasures of the world. In other words, you really get all that is in reality necessary for you, everything. If food is necessary, food will come to you.

Many of you are men and women of prayer, and I have heard from very many Christian lips that they got their food in answer to prayer, that they get everything in answer to prayer. I believe it. But I want you to come with me a step further, and believe with me that those who voluntarily give up everything on earth, including the body, that is to say, have readiness to give up everything (and they must examine themselves critically, rigidly, and give always an adverse judgment against themselves) — those who will follow this out will really find that they are never in want.

And I will confess to you that, when I felt God had given me some portion of the riches of the earth and when I had many possessions, I had not the facilities for possessing things that I have at this time. I had not certainly one-millionth part of the ability to command money and everything that I need for service.

A spirit of service had come to me even when I was practising and earning money and was in possession of several things, but at that time I had certainly not the capacity for getting whatever I wanted for service. But today (whether it is good for me or bad for me I do not know, God alone knows) I can give you this evidence, that I have never been in want.

After a period when I had really dispossessed myself by intention and had no hankering after anything that I could call my own, and began to share everything I possessed in common with my neighbours (I cannot share everything with the whole world; if I share with my neighbours, I do share with the whole world, my neighbours also doing likewise; if we do that, it is all a limited human being can do) but immediately I came to that state to a fair extent, I found that I was never in want.

Want must not, again, be taken literally. God is the hardest task-master I have known on this earth, and He tries you through and through. And when you find that your faith is failing or your body is failing you and you are sinking, He comes to your assistance somehow or other and proves to you that you must not lose your faith and that He is always at your beck and call, but on His terms, not on your terms. So I have found. I cannot really recall a single instance when, at the eleventh hour, He has forsaken me. And I have got this reputation, which I can repeat to you, a reputation for being one of the best beggars in India. And, as my critics will tell you, at one time I collected one crore of rupees; in pounds, shillings and pence I cannot count it for you, but it is some horribly large sum (about £750,000), but I had no difficulty in collecting it. And since then, whenever any emergency has arisen, not for any consideration, not in the soul of my fundamental being, can I recall a single instance of my failing to obtain whatever was necessary for service.

But you will say: this is in answer to prayer. It is not just an answer to prayer, it is a scientific result of this vow of non-possession or vow of voluntary poverty. You do not want to possess anything whatever: and the more therefore you simplify your life, dispossess yourself, the better it is for you.

Immediately you come to that, you can command anything. You can command vanities, but if you only once take possession of these, this power will immediately be gone; you must not take for yourself. If you do, you are done for. I have known this happen in so many instances. Many a man has said: 'Oh yes, God has now answered my prayer for money or possessions. I will now keep this—this Koh-i-noor diamond, or whatever it may be.' That will be the last time. He won't be able to defend that diamond.

Therefore, all I am just now holding out before you as a grand thing is that you can command all the resources of the world for service. To one who does not believe, that may seem an arrogant statement to make. But, as I believe, it is not an arrogant thing to say that you can command all the resources of the earth for service—to the extent of your ability to serve. If you want to command the whole services of the world, it is not enough to go down to some of those houses in the East End, find out the distress of those who live there and fling in their faces a few coppers; you will not have all the resources of the world for that; God will fling in your face also a few coppers.

But if you surrender yourself, body, soul and mind, and give yourself up to the world, then I say: the treasures of the world are at your feet, not for your enjoyment, but for the enjoyment of that service, only yours for that service.

The moral that I would have us to draw from this talk that I have given to you is really very apposite at this time. I want you to believe me when I tell you that my whole heart goes out to this nation in its distress. I cannot possibly present my solution of your financial difficulty. You are great enough, resourceful enough, to find out your own remedies. But I would ask you to elaborate this thought in your own minds in connection with the present distress.

Mr. C.F. Andrews brought to my notice a letter that was written by the Prime Minister to a correspondent and which he told me yesterday was being used as an advertisement throughout the District Railway, probably in the Tubes also; it runs somewhat like this: 'You must buy only British goods; must employ only British labour, and try to buy as much as you can.' That is one remedy I know. But I want to suggest to you that, in order to solve the problem of distress in the world, this idea of voluntary poverty is a root idea. No doubt, with your resourcefulness, you will tide over the difficulty and feel that there was nothing wrong. If you will permit me to say so, that would be perhaps short-sighted, for a time perhaps has come for a revision of values.

But again I must not go into deep waters. I can only throw out this hint to those who can appreciate the necessity of voluntary poverty for service. I have not tonight presented this blessed thing for the acceptance of all: though let me add that, in the innermost recesses of my heart, I feel that the world would not go all wrong, would not become a world of idiots, if all of us took the vow of voluntary poverty. But I know that this is almost an impossible thing. Everything is possible for God but, humanly speaking, it is wise to say that it is an impossible thing. But it is not an impossible thing; indeed, I hold it to be absolutely indispensable that those who give themselves wholly to the service of their fellow-beings must take the vow of voluntary poverty.

Try to find out for yourself whether you are not thereby assisting very materially in solving this great national problem that today faces you.

You will not have solved the problem if the people, who do not want to give up their salaries or whatever they are required to give up, are compelled to give them up by law. While they say: 'What can we do? We do not want to resist; we cannot resist', their minds are still hankering after these things.

But imagine that, in the midst of this hankering, there is a body of servants arising, who will themselves become voluntarily poor. They would be like lighthouses to guide the paths of those who do not know what voluntary poverty is because they know only involuntary poverty. I do not go among my fellows who starve and talk of voluntary poverty; I do not tell them how blessed they would be if they changed that involuntary poverty into voluntary. There is no such thing as magic of that character on this earth. It is a painful process, and these men have first of all to have the necessities of life before I can talk to them of voluntary poverty.

What does happen is this: that a man like me going among

them, living in their midst as best he can their life, can bring a ray of hope into their hearts. They will accept remedies that a man like me may suggest to them. At least, if I cannot suggest any immediate remedy, they would find in a man like me a friend. They would say: 'He is happy although he possesses nothing; how is it?' I do not need to argue with them; they begin to argue for themselves.

How can I share these richest treasures from my experience with everybody on earth? I could not. But today, having undertaken to speak on voluntary poverty, I am sharing, to a certain extent only, these treasured experiences of mine not amongst a few hundred people here but amongst millions of people. I tell you that it is beyond description, the bliss, the happiness, and the ability that this voluntary poverty gives one. I can only say: try it and experiment with it, test it for yourselves.

I thank you for giving me your undivided attention. There are still exactly ten minutes left before the hour of prayer, and if any of you wish to ask me any questions, I shall be glad. You need not hesitate to ask anything that is in your mind; you will never offend me by asking any questions, let them be as awkward as they may be.

Q. Can the Mahatma tell us how he can justify collecting large sums of money when Jesus, the Buddha and other great religious teachers who have practised voluntary poverty have never asked for or received large sums of money? I cannot reconcile this with the rest of what he told us.

A. Did these great teachers never ask for or receive moneys? After Jesus many Christians, who believed in poverty also, took moneys and used them for service. And I can speak with better confidence about the Buddha, who is reported in his own lifetime to have founded institutions. He could not possibly found institutions without money. And it is said that they who gave themselves body, soul and mind gave their riches also and placed them at the feet of the Buddha, who gladly accepted them—but not for himself.

Q. Why should we serve our fellow-beings?

A. In order that we may see a glimpse of God through them; because they have got the same spirit as we have, and unless we learn that, there is a barrier drawn between God and ourselves; if we want to demolish that barrier, the beginning is made by complete identification with our fellow-beings.

Speech at Guildhouse Church, under the auspices of the Franciscan Society
The Guildhouse, 23 Sept. 1931

[1] Dr. Maude Royden, progressive social reformer, was in the chair.

136. RELIGION AND IRRELIGION IN POLITICS

May 30, 1932

In my opinion unity will come not by mechanical means but by change of heart and attitude on the part of the leaders of public opinion. I do not conceive religion as one of the many activities of mankind. The same activity may be either governed by the spirit of religion or irreligion. There is no such thing for me therefore as leaving politics for religion. For me, every, the tiniest, activity is governed by what I consider to be my religion.

A Letter
Mahadevbhaini Diary, Vol. i, p. 189

137. TOWARDS UNIVERSAL BROTHERHOOD

Q. In your autobiography you have said that you cannot think of politics apart from religion. Do you still hold that view? If so, how is it that in a country of many diverse religions like India you expect a common political policy to be adopted?

A. Yes, I still hold the view that I cannot conceive politics as divorced from religion. Indeed, religion should pervade every one of our actions. Here religion does not mean sectarianism. It means a belief in ordered moral government of the universe. It is not less real because it is unseen. This religion transcends Hinduism, Islam, Christianity, etc. It does not supersede them. It harmonizes them and gives them reality.

Q. Is it true that you advised some Sikhs, who came to seek your advice on certain matters, that Guru Govind Singh taught the use of the sword while you stood for non-violence, and therefore the Sikhs must be ready to choose the one or the other?

A. The question is badly, if not mischievously, put. What I did say was that, if they thought the teaching of Guru Govind Singh excluded implicit belief in non-violence, they could not be consistent Congressmen so long as the Congress creed remained what it was. I added that they would be compromising themselves if they joined or remained in the Congress and might even damage their own cause.

Q. By teaching respect for all religions you want to undermine the power of Islam. You want to emasculate the Pathans by taking away the rifle from them. There can be no meeting ground between us and you.

A. I do not know what you thought during the Khilafat days. Let me give you a bit of the history of our own times. The foundation of the Khilafat struggle was laid by me. I had a hand in the agitation for the release of the Ali Brothers. So when they came out, they, together with Khwaja Abdul Majid, Shuaib Qureshi and Moazam Ali, and I met together and devised the plan of action which the world knows. I discussed with them the implications of non-violence and told them that, if they could not accept non-violence as true Muslims, I should be out of the picture. Their reason was satisfied, but they said they could not act without endorsement from Muslim divines, and so there was a conference of the *ulemas* at the late Principal Rudra's house where I used to stay, when in Delhi, during his lifetime. This learned company included, among several others, Maulana Abul Kalam Azad and the late Maulana Abdul Bari. Led by Maulana Abul Kalam Azad they held that belief in non-violence was not only not inconsistent with Islam, but it was obligatory in the sense that Islam had always preferred it to violence. It is noteworthy that this took place before the acceptance of non-violence by the Congress in 1920. Many were the discourses given by learned Muslims on non-violence before crowded Muslim meetings. The Sikhs too came in later without any scruples and they listened with rapt attention to my exposition of non-violence. Those were great and glorious days. Non-violence proved infectious. Under its spell there was a mass awakening such as had never been seen before in this land. All communities felt as one, and they thought that non-violence had clothed them with an irresistible power. Those halcyon days are gone and now I am obliged seriously to answer questions like the above.

I cannot give you the faith in non-violence you do not possess. God alone can give it to you. Mine remains unshaken. I do maintain, in spite of you and others like you suspecting my motives, that mutual respect for one another's religions is inherent in a peaceful society. Free impact of ideas is impossible on any other condition. Religions are meant to tame our savage nature, not to let it loose. God is only one though He has countless names. Don't you expect me to respect your faith? If you do, may I not expect the same respect from you for mine? You say Muslims have nothing in common with Hindus. In spite of your separatism, the world is moving towards universal brotherhood when mankind will be one nation. Neither you nor I can stop the march towards our common destiny. As for the emasculation of Pathans, let Badshah Khan answer. He had accepted non-violence before we met. He believes that the Pathan has no future save through non-violence. Without it, if nothing else, his blood-feuds will keep him from going forward. And he thinks that the Pathan found his feet in the Frontier Province after he accepted non-violence and became servant of God—*Khudai Khidmatgar*.

Q. You did not hesitate to join the Ali Brothers in their intrigue to invite Amanulla Khan to invade India and set up Muslim Raj. You drafted a wire for Maulana Mahomed Ali advising the then Amir not to enter into a treaty with the British. The late Swami Shraddhanandji is reported to have seen the draft. And now you want the Hindus of Sindh to make a present of their hearths and homes to their Mussalman oppressors instead of demanding the re-amalgamation of Sindh with the Bombay province, which alone can restore the reign of law to Sindh. Why won't you realize that in this age of enlightenment and progress what the minorities expect is effective protection of their due rights, not mere pious counsels of perfection?

A. I have several such letters. Hitherto I have ignored them. But now I see that the news has gone through a revised and enlarged edition in the Hindu Mahasabha. An angry correspondent threatens that persons like him will begin to believe what has been stated so authoritatively. For the sake of my reputation, therefore, I must answer the question. But my correspondents should know that life for me would be a burden if I were to make it a point of controverting every false report about me or distortion of my writing. A reputation that requires such a mud wall of protection is not worth keeping. So far as the charge of my intriguing with the Amir is concerned, I can say that there is no truth whatsoever in it. Further, I know that the Brothers stoutly denied the charge when it was brought to their notice. And I believed them implicitly. I do not remember having drafted any telegram on behalf of Maulana Mahomed Ali to the then Amir. The alleged telegram is harmless in itself and does not warrant the deduction drawn from it. The late Swamiji never referred the matter to me for confirmation. It is wrong to say anything against dead men unless one has positive proof and stating it is relevant.

The romance has been woven round my writings in *Young India*. Deductions drawn from them are wholly unjustified. I would not be guilty of inviting any power to invade India for the purpose of expelling the English. For one thing, it would be contrary to my creed of non-violence. For another, I have too great a respect for English bravery and arms to think that an invasion of India can be successful without a strong combination of different powers. In any case, I have no desire to substitute British Rule with any other foreign rule. I want unadulterated Home Rule, however inferior in quality it may be. My position remains today what it was when I wrote the *Young India* paragraphs now sought to be used against me. Let me further remind the readers that I do not believe in secret methods.

As for Sindh my advice stands. Reincorporation of Sindh in the Bombay Province may or may not be a good proposition on other grounds, but certainly it is not for the purpose of greater protection of life and property. Every Indian, be he Hindu or any other, must learn the art of protecting himself. It is the condition of real democracy. The State has a duty. But no State can protect those who will not share with it the duty of protecting themselves.

On the way to Delhi, February 4, 1940

'Question Box'
Harijan, 10 Feb. 1940

138. RELIGION AND THE STATE

[New Delhi,
Before 16 September 1946]

If I were a dictator, religion and State would be separate. I swear by my religion. I will die for it. But it is my personal affair. The State has nothing to do with it. The State would look after your secular welfare, health, communications, foreign relations, currency and so on, but not your or my religion. That is everybody's personal concern!

You must watch my life, how I live, eat, sit, talk, behave in general. The sum total of all those in me is my religion.

Asked which movement, e.g., women's, political, scientific or religious, would have had the most far-reaching influence in the world of tomorrow and would be considered 50 years hence as having had the greatest impact on world affairs as a whole and for the greatest good of mankind, he said it was wrong to bracket religious movement with the rest. He said:

It is the religious movement that will dominate the future. It would do so today but it does not, for religion has been reduced to a Saturday or a Sunday affair; it has to be lived every moment of one's life. Such religion, when it comes, will dominate the world.

Q. Do you feel there is any special significance in the increasing number and magnitude of labour strikes, especially in India of late? What do you think will be the outcome of this labour trouble in India?

A. Strikes have today become a universal plague. There are strikes everywhere, America and England not excepted. But in India they have a special significance. We are living under an unnatural condition. As soon as the lid is removed and there is a crevice letting in the fresh air of freedom, there will be an increasing number of strikes. The fundamental reason for this spreading strike fever is that life here as elsewhere is today uprooted from its basis, the basis of religion, and what an English writer has called 'cash nexus' has taken its place. And that is a precarious bond. But even when the religious basis is there, there will be strikes, because it is scarcely conceivable that religion will have become for all the basis of life. So there will be attempts at exploitation on the one hand and strikes on the other. But these strikes will then be of a purely non-violent character. Such strikes never do harm to anyone. It was such a strike perhaps that brought General Smuts to his knees. 'If you had hurt an Englishman', said Jan Smuts, 'I would have shot you, even deported your people. As it is, I have put you in prison and tried to subdue you and your people in every way. But how long can I go on like this when you do not retaliate?' And so he had to come to terms with a mere coolie on behalf of coolies as all Indians were then called in South Africa.

'Talk With a Christian Missionary'
Harijan, 22 Sept. 1946

§ 2. Political Power

139. THE CROWN OF THORNS

The Congress crown has ceased to be a crown of roses. The rose petals are year by year falling off and the thorns are becoming more and more prominent. Who should wear such a crown? Father or the son? Pandit Motilalji the weather-beaten warrior or Pandit Jawaharlal Nehru, the disciplined young soldier who by his sterling worth has captured the imagination of the youth of the country? Sjt. Vallabhbhai Patel's name is naturally on everybody's lips. Panditji says in a private letter that he as the hero of the hour should be elected and the Government should be made to know that he enjoys the fullest confidence of the nation. Sjt. Vallabhbhai is however out of the question just now. His hands are too full to allow of his attention being diverted from Bardoli. And before December comes upon us he may be a guest in one of His Majesty's innumerable prisons. My own feeling in the matter is that Pandit Jawaharlal should wear the crown. The future must be for the youth of the country. But Bengal wants Motilalji to guide the Congress barque through the perilous seas that threaten to overwhelm us during the coming year.

We are torn within and are encircled by an enemy that is as unscrupulous as he is powerful.

Bengal has special need of an older head and one moreover who has proved a tower of strength to her in the hour of her trial. If India as a whole has no easy time before her, Bengal has still less. There are a thousand reasons why Panditji should be chosen to wear the crown of thorns. He is brave, he is generous, he enjoys the confidence of all parties; Mussalmans acknowledge him as their friend, he commands the respect of his opponents and often bends them to his view by his forceful eloquence. He has moreover deep down in him a spirit of conciliation and compromise which makes him an eminently worthy ambassador of a nation that is in need of and is in the mood to take an honourable compromise. It is these considerations which actuate even the dare-all Bengal patriot to want Pandit Motilal Nehru as the helmsman for the coming year. Let the impatient youth of the country wait a while. They will be all the stronger for the waiting.

'Crown of Thorns'
Young India, 26 July 1928

140. SEEKING OFFICE

Marwar Junction,
February 16, 1929

Dear Swami,

I was sorry I was not able to have more time with you alone than I was able to have. I would now share with you my innermost thoughts through the more imperfect medium of correspondence.

I look to you to discharge your responsibility and shed sweetness around and about you, as you have promised. You have to stoop to conquer. You should become a real *sannyasi* without anger, without malice, without desire for self. I was not unobservant of the fact that you had no power behind you outside the ranks of your own immediate and very limited following. You are evidently unable to command money when you need it for public work. All this must be altered. You have sacrifice behind you. You have courage. What is it that keeps you from possessing greater influence over the people for their own good?

Why do you want office when the office does not want you? The narrow majority by which you may be able to retain office can bring no real satisfaction, can give you no real opportunity for service. If you will interpret office in terms of service, why will you not refuse to hold it unless your opponents too insist upon your holding it?

I have given effect as much as possible to your wishes in so far as I have been able to interpret them. But I would like you after the election is over or even before, if you have the humility, to confer with Jairamdas and others and think of another president whom you will unanimously elect. Anyway I look to you to run the Congress machinery in Sind smoothly and honestly. *Verb. sap.* You will not I hope misunderstand or misinterpret this letter.

Yours sincerely,

[P. S.] I am in Delhi between Sunday and Tuesday care Speaker Patel. Then Sabarmati for a week.

Letter to Swami Govindanand
SN 15339

141. POLITICAL POWER

Sjt. Satyamurti writes:

I write to you about the article 'Substance Not Shadow' by you, in the *Young India*, of the 18th June. The first sentence which causes me some doubt and anxiety is, 'My proposition therefore before the Working Committee was that agreed settlement failing, the Congress should give up the hope of winning a *swaraj* constitution by way of the present Round Table Conference and should wait till all the communities were satisfied to adopt a purely national solution.' Does this not mean a charter for extreme communalists to go on obstructing?...

But the sentence which causes me grave anxiety is that which ends as follows: 'We can gain our end without political power and by directly acting upon the powers that be.' I would add the following sentences also from your article in order to make my doubts clear to you. 'One form of direct action is adult suffrage. The second and more potent form is *satyagraha*. It can easily be shown that whatever is needful and can be gained by political power can perhaps be more quickly and more certainly gained by *satyagraha*.' I venture to join issue with you. I was always and am today under the impression that what the Congress wants is political power more than anything else. And, concretely speaking, prohibition can be more easily brought about by State action than by peaceful picketing. *Khaddar* and *swadeshi* cloth can be spread better and more quickly by State action than by peaceful picketing, and the necessary reforms embodied in the Karachi resolution on Fundamental Rights can be enforced only by a *swaraj* Government.

In any case, I do not see why the nation should not concentrate all its energies today, upon the gaining of political power.

To me, political power is the substance, and all other reforms can and ought to wait....

I am thankful for this letter. It enables me more clearly than I have been able to explain my position.

My implicit faith in non-violence does mean yielding to minorities when they are really weak. The best way to weaken communalists is to yield to them. Resistance will only rouse their suspicion and strengthen their opposition. A *satyagrahi* resists when there is threat of force behind obstruction. I know that I do not carry the Congressmen in general with me in this what to me appears as very sensible and practical point of view. But if we are to come to *swaraj* through non-violent means, I know that this point of view will be accepted.

Now for Sjt. Satyamurti's second difficulty. To me political power is not an end but one of the means of enabling people to better their condition in every department of life. Political power means capacity to regulate national life through national representatives. If national life becomes so perfect as to become self-regulated, no representation is necessary. There is then a state of enlightened anarchy. In such a state everyone is his own ruler. He rules himself in such a manner that he is never a hindrance to his neighbour. In the ideal state therefore there is no political power because there is no State. But the ideal is never fully realized in life. Hence the classical statement of Thoreau that that Government is best which governs the least.

If then I want political power, it is for the sake of the reforms for which the Congress stands. Therefore when the energy to be spent in gaining that power means so much loss of energy required for the reforms, as threatens to be the case if the country is to engage in a duel with the Mussalmans or Sikhs, I would most decidedly advise the country to let the Mussalmans and Sikhs take all the power and I would go on with developing the reforms.

If we were to analyse the activities of the Congress during the past twelve years, we would discover that the capacity of the Congress to take political power has increased in exact proportion to its ability to achieve success in the constructive effort. That is to me the substance of political power. Actual taking over of the Government machinery is but a shadow, an emblem. And it could easily be a burden if it came as a gift from without, the people having made no effort to deserve it.

It is now perhaps easy to realize the truth of my statement that the needful can be 'gained more quickly and more certainly by *satyagraha* than by political power'. Legislation in advance of public opinion has often been demonstrated to be futile. Legal prohibition of theft in a country in which the vast majority are thieves would be futile. Picketing and the other popular activities are therefore the real thing. If political power was a thing apart from these reforms, we would have to suspend latter and concentrate on the former. But we have followed the contrary course. We have everywhere emphasized the necessity of carrying on the constructive activities as being the means of attaining *swaraj*. I am convinced that whenever legal prohibition of drinks, drugs and foreign cloth comes, it will come because public opinion had demanded it. It may be said that public opinion demands it today but the foreign Government does not respond. This is only partly right. Public opinion in this country is only now becoming a vital force and developing the real sanction which is *satyagraha*.

'Power not an End'
Young India, 2 July 1931

142. POWER AND REPRESENTATION

[26 October 1934]

You have in your affection misread the whole of my position.[1] If you have given me the position of a general commanding an army, you must allow that general to judge whether he serves the army by being at its head or whether he serves the army by retiring and giving place to lieutenants who have served well.

It is always the wisest thing for the general to swing to power or office or withhold that power that came to him unsought, because there are occasions when generals have been found to consider themselves unequal to hold the reins, and in the interests of the army, the interests of the cause for which the army and its chief are fighting to give up the command. If you believe that I have been a fairly wise general, you must believe in my judgment even now when I seem to be deserting you in the hour of your defeat.

I have said times without number that I myself share no feeling of defeat. I am not deserting the cause for which I live, and for which I love to work. I go away from you in order, as I have said, to discover if there are still greater possibilities in experiments that we have been making, and for which you have given me such a generous support. I must express to you that I seem to have come to the end of my resources. I must confess to you that I have not lost faith in the efficacy of the means that I placed at the disposal of the Congress. But I feel there is a body of opinion rising in the country which has begun to question the efficacy of this means, and it is because they question the efficacy of this means that they have the sense of defeat in them.

In the dictionary of a *satyagrahi*, there is no such thing as defeat. To him, the very pursuit of his battle is its own reward. But when I find some of my best companions, who have believed in truthfulness and non-violence with all its implications, are filled with doubt and feeling of helplessness, when I find that I am not able to touch them with my faith, I see all around me an impenetrable darkness. I see no ray of light. I see I cannot infect them with the faith that is in me.

Therefore I would like you not to take me literally when I

say that I go, if I can get your blessing. I go with your blessing in search of greater power to discover means whereby I can give you the faith that is in me. It may be that the search will be in vain. It may be that my roaming in solitude will be in vain. But depend upon it that I shall not wait a moment longer to be called by you, when I feel the necessity. I shall come to you, and once more come to the Congress to register myself as a primary member, and do whatever I can in the interests of the Congress.

I ask you therefore not to feel dispirited, but to feel that you will be able to discharge your duty, that you will have leaders enough to lead you on to victory. When it becomes necessary for me to come to this House, I will be entirely at your disposal. Therefore, I would simply urge not to press the resolution further. You have unanimously passed that resolution. Let that be enough. Somehow or other I believe good, and only good, will come out of this decision. If you yourselves stand for the principle for which I stand, and for which you have given allegiance, which I acknowledge you have given out of your generosity, some of you no doubt blindly, some of you after having considered for yourselves, I have in all humility accepted all these allegiances. Let me now prove to myself, if it is possible, that I am worthy of it, and I invite you to prove it for yourselves also, whether I remain a member of the Congress or not, that the principle that you have more or less followed during all these long or short 14 or 15 years, you have followed because you have believed in it.

I tell you that it gives me still greater joy and pleasure, for I am able to say to myself, and if you are able to say to the world, that although I am gone you believe in the principle that you have fought for and lived for all these years and that you are going to follow them out not loosely, but still more truly and thoroughly than you have done before.

Immediately after his speech dealing with the question of his retirement Gandhiji proceeded to move formally a comprehensive resolution incorporating all the recommendations of the constitution sub-committee and the Congress Working Committee. In commending the whole of the resolution for the acceptance of the House *in toto*, Gandhiji at the outset referred to the previous day's decision of the House turning down the Working Committee's recommendation relating to the change of the Congress creed.

Mahatma Gandhi delivered a stirring address lasting 90 minutes in which he appealed to the Subjects Committee to adopt the revised constitution.

Mahatma Gandhi referred to Thursday's vote and congratulated the House on the frank way in which they had voted in referring the question of change of creed to provincial committees. He said, when he had read the public and Press criticism of his proposals, he had made up his mind not to take up those amendments, but members of the Working Committee had of their own accord unanimously taken the view that the Committee would sponsor those resolutions. Gandhiji added:

I was surprised to find that members of the Committee devoted two hours this morning to deliberating as to what was their duty in connection with these amendments. I told the Working Committee to carry out your resolution, but my feeling is that Mr. Sidhwa's amendment was wholly unnecessary. You could have rejected the Working Committee's proposal, but there was no use circulating it.

I listened carefully to Mr. Patwardhan's speech. He argued ably, but either he was deceiving himself or he was acting simply as an advocate. He said that we had descended from ideals to realism. But has not socialism its ideals and if I told him to remove even a comma from their ideals he would reject my proposition. Do you suppose you will realize complete independence at any time even after fifty generations? The socialist creed goes even further. I suggest we must have a measuring rod. The day humanity ceases to believe in ideals, it will descend to the level of the beasts. Today if you believe in truthfulness and non-violence say so. It does not mean you will be cent per cent truthful. I am not cent per cent truthful. If I were, my words will pierce you like arrows and come out clean, but I must aspire to attain my ideals. Did Marx live to see his ideal realized? Is it not undergoing changes and are not meanings put to it which Marx himself never dreamt of? If you do not mean by 'legitimate and peaceful' means that they are 'truthful and non-violent', then merely define your creed to be that of attaining your end by whatever means you think legitimate. Then that would be your measuring rod, but you must have a measuring rod. You must be clear about your aims and means. They are convertible terms.

You know that *swaraj* is your goal. It means '*purna swaraj*' but someone said it did not mean complete independence; so we defined it to mean '*purna swaraj*'. Just as you did that you must define your creed so that it may not be open to double interpretation. No one can draw a right angle, yet Euclid drew it up in imagination and gave the engineers a measuring rod by which the world has progressed. There is a right angle which is treasured in the Tower of London.

The Working Committee on Friday asked you to define your measuring rod. You sent it round for circulation. That was not the right course to take. Let us not tell the world that there is one thing on our lips and another in our thoughts.

Mahatma Gandhi next explained the main purpose of the amendments proposed in the constitution. He said:

Do not pass these changes buoyed up with the hope that I will reconsider my decision. If I began my unofficial leadership of the Congress by the constitution for which I was primarily responsible I am also tendering you a humble gift in the shape of this amended constitution so that you may get the benefit of the judgment and experience of one who has endeavoured to live that constitution in closest communication and communion with you and has discovered its faults. I want you, therefore, to pass these amendments after touching them up.

You can take a day for consideration after hearing me and do whatever lobbying you like and give your verdict. You must first make up your mind to reduce your delegates from 6,000 to 1,000. If you reject that, the entire scheme falls through.

The second point is that the delegates should be real representatives of the people. We are indirectly representative of the nation's dumb millions. We are their mouthpiece, their voice and their thought. That is what the Congress has stood for since 1885, but indirectly we are representatives only of our electors.

Can anyone among us say whom he represents, as to

whether he is in living touch with his constituents and knows their feelings? Even the tallest amongst us cannot claim that. Sardar Vallabhbhai is the uncrowned king of Gujarat, but which electorate does he represent, whom do I represent, I do not know. I challenge anyone to produce the Congress register of electors. We must have constituencies and electors and each member shall represent his constituency and be in living touch with it. Then alone you will have your measuring rod.

The third principle I have put forward is that three things be combined in one election. It will not merely mean greater convenience and saving of money; it is rich with promise if you adopt it. Delegates will then be elected by their electorate. They will not assemble as now for three days in a year and then disappear from the horizon. They will remain active Congressmen as members of the A.I.C.C. and will take up work all the year round.

Today only 350 out of 1,530 are members of the A.I.C.C. Then again a body of 1,000 can be accommodated in Sardar Vallabhbhai's Ras or Bardoli. I have got a complete scheme ready for a successful session being held in the village of that character. Those villages can slave for you but they cannot give you money for your Reception Committee. You must have people who know the business of running a Congress meeting and even importing provisions and the Reception Committee shall not waste lakhs on its arrangements.

I question the wisdom of the Reception Committee wasting lakhs on *tamashas* and in building triumphal arches. Where is our triumph? We are a slave nation and all we want the Reception Committee to do is to give us the hospitality, not of pudding and ice-cream, but ordinary food. Our scheme is to relieve the Reception Committees of their heavy burden. If I had been in Nariman's position I would have become a lunatic.

Mahatma Gandhi, reverting to his plea that the matter be not referred to Provincial Committees, said:

Why do you want to refer it to the provinces? Why shirk your duty? You are super-delegates because you are the Subjects Committee. Others are mere delegates, but mere delegates representing 350 millions. Then I say it is for them to hammer out a constitution and not shirk the duty. I promise you fair consideration of every criticism you have made. I am giving you an iron constitution from which not one single man can go away. If there is any corruption you can detect it unless everyone becomes a fraud. No constitution can be robber-proof and thief-proof. I have sufficient faith in my countrymen that they will never betray their country.

Sri Prakasa says our present constitution is rotten. Then are we to waste another year in waiting for what we shall do? I am sure that the position will remain where it is and we will not receive a single report during the next twelve months. Make such use as you can of the wisdom you say is possessed by me whom the world holds as a fool. But wisdom sometimes does come out of the mouth of fools. I would like you to accept my suggestions and I will be ready for compromise. Already, when Mr. Masani came to me, I assured him that I wanted Socialists to be represented. They are the advance wing. There is nothing to be feared from their activity. They call themselves Congressmen and while they are Congressmen they are supposed to believe in the Congress creed and loyally follow the discipline of the Congress. If they fail to follow the Congress discipline, they will fail in their own creed. I have agreed to their suggestion regarding the single transferable vote. This constitution has been revised by able lawyers like Mr. Bhulabhai Desai and Mr. K. M. Munshi and many others have been consulted. I may also make it clear that there will be plural constituencies in urban areas, but their number is not defined. The Working Committee is entirely in your hands in these matters of detail, but do not postpone these matters.

You are going to have a Parliamentary Board. They will look after the Assembly work. In the Assembly it is not a battle between men, but between principles; our members will go to represent the principles which will not require eloquence, but will be enunciated in unequivocal terms.

Hereafter you will have your register of electors and you will be in living touch with them. Mr. Nageshwara Rao says that there are 7,000 voters for one seat. My scheme is modest. It wants only 1,000 voters for one seat. I will be satisfied if you are the representatives of one million and then we can have more and more. But I do submit that we will lose one precious year in the life of the nation by referring these matters to Provincial Committees.

Proceeding, Mahatma Gandhi referred to the amendments seeking to incorporate the existing convention of allowing the President to select his colleagues of the Working Committee in the Congress constitution. Gandhiji said that there had been no single instance in the past wherein the President's choice in this matter had been overridden by the A.I.C.C. But he thought it would be better to incorporate this convention in the constitution so as to avoid difficulties attendant on selection of candidates. The President under this amendment would not have to stand the fire of cross-examination at the A.I.C.C.

There must be some occasions in the life of the Working Committee when they would have to say: 'If you want us to carry on then we ask you to arm us with certain powers and if you cannot trust us then you must look for better men to fill our place.' For the same reason he had sought that the Secretaries and the Treasurer should also be the President's choice.[2]

Speech at Subjects Committee Meeting, All India Congress Committee
Home Department, Political, File No. 4/27/36

[1] Earlier the Subjects Committee had unanimously passed a resolution expressing the country's confidence in Gandhi and requesting him to revise his decision to retire.
[2] This appeared in the *Bombay Sentinel*, 27 Oct. 1934.

143. SILENT SERVICE

New Delhi,
April 15, 1946

The suggestion for not accepting office greatly appeals to me. Carry on silently whatever constructive work, you can. These are difficult times. If all want power who will render silent service?

Note to Baba Raghavdas (H.)
Hindustan, 16 Apr. 1946

§3. Politics and Society

144. INCLUSIVENESS OF THE POLITICAL

The Ashram, Sabarmati,
May 28, 1926

Dear Friend,

I have your letter of the 26th ultimo forwarded through Mr. Vaze. I thank you for that letter. I appreciate it for I know that it is only by a frank exchange of views that we come nearer one another. For me the word 'political' is inclusive. I do not divide different activities—political, social, religious, economical—into water-tight compartments. I look upon them all as one indivisible whole each running into the rest and affected by the rest. I also believe with you that our political freedom properly so called will depend upon our ability to solve many of our domestic problems such as the communal trouble. In other words, it will depend upon internal reforms. The outward will be merely therefore a symptom of the inward. I do not at all hold that this communal problem is insoluble. It seems to defy human effort for the time being. But I have the fullest faith in our ability finally to solve it. It may be of course that before we reach a solution there will be bloody fights between the two. In spite of all effort sometimes they become as it were unavoidable.

I would however like to add a word of caution against thinking, if you did think, that these communal troubles are not due to our dependence upon the British rule. It is my conviction that that rule is based upon the policy of 'divide and rule' which policy sometimes English officials have been frank enough to admit. If the Government chose, it can certainly contribute much to a speedy and permanent solution. But if I mention this I do so not to excuse ourselves for our inability to solve the problem without the help of the rulers, but I mention it to explain our difficulty.

Your second point is too difficult for me to pronounce an opinion upon. I am not an attentive student of European politics or European history to be able to draw the distinction that you do between the European States where Roman Catholicism is predominant and where Protestantism is predominant. And, for that reason I would prefer to hold silence upon the third point which is equally interesting as the second. There is no doubt that mankind is affected largely by the way it looks upon God. So far as India is concerned the vast majority think of God as the Monitor within each one of us. Even the illiterate masses know that God is only one, that He is all-pervading, and, therefore, is the witness of all our actions.

If you wish to elucidate further the two points of your letter which at present I understand but partially and if you can find time, I shall read your elucidation with all the attention I am able to give to it. And I know that I shall profit by it.

Yours sincerely,

Letter to Dr Norman Leys
SN 12468

145. TRICKS AND TRADING

Satyagraha Ashram,
Sabarmati,
February 26, 1928

My dear Jawahar,

I have your letters. I am sensing all that is going on in Delhi and can understand every word of what you have said in your letter. I can't give you an adequate conception of my grief as I follow the Conference proceedings from day to day and read between the lines. Father's illuminating letter only confirmed my own reading from a distance. Then came Kripalani's letter yesterday to Krishnadas, and yours has come today to put the finishing touch.[1] What a miserable show we are putting up against the insolence of Lord Birkenhead and the crookedness of the Commissioners? I had not expected much from Sir John Simon, but I was not at all prepared for his resorting to all the known tricks of bureaucracy, and this the latest trade on untouchables adds to the ugliness of the whole picture. However, we have to be patient. You must therefore patiently go through the agony and mend where you can.

Do come as early as possible. I hope Kamala is keeping up her strength, if not actually adding to it. I wonder if Father has told you that, before you came, when Father was with me in Bangalore, he and I had contemplated your stay in Bangalore because of its magnificent climate during summer. There are just four weeks of somewhat trying weather, but you could always go to Nandi Hill only 35 miles from Bangalore where you have delightfully cool weather. In no case should Kamala be allowed to lose what she gained in Switzerland.

Yours sincerely,

Letter to Jawaharlal Nehru
SN 13079

[1] In his letter, dated Allahabad, 23 Feb., Jawaharlal Nehru had written: 'I wrote to you a few hours ago and informed you that I hoped to be in Sabarmati on Monday or Tuesday night. Immediately after I received a summons from Delhi to go there and remain there for the next fortnight or more to assist in constitution drafting.... Personally I have had enough of this All-Parties Conference. After ten days of it, the strain was too great for me and I fled to avoid riot and insurrection! I feel better already after a three-day absence, but another dose of all the parties may go to my head.'

146. THE SANCTION OF PUBLIC OPINION

The Ashram,
Sabarmati,
March 3, 1928

Dear Motilalji,

I have your letter. I have discussed it with Jawaharlal, but he suggests, and I agree, that it will be better for me to reduce to writing the views I have expressed to him so that there may be no misunderstanding about the correct interpretation of my views, and so that he may also know whether he understood me correctly.

ELECTORATE: I am of the same opinion that I expressed years ago at Delhi that we should not be party to separate electorates or to reservation of seats, the latter should be by mutual voluntary arrangement if such is necessary. But unless the Mussalmans agree, there is no going back by us on reservation of seats. The Congress is committed to it. I think, therefore, that we must simply adhere to the Congress resolution and expect Hindus and Mussalmans to carry out that resolution. If the All-Parties Conference cannot discover another method acceptable to all, we must simply work out the Congress formula.

THE CONSTITUTION: Personally I am of opinion that we are not ready for drawing up a constitution till we have developed sanction for ourselves. Any constitution that we may arrive at must be a final thing in the sense that we may improve upon but we may not recede from it even by an inch. There seems to be no atmosphere for arriving at such a constitution. I would personally therefore prefer instead of a constitution, a working arrangement between all parties upon which all may be agreed. This would be not a constitution but chief heads of it, as for instance, the Hindu-Muslim arrangement, the franchise, the policy as to the Native States. If we are to make this thing popular, I should bring in total prohibition and exclusion of foreign cloth as an indispensable condition. Of course we should guarantee equality of treatment of all religions as also of the so-called untouchables. I am not exhaustive in the list of things on which there should be an agreement, but I have simply given a few things by way of illustration. I think that if we go beyond such a general agreement, we would be making a mistake. In any case, I do hope that the Conference will not break up without doing anything, and even if it does, the Working Committee should take the matter in its own hands and issue its own authoritative statement on behalf of the Congress on all the matters for which the Conference has been convened.

SANCTION: More important than the two foregoing things, in my opinion, is the sanction. Unless we have created some force ourselves, we shall not advance beyond the position of beggars, and I have given all my time to thinking over this one question, and I can think of nothing else but boycott of foreign cloth with the assistance of mills if possible, without if necessary. I hold it to be perfectly capable of attainment within a measurable distance of time if we can create sufficient public opinion in its favour. I would have exclusive concentration upon this thing if I had my way. Though I have said nothing in public, I do not at all like what is going on in Bengal. So far as I can see, it is doomed to failure and I can see much harm coming out of that failure; and unlike boycott of foreign cloth, it is valueless, unless it succeeds to the extent we want. Jawaharlal and I have given most of our time to a consideration of this question. And he will explain it all to you. As soon as he can be dispensed with, I would like you to send him back for further discussion of this problem if we do not finish before he leaves for Delhi.

I see that I am not to expect you here in the near future.

Yours sincerely,

Letter to Motilal Nehru
SN 13095

147. POLITICAL HARANGUES AND CAMOUFLAGE

[London
6 November 1931]

Mr. Bernard Shaw had long been wanting to see Gandhiji, and it was not without considerable hesitation that he came. He sat with Gandhiji for close on an hour, interrogating him on a bewildering variety of topics—ethnographical, religious, social, political, economic—and his talk was illumined by his sparkling wit and sardonic humour. 'I knew something about you and felt something in you of a kindred spirit. We belong to a very small community on earth', said he. Whilst his other questions were of universal importance, he could not help asking a question about the R.T.C. 'Does not the Round Table Conference try your patience?' he asked, and Gandhiji had to confess with sorrow:

It requires more than the patience of a Job. The whole thing is a huge camouflage and the harangues that we are treated to are meant only to mark time. Why not, I ask them, make a clean breast and announce your policy and let us make our choice? But it does not seem to be in the English political nature to do so. It must go by round about and tortuous ways!

Interview with George Bernard Shaw
Young India, 19 Nov. 1931

148. POLITICAL POWER AND SOCIAL REFORM

Among the questions that a correspondent asked me for discussion in *Harijan* there was one which I have kept on my file for some time:

Don't you think that it is impossible to achieve any great reform without winning political power? The present economic structure has also got to be tackled. No reconstruction is possible without a political reconstruction and I am afraid all this talk of polished and unpolished rice, balanced diet and so on and so forth is mere moonshine.

I have often heard this argument advanced as an excuse for failure to do many things. I admit that there are certain things which cannot be done without political power, but there are numerous other things which do not at all depend upon political power. That is why a thinker like Thoreau said that 'that government is the best which governs the least'. This means that when people come into possession of political power, the interference with the freedom of people is reduced to a minimum. In other words, a nation that runs its affairs smoothly and effectively without much State interference is truly democratic. Where such a condition is absent, the form of government is democratic in name.

There is certainly no limit or restraint on the freedom of thought. It may be remembered that many reformers are nowadays laying the greatest emphasis on a new ideology. How few of us are going in for any reform in our opinions? Modern scientists recognize the potency of thought and that is why it is said that as a man thinks so does he become. One who always thinks of murder will turn a murderer, and one who thinks of incest will be incestuous. On the contrary he who always thinks of truth and non-violence will be truthful and non-violent, and he whose thoughts are fixed on God will be godly. In this realm of thought political power does not come into play at all. Even so it must be obvious that political power or want of it is of no consequence in many of our activities. I would make a humble suggestion to the correspondent. Let him make a detailed note of all his daily activities and he is sure to find that many of them are performed independently of any political power. Man has to thank himself for his dependence. He can be independent as soon as he wills it.

The correspondent has raised the bugbear of 'great' reform and then fought shy of it. He who is not ready for small reforms will never be ready for great reforms. He who makes the best of his faculties will go on augmenting them, and he will find that what once seemed to him a great reform was really a small one. He who orders his life in this way will lead a truly natural life. One must forget the political goal in order to realize it. To think in terms of the political goal in every matter and at every step is to raise unnecessary dust. Why worry one's head over a thing that is inevitable? Why die before one's death?

That is why I can take the keenest interest in discussing vitamins and leafy vegetables and unpolished rice. That is why it has become a matter of absorbing interest to me to find out how best to clean our latrines, how best to save our people from the heinous sin of fouling Mother Earth every morning. I do not quite see how thinking of these necessary problems and finding a solution for them has no political significance and how an examination of the financial policy of Government has necessarily a political bearing. What I am clear about is that the work I am doing and asking the masses to do is such as can be done by millions of people, whereas the work of examining the policy of our rulers will be beyond them. That it is a few people's business I will not dispute. Let those who are qualified to do so do it as best as they can. But until these leaders can bring great changes into being, why should not millions like me use the gifts that God has given them to the best advantage? Why should they not make their bodies fitter instruments of service? Why should not they clear their own doors and environments of dirt and filth? Why should they be always in the grip of disease and incapable of helping themselves or anyone else?

No, I am afraid the correspondent's question betrays his laziness and despair and the depression that has overtaken many of us. I can confidently claim that I yield to none in my passion for freedom. No fatigue or depression has seized me. Many years' experience has convinced me that the activities that absorb my energies and attention are calculated to achieve the nation's freedom, that therein lies the secret of non-violent freedom. That is why I invite everyone, men and women, young and old, to contribute his or her share to the great sacrifice.

'A Fatal Fallacy'
Harijan, 11 Jan. 1936

149. TRUE POLITICS AND POWER POLITICS

[Malikanda,
21 February 1940]

Let me reply to what Gokulbhai has said. There is some misunderstanding about the two things he mentioned. Let me clear those misunderstandings so as to avoid needless discussion.

I did not say that members of the Gandhi Seva Sangh should withdraw from politics. What I said was that they should not be in politics in their capacity as members of the Sangh. I said in the morning that no member of the Sangh should give up politics. I cited the examples of Sardar and Rajendra Babu and said that I would not ask them to leave politics. But I said that they should keep out of the Sangh. I wish politics had no place in the Sangh.

Gokulbhai's second question is: When our whole life is one and indivisible, how can we specify whether a particular thing comes under the category of politics or not? I did say that we should not compartmentalize our lives since things are interwoven. That is why life is one and indivisible. But there is also another point of view. Our body is one whole, but it has several organs such as eyes, ears, nose, etc. When we consider the body as a whole, we do not think of the separate organs. But when we have particular organs examined we think of each of them separately. They are parts of one and the same body but, none the less, we do think of them separately and make distinctions such as organs of perception and organs of action. Today we are looking at this thing from such an analytical point of view. Politics pervades all our activities. But I am not talking of retirement from politics in this broad sense. I am referring to the politics of the Congress and elections and to groupism. I have explained why power politics should be kept out of the Sangh. We are taking that step not out of cowardice but for the sake of self-purification. That is the way of non-violence. I have not forbidden all political activity. I know that in this country all constructive activities are part of politics. In my view that is true politics. Non-violence can have nothing to do with the politics of power.

The existence of the Sangh is not indispensable for non-violence. This is not to say that there can be no Sangh at all for a non-violent purpose. But our Sangh today is not such a one. We formed a Sangh and tasted its benefit. I at any rate have done so. We have seen that a non-violent Sangh cannot and should not function like other organizations. A non-violent Sangh should have some special features. Hence, from the experience I have of the Sangh, I suggest that we keep away from politics. If we do not, Gandhism is sure to perish.

We have adopted a novel policy. We must adopt novel means to follow that policy. I have been trying to find out what these means could be. I am only experimenting. I have to change my methods as situations change. But I have no ready-made formula. Our experiment is absolutely new. The order in which the steps may be taken is not fixed. I am a person with an inquiring mind. With great patience I am discovering and developing the science of *satyagraha*. In the course of this search I am acquiring new knowledge and new light every day.

It was because I was keen on this experimenting that I suggested at Hudli that we should appear on the political stage and gain experience in that role too, and put to test the strength of our truth and non-violence. It is possible that I was mistaken in giving this advice; but I do not regret having done it. It was just as well that we entered politics. We have had a very valuable experience. Had we not had this experience, I would have been in two minds. I would have been always conscious of the fact that we had had no political experience. Now, after that experience, I can positively advise you as members of the Sangh to retire from politics.

One more thing you should not forget. The Sangh has never accepted politics as its field. Remember, have we ever discussed politics during the conventions of the Sangh? You will find no trace of politics in its reports and resolutions. This is the subject that matters to those members of the Sangh who are interested in politics. This is not a matter of interest to the Sangh. It is meant for Sardar who is involved in politics. For other members of the Sangh too it is an activity outside the purview of the Sangh. When do you ever discuss politics here? Do you ever trouble Sardar and insist that he should explain to you how he conducts the affairs of the Parliamentary Board? Do you ever ask me what I convey to the Viceroy? We are no doubt curious to know about these things; but it is not as members of the Sangh, but in some other capacity. At Hudli we passed a resolution to take part in politics. But even after that we did not discuss politics during our conventions. We come here with an altogether different mind. Ours is a group of seekers of knowledge. We come here with a view to doing some introspection and correcting our mistakes. Hence, our conventions have quite a different atmosphere. Politics is an outer activity. Hence, we do not discuss it here. That is not the sphere of the Sangh.

Even after Hudli the inner character of the Sangh did not change. It is clear from this that the misunderstanding on the part of people that we are forming a political group is totally unfounded. For no reason are we dragging ourselves down. That is why I suggest we wind up the Sangh. I am not winding up political activity. But I am putting a stop to it as far as the Sangh is concerned; because that is none of our business. Even at Hudli we did not change the original character of the Sangh. The only mistake we made was that we permitted members of the Sangh to take part in parliamentary activities.

But even after this I took very little direct interest in parliamentary work. As far as I am concerned I devote little time even to reading newspapers. If you ask Kishorelalbhai he would say he knows nothing about it and that you had better ask Sardar. The obvious reason for this is that the Sangh never considered politics as its sphere. We did not form the Sangh for that purpose. Then, why should we needlessly jump into that fire? Let us withdraw from it.

This distinction was not clear to Gokulbhai. He thought I was forbidding political activity. But that is not the case. What I say is that the Sangh should not get involved in this.

The question now remains why we may not try to introduce truth and non-violence into politics. Why should the Sangh leave this field untouched? I have already answered this question. The method of non-violence implies that when we do not have the capacity to undo an evil, we should keep away from it. This is non-co-operation. I have placed before India the great principle of non-co-operation. I am applying that very principle here.

Let me give an illustration. There is a protest demonstration going on here. Should we force ourselves to stand before the demonstrators and say: 'Well, here we are! Do with us what you will.' This would be sheer folly. That is why the Shastras say that we need not rush to hear abuse of ourselves.

I do not know if you have heard about the figure of the three monkeys. It is in Kobe, Japan. Somebody had sent me a small figure—a replica of the original. It consists of the statuettes of the three monkeys. One of them has its mouth closed, the second its eyes and the third its ears. These monkeys teach the world to speak no evil, to see no evil, to hear no evil. This is the secret of non-co-operation. There is a protest demonstration going on here. If the demonstrators enter this *pandal* and attack us, I would ask you to stick to your places and take their blows. But I would certainly not ask you to go to the place of demonstration and invite their blows. This would mean deliberately inciting them. This is not non-violence. It is self-assertion.

I propose to adopt here this pattern of non-co-operation. Our entering politics will provoke them. It will be like feeding their wrath. Therefore, non-violence tells me that we should withdraw ourselves from politics. If people criticize, oppose or attack us even after we quit politics, we should put up with all this. After we have quit politics if anyone wishes to destroy us, let him do so.

In spite of all this, those who do not need the protection of the Sangh for their political work may continue there. For instance Vallabhbhai. What does he need the Sangh for? He does not enhance his prestige by remaining with the Sangh. He had established his political prestige even before the Sangh was born. Thus, it is the Sangh that acquires prestige by his being in it. Then he happens to be a born politician. Politics is in his blood. He is not born for the constructive programme. In a way, he has accepted the constructive programme out of a sense of compulsion. It is not an inseparable part of his temperament as I claim it is in my case. Unlike me, he is not absorbed in the constructive programme. I was born for the constructive programme. It is part of my soul. Politics is a kind of botheration for me. I would dance for joy if I could shake it off. Sardar would hardly do any such thing. This is the main difference between us. He hears what I say and if I am making any mistake he can put me right.

But if there are many other members who derive prestige from their association with the Sangh and if this prestige proves helpful to them for political purposes, the only way out is that we should not let them have such prestige. They too ought to give up such borrowed prestige. If we lend them such prestige and they accept it, we deserve the charge that we are competing with the Congress socialists or the communists. When this is not actually the case, why should we lay ourselves open to such a charge?

We do not wish to compete with anyone. We want to absorb even our opponents. If they are opposing us, it is because they do not understand us. But we know that we belong to them and they belong to us. Hence, let us take no active interest in politics so long as the people do not invite us. Let us carry on quietly our constructive programme. And let us, by withdrawing ourselves from politics, lend glory to non-violence. These are the words of an experienced person. You must understand and grasp its significance and pay attention to what it implies. Thus, your winding up the Sangh would be an object-lesson in non-violence. It is quite simple. There is no harm in it.

Now, why should Gokulbhai feel hurt by my saying that I gained nothing from the Sangh? Actually being in this place is like being in my home. You all love me. We form a family here. In a family, the children, wife, parents, brothers, sisters—all love one another. There is very little scope in a family to carry on an experiment in non-violence. Similarly, my non-violence cannot be tested in this Sangh. What can you teach me here? My non-violence is tested in the Congress where I am opposed, censured and criticized. My proclivity to anger is nurtured and aroused. I put up with everything very respectfully. I have an opportunity to practise love and non-violence. There I grow. Hence I say that I had in the Congress opportunities to develop my abilities. I derived all my [training] from the Congress. I learnt very little in the Sangh. I do not mean to suggest that the Sangh is something useless and hence deserves to be discarded. As a matter of fact I have only praised you. I have not criticized you. What can even Gokulbhai learn in a family where he finds only abundant love? He can learn more outside. It is in the outside world that we are tested. We have to learn to keep smiling in the face of abuse. Whatever the provocation, there should be no anger in our hearts. The field for our test lies outside. All we have to do here is to charge our batteries. How we can charge our batteries is another question. For that purpose we shall have to change the character of the Sangh. You may, if you wish, express your view on this matter at a later stage. For the time being we are only considering the question of separating the Sangh from politics. I have cleared the misunderstanding in Gokulbhai's mind. I have clarified all the points. Now, if he wants, he may confess his fault or ask any other question if he so desires.

I know Jajuji's temperament. He believes in doing everything methodically. He has said that it is easy to consider a matter when it is presented in definite terms. I too felt that it would be better if we could come to some decision. That is why a resolution has been drafted. Do wind up the Sangh if that is what you want. I still hold fast to my opinion in this regard and I have already offered my advice. If my suggestion is not acceptable to you, you can decide the manner in which the members of the Sangh may take part in politics. I have drafted this resolution because I have a feeling that you would prefer such an arrangement. The draft will be presently read out to you. Think over it during the night. We must come to a final decision by tomorrow. Kishorelalbhai has drafted two resolutions twice over. I could not accept the points when he presented the resolutions for the first time. He had given fresh drafts today. I could not accept the points in these drafts either. But the spirit of the resolutions has been brought in.

Speech at Gandhi Seva Sangh (H.)
Gandhi Seva Sanghke Chhathe Adhiveshan (Malikanda-Bengal) ka Vivaran, pp. 25-33

150. THE CONSTRUCTIVE PROGRAMME

[Malikanda,
22 February 1940]

Brothers and Sisters,

Are those sitting at the far end able to hear me? If not, please say so.

I was up by 3 o'clock this morning [and] started thinking about the Sangh. I am placing before you my conclusions. Later on you may express your views today itself if you so desire. Some sort of a tentative resolution was placed before you yesterday. I think I should say something about it and also something apart from it.

The views I had expressed on [taking part in] politics have been further confirmed. As I thought about it I felt that unwittingly we had acted wrongly. What I said last night was an indication of this. Hence it is that we have deliberately abjured power politics. If while remaining in the Sangh we wanted to enter into power politics we should have done so openly. For that matter we ought to have changed even the character of the Sangh. But we are not at all equipped for taking part in politics. In politics it is necessary to know what others also are doing and what they intend to do. It would also be necessary to be acquainted with the ideology and politics of the socialists. We do not have so much as a library of political literature.

I have not studied socialism. I have read no books on the subject; I have read one[1] by Jayaprakash. I have also read one book given by Masani.[2] Sampurnanandji has written a very good book.[3] He sent it to me very lovingly and I read that too. This is all my study of socialism. They say countless books have been written on socialism and communism. I have read nothing at all about communism. I do not know how many of you have read books about it.

This thing does not form part of my activity. My mental make-up is quite different. They do not even accept my claim to intelligence. My intelligence is not recognized in the field of politics. Those who are deeply involved in politics laugh at me. They wonder if this is politics. I acquired some status in politics because I had been a fighter, not because of my intelligence. They don't credit me with intelligence capable of winning over others. I believe in socialism and also in communism. I believe in everything, but from my own point of view. I belong to everybody and I belong to none. A believer in non-violence cannot be anybody's opponent. He would wish to build up the science of non-violence with the help of everyone. His is not the politics of opposing or vanquishing anyone.

I do not wish to qualify myself or others for what goes by the name of politics. I had suggested at Hudli that we should enter politics. Unwittingly I made that mistake. You may as well say that unwittingly I followed untruth. Instead of doing well the task for which we were born, we took in hand something else. But whatever happened was for the good. We gained some experience. We discovered that we were not capable of engaging in politics. We have realized our incompetence. Now we are withdrawing our hands. We did commit a mistake, but now, the moment we have become conscious of our faults, we are being cautious. A mistake, when it is corrected, does not remain a mistake. Our strength grows if we acknowledge our mistakes. I am asking you to realize your limitations, and carry on thoroughly the work for which the Sangh was formed.

Some friends from Noakhali visited me today. They told me: 'We accept everything you say. But we are unable to understand what your followers here say. We can agree to what you say. Call it Gandhism, spinning, village industries— we would accept anything. We are your followers. But we are not followers of your followers. Your followers have nothing [to offer].' Whatever these friends said sprang from sheer love. It is a matter over which we should ponder. We entered politics but we could do nothing in that field. Nor could we do our own work. We belonged neither here nor there. Now that we are aware of our folly let us try to remove it.

We used politics to put our principles into practice. Now after some experience we are renouncing politics. The politics which we are renouncing is the politics of acquiring positions of power within the Congress. We cannot take part in this politics. I am not talking about an individual; I am talking about the Sangh. Power politics has no place in the Sangh. An individual with a natural inclination and qualification may remain in politics. But this power politics is such a dreadful snare that even individuals may have to quit it. Their non-violence would be put to the severest test there. They too will quit when they have some bitter experiences. But what I am saying today applies only to the Sangh. The Sangh must definitely renounce the Congress Committees, that is, elections and power politics. My opinion in this matter has hardened. This Sangh is not meant for this kind of politics. I am myself not qualified for it. Your Chairman is much less so. He is a philosopher, a moralist and a writer.

Let me mention the third point in just one sentence. As a matter of fact, you must give up the very name of Gandhism. If not, you will be falling into a blind well. Gandhism is sure to be wiped out. I love to hear the words: 'Down with Gandhism.' An 'ism' deserves to be destroyed. It is a useless thing. The real thing is non-violence. It is immortal. It is enough for me if it remains alive. I am eager to see Gandhism wiped out at an early date. You should not give yourselves over to sectarianism. I did not belong to any sect. I have never dreamt of establishing any sect. If any sect is established in my name after my death, my soul would cry out in anguish. What we have carried on for all these years is not a sect. We do not have to subject ourselves to any 'ism'. Rather, we have to serve in silence in keeping with our principles.

Whatever people might say, service can never be made into a sect. It is meant for everybody. We shall accept everyone. We shall try to fall in line with everyone. This is the way of non-violence. If we have any 'ism', it is just this. There is nothing like Gandhism. I have no followers. I am my own follower. No, no, even I have not succeeded in being my true follower, have I? Do I put my own thoughts into practice? How then can others be my followers? I would like others keeping me company and walking the same way. But how can I know who would walk ahead and who would fall behind? You are all my fellow-students and co-workers, fellow-servants and fellow-researchers. Forget the idea of being followers. Nobody is leading and nobody is following. Nobody is a leader and nobody is a follower. We are all going together in one line. I have said this quite often but I am repeating it in order to remind you.

We have to give up offices in the Congress. Let this point be absolutely clear in your minds. Then you would grasp the significance of what I am going to say. I am asked: 'Should we withdraw ourselves even from municipal bodies and similar organizations?' I would say that we should withdraw even from municipal bodies. I know what had happened in the case of Nagpur Municipality. I was amazed to find how much bitterness and hatred had crept into the Congress Municipal Party. I have very little information about it. Gopalrao knows the inside story. There are three factions in the Congress Municipal Party of Nagpur. They are always at daggers drawn. People belonging to all the three factions came to me and narrated their own versions. I was very deeply pained. There is a lot of animosity prevailing in the Provincial Congress Committee also. I told them that the Provincial Congress Committee was under the control of the A.I.C.C. Hence they should approach either Sardar or Rajendra Babu. I would be infinitely pained if any member of the Sangh chose to go into any such organization. What can he attain by going there? We wish to establish rapport with 300 million people. This would happen only when we reduce ourselves to a cipher. What have we to do with rights? Politics of power is unreal.

We must tell the people what is real politics. We shall concentrate only on the constructive programme which others not only shun but regard with contempt. The Gandhi Seva Sangh would justify its existence by persistently devoting itself to the constructive programme. There are eighty-four persons among the members of the Sangh taking active part in politics. They would of course leave the Sangh. But I shall presently explain to you what shape the Sangh would take after that.

The Gandhi Seva Sangh has come into existence for the purpose of carrying on the constructive programme. That alone is real politics. We have to lend glory to this real politics by renouncing our rights. What do we care if they do not call it politics? We shall remain within the Congress fold, but keep ourselves away from power and elections.

There is no need to have the Gandhi Seva Sangh for the sake of keeping a register of the list of persons who believe as I do in truth and non-violence. I see no need for such a list. I was thinking yesterday about the future form and functions of the Sangh. I shall now place before you my conclusions. In my view, the Gandhi Seva Sangh has now to become something like an organization for post-graduate studies. All the organizations in the country which carry my name or are run under my supervision or guidance are meant only for the constructive programme. The Charkha Sangh, Gramodyog Sangh, Harijan Sevak Sangh, Talimi Sangh—I guide the functions of all these organizations. I was responsible for the propagation of Hindi throughout the country. For me, these are inseparable aspects of real politics. I also control the labour union of Ahmedabad. The Gandhi Seva Sangh can undertake a great deal of post-graduate study and research for these organizations. All these organizations cannot take up such work to the required extent as their field of activity is limited. Take for instance the Charkha Sangh. I have laid down its policy which is aimed at giving the maximum wages to the hungry and poor who are idle for nearly six months in a year, and persuading others to buy the *khadi* produced by these poor people by appealing to their altruism. Self-sufficiency is also one of the functions of the Charkha Sangh, but as auxiliary to its primary objective. It does not include that aspect of self-sufficiency which Prafulla Babu had placed before you. Prafulla Babu had correlated it with non-violence and *swaraj*. Self-sufficiency in that sense of the term is not included among the functions of the Charkha Sangh. I cannot expect Shankerlal to go and explain to the poor women spinners all the things that the spinning-wheel implies.

All that is your work. Prafulla Babu had said the other day that you should spin at least 100,000 yards of yarn in a year. On calculation I realized that it would mean spinning nearly 300 yards every day to produce 100,000 yards in a year. It would take 45 minutes to spin 300 yards if the wheel and the slivers are good. There are people who do it even in half an hour. I would of course expect the maximum from you. For you the art of spinning should be child's play, for, after all, you are in the post-graduate stage. You would spin sincerely and with great concentration. There are experts who can produce 300 yards of yarn in half an hour. But you would not stop at being expert spinners. You would try, as I am doing, to realize God through the spinning-wheel. Then alone would you pass my test. This is my explanation of Prafulla Babu's scheme.

People from the Charkha Sangh and the Gramodyog Sangh come to you for acquiring the expertise in their respective fields. You would have to achieve perfection and specialize in such things. Everyone cannot become an expert in every field. But everybody can specialize in one thing. Among doctors too some are physicians and some are surgeons. Even in the field of surgery some specialize in the eye, some in the nose and throat. Similarly we should take up some field for specialization. This is not a matter of making money. Other specialists go in for invention and research for the purpose of earning money. We have to become specialists for the purpose of serving the poor and bettering their lot. The Gandhi Seva Sangh would justify its existence in future only if it undertakes such work. If you must continue the Sangh you should continue it in this form or wind it up. If this is not done, the whole world will laugh at us and Gandhism will be killed by our own hands. Let us not deceive ourselves.

We shall follow yet another programme. I have told the labour union people that they would be doomed if they meddled in politics. I have told the Charkha Sangh people that they have nothing to do with politics. I told them that if they got involved in the politics of *swaraj*, they would be spinning weak yarn as their attention would be divided. They would not be able to do their work properly and they would prove worthless. I told the Harijan Sevak Sangh that it had nothing at all to do with *swaraj*. If they stuck to their work, *swaraj* would issue from it automatically.

But what about you? You will have to understand what Gandhi means when he says he sees God in the spinning-wheel. Should we take it literally or grasp its significance? Or should we accept both? I would suggest that you accept both the meanings. It means that you should see God in every thread [you draw] from the spinning-wheel. Acts of service such as spinning take us nearer to God, and therein lies its significance. You will thus find that all our activities have a far-reaching meaning.

Even its literal meaning has a wider sense. You are all going to do research and make discoveries in this post-graduate laboratory. For that purpose you should have expert knowledge not only of the science of the spinning-wheel but also of the art of spinning. Knowledge and skill would be especially expected of the Gandhi Seva Sangh. Your spinning-wheel should work more efficiently than that of the Charkha Sangh. Your yarn should be fine, strong and may not snap. It would be a matter of shame if your thread snaps. For the expert, his tools should be of the highest quality. There should be something special about your slivers, your implements. I do not wish to make you just skilled labourers. I want to make you expert craftsmen and scientific researchers. I expect something unique from you. The difference between your spinning-wheel and that of the Charkha Sangh should be as much as that between my razor and a barber's. Your implements would be out of the ordinary.

Here you spin from the slivers made by others. But the slivers from which you would be spinning then would be no ordinary slivers. You would improve the process of making slivers. You would make the best slivers yourself and make others improve theirs. Thus you would study the smallest detail just as Vinoba has done. After persistent experiments he has built up a science round everything. He has discovered a new method of making slivers. In Andhra fish-bones are used for this purpose. Vinoba's is a modification of the Andhra method, but the yarn spun from these slivers can well compete with the Andhra yarn. He has sent me cloth woven from the yarn of 40 counts spun from *verum* cotton. It deserves to be displayed in the exhibition. It is made by a Muslim boy. Your carding process would also be such as to cause trouble to no one. Cotton fibres would not fly about from your carding-bow. Even one suffering from cough would be able to do the carding without a hitch. There would be something special even in your choice of cotton. How could you ever take cotton from the bales? Your ginning process too would be special. For this one needs patience, talent for research and perseverance. If you have achieved all this, you would have followed my suggestion to the very letter.

After this you would have to see if the *charkha* increases your non-violent powers. Vinoba has composed a couplet comprising the eleven vows. You recite it every day. You would have to find out if the spinning-wheel in any way helps you in observing these vows. There may not be politics in the spinning-wheel of the Charkha Sangh; but you would have to see if it is there in your spinning-wheel. In other words you would have to see if it increases the strength of the people and whether, in free India, the economic provisions of *swaraj* could be based on the spinning-wheel. Would it turn people into mere automatons capable of physical labour or would it make them non-violent soldiers of *swaraj*? You must think about all this. These things do not fall within the purview of the Charkha Sangh. The special task which lies beyond its scope would be your field.

You will have to consider if we can really win *swaraj* with the help of the spinning-wheel. Are you really convinced of it? Or do you believe in it just because Gandhi says so? Gandhi can see God through the spinning-wheel, or hope to find Him there. But it may as well be a personal matter with him. You will have to find out if this principle can be applied universally. Like Jagdish Chandra Bose you will have to become research scholars in your own field. He had even devised a post-graduate course. I have seen how he remained deeply engrossed in it. This had become the main objective of his life. I used to be his close associate. I had stayed in his house for several days. He had a group of some twenty chosen men. But a few such men, if they are firm in their conviction, can work like millions. This is how experts work. The Charkha Sangh, Gramodyog Sangh, cannot do such work. There, too, there are experts and they also do research. But your field would be much wider and much more specialized. Through them I am trying to serve *Daridranarayana* in particular. Their development will be only along those lines. Your work however would be quite unusual. You would not merely improve the tools and implements, but also see their conformity with our principles. I shall exercise my mind to help you as your co-student; but I want to get the work mainly out of you.

In this way research and discoveries are undertaken everywhere. Look at Germany. How many specialists are there in that country! There they are in need of specialists in the science of violence. We too could do research and make progress in the field of non-violence if we can have a small centre. We have to link the spinning-wheel and related activities with non-violence and ultimately with God.

You will have to consider if all these things are possible. You know that even a society based on violence functions only with the help of experts. We want to bring about a new social order based on truth and non-violence. We need experts to develop this into a science. The world as it functions today represents a mixture of violence and non-violence. The external surface of the world suggests its internal state. A country like Germany which regards violence as God is engaged only in developing violence and glorifying it. We are watching the efforts that the votaries of violence are making. We must also know that those given to violence are watching our activities. They are observing what we are doing for developing our science.

But the way of violence is old and established. It is not so difficult to do research in it. The way of non-violence is new. The science of non-violence is yet taking shape. We are still not conversant with all its aspects. There is a wide scope for research and experiment in this field. You can apply all your talents to it.

For me non-violence is something to be shunned if it is a private virtue. My concept of non-violence is universal. It belongs to the millions. I am here just to serve them. Anything that cannot reach the millions is not for me. It should be so for my colleagues as well. We were born to prove that truth and non-violence are not just rules for personal conduct. They can become the policy of a group, a community, a nation. We have not yet proved this, but that alone can be the aim of our life. Those who do not have this faith or those who cannot acquire it should be good enough to remove themselves. But I have only this dream. I have regarded this alone as my duty. I shall not leave it even if the whole world abandons me. So profound is my faith. I would live only to attain this and die only in that endeavour. My faith gives me new visions every day. Now, in my old age, I am not likely to do anything else. It would be another thing if my mind is corrupted or I have a new vision. But today I am seeing ever new miracles of non-violence.

Every day I have a new vision and I experience a new joy. I am certain that non-violence is meant for all time. It is an attribute of the *atman* and is, therefore, universal since the *atman* belongs to all. Non-violence is meant for everybody and for all time and at all places. If it is really an attribute of the *atman* it should be inherent in us. Nowadays it is said that truth cannot help in trade and politics. Then where can it be of help? Truth is not worth a dime if it cannot be of help in all the fields of life, in all worldly dealings. Of what use would it be in life then? I see every day that truth can be applied to life in all its practical aspects. I have been doing this *sadhana* for more than fifty years now. I have been conveying to you from time to time some of my experiences during this *sadhana*. You too can have glimpses of it.

If the Sangh has to continue it should be for this purpose. If you do not have this much inclination, earnestness or strength, disbanding the Sangh would itself be the pursuit of truth and non-violence. If we do not do so, we would prove guilty, we would be committing the sin of dishonesty. We would be committing yet another sin. We have in our midst a worker like Kishorelalbhai. He is a pure man given to uninterrupted hard work. God has given him a keen intelligence.

We would be misusing the services of such a man. Would we take his life even if he offered it himself? But how can I press him? What special work is now left for us? We must relieve him of his responsibility right from today.

The other day I narrated to you the circumstances that should spell the doom of Gandhism. I repeat today that if Gandhism encourages any wrong thing it deserves to be doomed. Truth and non-violence are not just flowers in the sky. They should be manifest in all our words, dealings and acts. Kishorelal does not wish to speak about all his bitter experiences in the Sangh. He is silent because he has forbearance. He conveys it to me just in passing when things go too far. This gives me some idea of the situation. Why is there such vicious estrangement amongst us? We all form one family. We have adopted truth and non-violence. But we are given to strife and dogma and jealousy. Are not truth and non-violence things of this earth then? Do they deserve only to be installed in the heavens? We have learnt our lesson by entering politics. It created antagonism even among good people in the Sangh. Oh, we are far away from Rama and Bharata. Kishorelalbhai is a poet too. He gave a poetic interpretation in terms of Rama and Bharata. But here no one is like Rama or Bharata. If we had Rama and Bharata amongst us would such things have happened? What a world of difference between the love of Rama and Bharata and these feuds! These people are miles away from Rama and Bharata.

Why should such difference arise among us when we all belong to the same Sangh and have come here with the sole idea of service? There ought to be only love among us. Are all the people who have come here like this? Do they have cordial relations with each other? If your reply is in the affirmative, I shall have to put you to a very hard test. I am afraid we shall not pass that test. Ask the honest opinion of one another among yourselves. Ask yourselves, and you will realize that we are still far removed from non-violence. Why should we meddle with the Congress when we are so weak? What service can we render there? Why do we not withdraw when we are faced with a rival? What have we to do with positions in the Congress? Why should we think of defeating anyone in an election?

If we are not possessed by the desire to power and a feeling of jealousy, why has so much bitterness spread round us? Why do these people come to Malikanda and shout violent slogans? Why is there so much bungling? I would like to ask the Bengali workers present here: 'What gave rise to so much bitterness? Why don't you try to win them over with love? Why don't the leading workers among you go to them and talk to them lovingly?' There are Prafulla Babu and Satis Babu. Let them go to the people who are raising the slogans, mix with them and talk to them with love. I have tried this myself. It is nothing new for me. Surely I am not their rival. I do not want what they desire. They want power, and I have no desire for power at all. Rivalry is possible only when two persons want the same thing. We have to qualify ourselves for their love. That is why we must give up politics, and leave the field open for them. That is the way non-violence functions. For that very reason we must quit politics.

But are we going to remain idle, doing nothing? We shall continue to spin and see if the spinning-wheel dulls our intelligence or sharpens it. We shall study the spinning-wheel. We shall find out what the spinning-wheel signifies, why and how it came into being and why it went out of use and in what way it can be useful to us. The Gandhi Seva Sangh should maintain a special library for this purpose. There will be no other library of its kind anywhere in the world. We shall be free from the bother of Gandhism only when we start working along these lines. Our present opponents will admire us for having such a programme and bless us, although today they are cursing us. Let me know if this thing is beyond your intelligence, strength or wishes. It would only prove that we are not yet qualified to form a new Sangh. But does this mean that what I have said is proved wrong thereby?

I had not imagined this even when this Sangh was formed. Jamnalalji had set up the Sangh when I was in jail. Jamnalalji is a worker with a pure heart. He opened his coffers when I started the non-co-operation movement. I had appealed to lawyers to give up their practice. Jamnalalji offered to support one hundred lawyers who would give up their practice and devote themselves to the service of the nation. This was not because he wanted to dominate the Congress but because he wanted to awaken the masses to the power of *satyagraha*. It was not his intention to send these gentlemen to the Congress. Every time he referred that matter to me he said that these people should be kept away from politics. Subsequently it was decided to use the money for the maintenance of constructive workers instead of the lawyers who had joined the non-cooperation movement. Out of this the Sangh was born. Jamnalalji had never intended to do political work through the Sangh. He was not agreeable again when I expanded the Sangh in 1934. Later on he also opposed my resolution at Hudli. If I am accused of dragging the Sangh into politics I would plead guilty. It would not be right to levy such a charge against Jamnalalji.

Today I gave deep thought to the policy of the Sangh. I had not been able to do this with so much faith. Age has sapped my physical energy. I cannot observe things in all directions. I cannot be alert to all things and always. My body has become weak. I have somehow shouldered my responsibilities to this day. I even wanted to avoid coming to this convention; but I came because Prafulla Babu insisted. After coming here I concentrated my thoughts on the affairs of the Sangh. I placed the conclusions before you. Today my mind is saturated with these thoughts. Kishorelal carried this burden all these years. It was I who brought him into the Sangh. What is his condition today? His body has become weak. I have read his speech. Now Kishorelal cannot work here. Hence I am placing this thing before you with so much concern.

In a way I am only casually mentioning this thing here. If you are not going to have in the Sangh any members who want to take part in politics independently, the Sangh would become quite small. But if it is genuine, it would grow into a huge tree, otherwise it would perish.

Because of politics, 84 out of 243 members go out of the Sangh. Not because they are worthless, but because they are useful in that field and are qualified for the task. How many from the remaining members can accomplish this ideal? For this purpose you should form a committee with three or four members. This committee would prepare, with God as their witness, a list of persons who are qualified to carry on this work. This committee should also be empowered to decide on the future character of the Sangh. On this point you may also place your views before the committee in brief. But do not make speeches. In any case, the Sangh as it is functioning today has to be disbanded. It need not be considered whether or not it should function in future and if it should, in what shape. I have given you an outline of the shape it can take.

The question that now remains is: What would happen to the paid members who number nearly thirty? It is a matter of settling accounts. But the Sangh does not exist for settling accounts, does it? It is Jamnalalji's trust and this is a small matter. Dhotre and Jamnalalji can attend to it. It is not a matter for others to worry about. Let Jamnalalji and others concerned with him think about it. With this the matter comes to a close.

I am asking the rest of you if, in your view, the Gandhi Seva Sangh should continue to exist. If you want it to continue, it will have to be in the form I have indicated. You will have to study, do research and conduct experiments. And life will then be a grand thing. For that purpose all intellectual, physical and spiritual forces will have to be brought together. It will be a unique thing for the country, and for the world at large, which none may possibly envy. You will be placing before the country and the world the ideal of a new culture by this unity of body, mind and spirit. Can there be a goal loftier than this?

I have placed this serious point before you. If this is beyond your capacity, it is only wise to wind up the Sangh. Only those who can pass the test of truth and non-violence and who are capable of working efficiently with a sense of complete dedication can manage the Sangh in its changed form.

I have explained my points at great length. Now you may express your views.

SWAMI ANAND: ... Will communal unity also form part of the activities of the Sangh? Or has it been kept out?

GANDHIJI: It has not been kept out. It is there all right. Non-violence has no meaning without it. We have no programme for it right now, hence I deliberately omitted to mention it.

Politics and Society

SWAMI: I am afraid if we get involved in the work of labour unions and communal unity it also may arouse jealousy and anger in other people. There too we shall come in conflict with them, and we shall have to withdraw from that field as well.

GANDHIJI: If feelings of jealousy and anger oblige us to withdraw it would be proved that we who are experimenting with non-violence are good for nothing. Some people may as well say that non-violence itself is meaningless. From their point of view non-violence would prove to be a great *adharma* instead of a great *dharma*, for they say that the world functions because of the mixture of violence and non-violence. In their view, it is necessary that these two go hand in hand, otherwise the world would come to a standstill. Labourers have become cowards, they say, and it is necessary to train them in violent protest to build up their self-confidence. The Hindus are also cowards and for them non-violence is not the supreme *dharma*. I am still receiving letters to this effect. But we have to prove that we can solve labour disputes and communal problems by adopting non-violent methods.

RAM RATAN SHARMA: I wish to place before you one of my difficulties. I requested Kishorelalbhai to arrange for sending me somewhere to be given some training. He said that he had no provision for such training for me. I had also wanted to shape the Sangh in the manner Bapu has suggested. For that purpose it is necessary that I get some expert training somewhere.

GANDHIJI: The Sangh about which I am thinking can come about after specialized training. It is a different question what we should do for training experts.

SUDHAKAR: Are we giving up politics for good or for some time?

GANDHIJI: Who can say it is for good? We are not God, after all. We are talking only about the present.

KRISHNAN NAIR: People attached to the Charkha Sangh, the Gramodyog Sangh, etc., have their own organizations to lean on. But how should we organize those who do not belong to any of these five organizations? Is it not necessary to have for them some kind of an old students' association? Are all our members covered by these five organizations?

GANDHIJI: If you have read the report carefully, you should know how many of our members are working in organizations devoted to constructive programme. This report is a thing which deserves to be studied. It is not an ordinary report. It indicates the number of the members in several ways. They are classified according to their Provinces, their activities and the organizations to which they belong. I was fascinated by the report. Look at the appendix. You find in the report an answer to every point. After reading the report you will realize that we are not going to achieve anything great by having an independent organization like the Gandhi Seva Sangh. Those who would work with complete dedication would not need frequent consultation or guidance at every step. Attaining freedom is not an ordinary thing. For that purpose we would need workers who could carry on their tasks independently and with a sense of dedication. Those who require supervision would work under the aegis of some organization or other. The Sangh does not have to exist for that purpose. We have five organizations devoted to the constructive programme. Most of the members are covered by these organizations. What more can we offer them by having them to continue in the Sangh? The Sangh remains like a sixth finger. And how can we carry on supervision or inspection? We have no means to do it.

KRISHNAN NAIR: This means that the Sangh ceases to exist physically from today.

GANDHIJI: Yes, that is so. Today we are pushing Sita down into the nether world.

PREMABEHN: At the time of the Dandi March[4] also you said that we were all worthless and disbanded the Ashram. Has a similar situation arisen now? Is this also a preparation for some struggle? Or is it merely an attempt at purification?

GANDHIJI: I did not say that anyone was worthless. And how could I say such a thing so long as people like you are around? But this is only by way of a joke. The fact is that we want to become purer. I do not consider anyone worthless nor call him so. I did not describe anyone in the Sabarmati Ashram as worthless. You would be doing me an injustice by putting such an interpretation on my speeches. All I say is that applying the yardstick we have chosen we have not passed the test. And if we have really followed the principles of truth and non-violence why have these people come here to shout slogans like 'Down with Gandhism'? They teach us a lesson that we are yet impure. Otherwise why would they raise such slogans? Neither you nor I know if they are being paid. But even if it is true it is a matter for us to think over. Why do people pay them to shout these slogans? Why are they so bitter? We must learn something from this. As far as I am concerned, I have learnt from these demonstrations what I could not learn by seeing you. They have set me thinking. Very few among us would have practised non-violence. In this matter we can have no guidance even from history, for we do not find there instances of such experiments. But if you have a sharp mind like mine, you would find proof of non-violence everywhere in the universe. This world changes every moment. There are so many forces of destruction that nothing remains steady. But the human race is not destroyed in spite of this. It only suggests that non-violence prevails everywhere. I have vision of it. Like the force of gravitation, non-violence is drawing everything in the universe towards itself. Love has this power. Do I not call myself a scientist of non-violence? That is why I am aware of its laws and see them [in action]. We cannot be votaries of non-violence if we do not cherish a feeling of equality for all. If you have such a feeling of equality, you would not resort to protest or violence when anybody raises slogans invoking doom on you. Let us abjure any such practice which creates violent reactions in others. We are changing the character of the Sangh from this point of view. So many people come here to have a glimpse of me. They think I am an incarnation of God. But nobody is an incarnation of God. Or, if anybody is, everybody is. If I am such an incarnation, all of them are also such incarnations. Then why are they drawn towards me like this? It is the law of non-violence at work. There is in me some non-attachment also. Hence I am able to take the detached view that it is not my own power but the power of non-violence. What am I? I am but an embodiment of a thousand faults. When I search my heart I discover nothing but faults. If you take to such introspection you too would go crazy. We must try to control every thought of ours. I try it myself. I have a glimpse of God even in my opponents. You should also have that glimpse of God.

PRABHUDAS: Kishorelalbhai has said in *Sarvodaya* that there should be an organization even of imperfect followers of non-violence. They too must acquire the strength of organization. But from your speeches it seems that there can be no organization of those who practise non-violence imperfectly.

GANDHIJI: I shall not accept your point in your own words. An organization like the Gandhi Seva Sangh cannot be an organization merely of non-violent people The Sangh was formed under specific circumstances. It was formed with the idea of supervising the work of some twenty or thirty people by giving them financial assistance. Then it expanded, and it was necessary to make improvements in it. But that special purpose is no more. There is no need to have such organization for the progress of people who follow non-violence imperfectly. An institution is formed only with imperfect people. If men were perfect, no organization would be needed. Non-violence is imbibed even in the Charkha Sangh and the Harijan Sevak Sangh. These also want to follow the path of peace. The Gandhi Seva Sangh can become an organization of non-violent people only if it has some special programme. But what is the meaning of a mere organization of non-violent people? It is an implied *dharma*. When we wish to attain non-violence through certain means, an organization is formed. The Gandhi Seva Sangh has adopted no such specific means.

We have our institutions for each particular department, which organize various constructive activities. But it is not the function of these institutions to examine how far their activities have helped develop non-violence. I would ask Shankerlal[5] how many persons were paid wages. I would not ask how much progress he made in the direction of non-violence. I would ask a labour union how far it had improved the lot of the workers through peaceful methods. I shall not ask how the labour movement was related to non-violence. The same is true of the Gramodyog Sangh. The same is true of Satis Babu. He does not have to take the test about the progress of non-violence in the Khadi Pratishthan. He can be asked only about how much of good quality oil, paper, *khadi* or leather he has produced. After all this the point remaining to be considered is to find out in what way we can evolve through them a non-violent culture. We can have an organization for that purpose. But where is the need for an organization merely for those who believe in non-violence? Whatever strength or guidance they need would be available to them from my actions. God Himself would guide them. They would be organized even without an institution.

JAJU: It should be realized that those who are in politics will definitely cease to be members. Now we have to think about the future of the Sangh.
DADA DHARMADHIKARI: Let us wind up this Sangh. The new committee which is to be formed can form a new Sangh if it thinks it necessary.
SHANKARRAO DEO: I support Dada's suggestion. It is not proper to discriminate and say that those involved in politics cease to be members of the Sangh and the Sangh continues with the rest of the members.
GANDHIJI: Let it not be forgotten that we are not discriminating when we keep the politically inclined people out of the Sangh. All that we want to do is to eschew politics in the name of the Sangh. Had we been regarding men who are active in politics as inferior persons, we would not allow those persons even to remain in politics. But we are not asking them to quit politics.

SHANKARRAO: We are confronted with this whole moral dilemma because we take part in politics even as we remain members of the Sangh. That is why Bapuji has suggested that we change the present character of the Sangh. The present form of the Sangh is not its original form. That is why I suggest we wind up the Sangh as it is functioning today, and restore it to its original form.
GANDHIJI: You may do that.

MAGANBHAI: Politics involves struggle for power which breeds jealousy. Hence the members of the Sangh are being asked to retire from politics or those in politics are being asked to leave the Sangh. But those who are inclined to pursue power will do the same in the field of the constructive programme. Even today it cannot be said that there is no power struggle or jealousy or bitterness in the field of constructive work. Nor do I see any need of even a research institute for those devoted solely to constructive work. People working in different fields would continue research in their respective fields while remaining with their own organizations. If they have any difficulty, they would consult their departmental head. Or, they would seek guidance from someone in whom they have a special faith.... For that no institution is needed. If you set up a research institute, it would undertake some stray research projects and exercise unnecessary control over other organizations. Apart from conducting research, its function would also be to correlate it with truth and non-violence. Hence it would become an organization exercising superior moral authority. I see no need for such a superior authority. In my view, you are discarding the shape you gave to the Sangh in 1934. And now when we clearly find that it can have no specific function under these conditions, why should we not disband it altogether?

GANDHIJI: Maganbhai, I have understood your point. What you are saying is not quite relevant here. Krishnan Nair correctly observed the other day that this would be purely a research institute. It would place before the people the conclusions of its research. Anyone who is keen would be free to benefit from them. Those who do not find them useful can ignore them. Here the question of dominating others does not arise at all. There is nothing like 'superior moral authority' at the back of it. When there is no question of domination how can it give rise to quarrels?

It is possible that organizations devoted to the constructive programme, such as the Charkha Sangh, etc., may to some extent suffer from bickerings, misuse of authority or arrogance resulting from power. But where do you find in it what we describe as power politics? It is not a representative body formed after an election. It is an organization like a bank. How could power politics get into it? The Congress is a vast organization. It belongs to the millions. It is a representative organization. There is plenty of scope in it for the politics of power. Today the Gandhi Seva Sangh has neither the equipment nor the strength to enter power politics in its own capacity. That is why we are suggesting that those who are taking active part in politics should not remain in the Sangh. Did we ever say that they should give up politics? If they find that by their remaining in politics they are not serving any useful purpose but are instrumental in spreading poison, they would have to get out of it. But today we only ask them that they should not enter politics as members of the Sangh. We must deny them this much.

The question now is: What form should the Sangh take? I have given an outline. I cannot think of an alternative method. Those who want to engage in other activities like the study of politics can do it independently. This does not fall within my purview. For me, true politics consists in linking up the constructive programme with *swaraj*. I have never studied what people describe as politics. I carried on agitation even in South Africa; but I did not study politics. For me whatever I did was politics. I do not say that I have never taken part in politics. On my arrival in India, as chance would have it, the reins of the Congress fell into my hands. I remained with the Congress so long as I could be of service and then retired from it.[6] It seems to me that I have rendered a great service by going out of the Congress. It was my attitude of non-violence. My non-violent attitude served its purpose well. Non-violence had its place in the Congress even after I went out of it. Now I leave the politics of the Congress to Rajendra Babu and Vallabhbhai. If they find poison spreading by their remaining there, they too should quit. But it is not easy to quit. If the people insist on their remaining in politics and if the Congress and its President find it necessary that they should be there, then it becomes difficult for them to go out. But in that case they should not remain with the Sangh.

We have to keep the Sangh untouched by the politics of power and groupism. We have to work in studied silence. That alone would be the beginning of real politics. The field of constructive work is very vast. Let us study it. Let us do research and make discoveries in that field. Shankerlal alone cannot shoulder this responsibility. This is the only programme I have for the Sangh. I can only give you what I have. A new vista is opening before you for an experiment in non-violence. You can expand it as much as you choose. Even if you cannot expand it, its present scope itself is sufficient. Your work will inspire others in their research. I am not talking of ordinary inventions. Am I going to be satisfied by haphazard discoveries? Only when someone has something great should he place it before the world. That thing should be so extraordinary that the world should be wonder-struck by it. For having such a Sangh we should also have people who would be crazy about it.

I have put before you everything in simple and straight-forward words. Whatever work we do, we have to keep away from power politics and groupism. The Congress is guiding the politics of the whole nation. Those among us who want to render service by being on Congress committees will not be with the Sangh from now. It is possible to prepare for civil disobedience even by remaining out of the Congress. The Congress represents the whole nation. Hence it has to reflect its views in its programmes and resolutions. But we have to do silent service in our independent field.

KRISHNAN NAIR: If the Sangh is continued, there is the danger of the emergence of a sect in your name. Just as we are scared of groupism in politics, we should also be scared of a sect emerging in your name in the form of the Sangh. This Sangh which has carried your name has done no extraordinary work in your own lifetime. Who knows what would happen after you are gone? Hence it is just as well that the Sangh be disbanded.

GANDHIJI: There is, no doubt, the danger of forming a sect. In this matter I take as much precaution as possible.

MULCHAND AGRWAAL: In these two or three days Bapuji has said the same things that he has been saying for the last ten or fifteen years. He has said nothing new. All this trouble has arisen because the members of the Sangh were permitted to take part in politics. Hence it would be enough to correct this. It seems impracticable to wind up this Sangh and form another. Let there be some improvements in the present set-up. We need not have a new set-up.

JAJU: It is not a question of breaking up. We are reorganizing it. Even those who are in favour of closing it down desire its reconstitution. We had reconstituted the Sangh in 1934. We are now doing it again in 1940.

BABALBHAI: Inventors would be produced by organizations like the Charkha Sangh, etc. They alone would be the experts in their respective fields. No special research institute is needed for them. They can correlate their work with truth and non-violence.

GANDHIJI: You do not expect me to answer this point, do you? The answer is contained in whatever I have said.

PUNDALIK: Does withdrawal from politics as members of the Sangh indicate that politics of power is a bad thing? Why should we not take part in politics of power after remaining out of the Sangh? Why should we not prove for the world that the best possible means can be employed even in politics of power?

GANDHIJI: Why should we meddle with politics of power while we remain in the Sangh? I do not wish to give my view even about one individual doing it or not doing it. I do not have enough material about it. I do not know how far those who accept power observe non-violence and stand for truth. You would be put to test if you take part in politics of power. I cannot say to what extent you would pass that test. Vallabhbhai is conducting the experiment by remaining in politics. He would leave it when he finds that it is not proper for him to be there. The Sangh does not wish to take any responsibility in this matter. It is left to the conscience of each individual. Each should regard his own conscience as the authority. No outsider can take any decision in the matter. For instance, Vallabhbhai is engaged in [political] work. But I do not know his mind. I do not know Vallabhbhai's feelings, nor yours. After all, I am no God, am I?

SWAMI ANAND: The Government has set up Rural Reconstruction Boards. Should we withdraw from these Boards as well? After two years these Boards will be elected bodies. The Bombay Board will be elected after six months.

GANDHIJI: We may remain with the Boards so long as they are not infected by power politics. If the politics of power starts affecting them, we should quit. This is a matter to be decided independently for one's own self.

It is quite likely that the committee which we are going to form may not be able to finish its work here in Malikanda. But it should nevertheless fix a time-limit for submitting its conclusions. If people are not offering their names, Kishorelal is of course there. I am there too.

Speech at Gandhi Seva Sangh (H.)
Gandhi Seva Sanghke Chhathe Adhiveshan (Malikanda-Bengal) ka Vivaran, pp. 34-52

[1] Why Socialism? (1936)
[2] M.R. Masani.
[3] *Samajvad* (1936).
[4] In March 1930.
[5] Shankerlal Banker, Secretary, Akhil Bharatiya Charkha Sangh.
[6] In 1934.

151. COMPROMISE ON NON-ESSENTIALS

[Malikanda,
24 February 1940]

All incompatible mixtures are bound to explode. You must resolve to act on the square, and whilst you should be prepared to compromise on non-essentials you should never be in the uncomfortable position of having to compromise truth. You should retire from all such positions. That is the essence of compromise. Let service without near or distant objective be your motto. You are surrounded by poverty on all sides. Serve those that are afflicted, whether they are Muslims, *Namasudras* or others. *Satyagraha* transcends parties and divisions of class and creed. It should permeate the whole of our being and society. There is no question before you of enlisting members for the Congress. Give up all thought of gaining members for the sake of swelling your register. That is power politics. I would rather have no register than blacken it with bogus members. If you will thus become silent workers, even one of you will lead the Congress in the province without being in it.

I hope you will not now say, 'What will happen if the Congress is captured by the opponents?' You know the Upanishad precept: *Enjoy by means of renunciation*.[1] Give up the Congress in order to 'enjoy' or have it. The moment I set my heart on some kind of capturing I am done for. No manoeuvring to keep your hold on the Congress, no descending from the right path, and you will disarm all opposition. A bogus Congress register can never lead you to *swaraj* any more than a paper boat can help you to sail across the Padma.

'Talk with Workers'
Harijan, 9 Mar. 1940

[1] *Ishopanishad*, 1.

152. POLITICS AND SOCIAL WORK

[New Delhi,
Before 24 September 1946]

Q. Are you full of the joy of life? Why do you want to live for 125 years?

Gandhiji told him that his desire to live up to 125 years was not for enjoyment but service. He explained that both were not the same and proceeded to explain to the puzzled interviewer the doctrine of 'enjoyment through renunciation' as set forth in the *Ishopanishad*.

Q. When did your real enjoyment of life begin?

A. When I was born.

Q. No, I mean when did that pattern of life begin when service became a joy for ever?

A. When I understood the inner meaning of life.

Q. Is that India's speciality?

A. The only speciality of India is her poverty as America's is her glamour of riches.

Q. May not there be occasions when one may have to compromise ideals with expediency?

A. No, never. I do not believe that the end justifies the means.

Q. Is it possible that your activities may some day be removed from the political field?

A. Perhaps you do not know that I felt compelled to come into the political field because I found that I could not do even social work without touching politics. I feel that political work must be looked upon in terms of social and moral progress. In democracy no part of life is untouched by politics. Under the British you cannot escape politics in the good sense. It embraces the whole life. All who breathe must pay a tax. That is British rule in India. Take the salt tax for instance. It concerns everybody. The collector of revenue and the policeman are the only symbols by which millions in India's villages know British rule. One cannot sit still while the people are being ravaged.

Q. Then your job will never be finished?

A. It will be finished only with my death. I must be watchful, whether it is the foreign government that is in power or indigenous, if I am a social reformer in the true sense of the term. This is applicable to all.

Q. When people attain power they grow away from the people. What about here?

A. Let us hope and pray that this will never happen here. I have likened our people's office-acceptance to wearing a crown of thorns and pretty sharp thorns at that.

Q. What do you think of the students' strikes?

A. It seems to be a universal malady, an epidemic.

Q. Do you ever feel depressed?

A. I believe in an over-ruling Power as I believe I am talking to you just now. This may be unreal, but that *is* real. It dominates me and enables me to remain calm even in the midst of storm.

Gandhiji's questioner next asked his opinion about predestination.

A. It is a much-abused word. It is true that we are not quite as free as we imagine. Our past holds us. But like all other doctrines this may well be ridden to death.

Q. This provoked the question as to how one could overcome the unpleasant effects of one's predestination since predestination was a reality.

A. By taking the pleasant with the unpleasant in perfect detachment and thereby sterilizing the unpleasantness of its sting, even as you have tackled the problem of the prickly pear by removing its thorns through judicious selection and cultivation and converting it into edible fodder for cattle.

Q. How to prevent the next war?

A. By doing the right thing, irrespective of what the world will do. Each individual must act according to his ability without waiting for others if he wants to move them to act. There comes a time when an individual becomes irresistible and his action becomes all-pervasive in its effect. This comes when he reduces himself to zero.

If the third war comes, it will be the end of the world. The world cannot stand a third war. For me the second war has not stopped, it still goes on.

'Talk with an American Journalist'
Harijan, 6 Oct. 1946

VI

Religion

§ 1. Religion, Mysticism, and Society

153. RELIGIOUS REFORM

August 20, 1928

I do not think I am at all fitted to say anything on this great occasion, but I consented to come because of the great esteem in which I held the late Ramanbhai[1] and because I could not resist Shrimati Vidyagauri.[2] I am unfit to say anything today for many reasons. I have read nothing of or about Raja Ram Mohan Roy. What I know about him is based on what I have heard from his admirers. I do not claim to have studied the history of the Brahmo Samaj[3] either. I made a desperate effort to read something today, something from a brochure on Ram Mohan Roy that we have in our library, but I could not find a moment for it. I therefore contented myself with the prayer that God might give me the right word to acquit myself of the task.

Although, then, I cannot claim to have studied the history of the Brahmo Samaj, I do claim to have a close connection with the members of the Brahmo Samaj extending over several years. It dates as far back as 1896 when I first visited Calcutta. In 1901 I came in contact with several eminent members of the Brahmo Samaj through Gokhale and Dr. P.C. Ray. I used to visit occasionally the Samaj Mandir, and listen to the sermons of the late Pratap Chandra Mozumdar. I also came in touch with the late Pandit Shivnath Shastri.

These experiences were enough to convince me that the Brahmo Samaj has rendered a great service to Hinduism, and that it has rescued the educated classes of India, especially Bengal, from unbelief. I have always regarded it as a movement essentially designed for the educated classes. Though religion in India sometimes takes the form of superstition, wooden formalism and hysterics, I cannot escape the conviction that a man cannot long remain an unbeliever in India. But the faith of the educated class in the early part of the nineteenth century was in danger and it was Ram Mohan Roy who came to their rescue. I have heard that he had come under the influence of Christian missionaries and Kakasaheb told me that he had made a deep study of Persian and Arabic. There can be no question about his scholarship or about his catholicity. He made a deep study of Hinduism, especially the religion of the Vedas, and then allowed himself to be influenced by the essential principles of Christianity and Islam. As a result he saw that there was for him no escape from inaugurating a new movement to liberalize the existing Hinduism which had been overgrown with superstitious weeds. Sacrifice of animals and social evils were flourishing in the name of Hinduism. How could the educated class tolerate it? It was open to Ram Mohan Roy to rest content with individual dissociation from the evils. But he was a reformer. He could not afford to hide his light under a bushel, he gave public expression to his views, secured a following and founded the Brahmo Samaj in 1828.

But it would have languished, had not a man of great spiritual gifts like Maharshi Devendranath Tagore joined it. It is for the future historian to estimate the Tagores' contribution to the intellectual and spiritual life of Bengal, India, and even the world. Rabindranath Tagore's contribution in the direction has been stupendous. We cannot adequately estimate it, even as men living in the valleys of the Himalayas cannot adequately visualize their sublimity. And the Tagores received their inspiration from the Brahmo Samaj. The Brahmo Samaj liberated Reason, and left room enough for Faith. There was once a danger of the Brahmo Samaj severing its tie with Hinduism or the religion of the Vedas but the Maharshi's *tapasya* and knowledge rescued the Brahmo Samaj from that catastrophe. It is due to him that the Samaj remained part of Hinduism.

One may not measure the contribution of the Brahmo Samaj from the number of its adherents. The Brahmos are indeed very few but their influence has been great and good.

The service of the Brahmo Samaj lies in its liberalizing and rationalizing Hinduism. It has always cultivated a toleration for other faiths and other movements, it has tried to keep the fountain source of religion pure and to hold up the ideal of pure worship of the Supreme Being.

Not that there is nothing for me to criticize in the Samaj, but this is not the occasion for it. My desire is to place before you whatever is best in the Brahmo Samaj. Let this celebration awaken the religious instinct in you. True religion is not narrow dogma. It is not external observance. It is faith in God, and living in the presence of God; it means faith in a future life, in truth and *ahimsa*. There prevails today a sort of apathy towards these things of the spirit. Our temples appear today to be meant only for the simple and the ignorant. Few visit real temples of God. Let the educated class take up the work of reform in this direction.

We have rightly honoured Vallabhbhai over the Bardoli victory. But you may not know his greater victory. Vallabhbhai realized his 'Vallabha' (God) in Bardoli. He saw that nothing but faith in God could keep together the thousands of men and women bound to their pledge. It is through religion that he found his approach to the hearts of those simple illiterate men and women.

I am inundated with letters from young men who write frankly about their evil habits and about the void that their unbelief has made in their lives. No mere medical advice can bring them relief. I can only tell them that there is no way but that of surrender to and trust in God and His grace. Let us all utilize this occasion by giving the living religion in our lives the place it deserves. Has not Akho Bhagat said:

> Live as you will, but so
> As to realize God.

'Speech at Brahmo Samaj Centenary Meeting', Prarthana Samaj, Ahmedabad (G.)
Young India, 30 Aug. 1928

[1] Ramanbhai Nilkanth.
[2] Vidyagauri Nilkanth.
[3] A reformist Hindu movement, started in Bengal in the nineteenth century.

154. RELIGION AND CULTURE

A student of the Gujarat Vidypith[1] writes:

What concrete form ought religious instruction to take in the Vidyapith?

To me religion means truth and *ahimsa* or rather truth alone, because truth includes *ahimsa*, *ahimsa* being the necessary and indispensable means for its discovery. Therefore anything that promotes the practice of these virtues is a means for imparting religious education and the best way to do this, in my opinion, is for the teachers rigorously to practise these virtues in their own person. Their very association with the boys, whether on the playground or in the class-room, will then give the pupils a fine training in these fundamental virtues.

So much for instruction in the universal essentials of religion. A curriculum of religious instruction should include a study of the tenets of faiths other than one's own. For this purpose the students should be trained to cultivate the habit of understanding and appreciating the doctrines of various great religions of the world in a spirit of reverence and broad-minded tolerance. This, if properly done, would help to give them a spiritual assurance and a better appreciation of their own religion. There is one rule, however, which should always be kept in mind while studying all great religions and that is that one should study them only through the writings of known votaries of the respective religions. For instance, if one wants to study the *Bhagavata* one should do so not through a translation of it made by a hostile critic but through one prepared by a lover of the *Bhagavata*. Similarly to study the Bible one should study it through the commentaries of devoted Christians. This study of other religions besides one's own will give one a grasp of the rock-bottom unity of all religions and afford a glimpse also of that universal and absolute truth which lies beyond the 'dust of creeds and faiths'.

Let no one even for a moment entertain the fear that a reverent study of other religions is likely to weaken or shake one's faith in one's own. The Hindu system of philosophy regards all religions as containing the elements of truth in them and enjoins an attitude of respect and reverence towards them all. This of course presupposes regard for one's own religion. Study and appreciation of other religions need not cause a weakening of that regard; it should mean extension of that regard to other religions.

In this respect religion stands on the same footing as culture. Just as preservation of one's own culture does not mean contempt for that of others, but requires assimilation of the best that there may be in all the other cultures, even so should be the case with religion. Our present fears and apprehensions are a result of the poisonous atmosphere that has been generated in the country, the atmosphere of mutual hatred, ill-will and distrust. We are constantly labouring under a nightmare of fear lest someone should stealthily undermine our faith or the faith of those who are dear and near to us. But this unnatural state will cease when we have learnt to cultivate respect and tolerance towards other religions and their votaries.

'Religious Education'
Young India, 6 Dec. 1928

[1] An educational institution.

155. THE SOUL AND ITS SOURCE

[5 August 1927]

Q. 1. Do you believe in the distinct existence of Trinity—Nature, Soul and God?

A. I do believe but the word 'distinct' sticks in the throat, for though the three are separate in name they are one in substance.

2. Whether soul is one or numerous? Has it entirely its separate existence or is it a portion of the Great *Atman*—God?

Souls seem to be many; but underneath the seeming variety, there is an essential oneness.

3. If it is a particle [of God] it ought to be all-knowing, quite free from all evils and should possess all the qualities of God, as a particle

of fire, water or gold possesses all the talents of the whole from which it is separated.

In its essence the soul is free from all evil, but torn from its source it partakes of evil and all other limitations even as a drop of water torn from its source and found as part of a dirty pool seems for the time being to partake of the pool's dirt.

4. We often see that a man does evil deeds. Whence that evil comes?

We should be satisfied to know that there is evil in the world and that it has got to be avoided. If we knew its origin we should be God Almighty. But at the present we are in the unfortunate position of that drop of water in the dirty pool, and whilst we are doers of evil deeds, we must suffer the consequences thereof

5. In the world we see different men. Those who work and do right deeds are gifted; those who do wrong or bad deeds are punished and have to suffer. This is all in accordance with the theory of *karma*. When one suffers—as a peasant or a labourer in India does at present, he suffers through his own past deeds, that is the Will of God. Why should we interfere in the Will of God by helping him so that he may not suffer?

If the Indian peasant was isolated and was seated on some mountain top, we shall not be perhaps responsible for his state. But seeing that he is part of the same society that we are, we cannot divest ourselves of responsibility for his state even as that drop of water in the pool cannot. Why God suffers this evil to exist, I am thankful to feel that He alone knows.

6. Instead of devoting our energies for the good of *others* should we not devote our energies in acquiring knowledge or earning money for ourselves or at the most for our relatives and neighbours and that too because they may help us in time of need?

Precisely for the reason that we may help our relatives, must we help all others whom in our ignorance we refuse to recognize as relatives. We are like the drop of water. It enjoys even that dirty existence by the help of its neighbours at the farthest end; for, its near neighbours have to depend upon theirs and so on to the end.

7. Yourself as well as others define *swarajya* as control over the self, reformation, betterment or improvement of the *self*, and also it is evident, that if one controls and improves or reforms himself, the society or the nation will be reformed by itself as the *nation* is a composition of *individuals*.

It is a correct statement; self-rule is home-rule.

Answers to Questions from Moolchand Agrawal
GN 765

156. RELIGION AT ITS HIGHEST

A friend sends me a number of *Spiritual Life* which contains a beautiful paragraph under the heading 'Broken' which reminds one of so many hymns by Tulsidas, Surdas and the other saints that I am tempted to give the following condensation of the paragraph:

'God uses most for His Glory those people and things which are most perfectly broken. The sacrifices He accepts are broken and contrite hearts. It was the thorough breaking-down of Jacob's natural strength at Peniel that got him where God could clothe him with spiritual power. It was by breaking the surface of the rock at Horeb by the strokes of Moses' rod, that it let out the cool waters to thirsty people. It was when the three hundred elect soldiers under Gideon broke their pitchers, a type of breaking themselves, that the hidden lights shone forth to the consternation of their adversaries. It was when the poor widow broke the seal of the little pot of oil, and poured it forth, that God multiplied it to pay her debts and supply means of support. It is when a beautiful grain of corn is broken up in the earth by death, that its inner heart sprouts forth and bears hundreds of other grains. And thus on and on through all history, and all biography, and all vegetation, and all spiritual life, God must have *broken things*. Those who are broken in wealth and broken in heart, broken in their ambitions, broken in their beautiful ideals, broken in worldly reputation, broken in their affections, and broken sometimes in health, and those who are despised, and seem utterly helpless and forlorn, the Holy Ghost is seizing upon and using for God's glory. It is "the lame that take the prey", Isaiah tells us. It is the weak that overcome the devil. God is waiting to take hold of our failures and cleanse away the cause and shine through us in victory.'

The paragraph shows how religion at its highest is one.

'Broken'
Young India, 3 Jan. 1929

157. TRUE SPIRITUAL TRANSFORMATION

[Before 1 March 1929]

DR. MOTT: What do you consider to be the most valuable contribution that India can make to the progress of the world?

GANDHI: Non-violence, which the country is exhibiting at the present day on a scale unprecedented in history. But for it, there might have been a blaze, for provocation of the gravest kind has not been wanting on the side of the Government. There is no doubt a school in the country that believes in violence, but it is a mere excrescence on the surface and its ideals are not likely to find a congenial soil in the country.

What causes you solicitude for the future of the country?

Our apathy and hardness of heart, if I may use that Biblical phrase, as typified in the attitude towards the masses and their poverty. Our youth are full of noble feelings and impulses but these have not yet taken any definite practical shape. If our youth had a living and active faith in truth and non-violence, for instance, we should have made much greater headway by now. All our young men, however, are not apathetic. In fact without the closest co-operation of some of our educated young men and women, I should not have been able to establish contact with the masses and to serve them on a nationwide scale; and I am sustained by the hope that they will act as the leaven, and in time transform the entire mass.

From this they passed on to the distinctive contributions of Hinduism, Islam and Christianity to the upbuilding of the Indian nation.

The most distinctive and the largest contribution of Hinduism to India's culture is the doctrine of *ahimsa*. It has given a definite bias to the history of the country for the last three thousand years and over and it has not ceased to be a living force in the lives of India's millions even today. It is a growing doctrine, its message is still being delivered. Its teaching has so far permeated our people that an armed revolution has almost become an impossibility in India, not because, as some would have it, we as a race are physically weak, for it does not require much physical strength so much as a devilish will to press a trigger to shoot a person, but because the tradition of *ahimsa* has struck deep roots among the people.

Islam's distinctive contribution to India's national culture is its unadulterated belief in the oneness of God and a practical application of the truth of the brotherhood of man for those who are nominally within its fold. I call these two distinctive contributions. For in Hinduism the spirit of brotherhood has become too much philosophized. Similarly though philosophical Hinduism has no other god but God, it cannot be denied that practical Hinduism is not so emphatically uncompromising as Islam.

What then is the contribution of Christianity to the national life of India? I mean the influence of Christ as apart from Christianity, for I am afraid there is a wide gulf separating the two at present.

Aye, there's the rub. It is not possible to consider the teaching of a religious teacher apart from the lives of his followers. Unfortunately, Christianity in India has been inextricably mixed up for the last one hundred and fifty years with the British rule. It appears to us as synonymous with materialistic civilization and imperialistic exploitation by the stronger white races of the weaker races of the world. Its contribution to India has been therefore largely of a negative character. It has done some good in spite of its professors. It has shocked us into setting our own house in order. Christian missionary literature has drawn pointed attention to some of our abuses and set us athinking.

What has interested me most is your work in connection with the removal of untouchability. Will you please tell me what is the most hopeful sign indicating that this institution is as you say on its last legs?

It is the reaction that is taking place in orthodox Hinduism and the swiftness with which it has come about. As a most illustrious example I will mention Pandit Malaviyaji. Ten years back he was as punctilious in the observance of the rules with regard to untouchability as perhaps the most orthodox Hindu of that day. Today he takes pride in administering the *mantra* of purification to the untouchables by the bank of the Ganges, sometimes even incurring the wrath of unreasoning orthodoxy. He was all but assaulted by the diehard section in Calcutta in December last for doing this very thing. In Wardha a wealthy merchant Sheth Jamnalal Bajaj recently threw open his magnificent temple to the untouchables and that without arousing any serious opposition. The most remarkable thing about it is that from the record kept in the temple of the daily visitors it was found that the attendance had gone up instead of declining since the admission of the untouchables to it. I may sum up the outlook by saying that I expect the tide against untouchability to rise still more swiftly in the near future, astonishingly swift as it has already been.

Where do you find your friends? Do you get the backing of the Mussalmans and the Christians in this work?

The Mussalmans and the Christians can from the very nature of the case render little help in this matter. The removal of untouchability is purely a question of the purification of Hinduism. This can only be effected from within.

But my impression was that Christians would be a great help to you in this connection. The Rev. Whitehead, Bishop of the Church of England Mission, made some striking statements about the effect of Christian mass movement in ameliorating the condition of the untouchables in the Madras Presidency.

I distrust mass movements of this nature. They have as their object not the upliftment of the untouchables but their ultimate conversion. This motive of mass proselytization lurking at the back in my opinion vitiates missionary effort.

There are conflicting opinions on this point. There are some who seriously believe that the untouchables would be better off if they turned Christians from conviction, and that it would transform their lives for the better.

I am sorry I have been unable to discover any tangible

evidence to confirm this view. I was once taken to a Christian village. Instead of meeting among the converts with that frankness which one associates with a spiritual transformation, I found an air of evasiveness about them. They were afraid to talk. This struck me as a change not for the better but for the worse.

Do you then disbelieve in all conversion?

I disbelieve in the conversion of one person by another. My effort should never to be undermine another's faith but to make him a better follower of his own faith. This implies belief in the truth of all religions and therefore respect for them. It again implies true humility, a recognition of the fact that the divine light having been vouchsafed to all religions through an imperfect medium of flesh, they must share in more or less degree the imperfection of the vehicle.

Is it not our duty to help our fellow-beings to the maximum of truth that we may possess, to share with them our deepest spiritual experiences?

I am sorry I must again differ from you, for the simple reason that the deepest spiritual truths are always unutterable. That light to which you refer transcends speech. It can be felt only through the inner experience. And then the highest truth needs no communicating, for it is by its very nature self-propelling. It radiates its influence silently as the rose its fragrance without the intervention of a medium.

But even God sometimes speaks through His prophets.

Yes, but the prophets speak not through the tongue but through their lives. I have however known that in this matter I am up against a solid wall of Christian opinion.

Oh, no, even among Christians there is a school of thought—and it is growing—which holds that the authoritarian method should not be employed but that each individual should be left to discover the deepest truths of life for himself. The argument advanced is that the process of spiritual discovery is bound to vary in the case of different individuals according to their varying needs and temperaments. In other words they feel that propaganda in the accepted sense of the term is not the most effective method.

I am glad to hear you say this. That is what Hinduism certainly inculcates.

What counsel do you give to the young men who are fighting a losing battle with their lower selves and come to you for advice?

Simply prayer. One must humble oneself utterly and look beyond oneself for strength.

But what if the young men complain that their prayer is not heard, that they feel like speaking to brass heavens as it were?

To want an answer to one's prayer is to tempt God. If prayer fails to bring relief it is only lip prayer. If prayer does not help nothing else will. One must go on ceaselessly. This then is my message to the youth. In spite of themselves the youth must believe in the all-conquering power of love and truth.

The difficulty with our youth is that the study of science and modern philosophy has demolished their faith and so they are burnt up by the fire of disbelief.

That is due to the fact that with them faith is an effort of the intellect, not an experience of the soul. Intellect takes us along in the battle of life to a certain limit but at the crucial moment it fails us. Faith transcends reason. It is when the horizon is the darkest and human reason is beaten down to the ground that faith shines brightest and comes to our rescue. It is such faith that our youth require and this comes when one has shed all pride of intellect and surrendered oneself entirely to His will.

Interview with Dr John Mott
Young India, 21 Mar. 1929

158. RELIGION AND ART

[London,
14 October 1931]

The movement of artists and poets who are endeavouring to free themselves from the shackles of commercial and industrial influences of this age is a most laudable venture if only they have strength enough to do it. Religion is the proper and eternal ally of art. What religion teaches people the artist brings near to them in form on the plastic plane. I hate 'art for art's sake', which I think is a lamentable aberration of the human mind. Art has a profound similarity with religion inasmuch as the fundamental experience in both of them belongs to the domain of man's relationship with God. Indian art symbolizes this relationship and at the same time expresses the ritual of religious worship. If an artist who thinks he is surrounded by people without any religious sentiment chooses to become a scoffer, he will inevitably frustrate his own vocation. On the other hand if he feels that his is a mission, then a poet or artist has a right to oppose the prevalent creed or lack of creed and he will be justified by the greater value of his own revelation. I do not pretend to know anything about art, but I believe firmly that both religion and art have to serve the identical aims of moral and spiritual elevation.

The central experience of life will for ever remain the relationship which man has to God and it will never be superseded or replaced by anything else, just as human bodies will never free themselves from the law of gravitation. In this relationship of man to God it is the mysterious forces which matter, not the meagre texts expressed in words. There may be changes in this relationship of man to God as represented by the various and successive religions of mankind; but to quote Cardinal Newman: 'One step enough for me.'

Statement to *The Island*
GN 1055a

159. THE UNFAILING TEST

April 25, 1932

I wish you will not take to heart what the Bishop[1] has been saying. Your church is in your heart. Your pulpit is the whole earth. The blue sky is the roof of your church. And what is this Catholicism? It is surely of the heart. The formula has its use. But it is made by man. If I have any right to interpret the message of Jesus as revealed in the Gospels, I have no manner of doubt in my mind that it is in the main denied in the churches, whether Roman or English, High or Low. Lazarus has no room in those places. This does not mean that the custodians know that the Man of Sorrows has been banished from the buildings called Houses of God. In my opinion, this excommunication is the surest sign that the truth is in you and with you. But my testimony is worth nothing, if when you are alone with your Maker, you do not hear His voice saying, 'Thou art on the right path.' That is the unfailing test and no other.

Letter to Dr Verrier Elwin
The Diary of Mahadev Desai, Vol. 1, p. 87

[1] A Bishop had called Dr. Elwin a traitor to Christ and had prohibited him from preaching in churches.

160. EVER-GROWING INWARDNESS

November 25, 1932

What a joy it would be when people realize that religion consists not in outward ceremonial but an ever-growing inward response to the highest impulses that man is capable of.

From a Letter to Samuel E. Stokes
Mahadevbhaini Diary, Vol. II, p. 279

161. RELIGION AND SOCIAL SERVICE

[About 23 January 1935]

I have been asked by Sir S. Radhakrishnan to answer the following three questions:
 (1) What is your religion?
 (2) How are you led to it?
 (3) What is its bearing on social life?

My religion is Hinduism which, for me, is the religion of humanity and includes the best of all the religions known to me.

I take it that the present tense in the second question has been purposely used instead of the past. I am being led to my religion through Truth and Non-violence, i.e., love in the broadest sense. I often describe my religion as religion of Truth. Of late, instead of saying God is Truth I have been saying Truth is God, in order more fully to define my religion. I used at one time to know by heart the thousand names of God which a booklet in Hinduism gives in verse form and which perhaps tens of thousands recite every morning. But nowadays nothing so completely describes my God as Truth. Denial of God we have known. Denial of Truth we have not known. The most ignorant among mankind have some truth in them. We are all sparks of Truth. The sum total of these sparks is indescribable, as-yet-Unknown Truth, which is God. I am being daily led nearer to it by constant prayer.

The bearing of this religion on social life is, or has to be, seen in one's daily social contact. To be true to such religion one has to lose oneself in continuous and continuing service of all life. Realization of Truth is impossible without a complete merging of oneself in and identification with this limitless ocean of life. Hence, for me, there is no escape from social service; there is no happiness on earth beyond or apart from it. Social service here must be taken to include every department of life. In this scheme there is nothing low, nothing high. For all is one, though we *seem* to be many.

'Questions and Answers'
S. Radhakrishnan, *Contemporary Indian Philosophy*, p. 21

162. LIMITED THINKING AND LIMITLESS SERVICE

Sevagram,
[Before 12 March 1940]

Q. Could you tell me the things one should avoid in order to present the gospel of Christ?

A. Cease to think that you want to convert the whole world to your interpretation of Christianity. At the end of reading the Bible, let me tell you, it did not leave on my mind the impression that Jesus ever meant Christians to do what the bulk of those who take his name do. The moment you adopt the attitude I suggest, the field of service becomes limitless. You limit your own capacity by thinking and saying that you must proselytize.

I see what you mean. We have been cumbered by creeds and man-made things. We feel that we should be in a place where all barriers have broken down.

Gandhiji instanced a few Christians who, he said, saw the central fact that, if they wanted to live this Christian life, they should literally follow the words: 'Not he that sayeth "Lord, Lord", but he that doeth His will.'[1]

You are living a guided life. Could you kindly tell me your experience of guidance?

I do not regard God as a person. Truth for me is God, and God's Law and God are not different things or facts, in the sense that an earthly king and his law are different, because God is an Idea, Law Himself. Therefore it is impossible to conceive God as breaking the Law. He therefore does not rule our actions and withdraw Himself. When we say He rules our actions, we are simply using human language and we try to limit Him. Otherwise He and His Law abide everywhere and govern everything. Therefore I do not think that He answers in every detail every request of ours, but there is no doubt that He rules our action, and I literally believe that not a blade of grass grows or moves without His will. The free will we enjoy is less than that of a passenger on a crowded deck.

Do you feel a sense of freedom in your communion with God?

I do. I do not feel cramped as I would on a boat full of passengers. Although I know that my freedom is less than that of a passenger, I appreciate that freedom as I have imbibed through and through the central teaching of the *Gita* that man is the maker of his own destiny[2] in the sense that he has freedom of choice as to to the manner in which he uses that freedom. But he is no controller of results. The moment he thinks he is, he comes to grief.

'Discussion with a Missionary'
Harijan, 23 Mar. 1940

[1] Matthew 7: 21.
[2] *Bhagavad Gita*, VI. 5.

163. THE HEART AND THE EGO

An Arya Samajist[1] writes:

How can the Rama whom you believe to be undying, be the Rama who is the son of Dasharatha and the husband of Sita? Tormented by the doubt, I do join your prayers but take no part in singing the *Ramdhun*. This irks me for you say that everyone should join in singing and you are right. Could you not so modify the *Ramdhun* that all can join in it.

I have already explained what I mean by everyone. It means everyone who can join heartily and sing in unison. The others should remain silent. But this is of small importance. The more important question is how Dasharatha's son can be imagined as undying. Tulsidas himself has raised the question and answered it. Such questions cannot be answered by the intellect or to the satisfaction of the intellect. This is a matter of the heart and the heart alone knows the ways of the heart. I first worshipped Rama as Sita's Lord but, as realization and experience grew, my Rama became undying and all-pervasive. This means that He continued to be Sita's Lord but the content of that description was enlarged. This is how the world goes on. The Rama of the man who conceives him merely as Dasharatha's son cannot be all-pervasive. But to the man for whom Rama is all-pervasive, Dasharatha also becomes all-pervasive. It may be said that this is all arbitrary, 'to each man according to his faith'. But I see no other way.

If all religions are essentially one, we have to harmonize them. Today they are kept separate and that is why we kill each other. When we are tired of religion, we become atheists and then our ego alone is left and nothing else, not even God. But when we acquire true understanding, the ego perishes and God alone remains. Rama then is and is not the son of Dasharatha, the Lord of Sita, the brother of Bharata and Lakshamana. All honour then to those who not believing in Rama, the son of Dasharatha, still join the prayers. This is not rationalism. I have merely outlined what I do and what I believe.

New Delhi, September 16, 1946

'Rama the Son of Dasharatha' (H.)
Harijan Sevak, 22 Sept. 1946

[1] Member of the Arya Samaj, a reformist Hindu movement, started in Punjab in the nineteenth century.

164. MAN WITHOUT ROOTS

Q. Is it possible during prayers, for thousands who assemble at your prayer gatherings, to concentrate their minds on anything whatever?

A. I can only answer yes. For, if I did not believe in mass prayer, I should cease to hold public prayers. My experience confirms my belief. Success depends upon the purity of the leader and the faith of the audience. I know instances in which the audience had faith and the leader was an impostor. Such cases will continue to happen. But truth like the sun shines in the midst of the darkness of untruth. The result in my case will be known probably after my death.

A wag asks three questions in this matter. One requiring an answer runs: 'Is not political education infinitely superior to the religious?' In my opinion, political education is nothing worth, if it is not backed by a sound grounding in religion by which is not meant sectional or sectarian belief. Man without religion is man without roots. Therefore, religion is the basis on which all life structure has to be erected, if life is to be real.

Q. What is the employer to do when an employee of his is addicted to stealing and is not amenable to correction whether it is by way of entreaty or the cane?

A. It may well be that the others too are addicts though they may not be found out. Observation would show that we are all thieves, the difference being that most of us are tolerant towards ourselves and intolerant towards those that are found out and are not of the ordinary run. What is a man, if he is not a thief, who openly charges as much as he can for the goods he sells? If the reply be that the buyer is a willing dupe, it begs the question. In reality the buyer is helpless rather than willing. The stealing referred to is one of the symptoms of a deep-seated disease of society. It is symptomatic of the eternal strife between the monied few and the many paupers. Therefore, my advice to the employer will be to remove all temptation in the way of the thief, to treat him as if he was his own brother and, when he refuses to yield to any treatment, however humanitarian it may be, to ask him to go his way. Let the employer always ask himself whether he would treat his own brother in the same way at the given stage.

Panchgani, July 14, 1946

'Question Box'
Harijan, 21 July 1946

§ 2. Spiritual Teachers and Religious Traditions

165. THE KEYNOTE OF HINDUISM

[Johannesburg,
4 March 1905]

Mr. Gandhi introduced his subject by remarking that the endeavours of the Johannesburg Lodge to promote interest in the study of different religious systems were most praiseworthy, tending, as they did, to widen people's sympathies, and enlarge their comprehension of the motives and beliefs underlying the actions of those who were strangers in creed and colour. He himself had endeavoured, during his eleven years' residence in South Africa, to remove the prejudice and ignorance that existed concerning his own people.

Continuing, the lecturer described what was meant by the title 'Hindu', referring it to the branch of the Aryan people that had migrated to the trans-Indus districts of India, and had colonised that vast country. As a matter of fact, Aryanism would have been a better descriptive word than Hinduism, in explanation of the faith accepted by so many millions of his countrymen.

One of the most remarkable characteristics of the religion professed by Hindus was self-abnegation, and this was obviously shown in the name of the religion itself, for, unlike most of the great world religions, it did not derive its name from any prophet or teacher, although it counted some of the greatest within its fold. The lecturer further instanced the historic siege of Arcot in support of this contention, when the Indian soldiers, at a time when starvation faced the whole British Army, waived their claim to the rice rations in favour of the British soldiers, they themselves being content to have the water in which the rice had been boiled, although it was customary to throw this water away; and also the case of Prabhu Singh, an indentured British Indian, who was chosen for the post of honour of warning the inhabitants of Ladysmith during the siege, whenever a Boer shell was coming, by ringing a bell when perched in a tree, at imminent risk to himself. This man was mentioned several times in despatches by Sir George White.

The Hindus themselves claimed that the date of their scriptures was veiled in the mist of antiquity, the scriptures themselves being God-given. As against that, some Europeans contended that the scriptures were not more than 3,000 or 4,000 years old. Mr. Tilak, a well-known Indian Sanskritist, has, however, calculated that, from certain astronomical observations made in these works, they were at least 10,000 years old, although they were only committed to writing some three hundred years after Christ. The Vedas, as these scriptures were known, consisted of separate hymns, each being held to cover a definite period, and quite independent of each other. And, typically, not one author's name had passed down to posterity! The Vedas had inspired the thought of many illustrious men of the West, amongst whom might be mentioned Arthur Schopenhauer and Professor Max Müller.

Over two hundred millions of people professed Hinduism, and this faith entered into their every act. The key-note of Hinduism on the spiritual side was *moksha*, or salvation; that is, the final absorption of the Soul in the Infinite Soul that pervades all things. In regard to religion, pantheism was the chief characteristic, whilst, on the ethical plane, self-abnegation was the most notable quality, with its corollary, toleration. In social matters, the characteristic of caste was predominant, whilst the ceremonial characteristic was the sacrificing of animals. At a time when the faith of the Hindus had become

Spiritual Teachers and Religious Traditions 467

more formal, Prince Gautama Buddha, having learnt the spiritual worth of things during a prolonged period of contemplation, commenced to teach that animal sacrifices were despiritualising, and that the highest form of love was expressed by extending that toleration, which was already a tenet of their faith, in the direction of refraining from killing or otherwise destroying living things.

Hinduism had never, as a religion, been missionary, as were Christianity and Mahomedanism, but, under King Asoka, the Buddhist priests were sent far and wide to propagate the new belief. Buddhism had a reforming effect upon Hinduism, somewhat similar to that of Protestantism upon Catholicism, but there was a great difference in the spirit underlying that reform. No Hindu bore the Buddhist any ill-will, a statement that could not be made in reference to the Protestants and Catholics. Buddhism was sometimes said to have declined in India. This was not really so. The Buddhist priests had endeavoured to propagate their faith too zealously, and had aroused the jealousy, at the time, of the Hindu priesthood, who had driven the Buddhists to the outskirts of the country, to Tibet, China, Japan, Burma and Ceylon. But the spirit of the Buddhists remained in India, and actuated every principle professed by the Hindus.

In this connection, the lecturer briefly referred to Jainism as a most interesting form of faith. The Jains claimed that it was altogether independent of Buddhism, not a growth from it. Unlike others, they did not claim for the faith that it was of Divine origin, recognising that its sacred writings were the results of human workmanship. Jainism was, perhaps, the most logical of all faiths, and its most remarkable characteristic was its scrupulous regard for all things that lived.

Lectures on Religion, Johannesburg Lodge, Theosophical Society
The Star, 10 Mar. 1905

166. THE ELASTICITY OF HINDUISM

[Johannesburg,
11 March 1905]

The lecturer, after giving a short resume of the previous lecture, said that the second lecture would be devoted to what might be termed the second epoch of Hinduism. After the reformation that took place from within, under the influence of the teachings of Buddha, Hinduism was very largely addicted to idolatry. Several explanations were offered in extenuation, but the lecturer could not deny the fact that the Hindus seemingly worshipped stocks and stones. The Hindu philosophers easily recognised and worshipped God as the purest spirit, and Pantheism took them to the highest flight. It similarly brought down the ignorant masses to the lowest depths. If the infant mind could not realise God as a pure spirit, it had no difficulty in worshipping Him through His various manifestations. Many worship Him through the sun, the moon and the stars, and many worship Him through stocks and stones also, a mode of worship to which philosophical Hinduism, with its tolerant spirit, had no difficulty in reconciling itself. So the wheel of Hindu life went merrily on, until there arose in the desert of Arabia a power that was destined to revolutionise ideas and to leave a permanent impress. Mahomet, whilst yet a boy, on seeing around him people given to idolatry, to lustful indulgences, and to drunkenness, burned within himself with rage. He saw also Judaism prostrate and Christianity debased. He felt, even as did Moses and Christ, that he was a man with a mission. He decided to deliver his mission to the world, and he chose the members of his own family as the first recipients. The lecturer was not one of these who believed that the religion of Islam was a religion of the sword.

Washington Irving, in his work on Islam, had asked a pertinent question, namely, 'Where had Islam, in its first stage, found men to wield the sword?' He believed that the success of Mahomedanism was largely due to its simplicity and to its recognition of human weaknesses. He [the Prophet] taught that God was One and only One, that he was His Messenger. He taught also that prayer was absolutely necessary as an uplifting influence, and in order to bring together his followers, if only once in a year, he instituted the pilgrimage to Mecca, for those who could afford it, and recognising that people would amass wealth, he enjoined upon his followers that they set apart a certain portion of it religiously for charitable purposes. The key-note of Islam was, however, its levelling spirit. It offered equality to all that came within its pale, in the manner that no other religion in the world did. When, therefore, about 900 years after Christ, his followers descended upon India, Hinduism stood dazed. It seemed to carry everything before it. The doctrine of equality could not but appeal to the masses, who were caste-ridden. To this inherent strength was also added the power of the sword. The fanatical raiders who, from time to time, found their way into India, did not hesitate to convert by the sword if they could not do so by persuasion. They more or less overran all parts of India, breaking idols after idols, and although the Rajput valour was at the disposal of Hinduism, it was powerless to afford protection against the Mahomedan inroad.

In keeping with the spirit of Hinduism, an attempt was made at first to bring about reconciliation between the two faiths, and in the city of Benares, there arose a holy man about the 13th century, by name Kabir, who endeavoured, by keeping intact the chief tenets of Hinduism and by borrowing somewhat from Mahomedanism, to bring about a fusion, but the attempt was not quite successful. The Punjab, through which the Mahomedan conquerors poured into India, and which bore the first brunt, produced Guru Nanak, the founder of the Sikh religion, who drew upon Kabir for his doctrines and added to it militant Hinduism. He offered the olive-branch by respecting Moslem susceptibilities, but if that were not accepted, he was equally ready to defend Hinduism from the Moslem aggression, and thus Sikhism was a direct result of Islam. It was well known how brave the Sikh was and what service he had rendered to the British Government. The influence of Mahomedanism on Hinduism was that it gave rise to Sikhism and it brought out one of the chief characteristics of the religion, namely, toleration, in its true light and fulness. When there were no political influences at work, there was no difficulty about the Hindus and the Mahomedans living side by side in perfect peace and amity, each respecting the prejudices of the other, and each following his own faith without let or hindrance. It was Hinduism that gave Mahomedanism its Akbar, who, with unerring insight, recognised the tolerant spirit and adopted it himself in ruling India. Hinduism, moreover, showed its elasticity in that, in spite of the fierce struggle, the classes as well as a large majority of the masses were left totally untouched, and Hinduism arose out of the struggle braced, as we would rise out of a cold bath, with a warm glow. The first shock was, no doubt, severely felt, but the Hindu religion soon asserted itself. The lecturer also mentioned the *fakirs* and the *yogis*, and said that they lived almost the same kind of life, although the former belonged to the Islamic faith and the latter to the Hindu faith.

Lectures on Religion, Johannesburg Lodge, Theosophical Society
The Star, 18 Mar. 1905

167. THE TRIALS OF HINDUISM

When the Theosophical Society invited me to deliver these speeches, I accepted the invitation on two considerations. It is now nearly twelve years that I have been living in South Africa. Everyone is aware of the hardships suffered by my compatriots in this land. People view with contempt the colour of their skin. I believe all this is due to a lack of proper understanding; and I have continued to stay in South Africa with a view to helping as much as possible in the removal of this misunderstanding. I, therefore, felt that it would to some extent help me in the fulfilment of my duty if I accepted the Society's invitation; and I shall regard myself very fortunate if I am able, through these lectures, to give you a better understanding of the Indians. I am to speak to you no doubt about the Hindus; but the ways and manners of the Hindus and other Indians are all but identical. All Indians have similar virtues and vices and are descended from the same stock. The other consideration was that there was, among the objects of the Theosophical Society, this one, viz., to compare the various religions, find out the truth underlying these and

Spiritual Teachers and Religious Traditions

show the people how those religions were only so many roads leading to the realisation of God, and how one ought to hesitate to dub any of them false. I thought that this object, too, would be realised to some extent if I said a few words on the Hindu religion.

Hindus are not considered to be the original inhabitants of India. According to Western scholars, the Hindus as well as most of the European peoples lived at one time in Central Asia. Migrating from there, some went to Europe, some to Iran, others moved south-eastwards down into India through the Punjab, and there spread the Aryan religion. The Hindu population in India exceeds two hundred millions. They are called Hindus because they once lived beyond the river Sindhu (Indus). The Vedas are their oldest scripture. Very devout Hindus believe that the Vedas are of divine origin and without beginning. Western scholars hold that these were composed before 2000 B.C. The famous Mr. Tilak of Poona has shown that the Vedas must be at least 10,000 years old. The main thing that distinguishes the Hindus is their belief that the *Brahman* or oversoul is all-pervading. What we all have to attain is *moksha* or liberation, *moksha* here meaning freeing oneself from the evil of birth and death and merging in the *Brahman*. Humility and even-mindedness are the chief qualities of their ethics, while caste reigns supreme in their temporal affairs.

The Hindu religion underwent its first trial on the advent of Lord Buddha. The Buddha was himself the son of a king. He is said to have been born before 600 B.C. At that time the Hindus were under the glamour of the outward form of their religion, and the *Brahmins* had, out of selfishness, abandoned their true function of defending the Hindu faith. Lord Buddha was moved to pity when he saw his religion reduced to such a plight. He renounced the world and started doing penance. He spent several years in devout contemplation and ultimately suggested some reform in the Hindu religion. His piety greatly affected the minds of the *Brahmins*, and the killing of animals for sacrifice was stopped to a great extent. It cannot, therefore, be said that the Buddha founded a new or different religion. But those who came after him gave his teachings the identity of a separate religion. King Ashoka the Great sent missionaries to different lands for the propagation of Buddhism, and spread that religion in Ceylon, China, Burma and other countries. A distinctive beauty of Hinduism was revealed during this process: no one was converted to Buddhism by force. People's minds were sought to be influenced only by discussion and argument and mainly by the very pure conduct of the preachers themselves. It may be said that, in India at any rate, Hinduism and Buddhism were but one, and that even to-day the fundamental principles of both are identical.

We have seen that Buddhism had a salutary effect on Hinduism, that the champions of the latter were aroused by its impact. A thousand years ago, the Hindu religion came under another influence more profound. Hazrat Mahomed was born 1300 years ago. He saw moral anarchy rampant in Arabia. Judaism was struggling for survival; Christianity was not able to gain a foothold in the land; and the people were given to licence and self-indulgence. Mahomed felt all this to be improper. It caused him mental agony; and in the name of God, he determined to make them realise their miserable condition. His feeling was so intense that he was able immediately to impress the people around him with his fervour, and Islam spread very rapidly. Zeal or passion, then, is a great speciality, a mighty force, of Islam. It has been the cause of many good deeds, and sometimes of bad ones too. A thousand years ago the army of Ghazni invaded India in order to spread Islam. Hindu idols were broken and the invasions advanced as far as Somnath. While, on the one hand, violence was thus being used, the Muslim saints were, on the other, unfolding the real merit of Islam. The Islamic principle that all those who embraced Islam were equals made such a favourable impression on the lower classes that hundreds of thousands of Hindus accepted that faith, and there was great commotion in the whole community.

Kabir was born in Benares. He thought that, according to Hindu philosophy, there could be no distinction between a Hindu and a Muslim. Both of them, if they did good works, would find a place in heaven. Idolatry was not an essential part of Hinduism. Reasoning thus, he attempted to bring about a synthesis between Hinduism and Islam; but it did not have much effect, and his became no more than a distinct sect, and it exists even to-day. Some years later, Guru Nanak was born in the Punjab; he accepted the reasoning of Kabir and made a similar attempt to fuse the two religions. But while doing so, he felt that Hinduism should be defended against Islam, if necessary with the sword. This gave rise to Sikhism, and produced the Sikh warriors. The result of all this is that, despite the prevalence of Hinduism and Islam as the two principal religions of India to-day, both the communities live together in peace and amity and are considerate enough not to hurt one another's feelings save for the bitterness caused by political machinations and excitement. There is very little difference between a Hindu *yogi* and a Muslim *fakir*.

While Islam and Hinduism were thus vying with each other, the Christians landed at the port of Goa about 500 years ago, and set about converting Hindus to Christianity. They also partly resorted to force and [converted] partly through persuasion. Some of their ministers were exceedingly tender-hearted and kind, rather one would call them saintly. Like the *fakirs* they made a deep impression on the lower classes of Hindu society. But later, when Christianity and Western civilisation came to be associated, the Hindus began to look upon that religion with disfavour. And to-day, we see few Hindus embracing Christianity in spite of the fact that the Christians are ruling over a vast kingdom. Nevertheless, Christianity has had a very considerable influence on Hinduism. Christian priests imparted education of a high order and pointed out some of the glaring defects in Hinduism, with the result that there arose among the Hindus other great teachers who, like Kabir, began to teach the Hindus what was good in Christianity and appealed to them to remove these defects. To this category belonged Raja Ram Mohan Roy,[1] Devendranath Tagore, and Keshab Chandra Sen.[2] In Western India we had Dayanand Saraswati.[3] And the numerous reformist associations like the Brahmo Samaj and the Arya Samaj that have sprung up in India today are doubtless the result of Christian influence. Again, Madame Blavatsky[4] came to India, told both Hindus and the Muslims of the evils of Western civilisation and asked them to beware of becoming enamoured of it.

Thus, we have seen how there have been three assaults on

Hinduism, coming from Buddhism, Islam and then Christianity, but how on the whole it came out of them unscathed. It has tried to imbibe whatever was good in each of these religions. We should, however, know what the followers of this religion, Hinduism, believe. This is what they believe: God exists. He is without beginning, immaculate, and without any attribute or form. He is omnipresent and omnipotent. His original form is *Brahman*. It neither does, nor causes to be done. It does not govern. It is bliss incarnate, and by it all this is sustained. The soul exists, and is distinct from the body. It also is without a beginning, without birth. Between its original form and the *Brahman*, there is no distinction. But it takes on, from time to time, a body as a result of *karma* or the power of *maya*, and goes on being born again and again into high or low species in accordance with the good or bad deeds performed by it. To be free from the cycle of birth and death and be merged in *Brahman* is *moksha* or liberation. The way to achieve this *moksha* is to do pure and good deeds, to have compassion for all living beings, and to live in truth. Even after reaching this stage, one does not attain liberation, for one has to enjoy embodied existence as a consequence of one's good deeds as well. One has, therefore, to go a step further. We will, however, have to continue to act, only we should not cherish any attachment to our actions. Action should be undertaken for its own sake, without an eye on the fruit. In short, everything should be dedicated to God. We should not cherish, even in a dream, the feeling of pride that we do or can do anything. We should look upon all equally. These are the beliefs or tenets of Hinduism, but there admittedly exist a number of schools. Also, there have arisen a few factions or sects resulting from [differences in] secular practices. But we need not consider them on the present occasion.

If, after listening to this, any one of you has been favourably impressed and has come to feel that the Hindus or the Indians, in whose country the religion expounded above prevails, cannot be altogether an inferior people, you can render service to my countrymen even without becoming involved in political matters.

All religions teach that we should all live together in love and mutual kindness. It was not my intention to preach you a sermon neither am I fit to do so. But if it has produced any favourable impression on your mind, I would appeal to you to let my brethren have its benefit and, as behoves the English people, to defend them, whenever they are maligned.

Lectures on Religion, Johannesburg Lodge, Theosophical Society
Indian Opinion, 15 April 1905

[1] Founder of the Brahmo Samaj.
[2] Raja Ram Mohan Roy's work was continued by Devendranath Tagore and Keshab Chandra Sen, the former on the lines of pure Hinduism and the latter along those of Christianity.
[3] Founder of the Arya Samaj.
[4] Founder of the Theosophical Society.

168. GOLDEN SAYINGS OF ISLAM

We discussed in previous issues some books in the series *The Wisdom of the East*. The same writers have brought out a book on the subject noted above and have sent us a copy for review. It is hardly necessary to say that the book is in English. It is priced at one shilling only. It contains extracts from the holy Koran, and reproduces the sayings of Arab thinkers on different matters. For instance, with reference to nobility, it is said that 'He who disregards his own honour gets no good from an honourable lineage.... Learning and high principles cover the shame of low origin.' The book is full of rich thoughts having a bearing on our struggle for honour. The poet says: 'Men see no fault in one who respects himself.' Then again: 'Be ashamed in your own sight more than in the sight of men.' Once more: 'He who respects not himself can have no respect for others.' And elsewhere it is said: 'Life has no worth and this world has no happiness for a man who has lost his self-respect and abandoned himself to shamelessness.' Under Character, we have: 'A man is truly religious when he is truly good.' Under Knowledge, we have: 'A man without education is like a brave man without arms.' 'Kings govern men and learned men govern kings.' 'A wise man is not he who considers how he may get out of an evil, but he who sees to it that he does not fall into it.' On Truthfulness, it is said: 'No man's religion can be right unless his heart becomes right, nor can his heart become right unless his tongue is right.... That man is a hypocrite who prays and fasts, but is untruthful in what he says, false to his word, and unfaithful in discharging a trust.' Such are the golden sayings contained in this little book. We advise everyone who can read English to buy this book.

'Arab Wisdom' (G.)
Indian Opinion, 28 Dec. 1907

169. THE BUDDHIST REFORMATION

[Bombay,
18 May 1924]

The only reason for inviting me to preside at this meeting is, I presume, that I am more than most people endeavouring to popularize the truth for which Gautama Buddha lived and died. For, my book-knowledge of Buddhism, I am sorry to have to confess, is of the poorest type. It is confined to the fascinating book of Sir Edwin Arnold, which I read for the first time now nearly thirty-five years ago, and one or two books during my brief incarceration in the Yeravda Jail. But that great Buddhist scholar, Professor Kausambi, tells me that *The Light of Asia* gives but a faint idea of Buddha's life, and that at least one incident in the beautiful poem is not to be found in any authoritative original Buddhist work. Perhaps some day the learned Professor will give us the results of his ripe scholarship in the shape of a reliable story of Buddha's life for the ordinary Indian reader.

For the moment, however, I would like to tell the meeting what I believe about Buddhism. To me it is a part of Hinduism. Buddha did not give the world a new religion; he gave it a new interpretation. He taught Hinduism not to take but to give life. True sacrifice was not of others but of self. Hinduism resents any attack upon the Vedas. It regarded the new interpretation as such attack. Whilst, therefore, it accepted the central truth of Buddha's teaching, it fought against Buddhism regarded as a new and anti-Vedic cult.

It has become the fashion nowadays in some quarters to say

that India's downfall dates from her acceptance of Buddha's teachings. It is tantamount to saying that love and piety, if sufficiently practised, will degrade the world. In other words, according to the critics, evil should triumph in the end. It is my unalterable belief that India has fallen not because it accepted Gautama's teaching, but because it failed to live up to it. The priest has ever sacrificed the prophet. Vedas to be divine must be a living word, ever growing, ever expanding and ever responding to new forces. The priest clung to the letter and missed the spirit.

But we need not despair. The reformation that Buddha attempted has not yet had a fair trial. Twenty-five hundred years are nothing in the life of the world. If the evolution of form takes aeons, why should we expect wonders in the evolution of thought and conduct? And yet the age of miracles is not gone. As with individuals, so with nations. I hold it to be perfectly possible for masses to be suddenly converted and uplifted. Suddenness is only seeming. No one can say how far the leaven has been working. The most potent forces are unseen, even unfelt, for long. But they are working none the less surely. Religion to me is a living faith in the Supreme Unseen Force. That Force has confounded mankind before, and it is bound to confound us again. Buddha taught us to defy appearances and trust in the final triumph of Truth and Love. This was his matchless gift to Hinduism and to the world.

He taught us also how to do it, because he lived what he taught. The best propaganda is not pamphleteering, but for each one of us to try to live the life we would have the world to live.

Speech at Buddha Jayanti, Bombay
CW 5176

170. INTERPRETATIONS OF THE KORAN

I gladly print the following from Maulana Zafar Ali Khan written by him in his capacity as President of the Punjab Khilafat Committee:

I have read with feelings of mingled amazement and pain your pronouncement, in *Young India* of the 26th instant, on stoning incidents in Kabul. You say that 'this particular form of penalty cannot be defended on the mere ground of its mention in the Koran'. You, moreover, declare that 'every formula of every religion has in this age of reason to submit to the acid test of reason and universal justice if it is to ask for universal assent'. Finally you maintain that error can claim no exemption even if it can be supported by the scriptures of the world.

I have always paid unstinted homage to your greatness and have all along looked upon you as one of the few men who are making modern history; but I would be failing in my duty as a Mussalman if I refrained from pointing out to you that by challenging the right of the Koran to regulate the life of its followers in its own way you have shaken the belief of millions of your Muslim admirers in your capacity to lead them.

You are at perfect liberty to express your opinion one way or the other as to whether renegades can be stoned to death under the law of Islam. But to hold that even if the Koran supported such form of penalty, it should be condemned outright as an error, is a form of reasoning which cannot appeal to the Mussalmans.

Error is after all a relative term and Mussalmans have their own way of interpreting it. To them the Koran is an unalterable law which transcends the ever changing policies and expediencies of puny humanity. Would to God that to your multifarious activities as leader of India you had not added the rather delicate task of adversely criticizing the teaching of the Holy Koran.

The Maulana has put an interpretation upon my note which it does not bear. I have not adversely (or otherwise) criticized the teachings of the Holy Koran. But I have criticised the teachers, that is, the interpreters, in anticipation of their defending the penalty of stoning to death. I claim to understand enough of the Koran and the history of Islam to know that a multitude of interpreters have interpreted the Koran to suit their preconceived notions. My purpose was to issue a warning against the acceptance of such interpretations. But I would like to say that even the teachings themselves of the Koran cannot be exempt from criticism. Every true scripture only gains by criticism. After all we have no other guide but our reason to tell us what may be regarded as revealed and what may not be. The early Mussalmans accepted Islam not because they knew it to be revealed but because it appealed to their virgin reason.

I fully endorse the Maulana's statement that error is a relative term. But we know as a matter of fact that some things are universally accepted as errors. Death by torture is, I expect, such an error. In making the three statements the Maulana has quoted, I have simply mentioned three canons of interpretations which I think are incapable of challenge. Anyway, I abide by them. And if I am at perfect liberty to express my opinion 'as to whether renegades can be stoned to death under the law of Islam' why may I not express an opinion as to whether penalty of stoning to death can be imposed at all under the law of Islam? The Maulana has betrayed intolerance of criticism by a non-Muslim of anything relating to Islam. I suggest to him that intolerance of criticism even of what one may prize as dear as life itself is not conducive to the growth of public corporate life. Surely Islam has nothing to fear from criticism even if it be unreasonable. I therefore suggest to the Maulana the advisability in the light of my criticism of applying himself to an elucidation of the tremendous issues involved in the incidents reported from Kabul.

'My Crime'
Young India, 5 Mar. 1925

171. BUDDHA AND THE MASSES

[Calcutta,
7 May 1925]

Friends, it is now my pleasant duty to perform this service. I shall not say anything of these proceedings. Dr. Dharmapala has added a pathetic touch to this service. And he has laid on my shoulders a burden which I consider I am ill-fitted to carry. I hesitated last year when Mr. Natarajan drew me out of my convalescent bed and asked me to preside at the anniversary last year, but I could not resist Mr. Natarajan, for I have very great and deep affection for him. I know that from that time I would be perhaps called upon to take part at such functions somewhere in India from year to year. And so it happened even when I came to Calcutta. It is a very strange thing that almost all the professors of great religions of the world claim me as their own. The Jains mistake me for a Jain. Scores of Buddhist friends have taken me for a Buddhist. Hundreds of Christian friends still consider that I am a Christian and some Christian friends do not even hesitate to ascribe, by implication, cowardice to me and say: 'We know, you are a Christian, but you are afraid to own it. Why don't you come forward boldly and say you believe in Jesus and his salvation?' Many of my Mussalman friends consider that, although I do not call myself a Mussalman, to all intents and purposes, I am one of them; and some Mussalman friends consider that I am on the road to it, very near, but still far short of it.

All this is extremely flattering to me and I take it as a mark of their affection and their esteem. For me, however, I regard myself as one of the humblest of Hindus, but the deeper I study Hinduism the stronger becomes the belief in me that Hinduism is as broad as the Universe and it takes in its fold all that is good in this world. And so I find that with Mussalmans I can appreciate the beauties of Islam and sing its praises. And so simultaneously with the professors of other religions, and still something within me tells me that, for all that deep veneration I show to these several religions, I am all the more a Hindu, none the less for it.

Nearly 40 or 38 years ago, I went to England as a lad and the first religious book that was placed into my hands was *The Light of Asia*. I had read nothing of any religion in the world, nothing therefore of Hinduism. I knew of Hinduism what my parents taught me, not directly but indirectly, that is, by their practice, and I knew a little more of it from a *Brahmin* to whom they sent me in order to learn *Rama Raksha*. That was the stock with which I sailed for England. So, when I found myself in possession of *The Light of Asia*, I devoured it.

From page to page I went; I was really an indifferent reader of literature, but I could not resist the temptation that each page afforded to me and I closed the book with deep veneration for the expounding or teaching which has been so beautifully expressed by Sir Edwin Arnold. I read the book again when I had commenced the practice of my profession in South Africa. At that time, I had read something of the other great religions of the world, but the second study of that book did not diminish my veneration. Beyond that I have practically no acquaintance with Buddhism. I read some more literature in the Yeravda Jail, but I know that the reason why I am called upon to preside at such functions, whether they were in connection with Buddha or Mahavira or even with Jesus Christ, is that I endeavour to follow to the best of my ability such of these masters' teachings as my limited understanding enables me to appreciate. Many friends consider that I am expressing in my own life the teachings of Buddha. I accept their testimony and I am free to confess that I am trying my level best to follow these teachings. Unlike Buddhistic professors and unlike also many Hindu students—I was going to say philosophers—I draw no distinction between the essential teachings of Hinduism and Buddhism. In my opinion, Buddha lived Hinduism in his own life. He was no doubt a reformer of his terrible time, that is to say he was a reformer deeply in earnest and counted no cost or pain great for achieving the reform which he thought was indispensable for his own growth and for the uplift of the body.

If historical records are correct, the blind *Brahmins* of that period rejected his reform because they were selfish. But the masses were not philosophers who whiled away their time in philosophizing. They were philosophers in action, they had robust common sense and so they brushed aside the beast in the *Brahmins*, that is to say, selfishness, and they had no hesitation in recognizing in Buddha the true exponent of their own faith. And so, being myself also one of the masses living in their midst, I found that Buddhism is nothing but Hinduism reduced to practice in terms of the masses. And, therefore, sometimes the learned men are not satisfied with the incredibly simple teachings of Buddha. They go to it for the satisfaction of their intellect and they are disappointed. Religion is pre-eminently a matter of the heart and a man who approaches it with intellectual pride is doomed to disappointment.

I make bold to say that Buddha was not an atheist. God refuses to see any person, any devotee who goes in with his pride. He believes not in men rubbing their noses on the ground, he wants not to see the marks on the noses, and some of you may not know that many Mussalmans really carry these marks on their foreheads as they lie prostrate in their mosques, rub their foreheads day after day so that they have got the scar on their forehead about the circumference of a rupee, sometimes, even larger. God does not want the marks. He sees through and through. A man may cut his nose and rub it on the ground, but God will not recognize him who will turn his back upon a man with pointed nose, if his heart is not bruised and blood does not flow freely from his heart. He recognizes that as his own. And the masses, not knowing what pride is, approach him in all humility and become the splendid philosophers in action, and we can freely follow them. That, in my opinion, is the essential teaching of Buddhism. It is pre-eminently a religion of the masses. I do not despair. I do not for one moment consider that Buddhism has been banished from India.

Every essential characteristic of Buddhism, I see, is being translated into action in India, much more perhaps than in China, Ceylon and Japan, which nominally profess Buddhism. I make bold to say that we in India translate Buddhism into action far more and far better than our Burmese friends do. It is impossible to banish Buddha. You cannot deprive him of his birth in India. In his own life, he made out for himself an imperishable name. He lives today in the lives of millions of human beings. What does it matter whether we go to a little temple and worship his image or whether we even take his name. My Hinduism teaches me that, if my heart is pure, I may mispronounce the name of Sri Rama as Mara, still I can speak it with as much force as may, even more than, the learned *Brahmins*. So, I say to Dr. Dharmapala, what does it matter whether he can count upon the support of so many men, or whether a lady from Honolulu contributes a huge sum or not. Buddha has taught us, in my humble opinion, that it is not necessary for millions to associate themselves with one man who seeks for truth.

Let each one say for himself how much of the message of mercy and pity that Buddha came to deliver we have translated into our own lives, and in so much as we have translated that message in our own lives are we fit to pay our homage to that great Lord, Master and Teacher of mankind. So long as the world lasts, I have not a shadow of doubt that he will rank among the greatest of teachers of mankind. The thoughts that Buddha gave about 2,500 years ago will never vanish—thoughts leave also a mark, though going at snail's pace. It is still germinating, though one may find that Buddhism, like every other religion, at the present moment is really decadent. I am optimistic enough to feel that a day is dawning when all these great religions will be purged of all frauds, hypocrisy, humbug, untruthfulness, incredulity and all that may be described under the term 'degradation'. They will be purified of that fraud and we will see a day dawned when he who learns to see will find that truth and love, after all, are two faces of a coin. That and that alone is the only current coin and every other is a base coin.

May God help us to realize the message that the Lord Buddha delivered to mankind so many hundred years ago and may we, each one of us, endeavour to translate that message in our lives, whether we call ourselves Hindus or not.

Speech at Buddha's Birth Anniversary, Mahabodhi Society
Amrita Bazar Patrika, 9 May 1925

172. THE SIKH *GURUS*

During his visit to Patna for the A.I.C.C., Sardar Mangal Singh drew my attention to an article in *Young India* entitled 'My friend, the revolutionary' in the issue dated 9th April last. He told me that many Sikh friends were offended because they thought I have described Guru Govind Singh as a misguided patriot whereas I had glorified Krishna. The Sardarji asked me to take an early opportunity of explaining what I meant by the passages he drew my attention to. The careful reader will note that my language is most guarded. I have made no positive assertion. All that I have said is that believing every statement made about the heroes mentioned including Guru Govind Singh to be true, had I lived as their contemporary I would have called every one of them a misguided patriot. But, in the very next sentence, I have hastened to add that I must not judge them and that I disbelieve history as far as the details of the acts of the heroes are concerned. My belief about the Sikh *Gurus* is that they were all deeply religious teachers and reformers, that they were all Hindus and that Guru Govind Singh was one of the greatest defenders of Hinduism. I believe, too, that he drew the sword in its defence. But I cannot judge his actions, nor can I use him as my model so far as his resort to the sword is concerned. What I would have done had I lived in his times and held the same views that I hold now I do not know. Such speculation I regard as perfect waste of time.

I do not regard Sikhism as a religion distinct from Hinduism. I regard it as part of Hinduism and the reformation in the same sense that Vaishnavism is. I read in the Yeravda Prison all the writings that I could lay my hands upon regarding the Sikhs. I read portions of Granth Saheb. Its deeply spiritual and moral tone I found to be uplifting. In the collection of hymns we have at the Ashram, we have some of Guru Nanak's also. At the same time, I do not quarrel with the Sikhs for considering, if they wish, Sikhism as totally distinct from Hinduism. And when during my first visit to the Punjab, a few Sikh friends told me that my reference to Sikhism as part of Hinduism displeased them, I ceased to refer to it as such. But the Sikh friends will pardon me for avowing my belief when I am asked to express my opinion about Sikhism.

Now about Krishna. Whilst I have dealt with the *Gurus* as historical personages about whose existence we have trustworthy records, I have no knowledge that the Krishna of the *Mahabharata* ever lived. My Krishna has nothing to do with any historical person. I would refuse to bow my head to the Krishna who would kill because his pride is hurt, or the Krishna whom non-Hindus portray as a dissolute youth. I believe in Krishna of my imagination as a perfect incarnation, spotless in every sense of the word, the inspirer of the *Gita* and the inspirer of the lives of millions of human beings. But if it was proved to me that the *Mahabharata* is history in the same sense that modern historical books are, that every word of the *Mahabharata* is authentic and that the Krishna of the *Mahabharata* actually did some of the acts attributed to him, even at the risk of being banished from the Hindu fold, I should not hesitate to reject that Krishna as God incarnate. But to me the *Mahabharata* is a profoundly religious book, largely allegorical, in no way meant to be a historical record. It is the description of the eternal duel going on within ourselves, given so vividly as to make us think, for the time being that the deeds described therein were actually done by the human beings. Nor do I regard the *Mahabharata* as we have it now as a faultless copy of the original. On the contrary I consider that it has undergone many emendations.

'Sikhism'
Young India, 1 Oct. 1925

173. THE TYRANNY OF CUSTOM

In South Africa it is the crime of colour and race for which we are being punished. In India we Hindus punish our co-religionists for the crime of caste. The fifth caste man—the *Panchama*—is the greatest offender deserving the punishment of untouchability, unapproachability, invisibility and what not. An extraordinary case that was tried in a Madras Presidency court brings vividly to light the sad plight of our suppressed countrymen. A simple cleanly dressed *Panchama* entered a temple in a perfectly devotional spirit without the slightest intention of hurting anybody's feeling or insulting any religion. He had been in the habit of paying his respects at this temple every year though he did not enter it. But last year in his ecstatic mood he forgot himself and entered the temple. The priest in charge could not distinguish him from the others and therefore accepted his offering. But when he regained self-possession he was terrified to find himself in a prohibited place and ran away from the temple. But some who knew him caught him and handed him to the police. The temple authorities when they discovered the crime had the temple duly purified. Then followed a trial. A Hindu magistrate convicted him and imposed a fine of Rs. 75 or one month's rigorous imprisonment for insulting his own religion! An appeal was filed. There was an elaborate argument over it. Judgment had to be reserved! And when conviction was set aside, it was not because the court held that the poor *Panchama* had a right to enter the temple but because the prosecution in the lower court had forgotten to prove the insult. This is no triumph of justice or truth or religion or morality.

The only consolation to be derived from the successful appeal is that the *Panchama* will not have to suffer imprisonment for having in his zeal for worship forgotten that he was a prohibited entrant. If however he or his fellow-*Panchamas* again dare to enter the temple, it is highly probable that they would be severely punished if they are not lynched by those who look down upon them with contempt.

It is a curious situation. We resent, and properly, the treatment meted out to our countrymen in South Africa. We are impatient to establish *swaraj*. But we Hindus refuse to see the incongruity in treating a fifth of our own co-religionists as worse than dogs. For dogs are not untouchables. Some of us nowadays even keep them as drawing-room pets.

What place shall the 'untouchables' occupy in our scheme of *swaraj*? If they are to be free from all special restraints and disabilities under *swaraj*, why can we not declare their freedom now? And if we are powerless today, shall we be less powerless under *swaraj*? We may shut our eyes and stuff our ears to these questions. But they are of the highest importance to the *Panchamas*. Surely, judgment will be pronounced against Hinduism, if we as a body do not rise as one man against this social and religious atrocity.

Much has no doubt been done to remove the evil. But it is all too little so long as criminal prosecutions for temple-entry are possible and so long as the suppressed classes continue to be denied the right of entering temples, using public wells, and sending their children freely to national schools. We must yield to them the same rights as we would have the Europeans concede to our countrymen in South Africa.

But this case is not without its relieving features. The quashing of the conviction is no doubt some consolation. But the best consolation lies in the fact of so many *savarna* Hindus actively interesting themselves in the poor *Panchama*'s behalf. The appeal would not have been noted, if someone had not gone to the accused's assistance. Not the least interesting feature of the case was the fact of C. Rajagopalachari arguing the appeal—a fit application in my opinion of the principle of non-co-operation. Being in the court, when he got the opportunity, he would have been like a Pharisee if he had sat there stiff gloating over the sanctimonious satisfaction of non-co-operating whilst the accused could have been discharged by his intervention. The *Panchama* knew nothing of non-co-operation. He had appealed to avoid payment of fine or imprisonment. It is to be wished that every educated Hindu will constitute himself the untouchable's friend and regard it his duty to free him from the tyranny of custom masquerading under the name of religion. Not the entry of a *Panchama* into a temple but the brand of prohibition against him is an insult to religion and humanity.

'The Crime of Caste'
Young India, 14 Jan. 1926

174. HINDUISM A LIVING ORGANISM

A correspondent styling himself '*Sanatani* Hindu' writes:

Hinduism of today presents many a curious anomaly. No one cares to study it, excepting perhaps some European missionaries....
There is no definite body of doctrines or practices which may be called *sanatana* and should be respected....
A *Sudra*'s status in a province where the *Brahmins* eat meat or fish is different from that in any province where the *Brahmins* and *Vaishnavas* alike refrain from meat or fish. You have somewhere said: 'If you have no objection to drinking clean water given by a *Sudra*, you should have none to drink it at the hands of an untouchable.... some Hindus have religious scruples against having water at the hands of meat-eaters, some against doing so at the hands of beef-eaters.' ... I wish you could call a meeting of the untouchables of the whole province and take promises from them to abstain from beef or carrion as a preliminary to their being taken into the Hindu fold.

The letter presents only one side of the case. There is reason for the correspondent's complaint. But Hinduism is a living organism liable to growth and decay, and subject to the laws of Nature. One and indivisible at the root, it has grown into a vast tree with innumerable branches. The changes in the seasons affect it. It has its autumn and summer, its winter and spring. The rains nourish and fructify it too. It is and is not based on scriptures. It does not derive its authority from one book. The *Gita* is universally accepted, but even then it only shows the way. It has hardly any effect on custom. Hinduism is like the Ganges pure and unsullied at its source, but taking in its course the impurities in the way. Even like the Ganges it is beneficent in its total effect. It takes a provincial form in every province, but the inner substance is retained everywhere. Custom is not religion. Custom may change, but religion will remain unaltered.

Purity of Hinduism depends on the self-restraint of its votaries. Whenever their religion has been in danger, the Hindus have undergone rigorous penance, searched the causes of the danger and devised means for combating them. The Shastras are ever growing. The Vedas, Upanishads, *smritis*, Puranas and *itihasas* did not arise at one and the same time. Each grew out of the necessities of particular periods, and therefore they seem to conflict with one another. These books do not enunciate anew the eternal truths but show how these were practised at the time to which the books belong. A practice which was good enough in a particular period would, if blindly repeated in another, land people into the 'slough of despond'. Because the practice of animal-sacrifice obtained at one time, shall we revive it today? Because at one time, we used to eat beef, shall we also do so now? Because at one time, we used to chop off the hands and the feet of thieves, shall we revive that barbarity today? Shall we revive polyandry? Shall we revive child-marriages? Because we discarded a section of humanity one day, shall we brand their descendants today as outcastes?

Hinduism abhors stagnation. Knowledge is limitless and so also the application of truth. Every day we add to our knowledge of the power of *atman*, and we shall keep on doing so. New experience will teach us new duties, but truth shall ever be the same. Who has ever known it in its entirety? The Vedas represent the truth, they are infinite. But who has known them in their entirety? What goes today by the name of the Vedas are not even a millionth part of the real Veda—the Book of

Knowledge. And who knows the entire meaning of even the few books that we have? Rather than wade through these infinite complications, our sages taught us to learn one thing: 'As with the self, so with the Universe.' It is not possible to scan the universe, as it is to scan the self. Know the self and you know the universe. But even knowledge of the self within presupposes ceaseless striving—not only ceaseless but pure, and pure striving presupposes a pure heart, which in its turn depends on the practice of *yamas* and *niyamas*[1]—the cardinal and casual virtues.

This practice is not possible without God's grace which presupposes Faith and Devotion. This is why Tulsidas sang of the glory of *Ramanama*, that is why the author of the *Bhagavata* taught the *dwadashamantra (Om Namo Bhagavate Vasudevaya)*. To my mind he is a *sanatani* Hindu who can repeat this *mantra* from the heart. All else is a bottomless pit, as the sage Akho[2] has said.

But to come to the other part of the letter. The Europeans do study our manners and customs. But theirs is the study of a critic, not the study of a devotee. Their 'study' cannot teach me religion.

Boycott of beef-eaters may have been proper in the past. It is improper and impossible today. If you want the so-called untouchables to give up beef, you can do so only by means of love, only by quickening their intellects, not by despising them. Non-violent efforts to wean them away from their bad habits are going on, but Hinduism does not consist in eating and not-eating. Its kernel consists in right conduct, in correct observance of truth and non-violence. Many a man eating meat, but observing the cardinal virtues of compassion and truth, and living in the fear of God, is a better Hindu than a hypocrite who abstains from meat. And he whose eyes are opened to the truth of the violence in beef-eating or meat-eating and who has therefore rejected them, who loves 'both man and bird and beast' is worthy of our adoration. He has seen and known God; he is His best devotee. He is the teacher of mankind.

Hinduism and all other religions are being weighed in the balance. Eternal truth is one. God also is one. Let every one of us steer clear of conflicting creeds and customs and follow the straight path of truth. Only then shall we be true Hindus.

Many styling themselves *sanatanis* stalk the earth. Who knows how few of them will be chosen by God! God's grace shall descend on those who do His will and wait upon Him, not on those who simply mutter 'Ram Ram'.

'Hinduism of Today' (G.)
Young India, 8 April 1926

¹ *Yamas*, the cardinal virtues, according to Yoga Shastra are: *ahimsa* (non-violence), *satya* (truth), *asteya* (non-stealing), *brahmacharya* (celibacy), *aparigraha* (non-possession); and the *niyamas* or the casual virtues are, according to the same authority: *shaucha* (bodily purity), *santosha* (contentment), *tapa* (forbearance), *swadhyaya* (study of scriptures), *Ishwarapranidhana* (resignation to the Will of God).

² Akha Bhagat, a poet-seer of Gujarat.

175. THE PROCESS OF PURIFICATION

A friend thus writes . . .[1]

We have more Ganges and Jumnas than the two. It is true they bear other names. They are, however, watering our vast plains from the North to the South and the East to the West. The message that the Ganga and the Jumna gave to this friend is not the only message that the mighty rivers of India are giving to her inhabitants. They remind us of the sacrifice we must make for the sake of the land we are living in. They remind us of the process of purification that we must continuously go through as the rivers themselves are going through from moment to moment. I wrote nearly 10 years ago that the *Gayatri*, the common prayer of Hinduism, was one of the precious gifts of the Ganga. Her shining waters must have given the *Rishis* of old the inspiration. In the modern rush, the chief use we have for our rivers is to empty our gutters in them and to navigate our cargo vessels, and in the process make them dirtier still. We have no time, as the friend had, to stroll down to these rivers, and in silent meditation listen to the message they murmur to us.

'The Message of Ganga and Jumna'
Young India, 23 Dec. 1926

¹ The letter is not reproduced here. The friend had written of the feelings of ecstasy inspired by the river Jumna.

176. BUDDHA'S GREAT RENUNCIATION

[15 November 1927]

I thank you very sincerely for the address that you have given to me. I appreciate the courtesy, in that you have supplied me with a translation of your address in advance. I am equally grateful to His Holiness and the priests for the benediction that they have pronounced just now. I shall always esteem it as a great privilege that I have received this benediction this afternoon, and I can give His Holiness and the priests in the presence of this assembly the assurance that I shall always strive to deserve that benediction. Your address mentions it and His Holiness also just now mentioned the fact about the Buddha Gaya temple which is situated in India. I have been interesting myself in this great institution for a long time, and when I presided over the deliberations of the Indian National Congress at Belgaum, I had the privilege of doing what was possible on behalf of the Congress in this connection. I had sent to me by some unknown friend in Ceylon a report of the controversy that took place in connection with what I did at the Congress. I did not think it proper to take part in that controversy nor do I desire even now to go into it. I can only give you my assurance that everything that was humanly possible for me to do to advance your claim I did and I shall still do. I can only tell you, however, that the Congress does not possess the influence that I would like it to possess. There are several difficulties raised in connection with the proprietary rights. There are technical, legal difficulties also in the way.

The Congress appointed a Committee of the best men that were at its disposal to go into this matter and if possible even to come to terms with the Mahant who is at the present moment in possession of the temple. That Committee has already reported, and I take it that some of you have seen the report of that Committee. That Committee endeavoured to have an arbitration appointed, but it failed in its efforts to do so. But there is absolutely no reason to lose hope. However, I can tell you that all my personal sympathies are absolutely with you and, if the rendering of its possession to you was in my giving, you can have it today. In your address was mentioned another temple that is situated in Ceylon. I do not know anything about the controversy regarding this temple. I, therefore, like some of you to give me particulars about it, and tell me if there is anything that I can do in connection with it whilst I am in your midst. You may take it for granted that I should take a personal interest in it if I feel that there is anything that I can do, and I should do so not in order that I can oblige you, but in order to give myself satisfaction.

For, you do not know perhaps that one of my sons, the eldest boy, accused me of being a follower of Buddha, and some of my Hindu countrymen also do not hesitate to accuse me of spreading Buddhistic teaching under the guise of *sanatan* Hinduism. I sympathize with my son's accusations and the accusations of my Hindu friends. And sometimes I feel even proud of being accused of being a follower of the Buddha, and I have no hesitation in declaring in the presence of this audience that I owe a great deal to the inspiration that I have derived from the life of the Enlightened One. Indeed, at an anniversary celebration in the new Buddha temple that has been erected in Calcutta I gave expression to this view. The leader in that meeting was Angarika Dharmapala. He was weeping over the fact that he was not receiving the response that he desired for the cause which was close to his heart, and I remember having rebuked him for shedding tears. I told the audience that though what passed under the name of Buddhism might have been driven out of India, the life of the Buddha and his teachings were by no means driven out of India. This incident happened, I think, now three years ago, and I have seen nothing since to alter the view which I pronounced at that meeting.

It is my deliberate opinion that the essential part of the teachings of the Buddha now forms an integral part of Hinduism. It is impossible for Hindu India today to retrace her steps and go behind the great reformation that Gautama effected in Hinduism. By his immense sacrifice, by his great renunciation and by the immaculate purity of his life he left an indelible impress upon Hinduism, and Hinduism owes an eternal debt of gratitude to that great teacher. And if you will forgive me for saying so, and if you will also give me the permission to say so, I would venture to tell you that what Hinduism did not assimilate of what passes as Buddhism today was not an essential part of Buddha's life and his teachings.

It is my fixed opinion that Buddhism or rather the teaching of Buddha found its full fruition in India, and it could not be otherwise, for Gautama was himself a Hindu of Hindus. He was saturated with the best that was in Hinduism, and he gave life to some of the teachings that were buried in the Vedas and which were overgrown with weeds. His great Hindu spirit cut its way through the forest of words, meaningless words, which had overlaid the golden truth that was in the Vedas. He made some of the words in the Vedas yield a meaning to which the men of his generation were utter strangers, and he found in India the most congenial soil. And wherever the Buddha went, he was followed and surrounded not by non-Hindus but Hindus, those who were themselves saturated with the Vedic law. But the Buddha's teaching like his heart was all-expanding and all-embracing and so it has survived his own body and swept across the face of the earth. And at the risk of being called a follower of Buddha I claim this achievement as a triumph of Hinduism.

Buddha never rejected Hinduism, but he broadened its base. He gave it a new life and a new interpretation. But here comes the point where I shall need your forgiveness and your generosity, and I want to submit to you that the teaching of Buddha was not assimilated in its fulness whether it was in Ceylon, or in Burma, or in China or in Tibet. I know my own limitations. I lay no claim to scholarship in Buddhistic law. Probably, a fifth-form boy from Nalanda Vidyalaya would plough me in a Buddhist catechism. I know that I speak in the presence of very learned priests and equally learned laymen, but I should be false to you and false to myself if I did not declare what my heart believes.

You and those who call themselves Buddhists outside India have no doubt taken in a very large measure the teachings of the Buddha, but when I examine your life and when I cross-question the friends from Ceylon, Burma, China or Tibet, I feel confounded to find so many inconsistencies between what I have come to understand as the central fact of Buddha's life and your own practice, and if I am not tiring you out, I would like hurriedly to run through three prominent points that just now occurred to me. The first is the belief in an all-pervading Providence called God. I have heard it contended times without number and I have read in books also, claiming to express the spirit of Buddhism, that Buddha did not believe in God. In my humble opinion such a belief contradicts the very central fact of Buddha's teaching. In my humble opinion the confusion has arisen over his rejection, and just rejection, of all the base things that passed in his generation under the name of God.

He undoubtedly rejected the notion that a being called God was actuated by malice, could repent of his actions, and like the kings of earth could possibly be open to temptations and bribes and could possibly have favourites. His whole soul rose in mighty indignation against the belief that a being called God required for His satisfaction the living blood of animals in order that he might be pleased—animals who were his own creation. He, therefore, reinstated God in the right place and dethroned the usurper who for the time being seemed to occupy that White Throne. He emphasized and redeclared the eternal and unalterable existence of the moral government of this universe. He unhesitatingly said that the law was God Himself.

God's laws are eternal and unalterable and not separable from God Himself. It is an indispensable condition of His very perfection. And hence the great confusion that Buddha disbelieved in God and simply believed in the moral law, and because of this confusion about God Himself, arose the confusion about the proper understanding of the great word *Nirvana*. *Nirvana* is undoubtedly not utter extinction. So far as I have been able to understand the central fact of Buddha's life, *Nirvana* is utter extinction of all that is base in us, all that is vicious in us, all that is corrupt and corruptible in us. *Nirvana* is not like the black, dead peace of the grave, but the living peace, the living happiness of a soul which is conscious of itself, and conscious of having found its own abode in the heart of the Eternal.

The third point is the low estimation in which the idea of sanctity of all life came to be held in its travels outside India. Great as Buddha's contribution to humanity was in restoring God to His eternal place, in my humble opinion greater still was his contribution to humanity in his exacting regard for all life, be it ever so low. I am aware that his own India did not rise to the height that he would fain have seen India occupy. But the teaching of Buddha, when it became Buddhism and travelled outside, came to mean that sacredness of animal life had not the sense that it had with an ordinary man. I am not aware of the exact practice and belief of Ceylonese Buddhism in this matter, but I am aware what shape it has taken in Burma and China. In Burma especially the Burmese Buddhists will not kill a single animal, but do not mind others killing the animals for them and dishing the carcases for them for their food. Now, if there was any teacher in the world who insisted upon the inexorable law of cause and effect, it was inevitably Gautama, and yet my friends, the Buddhists outside India would, if they could, avoid the effects of their own acts. But I must not put an undue strain upon your patience. I have but lightly touched upon some of the points which I think it my duty to bring to your notice, and in all earnestness and equal humility I present them for your serious consideration.

One thing more and I shall have done. Last night the members of the Reception Committee asked me to speak at one of these audiences of the connection *khadi* had with Ceylon. I have not left much time for myself to expand this message before you, but I shall try to summarize it in two sentences. One thing is that you who regard Buddha as the ruler of your hearts owe something to the land of his birth, where millions of his descendants for whom he laboured and for whom he died are today living a life of misery, living in a state of perpetual semi-starvation. I venture, therefore, to suggest that *khadi* enables you to establish a living bond between yourselves and the ruler of your hearts. If you will follow the central fact of his teaching and regard life as one of renunciation of all material things, all life being transitory, you will at once see the beauty of the message of *khadi* which otherwise means simple living and high thinking. Taking these two thoughts with you, I suggest to every one of you to dot the 'i's and cross the 't's and make out your own interpretation of the message of *khadi*. I thank you again for the great kindness that you have shown, for the address and for the benediction, and

I hope that you have received the humble message that I have given to you in the same spirit in which it has been delivered. Regard it as a message not from a critic, but from a bosom friend.

Speech in Reply to Buddhists' Address, Vidyodaya College, Colombo
Young India, 24 Nov. 1927

177. TEACHERS OF MANKIND

[29 November 1927]

I am deeply grateful to you for the generous purse that you have given me on behalf of the semi-starving millions of India. You, Sir, sent me due notice yesterday of the very important question that you have repeated this morning.[1] I have many engagements between now and 10.30; therefore, and also for other reasons into which I do not want to enter, I would fain have avoided this question. But on the principle that has guided my life that I must take things as they come to me, unless I find it utterly impossible for me to cope with them, I propose to devote the very few minutes that I have at my disposal to answer that question.

I say in one sentence that for many many years I have regarded Jesus of Nazareth as one amongst the mighty teachers that the world has had, and I say this in all humility. I claim humility for this expression for the simple reason that this is exactly what I feel. Of course, Christians claim a higher place for Jesus of Nazareth than as a non-Christian and as a Hindu I have been able to feel. I purposely use the word 'feel' instead of 'give', because I consider that neither I, nor anybody else can possibly arrogate to himself the claim of giving place to a great man. The great teachers of mankind have had the place not given to them, but the place has belonged to them as a matter of right, as a matter of service that they have rendered, but it is given to the lowest and the humblest amongst us to feel certain things about certain people.

The relation between great teachers and ourselves is somewhat after the style of relation between a husband and wife. It would be a most terrible thing, a tragic thing, if I was to argue out intellectually for myself what place I was to give to my wife in my heart. It is not in my giving, but she takes the place that belongs to her as a matter of right in my heart. It is a matter purely for feeling. Then I can say that Jesus occupies in my heart the place of one of the great teachers who has had considerable part in finding that place in my heart. Leave the Christians alone for the present. I shall say to the 75 per cent Hindus receiving instruction in this College that your lives will be incomplete unless you reverently study the teaching of Jesus. I have come to the conclusion, in my own experience, that those who, no matter to what faith they belong, reverently study the teaching of other faiths broaden their own, instead of slackening their hearts. Personally, I do not regard any of the great religions of the world as false. All have served in embellishing mankind and are even now serving their purpose. A liberal education to all should include, as I have put it, a reverent study of other faiths, but I do not want to labour this point, nor have I the time to do so.

There is one thing which, as I am speaking to you, occurs to me, which comes to me from my early studies of the Bible. It seized me immediately I read the passage:

But seek ye first the Kingdom of God, and his righteousness; and all these things shall be added unto you.[2]

I tell you that if you will understand, appreciate and act up to the spirit of this passage, you won't even need to know what place Jesus or any other teacher occupies in your heart. If you will do the proper scavenger's work, clean and purify your hearts and get them ready, you will find that all these mighty teachers will take their places without invitation from us. That, to my mind, is the basis of all sound education. Culture of the mind must be subservient to the culture of the heart. May God help you to become pure!

Speech at Central College, Jaffna
The Hindu, 2 Dec. 1927

[1] The Principal of the College had asked what place Gandhi would give to Christ among the great world teachers, not as a divine instructor, but as a man and a teacher.

[2] Matthew 6:33

178. THE SERMON ON THE MOUNT

[15 November 1927]

There are some who will not even take my flat denial when I tell them that I am not a Christian.

The message of Jesus, as I understand it, is contained in his Sermon on the Mount unadulterated and taken as a whole, and even in connection with the Sermon on the Mount, my own humble interpretation of the message is in many respects different from the orthodox. The message, to my mind, has suffered distortion in the West. It may be presumptuous for me to say so, but as a devotee of truth, I should not hesitate to say what I feel. I know that the world is not waiting to know my opinion on Christianity.

One's own religion is after all a matter between oneself and one's Maker and no one else's, but if I feel impelled to share my thoughts with you this evening, it is because I want to enlist your sympathy in my search for truth and because so many Christian friends are interested in my thoughts on the teachings of Jesus. If then I had to face only the Sermon on the Mount and my own interpretation of it, I should not hesitate to say, 'Oh yes, I am a Christian.' But I know that at the present moment if I said any such thing I would lay myself open to the gravest misinterpretation. I should lay myself open to fraudulent claims because I would have then to tell you what my own meaning of Christianity is, and I have no desire myself to give you my own view of Christianity. But negatively I can tell you that in my humble opinion, much of what passes as Christianity is a negation of the Sermon on the Mount. And please mark my words. I am not at the present moment speaking of Christian conduct. I am speaking of the Christian belief, of Christianity as it is understood in the West.

I am painfully aware of the fact that conduct everywhere falls far short of belief. But I don't say this by way of criticism. I know from the treasures of my own experience that although I am every moment of my life trying to live up to my professions, my conduct falls short of these professions. Far therefore be it from me to say this in a spirit of criticism. But I am placing before you my fundamental difficulties. When I began as a prayerful student to study the Christian literature in

South Africa in 1893, I asked myself, 'Is this Christianity?' and have always got the Vedic answer, *neti neti* (not this, not this). And the deepest in me tells me that I am right.

I claim to be a man of faith and prayer, and even if I was cut to pieces, God would give me the strength not to deny Him and to assert that He is. The Muslim says He is and there is no one else. The Christian says the same thing and so the Hindu, and if I may say so, even the Buddhist says the same thing, if in different words. We may each of us be putting our own interpretation on the word God—God Who embraces not only this tiny globe of ours, but millions and billions of such globes. How can we, little crawling creatures, so utterly helpless as He has made us, how could we possibly measure His greatness, His boundless love, His infinite compassion, such that He allows man insolently to deny compassion, such that He allows man insolently to deny Him, wrangle about Him, and cut the throat of his fellow-man? How can we measure the greatness of God Who is so forgiving, so divine? Thus though we may utter the same words they have not the same meaning for us all. And hence I say that we do not need to proselytize or do *shuddhi* or *tabligh* through our speech or writing. We can only do it really with our lives. Let them be open books for all to study. Would that I could persuade the missionary friends to take this view of their mission. Then there will be no distrust, no suspicion, no jealousy and no dissensions.

Gandhiji then took the case of modern China as a case in point. His heart, he said, went out to young China in the throes of a great national upheaval, and he referred to the anti-Christian movement in China, about which he had occasion to read in a pamphlet received by him from the students' department of the Young Women's Christian Association and the Young Men's Christian Association of China. The writers had put their own interpretation upon the anti-Christian movement, but there was no doubt that young China regarded Christian movements as being opposed to Chinese self-expression. To Gandhiji the moral of this anti-Christian manifestation was clear. He said:

> Don't let your Christian propaganda be anti-national, say these young Chinese. And even their Christian friends have

come to distrust the Christian endeavour that had come from the West. I present the thought to you that these essays written by young men have a deep meaning, a deep truth, because they were themselves trying to justify their Christian conduct in so far as they had been able to live up to the life it had taught them and at the same time find a basis for that opposition. The deduction I would like you all to draw from this manifestation is that you Ceylonese should not be torn from your moorings, and those from the West should not consciously or unconsciously lay violent hands upon the manners, customs and habits of the Ceylonese in so far as they are not repugnant to fundamental ethics and morality. Confuse not Jesus' teachings with what passes as modern civilization, and pray do not do unconscious violence to the people among whom you cast your lot. It is no part of that call, I assure you, to tear the lives of the people of the East by its roots. Tolerate whatever is good in them and do not hastily, with your preconceived notions, judge them.

Do not judge lest you be judged yourselves. In spite of your belief in the greatness of Western civilization and in spite of your pride in all your achievements, I plead with you for humility, and ask you to leave some little room for doubt, in which, as Tennyson sang, there was more truth, though by 'doubt' he no doubt meant a different thing. Let us each one live our life, and if ours is the right life, where is the cause for hurry? It will react of itself.

To you, young Ceylonese friends, I say: Don't be dazzled by the splendour that comes to you from the West. Do not be thrown off your feet by this passing show. The Enlightened One has told you in never-to-be-forgotten words that this little span of life is but a passing shadow, a fleeting thing, and if you realize the nothingness of all that appears before your eyes, the nothingness of this material case that we see before us ever changing, then indeed there are treasures for you up above, and there is peace for you down here, peace which passeth all understanding, and happiness to which we are utter strangers. It requires an amazing faith, a divine faith and surrender of all that we see before us.

What did Buddha do, and Christ do, and also Mahomed? Theirs were lives of self-sacrifice and renunciation. Buddha renounced every worldly happiness, because he wanted to share with the whole world his happiness which was to be had by men who sacrificed and suffered in search of truth. If it was a good thing to scale the heights of Mt. Everest, sacrificing precious lives in order to be able to go there and make some slight observations, if it was a glorious thing to give up life after life in planting a flag in the uttermost extremities of the earth, how much more glorious would it be to give not one life, surrender not a million lives but a billion lives in search of the potent and imperishable truth? So be not lifted off your feet, do not be drawn away from the simplicity of your ancestors.

A time is coming when those who are in the mad rush today of multiplying their wants, vainly thinking that they add to the real substance, real knowledge of the world, will retrace their steps and say: 'What have we done?' Civilizations have come and gone, and in spite of all our vaunted progress I am tempted to ask again and again 'To what purpose?' Wallace, a contemporary of Darwin, has said the same thing. Fifty years of brilliant inventions and discoveries, he has said, has not added one inch to the moral height of mankind. So said a dreamer and visionary if you will—Tolstoy. So said Jesus, and Buddha, and Mahomed, whose religion is being denied and falsified in my own country today.

By all means drink deep of the fountains that are given to you in the Sermon on the Mount, but then you will have to take sackcloth and ashes. The teaching of the Sermon was meant for each and every one of us. You cannot serve both God and Mammon. God the Compassionate and the Merciful, Tolerance-incarnate, allows Mammon to have his nine days' wonder. But I say to you, youth of Ceylon, fly from that self-destroying but destructive show of Mammon.

Speech at YMCA, Colombo
Young India, 8 Dec. 1927

179. BUDDHA ON TRUE SACRIFICE

[25 November 1927]

There are some conditions laid down in Hinduism for a proper prayerful study of religions. They are of a universal character. Remember also that Gautama was a Hindu of Hindus. He was saturated with the spirit of Hinduism, with the Vedic spirit, he was born and bred up in those exhilarating surroundings, exhilarating for the spirit, and so far as I am aware, he never rejected Hinduism, or the message of the Vedas. What he did was therefore to introduce a living reformation in the petrified faith that surrounded him. I venture to suggest to you that *your study of Buddhism will be incomplete unless you study the original sources from which the Master derived his inspiration*, that is, unless you study Sanskrit and the Sanskrit scriptures. But your duty, if you are to understand the spirit of the Buddha and not the letter of Buddhism, does not end there. That study has those conditions which I am about to describe to you. Those conditions are that a man or a woman who approaches a study of religion has first of all to observe what are called the five *yamas*. They are the five rules of self-restraint and I will repeat them before you. First, *brahmacharya*, celibacy; the second is *satya*, truth; the third is *ahimsa*, absolute innocence, not even hurting a fly; the next condition is *asteya*, non-stealing, not merely not stealing in the ordinary sense in which the word is understood, but if you appropriate or even cast your greedy eyes on anything that is not your own, it becomes stealing. Lastly, *aparigraha*—a man, who wants to possess worldly riches or other things, won't be fit really to understand the spirit of the Buddha. These are the indispensable conditions. There are other conditions, but I am going into these because these are the fundamental ones, and Gautama before he attained his knowledge had conformed to all these rules, and conformed, as few of his contemporaries had ever done, to the spirit of those rules.

I humbly suggest to you that you will not understand the spirit of the Buddha unless you have also yourselves conformed to these rules and then prayerfully tried to ascertain what the Master meant. It makes no difference that you know of him through all the books that have been written, but even these very books, I make bold to assure you, you will understand and you will interpret with a new light, immediately you have gone, first of all, through these preliminary observances. Look what many critics of Islam have done—how they have torn the very book, that millions of Mussalmans swear by, to pieces and held up the teachings of Islam to scorn. They were not dishonest men who wrote this criticism, they were honest men, they were not men who were not trying to search the truth, but they did not know the conditions that they had to fulfil before they could make any religious study. Again look at what the critics of Hinduism have done. I read many of those criticisms, trying to enter into the spirit of the critics but came to the conclusion that they did not know the A B C of Hinduism and that they were grossly misinterpreting Hinduism.

Take Christianity itself. Many Hindus have misinterpreted Christianity. They approach the Bible, the Old Testament and the New Testament in a carping spirit, with preconceived notions. But why talk of the Hindus? Have I not read books written by Englishmen who, pretending to consider themselves atheists, have turned the Bible upside down, and put all the fiery writings into the hands of innocent men and women and thereby done grave injury to the simple people who read them? I have laid these points before the young men of this association because I am anxious that *you should be the pioneers of presenting Ceylon, and through Ceylon the world, with a real Buddhistic revival, that you should be the pioneers in presenting a living faith to the world, and not the dead bones of a traditional faith which the world will not grasp.*

The priests, whom I had seen by deputy, said they could not argue but that they could only say what the Master taught. It is all right, but today the spirit of enquiry is abroad. We have got to deal with that spirit. The world is trying to seek the truth, and thirsting for peace in the midst of this terrible strife. There is also the desire for knowing the truth, but as I have ventured to suggest to you, those, who made a scientific study of religion and those who gave their lives for arriving at the truth and those with whose bones the snows of the Himalayas are whitened, have left these treasures not merely for 300 millions of India, but they have left those treasures for everyone who cares to understand them, and they have said: 'We cannot deliver the truth to you.' It is incapable of being delivered through writings, it is incapable of being delivered with the lips, it is capable of being delivered only through life. It transcends reason. But it is not past experience. So they said: 'We tell you that such and such is the fact, but you will have to test it for yourselves. You will apply your reason, we do not want you to deaden your reason, but you yourselves, even as we, will come to the conclusion that reason which God has given is after all a limited thing, and that which is a limited thing will not be able to reach the limitless. Therefore, go through these preliminary conditions, even as when you want to study geometry or algebra, you have to go through preliminary processes, however trying and tiresome. Observe them and then you will find that what we tell you with our own experience will be also yours.'

I want to take you through only one illustration as to how the teaching of the Buddha is now not being observed. I have retained this part of my talk up to almost the very last moment except that I hinted at it in my speech at the Vidyodaya College.

You believe that Gautama taught the world to treat even the lowest creatures as equal to oneself. He held the life of even the crawling things of the earth as precious as his own. It is an arrogant assumption to say that human beings are lords and masters of the lower creation. On the contrary, being endowed with greater things in life, they are trustees of the lower animal kingdom. And the great sage lived that truth in his own life. I read as a mere youngster the passage in *The Light of Asia* describing how the Master took the lamb on his shoulders in face of the arrogant and ignorant *Brahmins* who thought that by offering the blood of these innocent lambs they were pleasing God and he dared them to sacrifice a single one of them. His very presence softened the stony hearts of the *Brahmins*. They looked up to the Master, they threw away their deadly knives and every one of those animals was saved. Was this message given to the world in order to falsify it, as it is being falsified here? I feel that you who are the repositories of this great faith are not true to the spirit of the Master's teachings so long as you do not regard all animal creation as sacred, and you cannot do so, so long as you do not abstain from meat and delude yourselves into the belief that you are not guilty of the crime of that slaughter because someone else killed the animals for you.

You entrench yourselves behind the wall of traditions. You say that the Master never prohibited meat-eating. I do not think so. If you would approach the teachings of the Master in the spirit indicated by me, and rub in the spirit of tradition, you will have a different vision and a different meaning. You will find that when the Master said, 'I do not prohibit you from meat-eating', he was preaching to a people who were, in Christian parlance, hard of hearts. It was because he wanted to make allowance for their weakness that he allowed them to eat it, and not because he did not know the logic of his own teaching. If animals could not be sacrificed to the gods above, how could they be sacrificed to the epicure in us? When he prohibited sacrifice he knew what he was saying. Did he not know that the animals were sacrificed to be ultimately eaten? Why do they sacrifice thousands of sheep and goats to the Goddess Kali in Calcutta—be it said to their discredit and the discredit of Hinduism—in spite of having received this message from the Hindu of Hindus—Gautama? Do they throw the carcases away in the Hooghly? No, they eat every bit of the meat with the greatest delight, thinking that it has been sanctified because of the presentation to Kali. So the Buddha said, if you want to do any sacrifice, sacrifice yourself, your lust, all your material ambition, all worldly ambition. That will be an ennobling sacrifice. May the spirit of the Buddha brood over this meeting and enable you to measure and assimilate the meaning of the words that I have spoken to you.

Speech at Young Men's Buddhist Association, Colombo
Young India, 8 Dec. 1927

180. WHICH JESUS?

The Ashram,
Sabarmati,
February 22, 1928

Dear Friend,

I have your letter.

Under British rule, millions of children are starving for want of nourishing food and they are shivering in winter for want of sufficient clothing. And this I say not of the cities of India, which contain but a microscopic minority of the population of India, but I say this without fear of contradiction about the seven hundred thousand villages of the country scattered over a surface 1,900 miles long and 1,500 miles broad.

I suppose your first question 'under non-Christian Religions' is included in the second. But, if your first question relates to India before British rule, I can only give you my inference that the little ones were infinitely happier than they are now under British rule.

Your third question is difficult to answer. Which Jesus have you in mind? The Jesus of history? Not being a critical student of history, I do not know the Jesus of history. Do you mean the Jesus whom Christian England and Christian Europe represent? If so, your question is, it seems to me, already answered. If you mean the mystical Jesus of Sermon on the Mount who has still to be found, I suppose the condition of India's children will be a trifle better than it is now when men conform to the precept of Love.

Yours sincerely,
M. K. GANDHI

Judge Henry Neil, Esq.
c/o American Express Co.
Rue Scribe
Paris—France

Letter to Henry Neil
SN 14248

181. AN ALL-EMBRACING RELIGION

Satyagraha Ashram, Sabarmati,
November 10, 1928

My dear Violet,

I have your letter. Hinduism should certainly be purged of all the evil and superstition that have crept into it. Nothing then need be a substitute for it. It is, in my opinion, an all-embracing and all-sufficing religion.

I hope your aunt is quite well now.

Yours sincerely,

[Mrs. Violet
c/o] Miss Baba Gunasekera
55 Hampden Lane, Wellawatte, Colombo

Letter to Violet
SN 12994

182. BUDDHA AND *AHIMSA*

[19 March 1929]

With this meeting ends what has been to me a most interesting and instructive tour in the interior of Burma. Delightful as my experiences of the people of Burma were when my visit some years ago was only confined to Rangoon and for a day only to Moulmein, that delight has been heightened by my experience during this tour even as far as Mandalay. It has been a great joy to me to see at all these meetings so many yellow-robed *phoongys* and so many Burman sisters and brothers. As this will be for many years to come, if not for ever, my last address to an audience such as this, I propose to submit a few remarks on what is to you and me dearest to our hearts. It cannot be without purpose and meaning that all your addresses, no matter where presented, have approved of and blessed my message of non-violence and of the spinning-wheel. I would therefore say a few words to you in explanation of what I understand by the message of *ahimsa*.

To me it is one of the most active forces in the world. It is like the sun that rises upon us unfailingly from day to day. Only if we would but understand it, it is infinitely greater than a million suns put together. It radiates life and light and peace and happiness. Why do we not see that light, that life, that peace and that happiness in a land that professes the law of *ahimsa*? As I said in Mandalay only yesterday, it has appeared to me that the message of the Buddha, the Enlightened One, has only touched but the surface of the heart of Burma. I would like to apply one or two tests. Now I hold that where the law of *ahimsa* reigns supreme, there should be no jealousy, no unworthy ambition, no crime. I read your criminal statistics and I find that you are not behindhand in the race for crime. Murder on the slightest pretext seems to me to be fairly common in Burma. I will therefore appeal to the friends on my left (the *phoongys*) who are supposed to be the repositories of the faith you have inherited from the Buddha. Having travelled in Ceylon and now fairly long enough in Burma, I feel that we in India have perhaps more fully, though by no means as fully as possible, interpreted the message of the Buddha than you have done.

We have it in our *shastras* that whenever things go wrong, good people and sages go in for *tapasya* otherwise known as austerities. Gautama himself, when he saw oppression, injustice and death around him, and when he saw darkness in front of him, at the back of him and on each side of him, went out in the wilderness and remained there fasting and praying in search of light. And if such penance was necessary for him who was infinitely greater than all of us put together, how much more necessary is it for us, no matter whether we are dressed in yellow or not? My friends, if you will become torch-bearers lighting the path of a weary world towards the goal of *ahimsa*, there is no other way out of it, save that of self-purification and penance. So many priests are sitting here today. If some of them will take upon themselves the work of interpreting the message of the Buddha, they will revolutionize life. You will not be guided by rigid traditions, but will search your hearts and your scriptures and tear the hidden meaning lying behind the written word and vivify your surroundings. You will then find upon searching your hearts that it is not enough not to take animal life, but you must see to it that it is not taken for the pleasures of the palate. You will then at once realize that it is inconsistent with the doctrine of love for all that lives to turn our mouths into chimneys. I understand that drink is on the increase in a people so simple-hearted as the Burmese and in a climate which does not necessitate the drinking of the fiery liquid. You will immediately see on further research that there is no room for one who loves everything that lives to have fear lurking in his breast. You will yourself cease to fear authority and you will teach all around you to cease to fear anybody.

I hope that these few words that I have spoken to you in all humility and from the bottom of my heart will be received by you in the same spirit in which they have been spoken. Since you have at all your meetings credited me with a spirit of non-violence and truth, I have endeavoured to interpret in the best way I could the message of non-violence and truth as I have understood it for an unbroken period of 40 years. May the words I have spoken find a lodgment in your hearts and may they bear abundant fruit, and if they do, there should be no difficulty in all factions and parties combining together for a common cause. I thank you for having listened to me with such patience and in perfect silence.

Speech at Public Meeting, Toungoo
Young India, 18 Apr. 1929

183. JESUS THE ANOINTED

[1 May 1929]

When I saw you, my heart actually leapt with joy and I really thought of Christ when I contemplated your consecration to the service of mankind. I have specially come to ask you what you think of Christ.

GANDHIJI: I consider Christ as one of the great teachers of the world. Beyond that I have not gone.

Is there any other kind of salvation apart from Christ's way?

These things are to be taken figuratively and not literally. Christ, no doubt, said: 'I am the way', etc., but he also said:

'The letter killeth'. Any teacher could say what Christ has said. After all, Christ is a common noun and Jesus Christ means Jesus the Anointed. Any teacher who has dedicated his life to the service of God and humanity and attained to complete purity can say that. The *Gita* said the same thing.

Have you any special plan of salvation?

My knowledge I derive from all the scriptures. What greater salvation does man need than the attainment of absolute purity in self-sacrifice and service?

Is not Jesus Christ the only sinless one?

What do we know of the whole life of Christ? Apart from the years of his life as given in the four gospels of the New Testament we know nothing of the rest of his life. As a man well versed in the Bible, you ought to have known that. Christ is, no doubt, one of those who attained sinlessness. These are, after all, not matters to be argued by the intellect but to be felt by the heart. All this is not for debate.

Interview with Abel[1]
The Hindu, 3 May 1929

[1] A Christian missionary from Vizagapatam.

184. JESUS AND NON-CO-OPERATION

An unknown English friend has thought it worth while cabling to me that in launching upon civil disobedience I am going against the teaching of Jesus: 'Render unto Caesar that which is Caesar's'. Another, an Indian Christian, writes from the Punjab somewhat in the same fashion and, forsaking charity, pours abuse upon my devoted head for my action. He says further that whereas he considered me to be a good man formerly, he is now utterly undeceived. I can reassure this friend that civil disobedience is no new thing with me. I began to preach and practise it in 1906. His regard for me therefore was evidently from ignorance, if his present dislike of me is wise. But I have learnt from the New Testament, as also from other sources, that if one wishes to walk in the fear of God, one should be indifferent about popular praise or blame.

Now for the question. As I hold my conduct to be in utter agreement with universal religion and as I hold the New Testament teaching in great esteem, I should not like it to be justly said of me that I was going against the teaching of Jesus. 'Render unto Caesar' was quoted against me before too. I have not read into the celebrated verse the meaning that my critics have sought to put into it. Jesus evaded the direct question put to him because it was a trap. He was in no way bound to answer it. He therefore asked to see the coin for taxes. And then said with withering scorn, 'How can you who traffic in Caesar's coins and thus receive what to you are benefits of Caesar's rule refuse to pay taxes?' Jesus' whole preaching and practice point unmistakably to non-co-operation, which necessarily includes non-payment of taxes. Jesus never recognized man's authority as against God's. He who disregarded the whole host of priesthood, which was in those days superior to kinghood, would not have hesitated to defy the might of emperors had he found it necessary. And did he not treat with supreme disdain the whole of the farcical trial through which he was made to pass?

Lastly, let me warn honest friends against running into the trap of literalism. The 'letter' surely 'killeth', it is the 'spirit' that 'giveth life'. In the present case I find no difficulty in reading into the text a satisfactory meaning. But it would matter little to me that some text should confound me, if there was no mistaking the spirit of the whole teaching of a book respected as among the world's religious scriptures.

'Render unto Caesar'
Young India, 27 Mar. 1930

185. THE MESSAGE OF *AHIMSA*

Sabarmati,
May 4, 1931

Dear Friend,

I thank you for your gift. I am sorry I cannot understand your language. My desire and hope is that Tibetans should understand and follow the secret of the message of *ahimsa* given by Lord Buddha.

Your friend,
M. K. GANDHI

His Highness the Lama
Tibet

Letter to the Thirteenth Dalai Lama (G.)[1]
CW 6208

[1] Tan Yun-Shan had brought a letter from the Thirteenth Dalai Lama of Tibet. It was written in Tibetan and no one could read it.

186. THE FEARLESS PROPHET

[Poona,
23 June 1934]

My friendship with the Mussalmans does not date from yesterday but from the time when I was a young man fifty years ago. My first visit to South Africa was in connection with the affairs of a Mussalman firm in that country. And there I had the privilege of coming into intimate touch with Muslim friends, for years together. In India too, you know, what close association there was between the Ali Brothers and myself. And although Maulana Shaukat Ali and I have appeared to have been estranged from each other, he knows that he has me always in his pocket.

Such being my relations with Mussalmans, I felt it to be my duty to study the life of the Prophet. I had tried to do this in South Africa but then I did not know sufficiently. In India imprisonment brought me good luck and thus I had the opportunity to read Maulana Shibli's biography of the Prophet which the late Hakim Saheb Ajmalkhan was good enough to send me at my request. At the same time I also read the volume on companions of the Prophet. I had read English works on Islam and the Prophet.

These studies led me to the conclusion that the Koran and the Bible were sacred books for me and not alone the Vedas and the *Gita*. Hazrat Mahomed was a great Prophet and so was Jesus Christ. From my reading I received the impression that the Prophet was a seeker of Truth. He was godfearing. In this I know I am not telling you anything new. I am only describing to you how I was impressed by his life. He suffered endless persecution. He was brave and feared no man but God alone. He did what he considered to be right in scorn of consequences. He was never found to say one thing and do another. He acted as he felt. If there was a change in his opinion, the next day he responded to the change without counting the cost and regardless of popular censure or opposition.

The Prophet was a *fakir*. He had renounced everything. He could have commanded wealth if he had so desired. Even as you would, I shed tears of joy when I read of the privations he, his family and companions suffered voluntarily. How can a Truth-seeker like me help respecting one whose mind was constantly fixed on God, who ever walked in God's fear and who had boundless compassion for mankind?

You all read the Koran. But how few put into practice what you read? You will perhaps retort that, if you do not live up to the precepts of the Koran, nor do the Hindus act according to the precepts of the *Gita*, and you will be right. It only comes to this that if both the communities followed the teachings of their respective faiths, communal quarrels would be a thing of the past. But at present some men in both communities appear to have taken leave of their senses, and are absorbed in slinging mud at one another. I will not have addressed this meeting in vain, if a single Mussalman will thereby come to a true understanding of the situation, and instead of picking holes learns to love other communities as well as his own.

Speech at Meeting to Commemorate the Prophet's Death,
Anjuman-i-Fide-e-Islam (H.)
The Hindu, 29 June 1934

187. HINDUISM EVER EVOLVING

Referring to my recent articles on the English peace movement led by Canon Sheppard, a friend writes:

I hold the view that independently of the *Gita* and the preliminary conversation between Arjuna and Shri Krishna, Hinduism does not stand decisively for non-violence in regard to organized invasion. It would be straining too much to interpret all our best scriptures in this way. Hinduism no doubt holds the spirit of compassion and love as the very highest duty for man. But it does not preach what you or the pacifists preach, and it is no good straining everything into an allegory for this object.

I have admitted in my introduction to the *Gita* known as *Anasaktiyoga* that it is not a treatise on non-violence nor was it written to condemn war. Hinduism as it is practised today, or has even been known to have ever been practised, has certainly not condemned war as I do. What, however, I have done is to put a new but natural and logical interpretation upon the whole teaching of the *Gita* and the spirit of Hinduism. Hinduism, not to speak of other religions, is ever evolving. It has no one scripture like the Koran or the Bible. Its scriptures are also evolving and suffering addition. The *Gita* itself is an instance in point. It has given a new meaning to *karma*, *sannyasa*, *yajna*, etc. It has breathed new life into Hinduism. It has given an original rule of conduct. Not that what the *Gita* has given was not implied in the previous writings, but the *Gita* put these implications in a concrete shape.

I have endeavoured in the light of a prayerful study of the other faiths of the world and, what is more, in the light of my own experiences in trying to live the teaching of Hinduism as interpreted in the *Gita*, to give an extended but in no way strained meaning to Hinduism, not as buried in its ample scriptures, but as a living faith speaking like a mother to her aching child. What I have done is perfectly historical. I have followed in the footsteps of our forefathers. At one time they sacrificed animals to propitiate angry gods. Their descendants, but our less remote ancestors, read a different meaning into the word 'sacrifice' and they taught that sacrifice was meant to be of our baser self, to please not angry gods but the one living God within. I hold that the logical outcome of the teaching of the *Gita* is decidedly for peace at the price of life itself. It is the highest aspiration of the human species.

The *Mahabharata* and the *Ramayana*, the two books that millions of Hindus know and regard as their guides, are undoubtedly allegories as the internal evidence shows. That they most probably deal with historical figures does not affect my proposition. Each epic describes the eternal duel that goes on between the forces of darkness and of light. Anyway, I must disclaim any intention of straining the meaning of Hinduism or the *Gita* to suit any preconceived notions of mine. My notions were an outcome of a study of the *Gita, Ramayana, Mahabharata,* Upanishads, etc.

'Teaching of Hinduism'
Harijan, 3 Oct. 1936

188. SALVATION FOR ALL

[16 December 1936]

The chief value of Hinduism lies in holding the actual belief that *all* life (not only human beings, but all sentient beings) is one, i.e., all life coming from the One universal source, call it Allah, God or Parameshwara. There is in Hinduism a scripture called *Vishnusahasranama* which simply means 'one thousand names of God'. These one thousand names do not mean that God is limited to those names, but that He has as many names as you can possibly give Him. You may give Him as many names as you like provided it is one God without a second, whose name you are invoking. That also means that He is nameless too.

This unity of *all* life is a peculiarity of Hinduism which confines salvation not to human beings alone but says that it is possible for all God's creatures. It may be that it is not possible, save through the human form, but that does not make man the Lord of creation. It makes him the servant of God's creation. Now when we talk of brotherhood of man, we stop there, and feel that all other life is there for man to exploit for his own purposes. But Hinduism excludes all exploitation. There is no limit whatsoever to the measure of sacrifice that one may make in order to realize this oneness with all life, but certainly the immensity of the ideal sets a limit to your wants. That, you will see, is the antithesis of the position of the modern civilization which says: 'Increase your wants.' Those who hold that belief think that increase of wants means an increase of knowledge whereby you understand the

Infinite better. On the contrary Hinduism rules out indulgence and multiplication of wants as these hamper one's growth to the ultimate identity with the Universal Self.

Interview with Miss Fitch
Harijan, 26 Dec. 1936

189. MYSTICS OF ISLAM

April 14, 1938

Sir Jogendra Singh is to be congratulated on having given us his rendering into English of the 'Sayings of the Mystic' by Abdullah Ansari.[1] Islam has given the world mystics no less than Hinduism or Christianity. In these days when irreligion masquerades as religion, it is well to remind ourselves of what the best minds of all the religions of the world have thought and said. We must not, like the frog in the well who imagines that the universe ends with the wall surrounding his well, think that our religion alone represents the whole Truth and all the others are false. A reverent study of the other religions of the world would show that they are equally true as our own, though all are necessarily imperfect.

M. K. GANDHI

Foreword
The Persian Mystics

[1] Sheikh Abdullah Ansari of Herat, A.D. 1005-90.

190. BEYOND THEOLOGY

[Sevagram,
12 March 1940]

Q. What started you on your career of leadership?

A. It came to me, unsought, unasked. I do not know, though, what sort of leader I am, and whether what I am doing is leadership or service. But whatever it is, it came to me unasked.

But the friends who came were sure that they were leaders, and they asked for guidance as leaders of Christian thought.

All I can say is that there should be less of theology and more of truth in all that you say and do.

Will you kindly explain it?

How can I explain the obvious? Amongst agents of the many untruths that are propounded in the world one of the foremost is theology. I do not say that there is no demand for it. There is a demand in the world for many a questionable thing. But even those who have to do with theology as part of their work have to survive their theology. I have two good Christian friends[1] who gave up theology and decided to live the gospel of Christ.

Are you sure that no great result has come through your own study of Jesus?

Why? There is no doubt that it has come, but not, let me tell you, through theology or through the ordinary interpretation of theologists. For many of them contend that the Sermon on the Mount does not apply to mundane things, and that it was only meant for the twelve disciples. Well, I do not believe this. I think the Sermon on the Mount has no meaning if it is not of vital use in everyday life to everyone.

Is there not to be found a solution to the present-day problems in the teaching of Jesus?

Well, you are now dragging me in deeper waters, and you will drown me.

What is the present trend of the thought of young India?

It would take a brave and knowing man to answer this question. But I must tell you that you have overstayed your time already. And if you go on questioning and cross-questioning me, I dare say you will floor me without being any the wiser for having done so.

Discussion with Christian Missionaries
Harijan, 23 Mar. 1940

[1] Samuel E. Stokes and C. F. Andrews.

191. THE EXAMPLE OF JESUS

Although I have devoted a large part of my life to the study of religion and to discussion with religious leaders of all faiths, I know very well that I cannot but seem presumptuous in writing about Jesus Christ and trying to explain what He means to me. I do so only because my Christian friends have told me on more than a few occasions that for the very reason that I am not a Christian and that (I shall quote their words exactly) 'I do not accept Christ in the bottom of my heart as the only Son of God', it is impossible for me to understand the profound significance of His teachings, or to know and interpret the greatest source of spiritual strength that man has ever known.

Although this may or may not be true in my case, I have reasons to believe that it is an erroneous point of view. I believe that such an estimate is incompatible with the message that Jesus Christ gave to the world. For He was, certainly, the highest example of one who wished to give everything asking nothing in return, and not caring what creed might happen to be professed by the recipient. I am sure that if He were living here now among men, He would bless the lives of many who perhaps have never even heard His name, if only their lives embodied the virtues of which He was a living example on earth; the virtues of loving one's neighbour as oneself and of doing good and charitable works among one's fellow-men.

What, then, does Jesus mean to me? To me He was one of the greatest teachers humanity has ever had. To His believers He was God's only begotten Son. Could the fact that I do or do not accept this belief make Jesus have any more or less influence in my life? Is all the grandeur of His teaching and of His doctrine to be forbidden to me? I cannot believe so.

To me it implies a spiritual birth. My interpretation, in other words, is that in Jesus' own life is the key of His nearness to God; that He expressed, as no other could, the spirit and will of God. It is in this sense that I see Him and recognize Him as the Son of God.

But I do believe that something of this spirit, that Jesus exemplified in the highest measure in its most profound human sense, does exist. I must believe this; if I do not believe it I should be a sceptic; and to be a sceptic is to live a life that is empty and lacks moral content. Or, what is the same thing, to condemn the entire human race to a negative end.

It is true that there certainly is reason for scepticism when one observes the bloody butchery that European aggressors have unloosed, and when one thinks about the misery and suffering prevalent in every corner of the world, as well as the pestilence and famine that always follow, terribly and inevitably, upon war.

In the face of this, how can one speak seriously of the divine spirit incarnate in man? Because these acts of terror and murder offend the conscience of man; because man knows that they represent evil; because in the inner depths of his heart and of his mind, he deplores them. And because, moreover, when he does not go astray, misled by false teachings or corrupted by false leaders, man has within his breast an impulse for good and a compassion that is the spark of divinity, and which some day, I believe, will burst forth into the full flower that is the hope of all mankind.

An example of this flowering may be found in the figure and in the life of Jesus. I refuse to believe that there now exists or has ever existed a person that has not made use of His example to lessen his sins, even though he may have done so without realizing it. The lives of all have, in some greater or lesser degree, been changed by His presence, His actions, and the words spoken by His divine voice.

I believe that it is impossible to estimate the merits of the various religions of the world, and moreover I believe that it is unnecessary and harmful even to attempt it. But each one of them, in my judgment, embodies a common motivating force: the desire to uplift man's life and give it purpose.

And because the life of Jesus has the significance and the transcendency to which I have alluded, I believe that He belongs not solely to Christianity, but to the entire world; to all races and people, it matters little under what flag, name or doctrine they may work, profess a faith, or worship a god inherited from their ancestors.

'What Jesus Means to Me'
Modern Review, October 1941

192. JESUS ON THE CROSS

Sevagram,
December 25, 1944

I had hoped that I would be able to speak a few words today. But God willed otherwise. Today is Christmas Day. We hold all religions in equal respect and all such festivals deserve respect. But our respect is different from the common kind. For us such festivals are for meditation and introspection. On such occasions we should search our hearts and cleanse it of all impurity. We should know that God is one, whether we call Him Ishwara or Khuda, and that His commands are the same for all. We should not harm others for what we regard as Truth or right. We should be prepared to die for Truth and when the call comes give our life for it and sanctify it with our blood. This in my view is the essence of all religions. On this day we should ponder upon this and remember that Jesus mounted the Cross for what he considered to be the Truth.

'A Christmas Message'
CW 4271

193. THE KINGDOM OF GOD

Sevagram, Wardha (India),
August 24, 1945

Dear Friend,

'The Kingdom of God is within you' is all-sufficient. Follow it out in action and you need nothing else. But if you will read anything of Hinduism, read Swami Vivekanand's works to be had there.

Yours sincerely,
M. K. GANDHI

Lawrence Mckenner Jr.
2132 High St.
Oakland 1, California

Letter to Lawrence McKenner
Pyarelal Papers

194. JESUS AN ASIATIC

[New Delhi,
26 June 1946]

Gandhiji asked about the rumours of war with Russia. I said there was a good deal of talk about war but perhaps it was only talk. 'You should turn your attention to the West', I added. He replied:

I? I have not convinced India. There is violence all around us. I am a spent bullet.

Since the end of the Second World War, I suggested, many Europeans and Americans were conscious of a spiritual emptiness. He might fill a corner of it.

But I am an Asiatic. A mere Asiatic.

He laughed, then after a pause:

Jesus was an Asiatic.

Interview with Louis Fischer
Louis Fischer, *The Life of Mahatma Gandhi*, p. 454

195. THE MEANING OF ISLAM

New Delhi,
September 7, 1946

The speeches being made by the Qaid-e-Azam and his followers cause me much pain. They say that they will take what they want by force. It is true that the reins of the Government are now in the hands of the Congress, but this only adds to their responsibilities and duties. During the struggle for freedom Congressmen were arrested, beaten and persecuted. They were even killed. These are now things of the past. Had the Congress taken to the path of violence it would have come to a bad end. In true suffering there is no room for revenge. Then alone can success be assured. The 40 *crores* of Indian people, that is to say the teeming millions in the villages, do not think of violence. They are slaves. Violence is lodged only in the hearts of a handful of men in the cities.

I am a villager. I belong with the villagers. The Congress has accepted power for the sake of these downtrodden villagers. I had taken up the cause of Hindu-Muslim unity long before I joined the Congress. I had a number of Muslim friends when I was at school. I went to South Africa to plead the case of some Muslim friends of my brother. I had gone there to gain my livelihood, but soon after my arrival there I gave the first place to service. As a coolie-barrister I served my friends of the labouring class. I had gone as an employee of a Muslim firm and I served the Hindus through them. My memory of those days is a happy one. It is a matter of deep regret that even in South Africa communal differences have arisen. Nevertheless they are unitedly fighting for the rights of Indians. I still remember those hefty Muslims, and especially Seth Cachalia, who participated in the *satyagraha* and who said they would rather die than live as slaves. When the Qaid-e-Azam and his followers describe Hindus as their enemies I am surprised and pained. I am not a Muslim but I venture to say that Islam does not preach enmity towards anyone. I think I am as much a Christian, a Sikh and a Jain as I am a Hindu. Religion does not teach one to kill one's brother however different his belief. No one can treat another as his enemy until the latter has become his own enemy. Muslim League leaders were not right when they said that they would compel the Congress, the Hindus and the British to accede to their demand.

I am reminded of an incident during the Khilafat[1] days. I was speaking at a meeting of Hindus. I said to them: 'If you want to protect the cow then protect Khilafat. If required even lay down your lives for it.' When I said this it brought tears of joy to the eyes of the Ali Brothers. But what a tragic change we see today. I wish the day may again come when Hindus and Muslims will do nothing without mutual consultation. I am day and night tormented by the question what I can do to hasten the coming of that day. I appeal to the League not to regard any Indian as its enemy. I appeal to the English not to nurse the thought that they can divide Hindus and Muslims. If they do they will be betraying India and betraying themselves. Hindus and Muslims are both born of the same soil. They have the same blood, eat the same food, drink the same water and speak the same language. The Qaid-e-Azam says that all the Muslims will be safe in Pakistan. In Punjab, Sind and Bengal we have Muslim League Governments. Can one say that what is happening in those provinces augurs well for the peace of the country? Does the Muslim League believe that it can sustain Islam by the sword? If it does it is committing a great error. The very meaning of the word 'Islam' is peace and I am certain that no religion worth the name can be kept alive except through peace.

Speech at Prayer Meeting (H.)
Hindustan, 8 Sept. 1946

[1] Khilafat, a Muslim movement protesting against the removal of Muslim sacred sites from the control of the Sultan of Turkey, *Caliph* or religious head for the majority of Indian Muslims.

§3. Religious Toleration and Equality

196. CHRISTIANITY AND OTHER RELIGIONS

The time has now passed when the followers of one religion can stand and say, ours is the only true religion and all others are false. The growing spirit of toleration towards all religions is a happy augury of the future. An article appeared recently in the columns of *The Christian World*, a London religious weekly, over the signature of 'J.B.', one of that journal's regular contributors, on this question, extracts from which I intend to quote.

The writer, in a most liberal and generous spirit, reviews the question from the Christian standpoint, and shows how the world's religions are linked one with the other, each having characteristics common to all others. The appearance of such an article in the Christian Press is worth noting, and shows that it is moving with the times. A few years back, such an article would have been classed as heretical teaching, and its author denounced as a traitor to the cause.

After remarking upon the new spirit which was changing the attitude of Christians to other religions, and pointing out how, a few years ago, the idea prevailed of the Christian religion standing out [as] the only true religion amongst a multitude of false ones, he goes on to say:

There has been an immense revulsion, and one of the features of it is the discovery, so vastly surprising to the average man, that the doctrine he was brought up on was not the earlier Christian teaching at all. The noblest of the old apologists thought very differently, he finds, of the outside races and faiths, from what he had been led to imagine. He hears of Justin Martyr, standing so close to the apostolic age, who regards the wisdom of Socrates as inspired by the 'Word'; of Origen, and Gregory of Nyssa, whose teaching is of the entire race of man as under the Divine tutorship; of Lactantius maintaining that belief in Providence was the common property of all religions.

The finer Christian minds have, in fact, in every age gone more or less along this line. It needed only that men should come into contact with these outside races, whether in their literature or face to face, to realise at once that the 'impassable gulf' theory between one religion and another was false to life and to the soul.

Religion, by a hundred different names and forms, has been dropping the one seed into the human heart, opening the one truth as the mind was able to receive it.

'J.B.' points out that many of the Christian institutions and doctrines were born of the knowledge of other religions. Many of the symbols are relics of ancient days.

How marvellous, too, in this connection, is that ancient cult of Mithras in Persia, where, as M. Cumont says: 'Like the Christians, the followers of Mithras lived in closely united societies, calling one another father and brother; like the Christians, they practised baptism, communion and confirmation; taught an authoritative morality, preached continence, chastity and self-denial, believed in the immortality of the soul and the resurrection of the dead.'

It is not surprising that the writer should claim for the Christian religion the premier position, but it is gratifying to find such a broadminded attitude taken up by Christian writers and the Christian Press.

To Europeans and Indians working together for the common good, this has a special significance. India, with its ancient religions, has much to give, and the bond of unity between us can best be fostered by a wholehearted sympathy and appreciation of each other's form of religion. A greater toleration on this important question would mean a wider charity in our everyday relations, and the existing misunderstandings would be swept away. Is it not also a fact that between Mahomedan and Hindu there is a great need for this toleration? Sometimes one is inclined to think it is even greater than between East and West. Let not strife and tumult destroy the harmony between Indians themselves. A house divided against itself must fall, so let me urge the necessity for perfect unity and brotherliness between all sections of the Indian community.

'The World's Religion'
Indian Opinion, 26 Aug. 1905

197. HINDU-MUSLIM UNITY

[8 April 1919]

In the huge mass meeting of Hindus and Mahomedans held in the Sonapur Masjid compound on Sunday, the 6th April, the day of humiliation and prayer, a vow of Hindu-Muslim unity was proposed to be taken as in the case of *swadeshi* proposed at the Chowpatty meeting and I had to utter a note of warning on both the occasions. At times in a fit of joyous passion we are spurred on to certain courses of action for which we have afterwards to repent. A vow is a purely religious act which cannot be taken in a fit of passion. It can be taken only with a mind purified and composed and with God as witness. Most of what I have said whilst writing about the *swadeshi* vow applies here. Acts which are not possible by ordinary self-denial become possible with the aid of vows which require extraordinary self-denial. It is hence believed that vows can only uplift us. If the Hindu and Muslim communities could be united in one bond of mutual friendship, and if each could act towards the other even as children of the same mother, it would be a consummation devoutly to be wished. But before this unity becomes a reality, both the communities will have to give up a good deal, and will have to make radical changes in ideas held heretofore. Members of one community when talking about those of the other at times indulge in terms so vulgar that they but acerbate the relations between the two.

In Hindu society we do not hesitate to indulge in unbecoming language when talking of the Mahomedans and *vice versa*. Many believe that an ingrained and ineradicable animosity exists between the Hindus and Mahomedans. In many places we see that each community harbours distrust against the other. Each fears the other. It is an undoubted fact that this anomalous and wretched state of things is improving day by day. The Time-Spirit is ceaselessly working on unchecked, and willy-nilly we have to live together. But the object of taking a vow is speedily to bring about, by the power of self-denial, a state of things which can only be expected to come in the fulness of time. How is this possible? Meetings should be called of Hindus—I mean the orthodox Hindus—where this question should be seriously considered. The standing complaint of the Hindus against the Mussulmans is that the latter are beef-eaters and that they purposely sacrifice cows on the *Bakr-i-Id* day. Now it is impossible to unite the Hindus and Mahomedans so long as the Hindus do not hesitate to kill their Mahomedan brethren in order to protect a cow. For I think it is futile to expect that our violence will ever compel the Mahomedans to refrain from cow-slaughter. I do not believe the efforts of our cow-protection societies have availed in the least to lessen the number of cows killed every day. I have had no reason to believe so. I believe myself to be an orthodox Hindu and it is my conviction that no one who scrupulously practises the Hindu religion may kill a cow-killer to protect a cow. There is one and only one means open to a Hindu to protect a cow and that is that he should offer himself as a sacrifice if he cannot stand its slaughter. Even if a very few enlightened Hindus thus sacrificed themselves, I have no doubt that our Mussulman brethren would abandon cow-slaughter. But this is *satyagraha*, this is equity; even as, if I want my brother to redress a grievance, I must do so by taking upon my head a certain amount of sacrifice and not by inflicting injury on him. I may not demand it as of right. My only right against my brother is that I can offer myself a sacrifice.

It is only when the Hindus are inspired with a feeling of pure love of this type that Hindu-Muslim unity can be expected. As with the Hindus, so with the Mussulmans. The leaders among the latter should meet together and consider their duty towards the Hindus. When both are inspired by a spirit of sacrifice, when both try to do their duty towards one another instead of pressing their rights, then and then only would the long-standing differences between the two communities cease. Each must respect the other's religion, must refrain from even secretly thinking ill of the other. We must politely dissuade members of both the communities from indulging in bad language against one another. Only a serious endeavour in this direction can remove the estrangement between us. Our vow would have value only when masses of Hindus and Mussulmans join in the endeavour. I think I have now made sufficiently clear the seriousness and magnitude of this vow. I hope that on this auspicious occasion and surely the occasion must be auspicious when a wave of *satyagraha* is sweeping over the whole country—we could all take this vow of unity. For this it is further necessary that leading Hindus and Mahomedans should meet together and seriously consider the question and then pass a unanimous resolution at a public meeting. This consummation will certainly be reached if our present efforts are vigorously continued. I think the vow may be taken individually even now and I expect that numerous people will do so every day. My warnings have reference to the taking of the vow publicly by masses of men. If it is taken by the masses, it should, in my humble opinion, be as follows:

'With God as witness we Hindus and Mahomedans declare that we shall behave towards one another as children of the same parents, that we shall have no differences, that the sorrows of each shall be the sorrows of the other and that each shall help the other in removing them. We shall respect each other's religion and religious feelings and shall not stand in the way of our respective religious practices. We shall always refrain from violence to each other in the name of religion.'

'The Vow of Hindu-Muslim Unity'
Young India, 7 May 1919

198. THE MEANING OF BROTHERHOOD

[Calcutta,
4 August 1925]

Mr. Chairman and Friends: You, Sir, have just said that probably this is for the first time I am privileged to address a meeting of Indian Christians only. If you refer to my present visit, you are perfectly correct. But if you refer or have referred to the whole of the time that I have been in India since my return from South Africa, then I have to inform you that I had such a privilege in 1915. But my connection with Indian Christian dates back to 1893. That was the time when I went to South Africa and found myself in the midst of a large Christian Indian community. I was agreeably surprised to find so many young men and young women who, whilst they were devoted Christians, were equally devoted to the motherland, and it gave me greater pleasure when I discovered that most of the young men and young women had never seen India. The majority of them were born in Natal; some of them in Mauritius, because it was from Mauritius that the first batch of free Indian settlers found their way to South Africa. They were most of them children of indentured parents. Indentured Indians were those who had gone to work on the sugar estates of Natal under an indissoluble contract to work on those estates for at least five years and, as they had gone under this contract, otherwise called indenture, they were called Indentured Indians. Their state was described during his lifetime by the late Sir William Hunter as a state very near to slavery.

I have mentioned this in order to show to you under what difficulties and disabilities these countrymen and countrywomen of ours laboured in South Africa and how they were able to overcome those difficulties, and, in the face of them, cut out for themselves honourable careers. Today, some of these men have even received a liberal education in England. Some of them are store-keepers, some of them occupying humbler walks of life. These brave lads offered their services to the Government at the time of the Boer War and the Zulu Rebellion. Some of them were brought up in my own home; two of them at least became barristers. So you understand what intimate relations I enjoyed with the Christian Indian community. I do not think there is in that land a single Indian Christian whom I do not know or who does not know me. It gives me, therefore, much pleasure to be able to come before you this evening to speak to you on 'Brotherhood of Man'.

It goes hard with people who have to suffer the disabilities that our countrymen, whom I have just now described to you, have to labour under, to understand that there can be any such thing as 'Brotherhood of Man'. If you are readers of newspapers and if you take any interest in what goes on outside the four corners of India, you may know that, today, in South Africa an attempt is being made by the Government of the country to drive away the Indians, or, as it has been well put by one of the newspapers here, English-owned, to starve them out of South Africa; and in this scheme of starvation are included some of these very men I have described to you. Whether ultimately this thing will come to pass, whether ultimately the Government of India will sanction or tolerate this thing, remains to be seen. But the connection in which I mention this thing to you is, as I have already told you, that it is difficult for such men to realize the meaning of brotherhood; and yet I have undertaken to speak to you on brotherhood at this time because it is in such times of stress and difficulty that one's spirit of brotherhood is really tested.

I receive compliments every often. They pass through my mind like water poured on to a duck's back. But you, Sir, have paid a compliment to me this evening which I feel inclined to accept. You think that if there is any person who has a right to speak on Brotherhood of Man, at least I should have that right, and I think so too. I have tried myself on many an occasion to find out whether it is possible for me to hate—I don't say love—my persecutor, and I must honestly but in all humility confess to you that I have not succeeded, I cannot recall a single occasion when I have felt constrained to hate a single human being. How I came to it I do not know. But I am simply giving to you a life-long practice and, therefore, it is really literally true that, if there is any person who has the right to speak on Brotherhood of Man, I at least have that right.

Brotherhood does not mean loving or sympathizing with those, extending the hand of fellowship to those who will in return love you. That is a bargain. Brotherhood is not a mercantile affair. And my philosophy, my religion teaches me that brotherhood is not confined merely to the human species; that is, if we really have imbibed the spirit of brotherhood, it extends to the lower animals. In one of the magazines issued in England by those great philanthropic societies 30 or 35 years ago, I remember having read some beautiful verses. I think the title of those verses was *My Brother Ox*. In them the writer beautifully described how on a man who loved his fellow men it was obligatory to love his fellow-animals also, taking the word animals to mean the sub-human species. The thought struck me most forcibly. At that time, I had learnt very little of Hinduism. All I knew about it was what I had imbibed from my surroundings, from my parents and others. But I realized the force of that writing. However, I do not intend to dwell upon this broadest brotherhood. I shall confine myself to 'Brotherhood of Man'. I have brought this thing in order to illustrate that our brotherhood is a mockery if we are not prepared to love even our enemies. In other words, one who has imbibed the spirit of brotherhood cannot possibly allow it to be said of him that he has any enemy at all. People may consider themselves to be our enemies, but we should reject any such claim. I have heard that claim made; that is the reason why I use the word 'claim'.

The question then arises: how is it possible to love those who consider themselves to be our enemies? Almost every week, I receive letters either from Hindus or from Mussalmans, sometimes from Christians, combating this fundamental position that I have taken up. If it is a Hindu who writes, then he asks me, 'How is it possible for me to love a Mussalman who kills the cow', which is dear to me as my life? Or if it is a Christian who writes to me, he asks, 'How is it possible to love Hindus who so ill-treat those whom they call untouchables, Hindus who have suppressed a fifth of their own numbers?' And if it is a Mussalman who writes, he asks, 'How is it possible to extend the hand of brotherhood or fellowship to Hindus who are worshippers of stock and stone?' I say to all these three: 'Your brotherhood is of no value to me if you cannot love the respective parties that you have described.' But what does the attitude signify after all? Does it not signify cowardly fear or intolerance? If all of us are God's creation, why should we fear one another or hate those who do not hold the same belief that we do? A Hindu will ask me, is he to sit or look on, while a Mussalman is doing something which is most repugnant to him? My brotherhood replies, 'Yes'. And

I add, 'You must sacrifice yourself, or in the language you have just listened to, you must bear the cross. If you want to defend one who is dear to you, you must die without killing.' I have personal experience of such occurrences.

If you have the courage to suffer lovingly, you melt the stoniest heart. You may raise your hand against one whom you regard as a ruffian, but how if he overpowers you? Will not the ruffian be more ferocious because of his victory over you? Does not history show that evil feeds on resistance? History also furnishes instances of men having tamed the fiercest men with their all-embracing love. But I admit that such non-resistance requires far greater courage than that of a soldier who returns two blows against one. I also admit that if a man has anger instead of love in him for the evil-doer, it is better for him to fight clean rather than, in a cowardly manner, to sit still for fear of dying. Cowardice and brotherhood are contradictory terms. I know that the world does not accept the fundamental position that I have endeavoured to place before you. I know that in Christian Europe, this doctrine of non-retaliation is pooh-poohed.

At the present moment, I am privileged to receive precious letters from friends all over Europe and America, some of them asking me to still further expound the doctrine of non-resistance. Some others are laughing at me and telling me: 'It is all right for you to talk these things in India, but you dare not do so in Europe.' Yet others tell me: 'Our Christianity is a whitewash, we do not understand the message of Jesus, it has got to be still delivered to us, so that we can understand it.' All these three positions are more or less right from the standpoint of the writers. But I venture to tell you that there is no peace for this world, and to take the name of brotherhood is a blasphemy, until we arrive at this fundamental position. Men there are who ask and so also women who ask: 'Is it human to refrain from retaliation?' I say it is human. Up to now we have not realized our humanity, we have not realized our dignity; we are supposed to be, if Darwin is to be believed, the descendants of monkeys, and I am afraid that we have not yet shed our original state.

The late Dr. Anna Kingsford in one of her books wrote once: 'As I walk about the streets of Paris, I seem to see before me diverse lions and snakes personified.' She says these animals have only the human form but no more. Man, to realize his full stature, has to become absolutely fearless. This he will do not by being armed from head to foot, but by generating force from within. A *Kshatriya* is one who does not fly from danger, he is not one who strikes a blow for a blow. The *Mahabharata* says also that forgiveness is the quality of a brave man. There is a statue erected, I am told, in the memory of the late General Gordon. The sculptor does not put a sword in his hands, he puts only a stick. It is considered to be a beautiful work of art. If I was born a sculptor and I had the order, I would not have put even a stick in the hands of General Gordon, but I would have pictured him as one with folded arms, with his chest put forward, in all humility telling the world: 'Come, all of you, who want to throw your darts, here is General Gordon to receive them without flinching, without retaliation.' That is my ideal of a soldier. Such soldiers have lived on the earth.

Christianity undoubtedly has given birth to such soldiers, and so has Hinduism, so has Islam. In my opinion, it is not true to say that Islam is a religion of the sword. History does not bear that out. But I am just now speaking to you of individual instances, and what is true of the individual can be true of nations or of groups of individuals; not all at once, I admit, but in the process of evolution, when men after men live this truth in their lives before our very eyes, they cannot but affect us. Such is the history of Quakers. Such is the history of Dukhobors whom Tolstoy has described. I do not know how far the latter, after having gone to Canada, are carrying out their original resolution, but the fact stands that they have lived this life of non-resistance as a community. I, therefore, feel that we are trifling with that sacred name, Brotherhood of Man, unless and until we are ruled by this fundamental fact in life.

What I am just now combating is the position that is taken up by some of the finest writers in Europe and by some of the finest writers even in India: that man, as a class, will never be able to arrive at a stage when he can do without retaliation. I have a fundamental quarrel with that position. On the contrary, I say that man, as man, will not realize his full destiny, and his full dignity, until he has been so far educated as to be able to refrain from retaliation. Whether we like it or whether we do not like it, we are being driven to it. It would be to our credit if, instead of being driven to the position, we will take ourselves to it, and I have come here this evening to ask you to exercise this privilege, the privilege of voluntarily taking up this idea in practice. Indeed, I ought not to have to be speaking to a Christian audience on this, because some of my friends tell me that I am really a Christian, when I talk about non-retaliation. Little do they know that I have got to strive with the Christians, as I have to with Hindus and my Muslim friends.

I do not know many Christians who have adopted this thing as a rule of their life. Some of the very best Christians that I know do not admit that this is the teaching of Christ. I do believe that it is the teaching of Christ. They say it was meant merely for his twelve disciples, not meant for the world, and they quote some passages from the New Testament in support of their contention. The opponents of non-violence as a rule of life say that it can only breed a race of cowards, and if India takes up this message of non-retaliation, she is a doomed country. On the contrary, the fundamental position that I place before you is, that unless India takes up this position, she is a doomed nation and with her all the nations of the world. India is a continent, and when India takes up the doctrine of force, as Europe today seems to have taken it up, then India becomes one of the exploiters of the weaker races of the world. Just imagine what it must mean to the world.

I call myself a nationalist and I pride myself in it. My nationalism is as broad as the universe. It includes in its sweep even the lower animals. It includes in its sweep all the nations of the earth, and if I possibly could convince the whole of India of the truth of this message, then, India would be something to the whole world for which the world is longing. My nationalism includes the well-being of the whole world. I do not want my India to rise on the ashes of other nations. I do not want India to exploit a single human being. I want India to become strong in order that she can infect the other nations also with her strength. Not so with the other nations of the world, not so with a single nation in Europe today. They do not give strength to the others. We are not receiving any strength. It is in the nature of things impossible for them to do so, and that is why I have taken the uncompromising position that I cannot possibly be a party to a constitution whose basis is brute force.

President Wilson mentioned his beautiful 14 points, and do you know what he wound up with? He said: 'After all, if this endeavour of ours to arrive at peace fails, we have got our armaments to fall back upon.' I want to reverse that position, and I say: 'Our armaments have failed already. Let us now be in search of something new, and let us try the force of love and God which is Truth.' When we have got that, we shall want nothing else. There is the story of the devotee, Prahlad. It may be a fable, but no fable for me. He was a lad of hardly 12 years. His father asked him not to take the name of God. Prahlad said: 'I can't do without it, it is my life.' Then his father asked him: 'Show me your God.' A red hot iron pillar was shown to Prahlad and he was asked to embrace it. Yes, there was God in that pillar. Prahlad embraced it in love and faith. He was unhurt. If we would realize brotherhood, we must have the love and the faith and the truth of Prahlad in us.

Speech at Meeting of Christians
Amrita Bazar Patrika, 15 Aug. 1925

199. TOLERANCE AND FAITH

Ashram, Sabarmati,
April 1, 1926

Dear Friend,

I have your letter. I do not think that it is necessary to forget one's own faith in order to be tolerant towards the rest. In fact, tolerance loses its value when one's own faith is forgotten. In my opinion, tolerance requires the same respect to be paid to another's faith as we claim for our own.

I am of opinion that God can be reached without a mediator.

Yours sincerely,

Sjt. S. V. Venkatanarasayyan
7, Miller Road
Kilpauk, Madras

Letter to S.V. Venkatanarasayyan
SN 19407

200. LOVE AND MUTUALITY

Ashram, Sabarmati,
April 6, 1926

Dear Friend,

I have your letter. I am afraid it is not possible for me to subscribe to the creed you have sent me. The subscriber is made to believe that the highest manifestation of the unseen reality was Jesus Christ. In spite of all my efforts, I have not been able to feel the truth of that statement. I have not been able to move beyond the belief that Jesus was one of the great teachers of mankind. Do you not think that religious unity is to be had not by a mechanical subscription and a common creed but by all respecting the creed of each other? In my opinion difference in creed there must be so long as there are different brains. But what does it matter if all these are... upon the common path of love and mutual judgment.

I return the stamp kindly sent by you. It cannot be used in India.

Yours sincerely,

Milton Newberry Frantz, Esq.
Collegeville

Letter to Milton Newberry Frantz
SN 12461

201. FELLOWSHIP AND TOLERATION

[Before 15 January 1928]

In order to attain a perfect fellowship,[1] every act of its members must be a religious act and an act of sacrifice. I came to the conclusion long ago, after prayerful search and study and discussion with as many people as I could meet, that all religions were true and also that all had some error in them, and that whilst I hold by my own, I should hold others as dear as Hinduism, from which it logically follows that we should hold all as dear as our nearest kith and kin and that we should make no distinction between them. So we can only pray, if we are Hindus, not that a Christian should become a Hindu, or if we are Mussalmans, not that a Hindu or a Christian should become a Mussalman, nor should we even secretly pray that anyone should be converted, but our inmost prayer should be that a Hindu should be a better Hindu, a Muslim a better Muslim and a Christian a better Christian. That is the fundamental truth of fellowship. That is the meaning of the wonderful passion, the story of which Andrews read out to you, of the song and verses that Khare, Shastri and Imam Saheb recited. If Andrews invited them to give their song and verses for mere courtesy or by way of patronizing toleration, he was false to the fellowship. In that case, he should not have done so, but I have known Charlie Andrews too well, and I know that he has given the same love to others as he has for his own, and thereby broadened his Christianity, as I broaden my Hinduism by loving other religions as my own.

If however there is any suspicion in your minds that only one religion can be true and others false, you must reject the doctrine of fellowship placed before you. Then we would have a continuous process of exclusion and found our fellowship on an exclusive basis. Above all I plead for utter truthfulness. If we do not feel for other religions as we feel for our own, we had better disband ourselves, for we do not want a wishy-washy toleration. My doctrine of toleration does not include toleration of evil, though it does the toleration of the evil-minded. It does not therefore mean that you have to invite each and every one who is evil-minded or tolerate a false faith. By a true faith I mean one the sum total of whose energy is for the good of its adherents, by a false I mean that which is predominantly false. If you, therefore, feel that the sum total of Hinduism has been bad for the Hindus and the world, you must reject it as a false faith.

Gandhiji's insistence on a member of the fellowship not even secretly wishing that a member of another faith should be converted to his own led to a general discussion on the question of conversion. Gandhiji again defined his position more clearly than before:

I would not only not try to convert but would not even secretly pray that anyone should embrace my faith. My prayer would always be that Imam Saheb should be a better Mussalman, or become the best he can. Hinduism with its message of *ahimsa* is to me the most glorious religion in the world—as my wife to me is the most beautiful woman in the world—but others may feel the same about their own religion. Cases of real honest conversion are quite possible. If some people for their inward satisfaction and growth change their religion, let them do so. As regards taking our message to the aborigines, I do not think I should go and give my message out of my own wisdom. Do it in all humility, it is said. Well, I have been an unfortunate witness of arrogance often going in the garb of humility. If I am perfect, I know that my thought will reach others. It taxes all my time to reach the goal I have set to myself. What have I to take to the aborigines and the Assamese hillmen except to go in my nakedness to them? Rather than ask them to join my prayer, I would join their prayer. We were strangers to this sort of classification—'animists', 'aborigines', etc.,—but we have learnt it from English rulers. I must have the desire to serve and it must put me right with people. Conversion and service go ill together.

The next day early morning the friends met for an informal conversation with Gandhiji when again the same question was asked by many of them.

'Would you have a ruling of such a character that those who had a desire to convert should not be eligible for membership?'

Personally, I think they should not be eligible. I should have framed a resolution to that effect as I regard it as the logical outcome of fellowship. It is essential for inter-religious relationship and contact.

'Is not the impulse to proselytize God-given?', inquired another friend.

I question it. But if all impulses are God-given, as some of our Hindus believe, He has also given us discrimination. He will say, 'I have given you many impulses so that your capacity to face temptation may be tested.'

'But you do believe in preaching an economic order?', inquired one of the fair sex.

I do, as I believe in preaching laws of health.

Then why not apply the same rule in religious matters?

It is a relevant question. But you must not forget that we have started with the fundamental principle that all religions are true. If there were different but good and true health laws for different communities, I should hesitate to preach some as true and some as false. I am positive that, with people not prepared to tolerate one another's religious belief, there can be no international fellowship.

Moreover, physical analogies when applied to spiritual matters are good only up to a certain point. When you take up an analogy from Nature, you can stretch it only to a certain point. But I would take an illustration from the physical world and explain what I mean. If I want to hand a rose to you, there is definite movement. But if I want to transmit its scent, I do so without any movement. The rose transmits its own scent without a movement. Let us rise a step higher, and we can understand that spiritual experiences are self-acting. Therefore, the analogy of preaching sanitation, etc., does not hold good. If we have spiritual truth, it will transmit itself. You talk of the joy of a spiritual experience and say you cannot but share it. Well, if it is real joy, boundless joy, it will spread itself without the vehicle of speech. In spiritual matters we have merely to step out of the way. Let God work His way. If we interfere, we may do harm. Good is a self-acting force. Evil is not, because it is a negative force. It requires the cloak of virtue before it can march forward.

Did not Jesus Himself teach and preach?

We are on dangerous ground here. You ask me to give my interpretation of the life of Christ. Well, I may say that I do not accept everything in the gospels as historical truth. And it must be remembered that he was working amongst his own people, and said he had not come to destroy but to fulfil. I draw a great distinction between the Sermon on the Mount and the Letters of Paul. They are a graft of Christ's teaching, his own gloss apart from Christ's own experience.

'Discussion on Fellowship'
Young India, 19 Jan. 1928

[1] Members of the Council of International Federation and their friends stayed in the Ashram and held discussions on 'the fundamental objective of the fellowship'.

202. CONVERSION AND HUMILITY

I gladly publish the foregoing ...[1] It was made clear at those meetings of International Fellowship that I had meant the principal religions of the world and I had maintained that all were true more or less and that all were necessarily imperfect. Here therefore there is agreement. But Mr. Ireland's letter leaves on the mind the impression that there is a fundamental difference between him and me regarding conversion, no matter by what name it is called. Let me extend the analogy of fragrance, faulty as all analogies are in their very nature. The rose imparts its fragrance not in many ways but only one. Those who have not the sense of smell will miss it. You cannot feel the fragrance through the tongue or the ear or the skin. So may you not receive spirituality except through the spiritual sense. Hence have all religions recognized the necessity of that sense being awakened. It is a second birth.

A man with intense spirituality may without speech or a gesture touch the hearts of millions who have never seen him and whom he has never seen. The most eloquent preacher if he has not spirituality in him will fail to touch the hearts of his audience. Therefore I venture to think that most of the effort of modern missions is not only useless but more often than not harmful. At the root of missionary effort is also the assumption that one's own belief is true not only for oneself but for all the world; whereas the truth is that God reaches us through millions of ways not understood by us. In missionary effort therefore there is lack of real humility that instinctively recognizes human limitations and the limitless power of God. I have no feeling that from a spiritual standpoint I am necessarily superior to the so-called savage. And spiritual superiority is a dangerous thing to feel. It is not like many other things which we can perceive, analyse and prove through our senses. If it is there, I cannot be deprived of it by any power on earth, and it will have its effect in its own due time. But if in matters of medicine and other natural sciences, I feel my superiority over others, a thing of which I may be legitimately conscious, and if I have love for my fellow beings, I would naturally share my knowledge with them. But things of the spirit I leave to God and thus keep the bond between fellow beings and myself pure, correct and within limits. But I must not carry this argument any further.

My first feeling was not to publish Mr. Irelands's letter but to send a brief reply to him privately. But my regard for him has prompted me to comply with his wish without any ado knowing full well that this is not a matter which admits of any conclusive argument especially from my side and in view of the position herein described by me.

'Difference Stated'
Young India, 22 Mar. 1928

[1] A letter from W. F. Ireland of the Cambridge Mission, not reproduced here.

203. CHANGE OF RELIGION

Satyagraha Ashram, Sabarmati,
September 1, 1928

My dear Boyd,

Your letter has been on my file for some days.

I think you have stated my position fairly correctly except that the way in which you have put it may cause a misunderstanding. I did not say that I would not desire that others should accept my viewpoint. But I did say that I would not desire that others should accept my religion. Evidently you have used the word viewpoint as synonymous with religion. I do not. Whilst I would not press my religion upon others, I would press my viewpoint upon others, as every one of us must. Religion is a matter of feeling or the heart and, therefore, not a matter for argument, and I would hold everybody's feeling as dear as my own, because I expect him to do so with reference to my feeling. Viewpoint is a matter of reasoning, the mind, the intellect. It may shift from time to time without touching the heart. Change of religion is a change of status. Change of viewpoint is an accident often due to external causes.

My feeling about the existence of God cannot be easily altered. My viewpoint regarding the connotation of the term may vary from time to time and expand with the expansion of my reason. Religion is beyond explanation and it seems to me to be impertinent for anyone to touch another's religion. A viewpoint must always be capable of explanation. I have entered upon this distinction because it enables me to explain my position about religion more clearly than by any other means. I do not want you to become a Hindu. But I do want you to become a better Christian by assimilating all that may be good in Hinduism and that you may not find in same measure or at all in the Christian teaching. I can't explain why I delight in calling myself and remaining a Hindu, but my remaining does not prevent me from assimilating all that is good and noble in Christianity, Islam and other faiths of the world.

I wonder if I have explained my position to your satisfaction. If not please ask.

All you say about Bardoli is quite true.

Yours sincerely,

Rev. B.W. Tucker
Principal, Collins High School
140 Dharmtala Street, Calcutta

Letter to B. W. Tucker
SN 13505

204. EQUALITY OF RELIGIONS

Tuesday morning, September 23, 1930

EQUALITY OF RELIGIONS. This is the new name we have given to the Ashram observance which we know as 'Tolerance'. *Sahishnuta* is a translation of the English word 'Tolerance'. I did not like that word, but could not think of a better one. Kakasaheb, too, did not like that word. He suggested 'Respect for all religions'. I didn't like that phrase either. Tolerance may imply a gratuitous assumption of the inferiority of other faiths to one's own and respect suggests a sense of patronizing whereas *ahimsa* teaches us to entertain the same respect for the religious faiths of others as we accord to our own, thus admitting the imperfection of the latter. This admission will be readily made by a seeker of Truth, who follows the law of Love. If we had attained the full vision of Truth, we would no longer be mere seekers, but would have become one with God, for Truth is God. But being only seekers, we prosecute our quest, and are conscious of our imperfection. And if we are imperfect ourselves, religion as conceived by us must also be imperfect.

We have not realized religion in its perfection, even as we have not realized God. Religion of our conception, being thus imperfect, is always subject to a process of evolution and re-interpretation. Progress towards Truth, towards God, is possible only because of such evolution. And if all faiths outlined by men are imperfect, the question of comparative merit does not arise. All faiths constitute a revelation of Truth, but all are imperfect and liable to error. Reverence for other faiths need not blind us to their faults. We must be keenly alive to the defects of our own faith also, yet not leave it on that account, but try to overcome those defects. Looking at all religions with an equal eye, we would not only not hesitate, but would think it our duty, to blend into our faith every acceptable feature of other faiths.

The question then arises: why should there be so many different faiths? The soul is one, but the bodies which she animates are many. We cannot reduce the number of bodies; yet we recognize the unity of the soul. Even as a tree has a single trunk, but many branches and leaves, so is there one true and perfect Religion, but it becomes many as it passes through the human medium.

All religions are divinely inspired, but they are imperfect because they are products of the human mind and taught by human beings. The one Religion is beyond all speech. Imperfect men put it into such language as they can command and their words are interpreted by other men equally imperfect. Whose interpretation is to be held to be the right one? Everybody is right from his own standpoint, but it is not impossible that everybody may be wrong. Hence the necessity for tolerance, which does not mean indifference towards one's own faith, but a more intelligent and purer love for it. Tolerance gives us spiritual insight, which is as far from fanaticism as the north pole is from the south. True knowledge of religion breaks down the barriers between faith and faith. Cultivation of tolerance for other faiths will impart to us a truer understanding of our own.

Tolerance obviously does not disturb the distinction between right and wrong, or good and evil. The reference here throughout is naturally to the principal faiths of the world. They are all based on common fundamentals. They have all produced great saints. There is some difference between tolerance towards other religions and tolerance towards their followers. We should have equal regard for all human beings—for the wicked as for the saintly, for the impious as for the pious—but we should never tolerate irreligion.

This idea may need elaboration. If you do not follow it easily, ask me.

Blessings from
BAPU

Tuesday morning, September 30, 1930

As I told you in my letter last week, I have translated into English the discourse on Equality of Religions and send the translation herewith. If Valjibhai has translated it and if his translation has already been printed, he should read the translation which I am sending. Anybody else who wishes to read it may do so and the last person should hand it over to Mirabehn. If Valjibhai's translation has not been printed, he should carefully go through my translation and then publish whichever he likes. Do you intend to publish the discourses in Gujarati only or their English translations also?

This subject is so important that I dwell a little further on it.

My meaning will perhaps become clearer if I describe here some of my experiences. In Phoenix we had our daily prayers in the same way as in Sabarmati, and Mussalmans as well as Christians attended them along with Hindus. The late Sheth Rustomji and his children too attended the prayer meetings. Rustomji Sheth very much liked the Gujarati *bhajan*. 'Dear, dear to me is the name of Rama.' If my memory serves me right, Maganlal or Kashi was once leading us in singing this hymn, when Rustomji Sheth exclaimed joyously 'Say the name of Hormazd instead of the name of Rama'. His suggestion was readily taken up, and after that whenever the Sheth was present, and sometimes even when he was not, we put in the name of Hormazd in place of Rama. The late Husain, son of Daud Sheth, often stayed at the Phoenix Ashram, and enthusiastically joined our prayers. To the accompaniment of an organ, he used to sing in a very sweet voice the song *Hai bahare bagh*, 'The garden of this world has only a momentary bloom'. He taught us all this song, which we also sang at prayers. Its inclusion in our *bhajanavali* is a tribute to truth-loving Husain's memory. I have never met a young man who practised Truth more devotedly than Husain. Joseph Royeppen often came to Phoenix. He was a Christian, and his favourite hymn was *Vaishnava jana*. He loved music and once sang this hymn saying 'Christian' in place of *Vaishnava*. The others accepted his reading with alacrity, and I observed that this filled Joseph's heart with joy.

When I was turning over the pages of the sacred books of different faiths for my own satisfaction, I became sufficiently familiar for my purpose with Christianity, Islam, Zoroastrianism, Judaism and Hinduism. In reading these texts, I can say that I felt the same regard for all these faiths although, perhaps, I was not then conscious of it. Reviving my memory of those days, I do not find I ever had the slightest desire to criticize any of those religions merely because they were not my own, but read each sacred book in a spirit of reverence and found the same fundamental morality in each. Some things I did not understand then, as I did not in Hindu scriptures. I do not understand those things even now, but experience has taught me that it is a mistake hastily to imagine that anything that we cannot understand is necessarily wrong. Some things which I did not understand first have since become as clear as daylight. Equimindedness helps us to solve many difficulties and even when we criticize anything, we express ourselves with a humility and courtesy which leave no sting behind them.

One difficulty still remains. As I stated last time, the acceptance of the doctrine of Equality of Religions does not abolish the distinction between religion and irreligion. We do not propose to cultivate tolerance for irreligion. That being so, some people might object that there would be no room left for equimindedness, if everyone took his own decision as to what was religion and what was irreligion. Such a question may be raised and one may even make a mistake in deciding what is religion and what is irreligion. If, however, we follow the law of love, we shall not bear any hatred towards the irreligious brother. On the contrary, though we see that he follows irreligion, we shall love him and, therefore, either we shall bring him to see the error of his ways or he will convince us of our error, or each will tolerate the other's difference of opinion. If the other party does not observe the law of love, he may be violent to us. If, however, we cherish real love for him, it will overcome his bitterness in the end. All obstacles in our path will vanish, if only we observe the golden rule that we must not be impatient with those whom we may consider to be in error, but must be prepared, if need be, to suffer in our own person.

As I have stated earlier in this letter, I think I will omit the subject of *swadeshi*. I have yet to think on what subject I shall write next.

Blessings from
BAPU

Letters to Narandas Gandhi (G.)
MMU/I (Microfilm)

205. THE ROOT OF ALL RELIGIONS

November 1, 1945

Panditji,

I have had your book read out to me from beginning to end. These days while I spin I have a friend read out to me some book which I think is worth reading. I am disappointed with your book. From what you had said I had expected that I would learn something new from it and at the same time some light would be thrown on the Hindu-Muslim question. Right from my childhood I have lived with Muslims and when I went to London Providence placed me in close association with Christians, Muslims and Parsis. Hindus of course were there. I came into contact with the intellectuals among them and that is how I read the holy books of all the four religions. I came to the conclusion that we could know the true facts about any religion only after reading its sacred books. I also read quite a few criticisms of them. I can say that I have also read the Muslim critics you mention in your letter. I had a talk with Muslims in South Africa about those books and also with an Englishman who had become Muslim. After coming here I read the works of Maulana Shibli. As a result I have realized that every religion contains both truth and untruth. The root of all religions is one and it is pure and all of them have sprung from the same source, hence all are equal. This equality of all religions has been included by Vinoba among the eleven vows in a Marathi *sloka*, which is daily recited in the prayers.

Non-violence, truth, non-stealing, *brahmacharya*, non-possession, body-labour, control of the palate, fearlessness on all occasions, equal respect for all religions, *swadeshi* and *sparshabhavana*[1]—these eleven vows should be observed in a spirit of humility.

I am trying to live in accordance with this and so are my associates.

As it is, the letter has become quite long but since you have taken so much trouble and have also sent to me extra copies of your book, I felt I should at least let you know my stand.

I do not wish to argue with you. I have an answer to what

you have suggested but I see no need to give it here. I therefore think that what I have said is enough.

Yours,
M. K. GANDHI

Letter to Mahadevshastri Divekar (H.)
Pyarelal Papers

[1] Refusal to treat anybody as untouchable.

§4. Prayer and Devotion

206. YEARNING OF THE HEART

A medical graduate asks:

What is the best form of prayer? How much time should be spent at it? In my opinion to do justice is the best form of prayer and one who is sincere about doing justice to all does not need to do any more praying. Some people spend a long time over *sandhya* and 95% of them do not understand the meaning of what they say. In my opinion prayer should be said in one's mother tongue. It alone can affect the soul best. I should say that a sincere prayer for one minute is enough. It should suffice to promise God not to sin.

Prayer means asking God for something in a reverent attitude. But the word is used also to denote any devotional act. Worship is a better term to use for what the correspondent has in mind. But definition apart, what is it that millions of Hindus, Mussalmans, Christians and Jews and others do every day during the time set apart for the adoration of the Maker? It seems to me that it is a yearning of the heart to be one with the Maker, an invocation for His blessing. It is in this case the attitude that matters, not words uttered or muttered. And often the association of words that have been handed down from ancient times has an effect which in their rendering into one's mother tongue they will lose altogether. Thus the *Gayatri* translated and recited in, say, Gujarati, will not have the same effect as the original. The utterance of the word 'Rama' will instantaneously affect millions of Hindus, when the word 'God', although they may understand the meaning, will leave them untouched. Words after all acquire a power by long usage and sacredness associated with their use. There is much therefore to be said for the retention of the old Sanskrit formulae for the most prevalent *mantras* or verses. That the meaning of them should be properly understood goes without saying.

There can be no fixed rule laid down as to the time these devotional acts should take. It depends upon individual temperament. These are precious moments in one's daily life. The exercises are intended to sober and humble us and enable us to realize that nothing happens without His will and that we are but 'clay in the hands of the Potter'. These are moments when one reviews one's immediate past, confesses one's weakness, asks for forgiveness and strength to be and do better. One minute may be enough for some, twenty-four hours may be too little for others. For those who are filled with the presence of God in them, to labour is to pray. Their life is one continuous prayer or act of worship. For those others who act only to sin, to indulge themselves, and live for self, no time is too much. If they had patience and faith and the will to be pure, they would pray till they feel the definite purifying presence of God within them. For us ordinary mortals there must be a middle path between these two extremes. We are not so exalted as to be able to say that all our acts are a dedication, nor perhaps are we so far gone as to be living purely for self. Hence have all religions set apart times for general devotion. Unfortunately these have nowadays become merely mechanical and formal, where they are not hypocritical. What is necessary therefore is the correct attitude to accompany these devotions.

For definite personal prayer in the sense of asking God for something, it should certainly be in one's own tongue. Nothing can be grander than to ask God to make us act justly towards everything that lives.

'What is Prayer?'
Young India, 10 June 1926

207. CLEANSING THROUGH PRAYER

The Ashram,
Sabarmati,
May 11, 1928

I must try to answer your questions today.

What you say about prayer at the Ashram is largely true. It is still a formal thing, soulless; but I continue it in the hope of it becoming a soulful thing. Human nature is much the same whether in the East or in the West. It does not therefore surprise me that you have not found anything special about prayers in the East and probably the Ashram prayer is a hotchpotch of something Eastern and something Western. As I have no prejudice against taking anything good from the West or against giving up anything bad in the East, there is an unconscious blending of the two. For a congregational life a congregational prayer is a necessity and, therefore, form also is necessary. It need not be considered on that account to be hypocritical or harmful. If the leader at such congregational prayer meetings is a good man the general level of the meeting is also good. The spiritual effect of an honest intelligent attendance at such congregational prayers is undoubtedly great. Congregational prayer is not intended to supplant individual prayer, which, as you well put it, must be heart-felt and never formal. It is there you are in tune with the Infinite. Congregational prayer is an aid to being in tune with the Infinite. For man who is a social being cannot find God unless he discharges social obligations and the obligation of coming to a common prayer meeting is perhaps the supremest. It is a cleansing process for the whole congregation. But, like all human institutions, if one does not take care, such meetings do become formal and even hypocritical. One has to devise methods of avoiding the formality and hypocrisy. In all, especially in spiritual matters, it is the personal equation that counts in the end.

The roll call is not the ordinary roll call. It is a note of the results of the daily *yajna*, that is, sacrifice. Everyone says what he has spun. Spinning has been conceived in a sacrificial spirit. The idea is to see God through service of the millions. The day must not close without every member of the congregation confessing whether he or she has or has not performed the daily sacrifice to the measure of his or her promise. It is therefore not business at the end of the prayer, but it is the finishing touch to the prayer. It is not done at the beginning of the meeting, because those who are late should have the opportunity of registering their sacrifice. Remember, too, this is a sacrifice not intended to be made in secret. It is designed to be done in the open.

In my opinion, Christianity or the message of Jesus is a response to the human want even as are the messages of Krishna, Buddha, Muhammad and Zoroaster. Though they were designed and delivered at different places and at different times, they have also a universal value. According to the needs of the time one message puts more emphasis on one thing than upon another. A man of religion will not hesitate to profit by all these messages and according to his predilection derive more comfort from one than from another.

I do believe that real art consists in seeing the hidden beauty of moral acts and effects and, therefore, much that passes for art and beauty is, perhaps, neither art nor beauty.

I think I have now answered all your questions. You will please remind me if I have missed any and you will not hesitate to write to me again if I am anywhere obscure or unconsciously evasive.

My love to both of you.

Yours sincerely,

Mrs. E. Bjerrum
United Theological College
Bangalore

Letter to Mrs E. Bjerrum
SN 13221 and 15365

208. THE SERVANT OF GOD

Under the above heading a gentleman has sent me the following article...[1]

I have no knowledge of what the writer has said about the

Marwari devotee. I am not acquainted with the three verses from *Siddhanta Rahasya*[2] whose purport he had given. But there is no doubt that a belief of the kind he has discussed does exist in Hinduism. I myself sing the following verse every morning:

> Guru is Brahma, Guru is Vishnu, Guru is God Siva,
> Guru verily is the Supreme Brahman; to that Guru I bow.

I am convinced there are strong reasons for the Hindu belief concerning the greatness of the *guru*. That is why I have been looking for the true meaning of the word '*guru*' and saying time and again that I am in quest of a *guru*. The *guru* in whom Brahma, Vishnu and Siva merge and who is the Supreme Brahman Himself cannot be an embodied man with his humours and diseases. He will possess the powers of Brahma, Vishnu and Shiva. In other words, He can only be an ideal being. This *guru*, our desired god, can only be God who is the embodiment of Truth. Hence the quest for such a *guru* is the quest for God. If we look at the matter thus, the meaning of all that the writer has said is easily understood. One who can show us God is certainly fit to be *guru* and may be said to be greater than God. We see God's creatures suffering in many ways. Anyone who can free us from this web would deserve a place superior to God's. This is also the meaning of the saying: 'The servant of Rama is greater than Rama.' The meaning of all these great utterances is so simple that if we examine them with a pure heart we shall not be led astray.

Every such great utterance has an indispensable condition attached to it. One who frees us from desire, anger and so on, initiates us into the religion of love, frees us from fear, teaches us simplicity, gives us not only the intelligence to establish identity with the poorest of the poor but also the heart to feel such identity, is certainly, for us, more than God. This does not mean that such a servant of God by himself is greater than God. If we fall into the sea we shall be drowned. However, if we drink, when we are thirsty, a jugful of water from the Ganga which flows into the sea, taking it from near the source, that Ganga water is more to us than the sea. But the same Ganga water is like poison if taken at the point where the Ganga meets the sea. The same is true with regard to the *guru*. To accept as *guru* one who is full of conceit and arrogance and hungering to be served is like drinking the poisonous water of the Ganga that carries all manner of filth into the sea.

Today we practise *adharma* in the name of *dharma*. We cherish hypocrisy in the name of truth and degrade ourselves as well as others by pretending to be possessed of spiritual knowledge and usurping all kinds of worship. At such a time *dharma* consists in refusing to accept anyone as *guru*. It is doubly sinful, when a true *guru* cannot be found, to set up a clay figure and make a *guru* of it. But so long as a true *guru* is not found there is merit in going on saying 'Not this. Not this', and it may one day lead to our finding a true *guru*.

There are many hazards in trying to go against the current. I have had, as I continue to have, many experiences—bitter and sweet—of this. I have learnt but one thing from these, viz., that whatever is immoral and must be opposed should be opposed, even if one is all alone in opposing it. And one should have the faith that if the opposition is truthful it will one day surely bear fruit.

A devotee who is after eulogy or worship, who is offended if not given honour, is no devotee. The true service of a devotee is to become a devotee oneself. Hence I oppose, wherever possible, the worship of human beings which is in vogue nowadays and urge others to do likewise.

'God or *Guru*—Who is Greater?' (G.)
Navajivan, 10 June 1928

[1] This is not translated here. The writer had pointed out that, according to Hindu tradition, God could be reached only by the grace of the *guru*.

[2] A work of Vallabhacharya.

209. THE SACRED ALLIANCE BETWEEN GOD AND MAN

A friend writes:

In the article entitled 'The Tangle of Ahimsa' appearing in *Young India* of October 11th, you have stated most forcefully that cowardice and *ahimsa* are incompatible. There is not an ambiguous syllable in your statement. But may I request that you tell us how cowardice can be exorcised from a man's character? I notice that all characters are but the sum total of habits formed. How are we to undo our old habits and build the new ones of courage, intelligence, and action? I am convinced that habits can be destroyed, and better and nobler habits can be formed giving birth to a new character in a person. It seems to me that you know prayers, discipline, and studies by which a man can attain a second birth. Won't you kindly tell us about them? Do give us your knowledge and advice in one of the numbers of *Young India*. Please help us by giving an account of the method of praying and working by which a man can recreate himself.

The question refers to the eternal duel that is so graphically described in the *Mahabharata* under the cloak of history and that is every day going on in millions of breasts. Man's destined purpose is to conquer old habits, to overcome the evil in him and to restore good to its rightful place. If religion does not teach us how to achieve this conquest, it teaches us nothing. But there is no royal road to success in this the truest enterprise in life. Cowardice is perhaps the greatest vice from which we suffer and is also possibly the greatest violence, certainly far greater than bloodshed and the like that generally go under the name of violence. For it comes from want of faith in God and ignorance of His attributes. But I am sorry that I have not the ability to give 'the knowledge and the advice' that the correspondent would have me to give on how to dispel cowardice and other vices. But I can give my own testimony and say that a heartfelt prayer is undoubtedly the most potent instrument that man possesses for overcoming cowardice and all other bad old habits. Prayer is an impossibility without a living faith in the presence of God within.

Christianity and Islam describe the same process as a duel between God and Satan, not outside but within; Zoroastrianism as a duel between Ahurmazd and Ahriman; Hinduism as a duel between forces of good and forces of evil. We have to make our choice whether we should ally ourselves with the forces of evil or with the forces of good. And to pray to God is nothing but that sacred alliance between God and man whereby he attains his deliverance from the clutches of the prince of darkness. But a heartfelt prayer is not a recitation with the lips. It is a yearning from within which expresses itself in every word, every act, nay, every thought of man. When an evil thought successfully assails him, he may know that he has offered but a lip prayer and similarly with regard to an evil word escaping his lips or an evil act done by him. Real prayer is an absolute shield and protection against this trinity of evils.

Success does not always attend the very first effort at such real living prayer. We have to strive against ourselves, we have to believe in spite of ourselves, because months are as our years. We have therefore to cultivate illimitable patience if we will realize the efficacy of prayer. There will be darkness, disappointment and even worse; but we must have courage enough to battle against all these and not succumb to cowardice. There is no such thing as retreat for a man of prayer.

What I am relating is not a fairytale. I have not drawn an imaginary picture. I have summed up the testimony of men who have by prayer conquered every difficulty in their upward progress, and I have added my own humble testimony that the more I live the more I realize how much I owe to faith and prayer which is one and the same thing for me. And I am quoting an experience not limited to a few hours, or days or weeks, but extending over an unbroken period of nearly 40 years. I have had my share of disappointments, uttermost darkness, counsels of despair, counsels of caution, subtlest assaults of pride; but I am able to say that my faith—and I know that it is still little enough, by no means as great as I want it to be—has ultimately conquered every one of these difficulties up to now.

If we have faith in us, if we have a prayerful heart, we may not tempt God, may not make terms with Him. We must reduce ourselves to a cipher. Barodada sent me a precious Sanskrit verse not long before his death. It means impliedly that a man of devotion reduces himself to zero. Not until we have reduced ourselves to nothingness can we conquer the evil in us. God demands nothing less than complete self-surrender as the price for the only real freedom that is worth having. And when a man thus loses himself, he immediately finds himself in the service of all that lives. It becomes his delight and his recreation. He is a new man never weary of spending himself in the service of God's creation.

'The Eternal Duel'
Young India, 20 Dec. 1928

210. INWARD COMMUNION

[On or after 17 January 1930]

I am glad that you all want me to speak to you on the meaning of and the necessity for prayer. I believe that prayer is the very soul and essence of religion, and therefore prayer must be the very core of the life of man, for no man can live without religion. There are some who in the egotism of their reason declare that they have nothing to do with religion. But it is like a man saying that he breathes but that he has no nose. Whether by reason, or by instinct, or by superstition, man acknowledges some sort of relationship with the divine. The rankest agnostic or atheist does acknowledge the need of a moral principle, and associates something good with its observance and something bad with its non-observance. Bradlaugh, whose atheism is well known, always insisted on proclaiming his innermost conviction. He had to suffer a lot for thus speaking the truth, but he delighted in it and said that truth is its own reward. Not that he was quite insensible to the joy resulting from the observance of truth. This joy however is not at all worldly, but springs out of communion with the divine. That is why I have said that even a man who disowns religion cannot and does not live without religion.

Now I come to the next thing, viz., that prayer is the very core of man's life, as it is the most vital part of religion. Prayer is either petitional or in its wider sense is inward communion. In either case the ultimate result is the same. Even when it is petitional, the petition should be for the cleansing and purification of the soul, for freeing it from the layers of ignorance and darkness that envelop it. He therefore who hungers for the awakening of the divine in him must fall back on prayer. But prayer is no mere exercise of words or of the ears, it is no mere repetition of empty formula. Any amount of repetition of

Ramanama is futile if it fails to stir the soul. It is better in prayer to have a heart without word than words without a heart. It must be in clear response to the spirit which hungers for it. And even as a hungry man relishes a hearty meal, a hungry soul will relish a heartfelt prayer. And I am giving you a bit of my experience and that of my companions when I say that he who has experienced the magic of prayer may do without food for days together but not a single moment without prayer. For without prayer there is no inward peace.

If that is the case, someone will say, we should be offering our prayers every minute of our lives. There is no doubt about it, but we erring mortals, who find it difficult to retire within ourselves for inward communion even for a single moment, will find it impossible to remain perpetually in communion with the divine. We therefore fix some hours when we make a serious effort to throw off the attachments of the world for a while, we make a serious endeavour to remain, so to say, out of the flesh. You have heard Surdas's hymn.[1] It is the passionate cry of a soul hungering for union with the divine. According to our standards he was a saint, but according to his own he was a proclaimed sinner. Spiritually he was miles ahead of us, but he felt the separation from the divine so keenly that he has uttered that anguished cry in loathing and despair.

I have talked of the necessity for prayer, and therethrough I have dealt with the essence of prayer. We are born to serve our fellowmen, and we cannot properly do so unless we are wide awake. There is an eternal struggle raging in man's breast between the powers of darkness and of light, and he who has not the sheet-anchor of prayer to rely upon will be a victim to the powers of darkness. The man of prayer will be at peace with himself and with the whole world, the man who goes about the affairs of the world without a prayerful heart will be miserable and will make the world also miserable. Apart therefore from its bearing on man's condition after death, prayer has incalculable value for man in this world of the living. Prayer is the only means of bringing about orderliness and peace and repose in our daily acts. We inmates of the Ashram who came here in search of truth and for insistence on truth professed to believe in the efficacy of prayer, but had never up to now made it a matter of vital concern. We did not bestow on it the care that we did on other matters. I awoke from my slumbers one day and realized that I had been woefully negligent of my duty in the matter. I have therefore suggested measures of stern discipline and far from being any the worse, I hope we are the better for it. For it is so obvious. Take care of the vital thing and other things will take care of themselves. Rectify one angle of a square, and the other angles will be automatically right.

Begin therefore your day with prayer, and make it so soulful that it may remain with you until the evening. Close the day with prayer so that you may have a peaceful night free from dreams and nightmares. Do not worry about the form of prayer. Let it be any form, it should be such as can put us into communion with the divine. Only, whatever be the form, let not the spirit wander while the words of prayer run on out of your mouth.

If what I have said has gone home to you, you will not be at peace until you have compelled your hostel superintendents to interest themselves in your prayer and to make it obligatory. Restraint self-imposed is no compulsion. A man, who chooses the path of freedom from restraint, i.e., of self-indulgence, will be a bondslave of passions, whilst the man who binds himself to rules and restraints releases himself. All things in the universe, including the sun and the moon and the stars, obey certain laws. Without the restraining influence of these laws the world would not go on for a single moment. You, whose mission in life is service of your fellowmen, will go to pieces if you do not impose on yourselves some sort of discipline, and prayer is a necessary spiritual discipline. It is discipline and restraint that separates us from the brute. If we will be men walking with our heads erect and not walking on all fours, let us understand and put ourselves under voluntary discipline and restraint.

Speech at Prayer Meeting, Sabarmati Ashram
Young India, 23 Jan. 1930

[1] Where is there a wretch
So loathsome and wicked as I?
I have forsaken my Maker,
So faithless have I been.

211. A GLIMPSE OF GOD

Abbottabad,
[On or before 26 July 1939]

GANDHIJI: It is a difficult thing to explain fully what I do when I pray. But I must try to answer your question.[1] The Divine Mind is unchangeable, but that Divinity is in everyone and everything—animate and inanimate. The meaning of prayer is that I want to evoke that Divinity within me. Now I may have that intellectual conviction, but not a living touch. And so when I pray for *swaraj* or independence for India I pray or wish for adequate power to gain that *swaraj* or to make the largest contribution I can towards winning it, and I maintain that I can get that power in answer to prayer.

FABRI: Then you are not justified in calling it prayer. To pray means to beg or demand.

Yes, indeed. You may say I beg it of myself, of my Higher Self, the Real Self with which I have not yet achieved complete identification. You may therefore describe it as a continual longing to lose oneself in the Divinity which comprises all.

And you use an old form to evoke this?

I do. The habit of a lifetime persists, and I would allow it to be said that I pray to an outside Power. I am part of that Infinite, and yet such an infinitesimal part that I feel outside it. Though I give you the intellectual explanation, I feel, without identification with the Divinity, so small that I am nothing. Immediately I begin to say I do this thing and that thing, I begin to feel my unworthiness and nothingness, and feel that someone else, some Higher Power, has to help me.

Tolstoy says the same thing. Prayer really is complete meditation and melting into the Higher Self, though one occasionally does lapse in imploration like that of a child to his father.

Pardon me, I would not call it a lapse. It is more in the fitness of things to say that I pray to God who exists somewhere up in the clouds, and the more distant He is, the greater is my longing for Him and I find myself in His presence in thought. And thought as you know has a greater velocity than light. Therefore the distance between me and Him, though so incalculably great, is obliterated. He is so far and yet so near.

It becomes a matter of belief, but some people like me are cursed with an acute critical faculty. For me there is nothing higher than what Buddha taught, and no great master. For Buddha alone among the teachers of the world said: 'Don't believe implicitly what I say. Don't accept any dogma or any book as infallible.' There is for me no infallible book in the world, inasmuch as all were made by men, however inspired they may have been. I cannot hence believe in a personal idea of God, a Maharaja sitting on the Great White Throne listening to our prayers. I am glad that your prayer is on a different level.

Let me remind you that you are again only *partially* true when you say my prayer is on a different level. I told you that the intellectual conviction that I gave you is not eternally present with me. What is present is the intensity of faith whereby I lose myself in an Invisible Power. And so it is far truer to say that God has done a thing for me than that I did it. So many things have happened in my life for which I had intense longing, but which I could never have achieved myself. And I have always said to my co-workers it was in answer to my prayer. I did not say to them it was in answer to my intellectual effort to lose myself in the Divinity in me! The easiest and the correct thing for me was to say, 'God has seen me through my difficulty.'

But that you deserved by your *karma*. God is Justice and not Mercy. You are a good man and good things happen to you.

No fear. I am not good enough for things to happen like that. If I went about with that philosophical conception of *karma*, I should often come a cropper. My *karma* would not come to my help. Although I believe in the inexorable law of *karma* I am striving to do so many things; every moment of my life is a strenuous endeavour which is an attempt to build up more *karma*, to undo the past and add to the present. It is therefore wrong to say that because my past is good, good is happening at present. The past would be soon exhausted, and I have to build up the future with prayer. I tell you *karma* alone is powerless. 'Ignite this match', I say to myself, and yet I cannot if there is no co-operation from without. Before I strike the match my hand is paralysed or I have only one match and the wind blows it off. Is it an accident or God or Higher Power? Well, I prefer to use the language of my ancestors or of children. I am no better than a child. We may try to talk learnedly and of books, but when it comes to brass tacks—when we are face to face with a calamity—we behave like children and begin to cry and pray and our intellectual belief gives no satisfaction!

I know very highly developed men to whom belief in God gives incredible comfort and help in the building of character. But there are some great spirits that can do without it. That is what Buddhism has taught me.

But Buddhism is one long prayer.

Buddha asked everyone to find salvation from himself. He never prayed, he meditated.

Call it by whatever name you like, it is the same thing. Look at his statues.

But they are not true to life. They are 400 years later than his death.

Well, give me your own history of Buddha as you may have discovered it. I will prove that he was a praying Buddha. The intellectual conception does not satisfy me. I have not given you a perfect and full definition as you cannot describe your own thought. The very effort to describe is a limitation. It defies analysis and you have nothing but scepticism as the residue.

What about the people who cannot pray?

'Be humble,' I would say to them, 'and do not limit even the real Buddha by your own conception of Buddha.' He could not have ruled the lives of millions of men that he did and does today if he was not humble enough to pray. There is something infinitely higher than intellect that rules us and even the sceptics. Their scepticism and philosophy does not help them in critical periods of their lives. They need something better, something outside them that can sustain them. And so if someone puts a conundrum before me, I say to him, 'You are not going to know the meaning of God or prayer unless you reduce yourself to a cipher. You must be humble enough to see that in spite of your greatness and gigantic intellect you are but a speck in the universe. A merely intellectual conception of the things of life is not enough. It is the spiritual conception which eludes the intellect, and which alone can give one satisfaction. Even monied men have critical periods in their lives; though they are surrounded by everything that money can buy and affection can give, they find at certain moments in their lives utterly distracted. It is in these moments that we have a glimpse of God, a vision of Him who is guiding every one of our steps in life. It is prayer.'

You mean what we might call a true religious experience which is stronger than intellectual conception. Twice in life I had that experience, but I have since lost it. But I now find great comfort in one or two sayings of Buddha: 'Selfishness is the cause of sorrow.' 'Remember, monks, everything is fleeting.' To think of these takes almost the place of belief.

That is prayer.

What would you say to the right of man to dispose of his life? Life as life I hold of very little importance.

I think that man has a perfect right to dispose of his life under certain circumstances. A co-worker, suffering from leprosy, knowing that his disease was incurable and that his life was as much an agony for those who had to serve him as it was for him, recently decided to end his life by abstaining from food and water. I blessed the idea. I said to him: 'If you really think you can stand the trial you may do so.' I said this to him for I knew how different it is to die by inches from, say, suddenly killing oneself by drowning or poisoning. And my warning was fully justified, for someone tempted him with the hope that there was one who could cure leprosy, and I now hear that he has resumed eating and put himself under his treatment!

The criticism seems to me to be that if one's mind is completely obscured by pain, the best thing for him would be to seek *Nirvana*. A man may not be ill but he may be tired of the struggle.

No, no. My mind rejects this suicide. The criterion is not that one is tired of life, but that one feels that one has become a burden on others and therefore wants to leave the world. One does not want to fly from pain but from having to become an utter burden on others. Otherwise one suffers greater pain in a violent effort to end one's agony. But supposing I have a cancer, and it is only a question of time for me to pass away,

I would even ask my doctor to give me a sleeping draught and thereby have the sleep that knows no waking....

No, according to you I should have no business to stay if I feel I have finished my task. And I do think I have finished mine!

No. I am convinced that you can serve humanity for many years. Millions are praying for your life. And though I can neither pray nor desire anything—

Yes, the English language is so elastic that you can find another word to say the same thing.

Yes, I can unselfishly opine that you have many years before you.

Well, that's it. You have found the word! Here too let me tell you there is the purely intellectual conception of a man being unable to live. If he has not the desire to live, the body will perish for the mere absence of the desire to live.

Discussion with Dr Charles Fabri ('A Dialogue with a Buddhist') *Harijan*, 19 Aug. 1939

[1] Dr. Fabri was a pupil of Professor Sylvain Levi and came out to India as assistant to the famous archaeologist, Sir Aurel Stein. He had in the past corresponded with Gandhi and even sympathetically fasted with him. He had come to Abbottabad specially to see Gandhi. He was particularly concerned about the form and content of prayer. Could the Divine Mind be changed by prayer? Could one find it out by prayer?

212. A CRY OF THE HEART

March 26, 1945

Prayer is a cry of the heart. It can be fruitful if it comes from within. But those who pray for an object do not know the meaning of prayer at all.

Blessing from
BAPU

Note to Gope Gurbuxani[1]
GN 1332

[1] He had asked: 'What is prayer and how can it be made fruitful?'

213. POTENCY OF PRAYER

There is little doubt that India is about to reach her cherished goal of political independence. Let the entrance be prayerful. Prayer is not an old woman's idle amusement. Properly understood and applied, it is the most potent instrument of action.

Let us then pray and find out what we have meant by non-violence, and how we shall retain the freedom gained by its use. If our non-violence is of the weak, it follows that we shall never be able, by such non-violence, to retain freedom. But it follows also that we shall not, for some length of time, at any rate, be able to defend ourselves by force of arms, if only because we have neither them nor the knowledge of their use. We have not even the requisite discipline. The result is that we shall have to rely upon another nation's help, not as equals, but as pupils upon their teachers, if the word 'inferiors' jars upon our ears.

Hence there is nothing but non-violence to fall back upon for retaining our freedom even as we had to do for gaining it. This means exercise of non-violence against all those who call themselves our opponents. This should not mean much for a man who has used himself to non-violence for nearly three decades. It is summed up in 'die for your honour and freedom' instead of 'kill if necessary and be killed in the act'. What does a brave soldier do? He kills only if necessary and risks his life in the act. Non-violence demands greater courage and sacrifice. Why should it be comparatively easy for a man to risk death in the act of killing, and almost super human for him to do so in the act of sparing life? It seems to be gross self-deception to think that we can risk death, if we learn and practise the art of killing, but cannot do so otherwise. But for the hypnotism induced by the repetition of an untruth we should not grossly deceive ourselves.

But the critic or the scoffer will ask, why bring in prayer if the matter is so simple as you put it. The answer is that prayer is the first and the last lesson in learning the noble and brave art of sacrificing self in the various walks of life culminating in the defence of one's nation's liberty and honour.

Undoubtedly prayer requires a living faith in God. Successful *satyagraha* is inconceivable without that faith. God may be called by any other name so long as it connotes the living Law of Life—in other words, the Law and the Law-giver rolled into one.

New Delhi, April 6, 1946

'Let Us Pray'
Harijan, 14 Apr. 1946

214. SELF-EXAMINATION

Q. You believe in mass prayer. Is congregational worship as practised today a true prayer? In my opinion, it is a degrading thing and therefore dangerous. Jesus said: 'When thou prayest, thou shalt not be as the hypocrites are, but enter into thine inner chamber and having shut thy door pray to the Father which is in secret.' Most people in a crowd are inattentive and unable to concentrate. Prayer then becomes hypocrisy. The *yogi* is aware of this. Should not the masses, therefore, be taught self-examination which is the true prayer?

A. I hold that congregational worship held by me is true prayer for a collection of men. The convener is a believer and no hypocrite. If he were one, the prayer would be tainted at the source. The men and women who attend do not go to any orthodox prayer house from which they might have to gain an earthly end. The bulk of them have no contact with the convener. Hence it is presumed, they do not come for show. They join in because they believe that they somehow or other, acquire merit by having common prayer. That most or some persons are inattentive or unable to concentrate is very true. That merely shows that they are beginners. Neither inattention nor inability to concentrate are any proof of hypocrisy or falsity. It would be, if they pretended to be attentive when they were not. On the contrary, many have often asked me what they should do, when they are unable to concentrate.

The saying of Jesus quoted in the question is wholly inapplicable. Jesus was referring to individual prayer and to hypocrisy underlying it. There is nothing in the verse quoted against collective prayer. I have remarked often enough that without individual prayer collective prayer is not of much use. I hold that individual prayer is a prelude to collective, as the latter, when it is effective, must lead to the individual. In other words, when a man has got to the stage of heart prayer, he prays always, whether in the secret or in the multitude.

I do not know what the questioner's *yogi* does or does not. I know that the masses when they are in tune with the Infinite, naturally resort to self-examination. All real prayer must have that end.

New Delhi, September 14, 1946

'Collective Prayer'
Harijan, 22 Sept. 1946

VII

God

215. THE OMNIPRESENCE OF GOD

September 19, 1924

Who can name Him, and knowing what he says,
Say, 'I believe in Him'? And who can feel,
And, with self-violence, to conscious wrong
Hardening his heart, say, 'I believe Him not!'?
The All-embracing, All-sustaining One,
Say, does He not embrace, sustain, include
Thee?—Me—Himself? Bends not the sky above?
And earth, on which we are, is it not firm?
And over us, with constant kindly smile,
The sleepless stars keep everlasting watch!
Am I not here gazing into thine eyes?
>   And does not All, that is,
>   —Seen and unseen—mysterious all—
>   Around thee, and within,
>   Untiring agency,
>   Press on thy heart and mind?
—Fill thy whole heart with it—and, when thou art
Lost in the consciousness of happiness,
>   Then call it what thou wilt
>   Happiness!—Heart!—Love!—God!
>   I have no name for it!
>   Feeling is all in all!
>   Name is but sound and reek,
>   A mist around the glow of Heaven.

Goethe's *Faust*

Last Thursday night, some Mussalman friends called on me by appointment. They appeared to me to be earnest and sincere. They had much to say against *shuddhi* and *sangathan*. I have said my say about these movements already. So far as possible I do not wish during the privileged weeks to say anything on matters of controversy. I wish to engage the attention of the reader on the solution they offered. They said: 'We believe in the divinity of the Vedas. We believe in Shri Krishnaji Maharaj and Ramachandraji Maharaj (the adjectives are theirs). Why cannot Hindus believe in the divinity of the Koran and say with us that there is no God but God and Mahomed is His Prophet? Ours is not an exclusive religion, but it is essentially inclusive.'

I told them that the solution was not quite so simple as they put it. The formula they suggested might be good enough for the cultured few, but it would prove ineffective for the man in the street. For the Hindus cow-protection and the playing of music even near the mosque was the substance of Hinduism, and for the Mussalmans cow-killing and prohibition of music was the substance of Islam. It was, therefore, necessary that the Hindus abandon the idea of compelling Mussalmans to stop cow-killing, and Mussalmans the idea of compelling the Hindus to stop music. The regulation of cow-slaughter and playing of music must be left to the goodwill of the respective communities. Each practice would assume a becoming proportion with the growth of the tolerant spirit. But I do not propose to elaborate here this ticklish question.

I wish to examine the attractive formula presented by the Mussalman friends and state what is at least acceptable to me; and as my instinct is wholly Hindu, I know that what I am about to say will be acceptable to the vast mass of Hindus.

In fact it is the average Mussalman who will not accept the divinity of the Vedas and the other Hindu scriptures, or Krishna or Rama as prophets or incarnations of the Deity. With the Hindu it is a new-fangled notion to revile the Koran and the Prophet. I have known the Prophet spoken of with reverence in Hindu circles. There are even Hindu songs paying tribute to Islam.

Take the first half of the formula. God is certainly One. He has no second. He is unfathomable, unknowable and unknown to the vast majority of mankind. He is everywhere. He sees without eyes, and hears without ears. He is formless and indivisible. He is uncreated, has no father, mother or child; and yet He allows Himself to be worshipped as father, mother, wife and child. He allows himself even to be worshipped as stock and stone, although He is none of these things. He is the most elusive. He is the nearest to us if we would but know the fact. But He is farthest from us when we do not want to realize His omnipresence. There are many gods in the Vedas. Other scriptures call them angels. But the Vedas sing of only one God.

I have no hesitation in regarding the Koran as revealed, as I have none in regarding the Bible, the Zend-Avesta, the Granth Saheb and any other clean scriptures as revealed. Revelation is the exclusive property of no nation, no tribe. If I know Hinduism at all, it is essentially inclusive and ever-growing, ever-responsive. It gives the freest scope to imagination, speculation and reason. I have found not the slightest difficulty in Hindu circles about evoking reverence for the Koran and the Prophet. But I have found difficulty in Mussalman circles about evoking the same reverence for the Vedas or the incarnations. I had a very good Mussalman client in South Africa. He is, alas, dead now. The relation of client and counsel developed into one of close companionship and mutual regard. We often had religious discussions. My friend, though not learned in any sense of the term, had an intellect as sharp as a razor. He knew everything of the Koran. He knew something of other religions also. He was interested in my accepting Islam. I said to him, 'I can pay full respect to the Koran and the Prophet, why do you ask me to reject the Vedas and the incarnations? They have helped me to be what I am. I find the greatest consolation from the *Bhagavad Gita* and Tulsidas's *Ramayana*. I frankly confess that the Koran, the Bible and the other scriptures of the world, in spite of my great regard for them, do not move me as do the *Gita* of Krishna and the *Ramayana* of Tulsidas.' The friend despaired of me and had no hesitation in saying that there must be something wrong with me. His, however, is not an exceptional case because I have since met many Mussalman friends who have held the same view. I do, however, believe that this is a passing phase.

I share Justice Ameer Ali's view that Islam in the days of Harun-al-Rashid and Maman was the most tolerant amongst the world's religions. But there was a reaction against the liberalism of the teachers of their times. The reactionaries had many learned, able and influential men amongst them and they very nearly overwhelmed the liberal and tolerant teachers and philosophers of Islam. We in India are still suffering from the effect of that reaction. But I have not a shadow of doubt that Islam has sufficient in itself to become purged of illiberalism and intolerance. We are fast reaching the time when the acceptance of the formula suggested by the friends will be a common thing among mankind.

The need of the moment is not one religion, but mutual respect and tolerance of the devotees of the different religions. We want to reach not the dead level, but unity in diversity. Any attempt to root out traditions, effects of heredity, climate and other surroundings is not only bound to fail, but is a sacrilege. The soul of religions is one, but it is encased in a multitude of forms. The latter will persist to the end of time. Wise men will ignore the outward crust and see the same soul living under a variety of crusts. For Hindus to expect Islam, Christianity or Zoroastrianism to be driven out of India is as idle a dream as it would be for Mussalmans to have only Islam of their imagination rule the world. But if belief in One God and the race of His Prophets in a never-ending chain is sufficient for Islam, then we are all Mussalmans, but we are also all Hindus and Christians. Truth is the exclusive property of no single scripture.

'God is One'
Young India, 25 Sept. 1924

216. DEFINING THE INDEFINABLE

A friend writes:

> 'There is one matter on which I have been longing to approach you for an explanation. It is about the term "God". As a national worker I would not have anything to say against such a passage as this occurring in a recent number of *Young India*: "I present it (*Ramanama*) to the reader whose vision is not blurred and whose faith is not damped by overmuch learning. Learning takes us through many stages in life, but it fails us utterly in the hour of danger and temptation. Then faith alone saves" (*Young*

India, 22-1-'25, p. 27). For it is a confession of your individual faith; and I know also that you have not failed on occasions to put in a word of praise about conscientious atheists where it was deserved. As witness the following sentences in your *Niti Dharma:* "We come across many wicked men, priding themselves on their religiosity, while doing the most immoral deeds. On the other hand, there are also men like the late Mr. Bradlaugh[1] who, while being extremely virtuous and moral, take pride in calling themselves atheists." As for the faith in *Ramanama* which "alone saves us in the hour of danger and temptation", I may mention the martyrdom of the rationalist Francisco Ferrer in 1909 at Barcelona in Spain at the hands of men who believed in Jesus' name, their *Ramanama*. I shall not dwell on the Holy Wars, the burnings and mutilations of heretics, and the torture and slaughter of animals and sometimes of men in sacrifice—all of which have been carried out "for the greater glory of God and in His name". This is by the bye [*sic*.].

'As a national worker, however, I feel, I must draw your attention to the objection which Mr.—raised (on behalf of a rationalist friend of his) to your saying that only "god-fearing" men can become true non-co-operators, and remind you of the assurance you then gave to all to the effect that the programme of national work does not require a man to declare his religious faith. (*Vide Young India*, May 4, 1921, pp. 138-9) That objection applies with greater force now than it did at that time, because "God" has now a place on pledges and vows such as that administered to Congress volunteers, which begins "with God as witness, I ..." Now you must be knowing that the Buddhists (like the Burmese ... now an "Indian" people, and your friend Prof. Dharmanand Kosambi) and the Jains, as well as many Indians who do not belong to these ancient recognized sects are agnostic in faith. Is it possible to these, if they wish, to enrol as Congress Volunteers conscientiously and with full understanding of a pledge which begins in the name of an Entity they ignore? If not, is it proper to exclude from Congress service any such merely because of their religious faith? May I suggest that a conscience clause be added to accommodate all such cases, allowing of solemn affirmation in place of the oath in the name of God (to which even some believers in a personal God object, as the Quakers), or else a substitution of "Conscience" in place of "God" by all conscientious objectors to the use of the latter, or—best of all—that a solemn affirmation without reference to God and with or without "Conscience" be required of *all* comers without distinction? I approach you as you are the author of that pledge and now the President of the Congress. I did so once before, but I am afraid, not in time for you to be able to attend to it before your historic arrest at Sabarmati in 1922.'

So far as the conscientious objection is concerned the mention of God may be removed if required from the Congress pledge of which I am proud to think I was the author. Had such an objection been raised at the time, I would have yielded at once. I was unprepared for the objection in a place like India. Though there is officially the Charvak[2] school, I do not know that it has any votaries. I deny that Buddhists and Jains are atheists or agnostics. The latter they cannot be. Those who believe in the soul as apart from and capable of life independent of and after the dissolution of the body cannot be called atheists. We may all have different definitions for 'God'. If we could all give our own definitions of God there would be as many definitions as there are men and women. But behind all that variety of definitions there would be also a certain sameness which would be unmistakable. For the root is one.

God is that indefinable something which we all feel but which we do not know. Charles Bradlaugh described himself as an atheist no doubt but many a Christian declined to regard him as such. He recognized in Bradlaugh a greater kinship with himself than many a lip Christian. I had the privilege of attending the funeral of that good friend of India. I noticed several clergymen at the function. There were certainly several Mussalmans and many Hindus in the procession. They all believed in God. Bradlaugh's denial of God was a denial of Him as He was known to Bradlaugh to have been described. His was an eloquent and indignant protest against the then current theology and the terrible contrast between precept and practice.

To me God is truth and love; God is ethics and morality; God is fearlessness. God is the source of Light and Life and yet He is above and beyond all these. God is conscience. He is even the atheism of the atheist. For in His boundless love God permits the atheist to live. He is the searcher of hearts. He transcends speech and reason. He knows us and our hearts better than we do ourselves. He does not take us at our word for He knows that we often do not mean it, some knowingly and others unknowingly. He is a personal God to those who need His personal presence. He is embodied to those who need His touch. He is the purest essence. He simply Is to those who have faith. He is all things to all men. He is in us and yet above and beyond us.

One may banish the word 'God' from the Congress but one has no power to banish the Thing Itself. What is a solemn affirmation if it is not the same thing as in the name of God. And surely conscience is but a poor and laborious paraphrase of the simple combination of three letters called God. He cannot cease to be because hideous immoralities or inhuman brutalities are committed in His name. He is long suffering. He is patient but He is also terrible. He is the most exacting personage in the world and the world to come. He metes out the same measure to that we mete out to our neighbours— men and brutes. With Him ignorance is no excuse. And withal He is ever forgiving for He always gives us the chance to repent. He is the greatest democrat the world knows, for He leaves us 'unfettered' to make our own choice between evil and good. He is the greatest tyrant ever known, for He often dashes the cup from our lips and, under cover of free will, leaves us a margin so wholly inadequate as to provide only mirth for Himself at our expense. Therefore it is that Hinduism calls it all His sport—*lila*, or calls it all an illusion—*maya*. We are *not*, He alone *Is*. And if we will be, we must eternally sing His praise and do His will. Let us dance to the tune of His *bansi*—lute, and all would be well.

Since reference has been made to a booklet of mine on ethics and religion, I must draw the attention of the reader to the fact that what the correspondent refers to is a translation of the original which is in Gujarati. But the Gujarati itself as the preface makes it clear is not an original effort but an adaption from an American publication called *Ethical Religion* by Mr. Salter. The translation came under my notice in Yeravda Jail and I regretted to notice the absence of any mention of the source from which I had borrowed. The translator himself I understand did not rely upon the original Gujarati but a Hindi translation. The English rendering therefore is a round-about thing. I owe this explanation to the author of the original book and I am glad that the mention of the work by my correspondent has reminded me of my obligation.

'God and Congress'
Young India, 5 Mar. 1925

[1] 1833-91; English free thinker and politician.
[2] Ancient materialist philosopher whose followers are called by his name.

217. THE SUM TOTAL OF *KARMA*

To The Editor, *Young India*
Sir,

With reference to your article 'God and Congress' I beg to say that while the Charvak school was materialistic out and out, Buddhism is silent on and Jainism doubts the existence of Ishvara or any supernatural Entity that may be said to correspond to God, although both faiths believe in the transmigration of the soul and the Law of Karma, in common with Hinduism. (Your friend Prof. Dharmanand Kosambi whom I mentioned may be consulted on this point.) Buddha with Karma, and Jina with Karma, respectively, may be said to take the place of God in the ritual practice of those two religions.

Of modern religious movements, the Deva Samaj of the Punjab which is mostly a humanitarian and social-service body and lays great store by *ahimsa* is (I believe) frankly atheistic in its creed, but *not* materialistic. I have read that it believes neither in God nor gods. In the light of this, its name of *Deva*[1] Samaj appears rather paradoxical. *Lucus a non lucendo!*[2]

Of Bradlaugh you say that his denial of God was a denial of Him as He was known to Bradlaugh to have been described. Was this denial inclusive, or was it exclusive, of that 'certain unmistakable sameness' behind all that variety of definitions which there would be if we could all give our own definitions of God, as you say? I presume, it cannot be the latter, for Bradlaugh was learned and observant enough. If the former is the case, what made Bradlaugh deny the existence of God even in the aspect of that 'unmistakable sameness'?

I doubt not but that the following excerpt will be of some interest to you in this connection:

The very idea of a god, as creating or in any way ruling the world, is utterly absent in the Buddhist system. God is not so much as denied; he is simply not known. Contrary to the opinion once confidently and generally held, that a nation of atheists never existed, it is no longer to be disputed that the numerous Buddhist nations are essentially atheist; for they know no beings with greater supernatural power than any man is supposed capable of attaining to by virtue, austerity, and science; and a remarkable indication of this startling fact is to be seen in the circumstance, that some at least of the Buddhist nations—the Chinese, Mongols, and Tibetans—have no word in their languages to express the notion of God. The future condition of the Buddhist, then, is not assigned him by the Ruler of the universe; the 'Karma' of his actions determines it by a sort of virtue inherent in the nature of things—by the blind and unconscious concatenation of cause and effect.—*Chamber's Encyclopaedia*, *sub.* Buddhism.

Let me conclude with a *shloka*:

> To the gods or angels we our homage pay;
> But to sorry Fate subject e'en them we find;
> Then, is our worship due to Fate?
> Sure *he* yields but the fruit our actions rate.
> And the fruit on actions of our own depends;
> —Hence small account of gods or angels or of Fate.
> Then hail, our actions small or great!
> Over whom not even Fate prevails!
>   (My own attempt at a free rendering from the Bhartrihari)
>   —*Nitishataka*.

I am, etc.,
S. D. Nadkarni

Karwar (N. Kanara)
10th March, 1925

I cannot refuse space to Mr. Nadkarni's clever letter. I must, however, adhere to my opinion that neither Jainism nor Buddhism are atheistic. I present Mr. Nadkarni with these definitions of God: The sum total of *karma* is God. That which impels man to do the right is God. The sum total of all that lives is God. That which makes man the plaything of fate is God. That which sustained Bradlaugh throughout all his trials was God. He is the Denial of the atheist.

'Is there God?'
Young India, 30 Apr. 1925

[1] God.
[2] 'The grove (*lucus*) is so named from its not shining (*lucendo*).'

218. SERVICE OF GOD

A Parsi gentleman has written to me from Persia. He has addressed to me some profound questions[1] which I quote below in his own words. At two or three places where he has used English words I have given their Gujarati equivalents instead.

If even a single leaf does not stir without the will of God, what is there left for man to do? This is a question dating back to times immemorial and it will always continue to be asked in the future; however, the answer is included in the question itself as it is God Himself who has given us the capacity to ask it. All our actions are governed by laws, and the same is true of God. As our laws and our knowledge are imperfect, we can violate these laws in a civil or uncivil manner. Being all-knowing and omnipotent God never violates His own laws. These admit of no improvements or additions. They are immutable.

Our freedom lies in the capacity bestowed on us to think, distinguish and choose between good and evil in various ways. This freedom is strictly limited. It is so limited that a learned person has said that it is even less than that enjoyed by a sailor on board a ship, to walk on the deck. However little it may be, there is this freedom and even though it is not much, it is sufficient, at any rate, to enable a human being to attain *mukti* through it. Destiny and man's effort to attain the aims of life go hand in hand. However, destiny does not thwart the purpose of anyone who treads the road of *mukti*.

Hence, all that now remains to consider is how to serve or worship God. He can be served in one way alone. To serve the poor is to serve God. By serving even an ant, one serves Him. It is He who gives the tiny particle [of food] to the ant and the six *maunds* to the elephant. Anyone who refrains from trampling an ant also serves it; one who does not intentionally harm it will not harm other animals or his fellow men. At each place and at each point of time service assumes a different form, although the sentiment involved in it is the same. In serving those who suffer, one serves God. Discretion should be exercised in this service. There is no reason to believe that one is doing nothing but service by giving grains to the hungry.

It is a sin to provide food for an idle person who makes no effort and depends on others for food. It is a meritorious act to provide him with an occupation and, if he refuses to work, to let him starve is to render service to him. God's name should be chanted and it is necessary to worship Him with rituals as this leads to self-purification, which in its turn enables man to find his own way. However, ceremonial worship by itself does not constitute the service of God. It is a means of doing that service. It is for this reason that Narasinh Mehta has sung: 'Nothing is achieved by bathing and worshipping, or counting one's beads and chanting His name.'

And from this reply we get the answer to the third question which is as follows. What is the aim of life? It is to know the Self. In the words of Narasinh Mehta, 'So long as the essence of the Self is not realized, all our efforts are in vain.' This realization of the Self, or Self-knowledge, is not possible until one has achieved unity with all living beings—has become one with God. To accomplish such a unity implies deliberate sharing of the suffering of others and the eradication of such suffering.

'How to Worship God' (G.)
Navajivan, 25 Oct. 1925

[1] Not given here.

219. GOD IS FORMLESS

Ashram, Sabarmati,
Sunday, June 13, 1926

Bhaishri Kasamali,

Your letter. God is one. I do not understand your difficulty in conceiving Him as formless. That which has a form cannot be all-pervading, it needs must be the subtlest. It can therefore be only formless. All must admit the need for a *guru* but one may not hastily accept someone as a *guru*. In this age to seek a *guru* is to believe in one, because to acquire a perfect guide one must perfectly qualify oneself. Moreover if we regard all faiths as true there is no need for us to relinquish our faith or make others change theirs, since all can adopt whatever satisfies them from all the religions.

Letter to Kasamali (G.)
SN 10932

220. REAL WORSHIP

Here in Ceylon where I am writing for *Young India* amid surroundings where Nature has bountifully poured her richest treasures, I recall a letter written by a poetically inclined friend from similar scenes. I share with the reader a paragraph from that letter.

A lovely morning! Cool and cloudy, with a drowsy sun whose rays are as soft as velvet. It is a strangely quiet morning—there is a hush upon it, as of prayer. And the mists are like incense, and the trees worshippers in a trance, and the birds and insects pilgrims come to chant *bhajans*. Oh! how I wish one could learn true abandonment from Nature! We seem to have forgotten our birthright to worship where and when and how we please. We build temples and mosques and churches to keep our worship safe from prying eyes and away from outside influences, but we forget that walls have eyes and ears, and the roofs might be swarming with ghosts—who knows!

Good gracious, I shall find myself preaching next! How foolish, on a lovely morning like this? A little child in the garden adjoining is singing as unconsciously and joyously as a bird. I feel inclined to go and take the dust of its little feet. And since I cannot pour out my heart in sound as simply as that little one, my only refuge is in silence!

Churches, mosques and temples, which cover so much hypocrisy and humbug and shut the poorest out of them, seem but a mockery of God and His worship, when one sees the eternally renewed temple of worship under the vast blue canopy inviting every one of us to real worship, instead of abusing His name by quarrelling in the name of religion.
'God's Temple'
Young India, 8 Dec. 1927

221. LIVING FAITH

As at the Ashram,
Sabarmati,
December 21, 1927

Dear Friend,

I have your letter. I can quite see that unless you have a living faith in God you cannot get rid of sinful thoughts much less sinful acts. The only way I can commend to you to have that belief is to realize man's littleness and therefore your own and persist in believing ... assuming that there must be a 'Being' who is perfect and who is responsible for the wonderful phenomenon, the world.

I have no capacity to argue with you about the origin of evil. It is enough for me humbly to recognize evil as evil and strive to combat it. For, I know that God in Whom I believe helps me always in the combat. Victory lies in striving. To find honestly a *guru* and the striving necessitates the living of a pure life.

Yours sincerely,

Sjt. Nirmal Chandra Dey
Engineering College Hostel, Shibpur
P.O. Botanical Garden
Howrath
Letter to Nirmal Chandra Dey
SN 12653a

222. BELIEF IN GOD

The Ashram,
Sabarmati,
April 8, 1928

My dear Shankaran,

I was thinking of you only early this morning, that is, just after prayer when talking to Pyarelal and here is your letter.

Suppose that there is a colony of orphans shipwrecked on an island, that they are all unmarried males, that they have never known that they ever had any parents, suppose further that they have a knowledge of letters and that from their reading they understood that they had all parents; suppose then that they in the course of their readings come upon a philosophical book called 'Our Spontaneous Origin', should the orphans feel convinced philosophically that they were all spontaneous creation? Just as the supposed philosophical book would not unsettle the conviction of the majority of the unsophisticated orphans, so should the philosophical book that you have read about the non-existence of God not unsettle your belief in God. If you will admit the fact of your having parents, how can you escape the fundamental fact of the First Cause? Having made sure of that, I am indifferent whether you call that First Cause God or some other thing. And having been also convinced of that fact it is wholly unnecessary to inquire how that First Cause disposes of justice or to inquire about the injustice that we seem to see around us.

There are endless theories. I believe in that of cause and effect, that is, of the law of *karma*. It seems to answer all a man's doubts. But if they do not answer yours, you must wait, watch and pray you will some day have the light. But if you do not believe in the First Cause, there is no hope. For to whom should you pray then? Therefore hold fast to your belief in God, never mind the reasoning. Can you reason out the existence of your parents? Will you not say, 'whether I can reason or not the existence of my parents is an absolute fact with me'? If you cannot prove it to the satisfaction of your inquirers, you will say 'my reasoning is at fault but not the fact'. Even so must you say to yourself, 'Though I may not be able to reason out the existence of God, I must accept the experience of and the belief of mankind in the First Cause'.

If even now you are not satisfied, you must ask me again.

Yours sincerely,

Letter to Shankaran
SN 13175 and 13180

223. THE WHY OF THE FIRST CAUSE

The Ashram,
Sabarmati,
April 21, 1928

My dear Shankaran,

I have your letter. So you are now president of a Congress Committee. This is very good. And I am glad that Girdharilal is taking such keen interest in *khadi*.

I am forwarding your letter to Sjt. Vithaldas Jerajani for attention.

I quite agree with you that in *khadi* organizations there should be no indifference, certainly never any cheating. I am asking Vithaldas what terms can be offered.

If you believe in the First Cause, you must regard the 'why' of the First Cause as a futile question. Whilst it is laudable and legitimate to bring everything under the dominion of reason we must be humble enough to recognize that there must be things beyond reason, seeing that man is an imperfect being.

I am glad you are making yourself serviceable all round. I entertain no fear about your being lax in the duty entrusted to you.

I have not been able to decide anything about the European visit.

Yours sincerely,

Letter to Shankaran
SN 13200

224. FUSION OF THOUGHT, WORD, AND DEED

Thanks to Shri Jaydayalji Goenka, an attempt is nowadays being made to create a spirit of devotion in the Marwari society. With this object in view *bhajan* groups have been formed and *bhajan bhavans* are also being run. One such *bhavan* called Govind Bhavan has been started in Calcutta. At Shri Jaydayalji's instance, a certain gentleman was put in charge of it. He indulged in debauchery in the name of devotion. He accepted *puja* from women; women regarded him as God and worshipped him; he gave them his left-overs to eat and debauched them. The simple-hearted women believed that having physical relations with one who had attained self-realization could not be regarded as sinful.

Although the incident is painful, it does not surprise me. All around us we find people who gratify their lust under the guise of devotion. And so long as the essence of devotion is not understood, is it surprising that robberies are committed in the name of religion? It would be surprising if false devotees did not bring evil.

I am a votary of *Ramanama* and the *dwadashamantra* but my worship is not blind. For anyone who is truthful, *Ramanama* is like a ship. But I do not believe that anyone who repeats *Ramanama* hypocritically is saved by it. Instances are cited of Ajamil and others; they are poetical creations and even there there is a hidden meaning. Purity of sentiments has been attributed to them. Anyone believing that *Ramanama* would calm his passions is rewarded by repeating it and is saved. The hypocrite repeating *Ramanama* in pursuit of his passions is not saved but is doomed—'A person will meet the fate conforming with his sentiment.'

Devotees should bear two things in mind:

One, that devotion does not merely consist in repeating the name but also in sacrificial activity that must constantly go with it. There is a belief nowadays that worldly activities have no connection with *dharma* or devotion. This is untrue. The truth is that all activities in this world are related to *dharma* or *adharma*. The carpenter who practises his trade merely in order to earn a living, steals wood and spoils his work is guilty of *adharma*. Another practises his trade for the good of others, say, for making a bed for a sick person, does not commit any theft and works to the best of his abilities and repeats *Ramanama* while working. This constitutes work done in pursuit of *dharma*. This carpenter is a true devotee of Rama. A third carpenter, whether deliberately or through ignorance, gives up his trade in order to repeat *Ramanama*, begs for himself and his children, and if asked to make something for a sick person says: 'For me there is only Rama. I would know no sick man and no happy man.' This carpenter is a degraded creature fallen into the well of ignorance.

Man does not pray to God through speech alone but through thought, word and deed. If any one of these three aspects is missing, there is no devotion. A fusion of these three is like a chemical compound. In the case of the latter, if a single ingredient is not present in its proper proportion, the expected result does not follow. The devotees of today appear to think that the limits of devotion are reached in the use of beautiful language and hence ceasing to be devotees become mere rakes and corrupt others too.

In the second place, how and where should man, who has a physical form, worship God? He is omnipresent. Hence the best and most understandable place where He can be worshipped is a living creature. The service of the distressed, the crippled and the helpless among living things constitutes worship of God. The repetition of *Ramanama* is also meant to help us learn to do so. If *Ramanama* does not thus result in service, it is both futile and a sort of bondage, as it proved in the case of the man in Govind Bhavan. Let this instance serve as a warning to all devotees.

Now a few words to the women. The man who makes others worship him necessarily becomes depraved; but why should the women become so? If they must worship human beings, why should they not worship an ideal woman? Moreover, why worship any living being? The saying of the learned Solon is worthy of being carved in one's heart—'No man can be called good while he is alive.' Those who were good one day have become wicked the following day. Moreover, we cannot even spot hypocrites. Hence God alone should be worshipped. If a human being has to be worshipped he should be worshipped only after his death. This is so because after his death we worship only his virtues, not his physical form. It is necessary that men repeatedly, insistently and courteously point this out to our gullible sisters.

'Indulgence in the Name of Devotion' (G.)
Navajivan, 6 May 1928

225. NEARNESS TO GOD

Correspondents often invite me to answer in these pages questions about God. That is the penalty I have to pay for what an English friend calls the God stunt in *Young India*. Whilst I am unable to notice all such questions in these columns, the following compels an answer:

I read your *Young India* of May 12, 1927, p. 149, where you write, 'I think it is wrong to expect certainties in this world, where all else but God that is Truth is an uncertainty.'

Young India, p. 152: 'God is long-suffering and patient. He lets the tyrant dig his own grave, only issuing grave warnings at stated intervals.'

I humbly beg to say that God is not a certainty. His goal ought to be to spread truth all round. Why does He allow the world to be populated by bad people of various shades? Bad people with their unscrupulousness flourish all round and they spread contagion and thus transmit immorality and dishonesty to posterity.

Should not God, omniscient and omnipotent as He is, know where wickedness is by His omniscience and kill wickedness by His omnipotence there and then and nip all rascality in the bud and not allow wicked people to flourish?

Why should God be long-suffering and be patient? What influence can He wield if He be so? The world goes on with all its rascality and dishonesty and tyranny.

If God allows a tyrant to dig his own grave, why should He not weed out a tyrant before his tyranny oppresses the poor? Why allow full play to tyranny and then allow a tyrant, after his tyranny has ruined and demoralized thousands of people, to go to his grave?

The world continues to be as bad as it ever was. Why have faith in that God who does not use His powers to change the world and make it a world of good and righteous men?

I know vicious men with their vices living long and healthy lives. Why should not vicious men die early as a result of their vices?

I wish to believe in God but there is no foundation for my faith. Kindly enlighten me through *Young India* and change my disbelief into belief.

The argument is as old as Adam. I have no original answer for it. But I permit myself to state why I believe. I am prompted to do so because of the knowledge that there are young men who are interested in my views and doings.[1]

There is an indefinable mysterious Power that pervades everything. I feel It, though I do not see It. It is this unseen Power which makes Itself felt and yet defies all proof, because It is so unlike all that I perceive through my senses. It transcends the senses.

But it is possible to reason out the existence of God to a limited extent. Even in ordinary affairs we know that people do not know who rules or why and how he rules. And yet they know that there is a power that certainly rules. In my tour last year in Mysore I met many poor villagers and I found upon inquiry that they did not know who ruled Mysore. They simply said some god ruled it. If the knowledge of these poor people was so limited about their ruler, I, who am infinitely lesser than God than they than their ruler, need not be surprised if I do not realize the presence of God, the King of kings. Nevertheless I do feel as the poor villagers felt about Mysore that there is orderliness in the Universe, there is an unalterable Law governing everything and every being that exists or lives. It is not a blind law; for no blind law can govern the conduct of living beings and, thanks to the marvellous researches of Sir J.C. Bose, it can now be proved that even matter is life. That Law then which governs all life is God. Law and the Law-giver are one. I may not deny the Law or the Law-giver, because I know so little about It or Him. Even as my denial or ignorance of the existence of an earthly power will avail me nothing, so will not my denial of God and His Law liberate me from its operation; whereas humble and mute acceptance of divine authority makes life's journey easier even as the acceptance of earthly rule makes life under it easier.

I do dimly perceive that whilst everything around me is

ever changing, ever dying, there is underlying all that change a living Power that is changeless, that holds all together, that creates, dissolves and recreates. That informing Power or Spirit is God. And since nothing else I see merely through the senses can or will persist, He alone is.

And is this Power benevolent or malevolent? I see It as purely benevolent. For I can see that in the midst of death life persists, in the midst of untruth, truth persists, in the midst of darkness light persists. Hence I gather that God is Life, Truth, Light. He is Love. He is the supreme Good.

But he is no God who merely satisfies the intellect if He ever does. God to be God must rule the heart and transform it. He must express Himself in even the smallest act of His votary. This can only be done through a definite realization more real than the five senses can ever produce. Sense perceptions can be, often are, false and deceptive, however real they may appear to us. Where there is realization outside the senses it is infallible. It is proved not by extraneous evidence but in the transformed conduct and character of those who have felt the real presence of God within.

Such testimony is to be found in the experiences of an unbroken line of prophets and sages in all countries and climes. To reject this evidence is to deny oneself.

This realization is preceded by an immovable faith. He who would in his own person test the fact of God's presence can do so by a living faith. And since faith itself cannot be proved by extraneous evidence, the safest course is to believe in the moral government of the world and therefore in the supremacy of the moral law, the law of truth and love. Exercise of faith will be the safest where there is a clear determination summarily to reject all that is contrary to Truth and Love.

But the foregoing does not answer the correspondent's argument. I confess to him that I have no argument to convince him through reason. Faith transcends reason. All I can advise him to do is not to attempt the impossible. I cannot account for the existence of evil by any rational method. To want to do so is to be coequal with God. I am therefore humble enough to recognize evil as such. And I call God long-suffering and patient precisely because He permits evil in the world. I know that He has no evil in Him, and yet if there is evil, He is the author of it and yet untouched by it.

I know too that I shall never know God if I do not wrestle with and against evil even at the cost of life itself. I am fortified in the belief by my own humble and limited experience. The purer I try to become, the nearer I feel to be to God. How much more should I be, when my faith is not a mere apology as it is today but has become as immovable as the Himalayas and as white and bright as the snows on their peaks? Meanwhile I invite the correspondent to pray with Newman who sang from experience:

> Lead, kindly Light, amid the encircling gloom,
>   Lead Thou me on;
> The night is dark and I am far from home,
>   Lead Thou me on;
> Keep Thou my feet, I do not ask to see
> The distant scene; one step enough for me.

'God Is'
Young India, 11 Oct. 1928

[1] What follows, excluding the last sentence and the stanza from Newman, was recorded on 20 October 1931, by the Columbia Broadcasting Company, London, during Gandhi's stay in Kingsley Hall.

226. GOD ELUDES ALL MEASURE

September 26, 1930

My dear Mathew,

The things you write about are ultimately decided by faith. Reason can take us only a little distance. Man is a person, God is not in the same sense. Man sins because he has the sense of right and wrong. Our difficulty arises through our effort to measure God by our little selves. And He eludes all measure.

Love.

BAPU

Letter to P. G. Mathew
GN 1553

227. SPIRITUAL SIGHT

March 28, 1932

Vanity is emptiness: Self-respect is substance. No one's self-respect is ever hurt except by self, vanity is always hurt from outside.

In the phrase 'Seeing God face to face', 'face to face' is not to be taken literally. It is a matter of decided feeling. God is formless. He can, therefore, only be seen by spiritual sight—vision.

Letter to Tilakam
Mahadevbhaini Diary, Vol. 1, p. 52

228. REALIZATION OF GOD

April 8, 1932

1 and 2. We must believe in God if we believe in ourselves. If living beings have existence God is the sum total of all life and this in my view is the strongest proof.

3. The denial of God is injurious in the same way as denial of ourselves. That is to say, to deny God is like committing suicide. The fact remains that it is one thing to believe in God and quite another to realize God emotionally and act accordingly. Truly, no one in the world is an atheist; atheism is merely a pose.

4. One can realize God only by ridding oneself totally of attachment, aversion, etc., and in no other way. I hold that one who claims to have realized God has not truly done so. Realization can be experienced, but is beyond description. Of this I have no doubt.

5. I can live only by having faith in God. My definition of God must always be kept in mind. For me there is no other God than Truth; Truth is God.

Letter to Hanumanprasad Poddar (H.)
Mahadevbhaini Diary, Vol. I, p. 82

229. GOD AS THE KNOWER

May 5, 1932

I got your letter. If you ask reason for proof of the existence of God, what proof can reason give since God is above all reason? For, if you say that nothing is beyond rational explanation, you will certainly run into difficulties. If we give the highest place to reason, we shall be faced with serious difficulties. Our own *atman* is beyond reason. People have indeed tried to prove its existence with logical arguments, as they have tried to prove the existence of God. But he who knows God and the *atman* with his intellect only does not know them at all. Reason may be a useful instrument of knowledge at one stage. But anybody who stops there will never enjoy the benefits of true spiritual knowledge, in the same way that intellectual knowledge of the benefits of eating food does not by itself help one to enjoy those benefits. God or the *atman* is not an object of knowledge. He Himself is the Knower. That is why we say that He is above reason.

There are two stages of knowledge of God. The first is faith, and the second and the last is first-hand experience to which faith leads. All great teachers of the world have testified to their having had such experience, and people whom the world would ordinarily dismiss as fools have also demonstrated their faith. If we have faith, like theirs, one day we may have first-hand experience too. If a person sees somebody with his eyes but, being deaf, cannot hear him and then complains that he had not heard him, he would not be justified. Similarly, to say that reason cannot know God betrays ignorance. Just as hearing is not the function of the eyes, so also knowing God is not the function of the senses or of reason. To know Him a different kind of faculty is required, and this is unshakeable faith. We know from experience that reason can be all too easily deceived. But we have not known anybody, howsoever mighty he may be, who could deceive true faith.

Letter to Ibrahimji Rajkotwala (G.)
Mahadevbhaini Diary, Vol. I, pp. 136-7

230. ACCEPTANCE OF GOD

August 15, 1932

I believe in both Gods,[1] the one that serves us and the one that we serve. It cannot be that we should render service and should not receive service of any kind. But both Gods are of our imagining. There is only one God who is real. The real God is beyond conception. He neither serves nor receives service. He cannot be described by any epithets, being not an external power but something dwelling in our heart. Since we do not understand the ways of God, we have necessarily to think of a power beyond our conception. And the moment we think of it, the God of our imagining is born. The fact is that belief in God is a function not of the intellect but of faith. Reasoning is of little help to us in this matter and once we accept God the ways of the world cease to bother us. Then we have to accept that no creation of God can be purposeless. Beyond this I cannot go.

Letter to Bhuskute (H.)
Mahadevbhaini Diary, Vol. I, p. 364

[1] 'There are two Gods. There is the God people generally believe in, a God who has to serve them sometimes in a very refined way; perhaps merely by giving them peace of mind. This God does not exist. But the God whom we all have to serve does exist and is the prime cause of our existence and of all we perceive.' Referring to these words of Tolstoy, the questioner had asked 'which of these two Gods Gandhiji believed in, for if a man believed in the second God, prayer had no meaning for him'.

231. THE LAW AND THE LAW-GIVER

March 9, 1945

God is not a person. He is the Law and also the Law-giver. Hence He is not like a human being. It means that man would reap as he sows. God neither does anything nor prompts one to do anything.

Note to Gope Gurbuxani[1] (H.)
GN 1325

[1] He had asked: 'Why should there be so much evil and unhappiness in the world in spite of God?'

232. GRACE THROUGH SILENCE

April 7, 1946

Mahatma Gandhi referred to his visit to the Quakers' silent prayer meeting this morning, and said that people at this meeting sat in absolute silence for half an hour. They were busy in worship of God and realized the sense of God's presence in their midst. Mahatma Gandhi said that he had himself felt the presence of God during silence. He asked the people to observe silence once a week as it played an important part in the individual's life.

Gandhiji described how his own experience tallied with that of the Quakers.

Emptying of the mind of all conscious processes of thought, and filling it with the spirit of God unmanifest, brings one ineffable peace and attunes the soul with the infinite.

The question may, however, be raised: Should not one's whole life be an unbroken hymn of praise and a prayer to the Maker? Why then have a separate time for prayer at all? Brother Lawrence testified that 'with him the set times of prayers were not different from other times; that he retired to pray according to the directions of his superior, but that he did not want such retirement, nor asked for it, because his greatest business did not divert him from God'. Gandhiji does not question that view. He observed:

I agree that if a man could practise the presence of God all the twenty-four hours, there would be no need for a separate time for prayer.

When the mind is completely filled with His spirit one cannot harbour ill will or hatred towards anyone and reciprocally the enemy will shed his enmity and become a friend. It is not my claim that I have always succeeded in converting enemies into friends, but in numerous cases it has been my experience that when the mind is filled with His peace all hatred ceases. An unbroken succession of world teachers since the beginning of time have borne testimony to the same. I claim no merit for it. I know it is due entirely to God's grace. Let us then in the Sacred Week seek His grace through the communion of silence and, maybe, the experience will abide us ever afterwards.

Speech at Prayer Meeting, New Delhi
Bombay Chronicle, 8 Apr. 1946
Harijan, 28 Apr. 1946

233. THE TRANSCENDENT AND THE IMMANENT

A friend from Baroda, who writes in English, says:

You ask us to pray to God to give light to the whites in South Africa and strength and courage to the Indians there to remain steadfast to the end. A prayer of this nature can only be addressed to a person. If God is an all-pervading and all-powerful force, what is the point of praying to Him? He goes on with His work whatever happens.

I have written on this topic before. However, if one keeps on saying the same thing again and again in different words some new words or phrases used are likely to help someone or other to understand the matter better. In my view, whether called Rama, Rahman, Ormuzd, God or Krishna, He is that Supreme Power that man is ever trying to find a name for. Man, though imperfect, strives after perfection and in so doing is caught up in the tides of thought. Then like a baby learning to toddle, he now stumbles, now stands up. Thus if we say that a reasoning man is only a few months' old child, we shall not in the least be exaggerating, judging by the immensity of cosmic time; we shall be stating a simple truth. Man can express himself only by means of language. But there can be no such medium as language for the Power that is God. Man however can describe this infinite Power only with his imperfect means. If one has grasped this there is nothing left further to ask. Then it would be right to pray to Him in the language of man, for one can comprehend Him somewhat by fitting Him into one's own mould.

One ought always to remember, while dwelling on Him, that one is but a drop, the tiniest of creatures of the ocean that is God. One may experience Him by being in Him, but one can never describe Him. As Madame Blavatsky[1] puts it, man in praying, worships the Great Power residing within. Only he who knows this may pray. He who does not, need not pray. God will not be offended by it, but I can say from experience that that man will be the loser by not praying. So it is immaterial if some worship God as a Person and some others as a Great Power. Both are right, each in his own way. Nobody knows what is intrinsically right and nobody is likely ever to know. The ideal, to be an ideal, must forever remain out of reach. All the other forces are static, while God is the Life Force, immanent and at the same time transcendent.

Sevagram, August 8, 1946

'Is God a Person or a Principle?' (G.)
Harijan, 18 Aug. 1946

[1] *The Key to Theosophy.*

Chronology
M. K. Gandhi (1869-1948)

1869	2 Oct.	Mohandas Karamchand Gandhi was born into a Vaishya family at Porbander in Kathiawar, Gujarat; the youngest of three sons of Karamchand Gandhi, Prime Minister in Porbander, Rajkot, and Vanakner States, and his fourth wife Putlibai.
1876		Goes to Rajkot with parents; attends primary school there until twelfth year. Betrothal to Kasturbai, daughter of Gokuldas Makanji.
1882		Marries Kasturbai Makanji.
1884		Experiments in meat-eating and agnosticism.
1888	Spring	Birth of Harilal.
	4 Sept.	Despite disapproval of caste elders, sails for England to study.
	6 Nov.	Enrols as law student at Inner Temple, London.
1889	Nov.	Meets H. P. Blavatsky and Annie Besant of the Theosophical Society. Becomes acquainted with the *Bhagavad Gita*, *The Light of Asia*, and the Sermon on the Mount.
1890	19 Sept.	Becomes executive member of the London Vegetarian Society.
1891	26 Mar.	Enrols as associate member of the London Theosophical Society.
	10 June	Called to the Bar and enrols in the High Court of London.
	12 June	Sails for India.
1892	Spring	Birth of Manilal.
	14 May	Receives permission to practise law in Kathiawar. Fails to establish successful practice. Settles in Rajkot as legal draughtsman.
1893	Apr.	Sails for South Africa as legal adviser to Dada Abdullah & Company.

1893	June	Ordered off train to Pretoria. Makes resolve to resist racial discrimination nonviolently.
	July	Kicked off footpath near President Kruger's house, but refuses to sue the assailant.
1894	Apr.	Studies religious literature, including the Bible, the Koran, and Tolstoy's *The Kingdom of God is Within You*.
	22 Aug.	Organizes Natal Indian Congress.
	3 Sept.	Enrolled as barrister in the High Courts of Natal and the Transvaal over opposition of European lawyers.
1895	Apr.	Visits Trappist monastery near Durban.
	May	Appeals to Natal Assembly and to Lord Ripon against re-indenture clause in Indian Immigration Bill.
	16 Dec.	Issues *The Indian Franchise: An Appeal to Every Briton in South Africa*.
1896	5 June	Sails for India. Addresses meetings on behalf of Indians in South Africa.
	30 Nov.	Sails for South Africa with his family.
1897	13 Jan.	Arrives in Durban and is attacked by a mob.
	20 Jan.	Declines to prosecute assailants.
	May	Birth of Ramdas.
1898		Petitions local and Imperial authorities regarding discriminatory laws.
1899	Dec.	Organizes Indian Ambulance Corps to serve in the Boer War.
1900	22 May	Birth of Devadas.
1901	18 Oct.	Sails with family to India.
	27 Dec.	Offers resolution on South Africa at Indian National Congress.
1902	Feb.	Stays with Gokhale for a month in Calcutta. Fails to establish successful law practice in Rajkot. Moves to Bombay where he pursues legal work.

	20 Nov.	Returns with family to South Africa in response to call to champion Indian cause against anti-Asiatic legislation in the Transvaal.
1903	Feb.	Enrols as Attorney of Supreme Court of the Transvaal. Opens law office in Johannesburg.
	4 June	Launches *Indian Opinion*.
1904	Oct.	Reads Ruskin's *Unto This Last*.
	Dec.	Founds the Phoenix Settlement near Durban.
1905	May	Begins learning Tamil.
	9 Aug.	Calls for revision of Bill levying poll tax against Natal Indians.
	19 Aug.	Calls for united opposition to Bengal partition and supports boycott of British goods.
1906	12 May	Advocates Home Rule for India.
	June-July	Engages in ambulance work during Zulu Rebellion. Vow of chastity.
	11 Sept.	Addresses mass meeting of Indians at Empire Theatre in Johannesburg calling for withdrawal of Asiatic Registration Bill.
	3 Oct.	Sails for England to seek redress from British government.
	7 Nov.	Addresses members of Parliament.
	Dec.	Returns to South Africa.
1907	Jan.-Feb.	Writes series of eight articles on 'Ethical Religion'.
	14 July	Calls upon Indians not to submit to re-registration.
	31 July	Explains significance of Passive Resistance. General Strike follows.
	28 Dec.	Conducts his own trial and appears in defence of pickets; ordered to leave Transvaal within forty-eight hours. Later, speaks at meeting in Government Square.
1908	10 Jan.	Adopts term '*Satyagraha*' in place of 'Passive Resistance'. Sentenced to two months' imprisonment. Released on 31 January along with all other *satyagrahis*.

1908	10 Feb.	Assaulted and nearly killed by Mir Alam Khan and other Pathans. Appeals from his sick-bed that assailants be forgiven, and asks Asiatics to give their finger-prints voluntarily.
	16 Aug.	Addresses mass meeting and encourages the burning of registration certificates.
	23 Aug.	Mass meeting in Johannesburg in which more registration certificates are burnt. Mir Alam, Gandhi's assailant, and other Pathans admit their error and resolve 'to fight to the end'.
	7 Oct.	Arrested at Volksrust for entering Transvaal without registration certificate. Sentenced to two months' hard labour. Released on 12 December.
1909	16 Jan.	Arrested again at Volksrust for failing to produce registration certificate. On deportation, returns and is re-arrested but released on bail.
	25 Feb.	Arrested at Volksrust on same charge. Sentenced to three months. Released on 24 May.
	23 June	Sails for England.
	10 July	Arrives in London. With assistance of Lord Ampthill, seeks to educate influential British leaders.
	13 Nov.	Returns to South Africa. *En route* writes *Hind Swaraj* and translates Tolstoy's 'Letter to a Hindu'.
1910	4 Apr.	Sends Tolstoy a copy of *Indian Home Rule* (*Hind Swaraj*).
	8 May	Tolstoy replies that Passive Resistance is of greatest importance for India and humanity.
	30 May	Establishes Tolstoy Farm Ashram on 1,100 acres provided by Herman Kallenbach.
1911	22 Apr.	Smuts agrees to assurances demanded by Indians in reciprocation of suspension of *Satyagraha* Movement.
1912	22 Oct.	Gokhale arrives in Cape Town. Gandhi accompanies him during a five-week tour.
		Gives up European dress and milk and restricts his diet to fresh and dried fruit.
1913	Apr.	Kasturbai joins the *Satyagraha* struggle.
	15 Sept.	*Satyagraha* is revived. Party of twelve men and four women, including Kasturbai Gandhi, leave Durban for Volksrust.

Chronology 597

	23 Sept.	Kasturbai is arrested along with other *satyagrahis*. Sentenced to three months' imprisonment at hard labour.
	28 Oct.	Leads march from Newcastle with 1,700 *satyagrahis*.
	6 Nov.	With 2,221 marchers, arrives at Volksrust border. Arrested at Palmford railway station. Others cross the border.
	7 Nov.	At Volksrust, released on bail and rejoins 2,037 marchers.
	8 Nov.	Arrested at Standerton, released on recognition. March continues.
	9 Nov.	Arrested at Teakworth and taken to Balfour.
	11 Nov.	Sentenced at Dundee to nine months' imprisonment at hard labour on charges of inducing a strike.
	18 Dec.	Released. From time of release till settlement takes only one meal a day and wears indentured labourer's dress.
1914	13 Jan.	Begins negotiations with General Smuts, resulting in a compromise on 22 January.
	22 Jan.	Suspends *Satyagraha* following agreement with Smuts.
	18 July	Sails for London *en route* to India, leaving South Africa for the last time.
	6 Aug.	Arrives in England two days after World War I begins.
	8 Aug.	Given reception at Hotel Cecil by English and Indian friends; Jinnah, Lala Lajpat Rai, Sarojini Naidu are among those present.
	13 Aug.	Organizes Ambulance Corps of Indian students in London.
	19 Dec.	Owing to ill health sails for India. Begins learning Bengali.
1915	9 Jan.	Arrival in Bombay. Awarded Kaiser-i-Hind Gold Medal for ambulance services.
	3 Mar.	At Poona meeting to mourn death of Gokhale.
	7 Apr.	Goes to Rishikesh, and visits Swargashram.
	20 May	Establishes Satyagraha Ashram (later known as Sabarmati Ashram) at Ahmedabad.

1915	Sept.	Admits untouchable family to Satyagraha Ashram.
1916		Tours India and Burma, travelling third class on railway.
	6 Feb.	Speaks at Benares University.
	21 Oct.	At Bombay Provincial Conference held at Ahmedabad, Gandhi proposes election of Jinnah as President.
	26 Dec.	Attends Indian National Congress at Lucknow.
	29 Dec.	Presides over All-India Common Script and Common Language Conference in Lucknow.
1917		Idea of using spinning-wheel to produce handmade cloth on large scale takes root in his mind.
	10 Apr.	Begins working with problems of indigo farmers in Champaran, leading to a workers' resolution in August.
	31 Aug.	Tells Mahadev Desai, 'I have got in you the man I wanted.'
	3 Oct.	Champaran Committee reaches compromise with planters.
1918	20 Feb.	Presides over annual gathering of Bhagini Samaj in Bombay, speaking on women's education.
	22 Feb.	Leads *Satyagraha* campaign on behalf of mill workers in Ahmedabad. Settlement reached 18 March.
	22 Mar.	Inaugurates Kheda *Satyagraha* in Nadiad. Successfully terminated 29 June.
	27 Apr.	Attends Viceroy's War Conference at Delhi, addressing it in Hindustani. Tours to raise recruits for British armed forces.
	14 Nov.	Opening of Gujarat Swadeshi Store.
1919	24 Feb.	Notifies Viceroy of *Satyagraha* Pledge.
	Mar.	Issues first '*Satyagraha* Leaflet', quoting Thoreau.
	19 Mar.	Speaks at meeting of Madras Labour Union, B. P. Wadia presiding.
	6 Apr.	Inaugurates all-India *Satyagraha* Movement: country-wide *hartal*.
	7 Apr.	First issue of *Satyagrahi* released without registration.

Chronology 599

	10–12 Apr.	Arrested on way to Delhi for refusal to comply with order not to enter the Punjab. Outbreaks of violence in several towns accompany his escort back to Bombay.
	13 Apr.	Massacre at Amritsar
	14 Apr.	Commencement of three-day penitential fast. Leads *Satyagraha* campaign against Rowlatt Act. Confesses his 'Himalayan miscalculation' regarding mass *Satyagraha*. Martial law declared in Punjab.
	18 Apr.	Suspends *Satyagraha*.
	Sept.	Assumes editorship of *Navajivan*.
	Oct.	Assumes editorship of *Young India*.
	4 Nov.	Received in Golden Temple at Amritsar.
	24 Nov.	Presides over all-India Khilafat Conference at Delhi.
1920	2 Apr.	Rabindranath Tagore visits Sabarmati Ashram.
	1 Aug.	Addresses letter to Viceroy, returning Kaiser-i-Hind, Zulu War, and Boer War medals.
	31 Aug.	Takes pledge to wear *khadi* for life.
	8 Sept.	Special session of Indian National Congress accepts his programme of non-co-operation to secure redress of Punjab and Khilafat wrongs.
	Dec.	Nagpur Congress session adopts his resolution declaring object of Congress to be attainment of *Swaraj* by legitimate and peaceful means.
1921	30 Mar.	In Vijayanagaram, pleads for Hindi to be made lingua franca of India.
	Apr.	Launches programme to set up twenty *lakhs* of *charkhas* in the country.
	31 July	Leads campaign for complete boycott of foreign cloth. Presides over huge bonfire at Bombay.
	31 Oct.	Takes vow of daily spinning.
	19 Nov.	Fasts for five days as protest against communal riots.
	Dec.	Mass *Satyagraha* campaign begins. Invested with full powers by Congress. Many Congress leaders arrested.

1922	4 Feb.	Riots at Chauri Chaura.
	12 Feb.	Commences five-day fast as protest against violence.
		Abandons plan of *Satyagraha* Movement.
	10 Mar.	Arrested for sedition at Sabarmati. Sentenced to six years' imprisonment.
1923	26 Nov.	In prison begins writing *Satyagraha in South Africa*.
1924	12 Jan.	Operated upon for appendicitis.
	4 Feb.	Ordered released from prison.
	12 Feb.	Requests Mahomed Yakub to desist from moving Assembly resolution recommending award of Nobel Peace Prize to him.
	18 May	In first public appearance after release from gaol, presides over Buddha Jayanti celebrations in Bombay.
	17 Sept.	Begins twenty-one-day fast on behalf of Hindu-Muslim unity. Ends fast on 8 October.
1925	15 Feb.	Inaugurates a national school and a Jain hostel at Rajkot.
	2 July	Bakr-i-Id day riots break out in Kidderpore, Calcutta. Gandhi, with Abul Kalam Azad, visits trouble spots, pacifies both communities.
	22 Sept.	Founds All-India Spinners' Association.
	7 Nov.	Madeline Slade (Mirabehn) joins the Ashram at Sabarmati.
	24 Nov.	Announces seven days' fast for misdeeds of Ashram inmates.
	29 Nov.	Begins writing *The Story of My Experiments With Truth*.
1927	Jan.–Nov.	Extensive *khadi* tour through North and South India.
	Nov.	Visits Ceylon.
1928	12 Feb.	Bardoli peasants refuse to pay taxes as an act of *Satyagraha*. Gandhi presides over successful settlement on 6 August.
	Dec.	Moves resolution at Calcutta Congress in favour of Independence if Dominion Status is not granted by end of 1929.

1929	3 Feb.	Completes *The Story of My Experiments With Truth*.
	4 Mar.	Arrested for burning foreign cloth. Released later on personal recognition.
	20 Aug.	Declines Congress Presidentship. Suggests Jawaharlal Nehru instead.
	27 Dec.	Declares for complete Indian independence at Lahore Congress.
1930	26 Jan.	Declaration of Independence prepared by him is proclaimed all over India.
	12 Mar.	Begins Salt March from Sabarmati to Dandi.
	6 Apr.	Breaks salt law on the beach at Dandi. Launches *Satyagraha* throughout India.
	18 Apr.	Riots at Chittagong.
	5 May	Arrested at Karadi and imprisoned at Yeravda gaol without trial. *Hartal* all over India. Over 100,000 are gaoled before close of year.
1931	26 Jan.	Released together with other Congress leaders.
	4 Mar.	Gandhi-Irwin Pact signed.
	8 Apr.	At Amritsar, discusses with Sikhs solutions to communalism.
	2 Aug.	Opens family temple of Chinubhai Madhavlal to untouchables in Ahmedabad.
	12 Sept.	Arrives in London to attend Round Table Conference, to meet with British leaders and elucidate the need for India's complete independence.
	26 Sept.	Talks with representatives of cotton industry.
	27 Sept.	Receives deputations from unemployed workers at Bradford.
	9 Oct.	Meets Madame Montessori.
	23 Oct.	Addresses a gathering at Eton College.
	24 Oct.	Delivers a talk to Oxford dons.
	6 Nov.	Mr and Mrs George Bernard Shaw call on Gandhi.
	14 Dec.	Sails to India after visiting Romain Rolland in Switzerland.
1932	4 Jan.	Arrested in Bombay after his draft resolution for resumption of *Satyagraha* is adopted by the Congress Working Committee. Detained at Yeravda gaol.

1932	20 Sept.	Begins fast unto death as protest against separate electorates for untouchables.
	24 Sept.	Yeravda Pact signed by high and low caste Hindus in the presence of Gandhi.
	26 Sept.	Concludes fast.
1933	Feb.	While in prison, founds the Harijan Sevak Sangh and *Harijan*.
	8 May	Begins fast 'for the purification of self and associates'. Is released from gaol.
	9 May	Announces suspension of *Satyagraha* Movement for six weeks and calls on Government to withdraw its Ordinances.
	29 May	Breaks fast after twenty-one days.
	26 July	Announces disbanding of Satyagraha Ashram, Ahmedabad. Prepares to march to Ras with thirty-three companions on 1 August.
	1 Aug.	Arrested and sentenced to one year's imprisonment for disobeying restraint order.
	16 Aug.	Begins fast because he is not allowed to work for untouchables while in prison. Four days later removed to hospital.
	23 Aug.	Released unconditionally from prison.
1934–4	Nov.–June	Extensive tour on behalf of Harijans in North and South India, last month of which is undertaken on foot.
1934	25 June	Escapes bomb attempt on his life.
	17 Sept.	Announces decision to retire from politics from 1 October to engage in development of village industries, Harijan service, and education through basic crafts.
	24 Oct.	Inaugurates All-India Village Industries Association.
	30 Oct.	Resigns from Congress.
1936	30 Apr.	Settles at Sevagram near Wardha in the Central Provinces, making it his headquarters.
1937	22 Oct.	Presides over Educational Conference at Wardha.
1938	3–5 Feb.	Attends Congress Working Committee at Wardha.

	Oct.	Tours North-west Frontier.
1939	3 Mar.	Commences fast unto death at Rajkot to secure local Ruler's adherence to his promise to reform administration. Ends 7 March on Viceroy's intervention.
	23 July	Writes letter to Hitler (undelivered).
1940		Attends frequent Congress Working Committee Meetings where he plays an active role.
	Oct.	Suspends *Harijan* and allied weeklies following official demand for pre-censorship on the subject of *Satyagraha*.
	17 Oct.	Launches limited civil disobedience campaign in protest against India's enforced participation in World War II.
1941	13 Dec.	Completes *Constructive Programme: Its Meaning and Place*.
1942	18 Jan.	Revives *Harijan* and weekly journals.
	27 Mar.	Meets Sir Stafford Cripps in New Delhi. Later declares his proposals to be a 'post-dated cheque'.
	8 Aug.	Launches 'Quit India' Movement.
	9 Aug.	Arrested and taken to Aga Khan's Palace, Poona.
	15 Aug.	Death of Mahadev Desai from heart failure in Aga Khan's Palace.
1943	10 Feb.	Begins twenty-one-day fast as an appeal for justice. Ends on 3 March.
1944	22 Feb.	Death of Kasturbai Gandhi while in prison at Poona.
	6 May	Released from prison because of ill health. Devotes himself to Constructive Programme.
	9 Sept.	Begins talks with Jinnah.
	27 Sept.	Announces breakdown of talks with Jinnah.
1945	17 Mar.	Declares Vinoba Bhave and Kishorelal Mashruwala as his successors in Sevagram Ashram.
	25 June	Attends Simla Conference.
	19 Dec.	Lays foundation stone of C. F. Andrews Memorial Hospital at Shantiniketan.
1945-6	Dec.-Jan.	Tours Bengal and Assam.

1946	Jan.–Feb.	Tours South India for anti-untouchability and the learning of Hindustani.
	10 Feb.	Once again revives *Harijan* and allied journals.
	Apr.	Participates in political talks with Cabinet Mission in Delhi.
	5–12 May	Attends Simla Conference.
	23 June	Advises Congress not to enter Interim Government proposed by Viceroy.
	24 June	Meets Cabinet Mission.
	29 June	Leaves Delhi for Poona by rail. Attempts made to derail train *en route*.
	7 July	Addresses Congress meeting at Bombay.
	16 Aug.	Four days' rioting starts in Calcutta as the consequence of 'Direct Action' called by Muslim League.
	27 Aug.	Cables warning to British Government against repetition of 'Bengal Tragedy'.
	15 Oct.	Muslim League enters Interim Government.
	Nov.	Tours riot-torn East Bengal on foot for four months.
1947	2 Jan.	Says: 'All around me is utter darkness.'
	3–29 Jan.	Leaves Srirampur on walking tour. Tours riot-affected areas in Bihar.
	29 Mar.	Lord Mountbatten, last Viceroy of India, arrives in India.
	1–2 Apr.	Addresses Asian Relations Conference in Delhi.
	15 Apr.	With Jinnah, issues joint appeal for communal peace.
	5 May	Denies that communal division of India is inevitable.
	2 June	Viceroy's Partition plan revealed. Congress Working Committee accepts.
	6 June	Writes to Mountbatten to persuade Jinnah to settle amicably all outstanding points with Congress.
	12 June	Addresses Congress Working Committee.
	15 Aug.	British India divided into two self-governing dominions; Gandhi rejoices for the deliverance from British rule, while deploring India's partition.

		Mass migration of Hindus and Muslims accompanied by widespread violence.
	1 Sept.	In Calcutta begins fast unto death. Fast broken after local peace is restored four days later.
1948	13 Jan.	Begins fast in New Delhi on behalf of communal unity.
	17 Jan.	Central Peace Committee formed and decides on 'Peace Pledge'.
	18 Jan.	Ends fast.
	20 Jan.	Bomb explosion at Birla House.
	30 Jan.	Struck by an assassin's bullet while on his way to evening prayer meeting. With hands folded in prayer and a gesture of forgiveness, he passed from this life with the words 'Hey Ram, Hey Ram' on his lips.

Bibliography

Primary Sources

COLLECTED WORKS

The Collected Works of Mahatma Gandhi (ninety volumes), New Delhi: Publications Division of the Government of India, Navajivan, 1958–84.

BOOKS BY GANDHI

Hind Swaraj, Ahmedabad: Navajivan, 1938.
Satyagraha in South Africa (translated by V. G. Desai), Ahmedabad: Navajivan, 1928.
The Story of My Experiments With Truth (translated by Mahadev Desai), Ahmedabad: Navajivan; Volume I, 1927; Volume II, 1929.
The Constructive Programme: Its Meaning and Place, Ahmedabad: Navajivan, 1941.
Ashram Observances in Action, Ahmedabad: Navajivan, 1955.
Discourses on the Gita, Ahmedabad: Navajivan, 1960.
A Guide to Health, Madras: S. Ganesan, 1921. Ahmedabad: Navajivan, 1967.

JOURNALS EDITED BY GANDHI

Indian Opinion, Natal, South Africa (1903–14).
Young India, Ahmedabad, India (1919–32).
Navajivan, Ahmedabad, India (1919–31).
Harijan, Ahmedabad, India (1933–48).

COLLECTIONS OF WRITINGS BY GANDHI

Bapu's Letters to Mira (1928–48), Ahmedabad: Navajivan, 1949.
Cent Per Cent Swadeshi, Madras; G. A. Natesan, 1933.
Conversations of Gandhiji (edited by Chandrashankar Shukla), Bombay: Vora & Co., 1949.

More Conversations of Gandhiji (edited by Chandrashankar Shukla), Bombay: Vora & Co., 1950.
Delhi Diary, Ahmedabad: Navajivan, 1948.
The Economics of Khadi, Ahmedabad: Navajivan, 1941.
Ethical Religion, Madras: S. Ganesan, 1922.
For Pacifists, Ahmedabad: Navajivan, 1949.
From Yeravda Mandir: Ashram Observances (translated by V. G. Desai), Ahmedabad: Navajivan, 1932.
Gandhiji's Correspondence with the Government, 1942-1944, Ahmedabad: Navajivan, 1945.
Gokhale: My Political Guru, Ahmedabad: Navajivan, 1958.
Hindu Dharma, Ahmedabad: Navajivan, 1950.
History of Satyagraha Ashram, Madras: G. A. Natesan, 1933.
India of My Dreams, Ahmedabad: Navajivan, 1947.
Letters to Manibehn Patel, Ahmedabad: Navajivan, 1963.
Letters to Rajkumari Amrit Kaur, Ahmedabad: Navajivan, 1961.
The Medium of Instruction (edited by Bharatan Kumarappa), Ahmedabad: Navajivan, 1954.
My Appeal to the British, New York: John Day Company, 1942.
'My Dear Child': Letters to Esther Faering, Ahmedabad: Navajivan, 1956.
Non-Violence in Peace and War, Ahmedabad: Navajivan; Part I, 1945; Part II, 1949.
The Rowlatt Bills and Satyagraha, Madras: G. A. Natesan, 1919.
Sarvodaya, Ahmedabad: Navajivan, 1951.
Satyagraha, Ahmedabad: Navajivan, 1951.
Selected Letters, Ahmedabad: Navajivan, 1962.
Self-Restraint v. Self-Indulgence, Ahmedabad: Navajivan, 1947.
Socialism of My Conception, Bombay: Bharatiya Vidya Bhavan, 1957.
Speeches and Writings, Madras: G. A. Natesan, 1933.
To a Gandhian Capitalist, Bombay: Hind Kitabs, 1951.
To Ashram Sisters, Ahmedabad: Navajivan, 1952.
Unto This Last, Ahmedabad: Navajivan, 1951.
Untouchability, Ahmedabad: Navajivan, 1954.
Women and Social Injustice, Ahmedabad: Navajivan, 1942.

Secondary Sources

ANDREWS, C. F., *Mahatma Gandhi's Ideas*, London: George Allen, 1929.

ANDREWS, C. F., *Mahatma Gandhi: His Own Story*, New York: The Macmillan Company, 1930.
—— *Mahatma Gandhi at Work*, New York: The Macmillan Company, 1931.
ASHE, GEOFFREY, *Gandhi: A Study in Revolution*, London: Heinemann, 1968.
BIRLA, G. D., *In the Shadow of the Mahatma*, Bombay: Orient Longmans, 1955.
BONDURANT, JOAN, *Conquest of Violence*, Berkeley: University of California Press, 1965.
BROWN, D. M., *The White Umbrella: Indian Political Thought From Manu to Gandhi*, Berkeley: University of California Press, 1958.
BROWN, JUDITH, M., *Gandhi's Rise to Power: Indian Politics 1915-1922*, Cambridge: Cambridge University Press, 1972.
—— *Gandhi and Civil Disobedience: The Mahatma in Indian Politics 1928-1934*, Cambridge: Cambridge University Press, 1977.
CATLIN, GEORGE, *In the Path of Mahatma Gandhi*, London: Macdonald & Co., 1948.
CHARPENTIER, MARIE VICTOIRE, *Gandhi*, Paris: Édition France-Empire, 1969.
DATTA, DHIRENDRA MOHAN, *The Philosophy of Mahatma Gandhi*, Madison: University of Wisconsin Press, 1961.
DESAI, MAHADEV, *Gandhiji in Indian Villages*, Madras: S. Ganesan, 1927.
—— *Gandhiji in Ceylon*, Madras: S. Ganesan, 1928.
—— *The Story of Bardoli*, Ahmedabad: Navajivan, 1929.
—— *The Nation's Voice*, Ahmedabad: Navajivan, 1932.
—— *The Gita According to Gandhi*, Ahmedabad: Navajivan, 1946.
DESAI, VALJI GOVINDJI (ed.), *The Diary of Mahadev Desai*, Ahmedabad: Navajivan, 1953.
DHAWAN, G., *The Political Philosophy of Mahatma Gandhi*, Ahmedabad: Navajivan, 1951.
DIWAKAR, R. R., *Satyagraha—Its Technique and Theory*, Bombay: Hind Kitabs, 1946.
DOKE, J. J., *M. K. Gandhi: An Indian Patriot in South Africa* (introduction by Lord Ampthill), London: The London Indian Chronicle, 1909.
ELWIN, VERRIER, *Mahatma Gandhi*, London: Golden Vista Press, 1932.
ERIKSON, ERIK H., *Gandhi's Truth*, New York: Norton, 1969.
FISCHER, LOUIS, *The Life of Mahatma Gandhi*, New York: Harper & Brothers, 1950.
GANDHI, MANUBEHN, *The Miracle of Calcutta*, Ahmedabed: Navajivan, 1959.

Bibliography

—— *Last Glimpses of Bapu*, Delhi: Shiva Lal Agarwala, 1962.
GEORGE, S. K., *Gandhi's Challenge to Christianity*, London: Allen & Unwin, 1939.
GREGG, RICHARD B., *The Power of Non-Violence*, Ahmedabad: Navajivan, 1938.
HORSBURGH, H. J. N., *Non-Violence and Aggression*, London: Oxford University Press, 1971.
HUNT, JAMES D., *Gandhi in London*, New Delhi: Promilla, 1978.
HUTTENBACK, ROBERT A., *Gandhi in South Africa: British Imperialism and The Indian Question, 1860-1914*, Ithaca: Cornell University Press, 1971.
IYER, RAGHAVAN N., *The Moral and Political Thought of Mahatma Gandhi*, New York: Oxford University Press, 1973. Galaxy Paperback, 1979. Second edition: Santa Barbara: Concord Grove Press, 1983.
—— *Utilitarianism and All That*, London: Chatto & Windus, 1960. Second edition: Santa Barbara: Concord Grove Press, 1983.
—— *Parapolitics: Toward the City of Man*, New York, Oxford: Oxford University Press, 1979. Second edition: Santa Barbara: Concord Grove Press, 1985.
KRIPALANI, J. B., *Gandhian Thought*, Bombay: Orient Longmans, 1961.
—— *Gandhi: His Life and Thought*, New Delhi: Publications Division of the Government of India, 1975.
KYTLE, CALVIN, *Gandhi, Soldier of Non-Violence: His Effect on India and the World Today*, New York: Grosset & Dunlap, 1969.
LAÑZA DEL VASTO, JOSEPH J., *Gandhi to Vinoba: The New Pilgrimage* (translated from the French by Philip Leon), London: Rider, 1956.
LEYS, WAYNE, and RAO, P. S. S. R., *Gandhi and America's Educational Future*, Carbondale: Southern Illinois University Press, 1969.
MAURER, HERRYMON, *Great Soul*, New York: Doubleday, 1948.
MUZUMDAR, HARIDAS T., *Gandhi Versus the Empire*, New York: Universal Publishing Co., 1932.
NAESS, ARNE, *Gandhi and the Nuclear Age*, Totowa: Bedminster Press, 1965.
NAG, KALIDAS, *Tolstoy and Gandhi*, Patna: Pustak Bhandar, 1950.
NAMBOODIRIPAD, E. M. S., *The Mahatma and the Ism*, New Delhi: People's Publishing House, 1958.
NANDA, B. R., *Mahatma Gandhi*, London: Allen & Unwin, 1958.
NIKAM, N. A., *Gandhi's Discovery of Religion: A Philosophical Study*, Bombay: Bharatiya Vidya Bhavan, 1963.
OSTERGAARD, G., and CURRELL, M., *The Gentle Anarchists*, Oxford: Clarendon Press, 1971.

PANTER-BRICK, SIMONE, *Gandhi Against Machiavellism: Non-Violence in Politics* (translated by D. Leon), London: Asia Publishing House, 1966.

PAYNE, ROBERT, *The Life and Death of Mahatma Gandhi*, New York: E. P. Dutton & Co., 1969.

POLAK, H. S. L., BRAILSFORD, H. N., PETHICK-LAWRENCE, FREDERICK, *Mahatma Gandhi*, London: Odhams Press, 1949.

POLAK, MILLIE GRAHAM, *Mr. Gandhi: The Man*, Bombay: Vora & Co., 1950.

POWER, PAUL F., *Gandhi on World Affairs*, Washington: Public Affairs Press, 1960.

PRABHU, R. K., and RAO, U. R. (eds.), *The Mind of Mahatma Gandhi*, Ahmedabad: Navajivan, 1967.

PRASAD, RAJENDRA, *Satyagraha in Champaran*, Ahmedabad: Navajivan, 1949.

PYARELAL, *The Epic Fast*, Ahmedabad: Mohanlal Maganlal Bhatt, 1932.

—— *Mahatma Gandhi: The Last Phase*, Ahmedabad: Navajivan; Volume I, February 1956; Volume II, February 1958.

—— *Mahatma Gandhi: The Early Phase*, Ahmedabad: Navajivan, 1965.

RADHAKRISHNAN, S. (ed.), *Mahatma Gandhi: Essays and Reflections*, London: Allen & Unwin, 1938.

—— *Mahatma Gandhi—100 Years*, New Delhi: Gandhi Peace Foundation, 1968.

RAMACHANDRAN, G., and MAHADEVAN, T. K. (eds.), *Gandhi: His Relevance for Our Times*, Bombay: Bharatiya Vidya Bhavan, 1964.

RAO, V. K. R. V., *The Gandhian Alternative to Western Socialism*, Bombay: Bharatiya Vidya Bhavan, 1970.

Reflections on 'Hind Swaraj' by Western Thinkers, Bombay: Theosophy Company, 1948.

REYNOLDS, REGINALD, *To Live in Mankind—A Quest for Gandhi*, London: Andre Deutsch, 1951.

ROLLAND, ROMAIN, *Mahatma Gandhi*, London: Allen & Unwin, 1924.

ROTHERMUND, INDIRA, *The Philosophy of Restraint*, Bombay: Popular Prakashan, 1963.

SHARMA, JAGDISH, *Mahatma Gandhi: A Descriptive Bibliography*, New Delhi: S. Chand & Co., 1955.

SHARP, GENE, *Gandhi As a Political Strategist*, Boston: Porter Sargent, 1979.

SHIRER, WILLIAM LAURENCE, *Gandhi: A Memoir*, New York: Simon & Schuster, 1979.

SHUKLA, C., *Gandhi's View of Life*, Bombay: Bharatiya Vidya Bhavan, 1954.

SPRATT, PHILIP, *Gandhism: An Analysis*, Madras: Huxley Press, 1939.
TENDULKAR, D. G., *Mahatma* (eight volumes), New Delhi: Publications Division of the Government of India, 1951–1954.
—— *Gandhi in Champaran*, New Delhi: Publications Division of the Government of India, 1957.
WATSON, FRANCIS, and BROWN, MAURICE (eds.), *Talking of Gandhiji*, Calcutta: Orient Longmans, 1957.

Glossary

abala, weak
adharma, amoral, unrighteous
advaita, non-duality, monism
ahimsa, non-injury, non-violence, harmlessness; renunciation of the will to kill and the intention to hurt; abstention from any hostile thought, word, or act; non-coercion
akhadas, centres for *sadhus* of particular sects
anasakti, selflessness; selfless action
anasakti yoga, yoga (discipline) of selfless action
angarakhun, a tight-fitting coat of relatively thin cloth
antyaja, untouchable
aparigraha, non-possession; renunciation
arya, literally 'the holy' or 'the noble'; originally the title of the *rishis*, those who walked the *aryamarga* or noble path of holiness
ashram, a spiritual fellowship or community
asteya, non-stealing
atman the Universal Self
avatar(a), literally a 'descent'—the incarnation of divinity

bahadur, brave, powerful, sovereign
bania, the caste of traders and farmers
bansi, a bamboo flute, such as that played by Krishna
bapu, literally 'father'; a term of affection and respect
bhajan, devotional hymn or chant
bhajan bhavan, a place where people gather to sing *bhajans*
bhajanavali, a collection of devotional hymns or chants
bhakti, devotion, faith, adoration of the divine
bhakti yoga, the path of faith, devotion, and adoration
bhang, Indian hemp, used as a narcotic
bhangi, member of a caste associated with sweeping and scavenging
bhavan, abode or lodging

bhumi, the land or the earth
brahmachari, a follower of *brahmacharya*, an exemplar of chastity
brahmacharya, fidelity, chastity; the first of the four stages of life
brahmin, member of the first of the four castes, whose chief duty is the study of the Vedas and the performance of sacrifices
buddhi, discrimination, moral perception

chapati, thin flat cake made of flour; unleavened bread
charkha, spinning-wheel
charpai, string cot
chawl, tenement building
chiku, Indian fruit
chitta, mind-stuff, pure perception

dacoit, bandit, robber
dana, gift-giving, charity
darshana, literally 'point of view'; school of philosophical thought
dastur, Parsee priest
deva, from the root *div*, 'to shine'; a celestial being
dharma, duty, righteousness, moral law; social and personal morality; natural law, natural obligation
dharmakshetra, the field of righteousness or *dharma*
dhoti, a single piece of cloth worn wrapped around the waist
duragraha, persistence in wrongdoing
dwadashamantra, twelve-syllabled *mantra*

ekadashi, eleventh day of each half of the lunar month, used for self-purification

fakir, a Muslim ascetic; a mendicant

gadi, cushion
gayatri, a most sacred verse, addressed to the Sun in the *Rig Veda*
gazal, a poetic composition of Persian origin in the style of lyrics
ghee, clarified butter
goonda, rowdy; vagabond

Glossary

grihastha, householder; family life
gunas, modes of cosmic energy; qualities or attributes; *sattva* (purity or clarity), *rajas* (restlessness), and *tamas* (torpidity)
guru, spiritual preceptor and guide

harijan, literally 'child of God', the name Gandhi gave to the untouchables
hartal, boycott, strike; cessation of work
himsa, injury; violence

id, a day sacred to Muslims
itihasa, a history; a record of events

jiva, individual soul
jnana, wisdom, knowledge
jnana yoga, the path of knowledge

kama, desire, pleasure; human affections and happiness
karma, moral law; the law of ethical causation and moral retribution; causality, action
karmabhumi, land of duty
karma yoga, spiritual realization through social action
karma yogin, practitioner of *karma yoga*
khadi (*khaddar*), hand-spun cloth
klesha, anguish
kosha, glossary
kshatriya, a member of the second or warrior caste

lakh, a hundred thousand
lila, sport, play

mahajan, leader
maharshi, great seer
mahatma, great soul
mahavakya, literally 'great utterance'
mantra, a sacred syllable or incantation
maulana, a scholarly and respected Muslim
maulvi, a learned Muslim divine
maund, unit of measure
maya, the veil of cosmic illusion, the façade of appearance
moha, primal ignorance and delusion
moksha, liberation, emancipation, enlightenment; spiritual freedom and redemption; salvation
mukta purusha, a liberated being
mukti, deliverance

mulla, a Muslim religious leader
mumuksku, seeker after *moksha*, deliverance from phenomenal existence
muni, silent sage

namasudras, a *harijan* caste from Bengal
neti, neti, literally 'not this, not this'—a philosophical exercise for negating relative truths
nirguna, without attributes
nirvana, emancipation from conditioned existence
niyamas, see *yama-niyama*
niyoga, insemination by a person other than one's husband

padarthakosha, a concordance and glossary
panchama, a member of the 'fifth caste'—an untouchable
papayoni, one born of sin, the most wicked of all sinners
pararaj, foreign rule
paricharya, service; attendance
phoongy, Buddhist monk of Burma
prabhatiyan, devotional songs with which the devout, getting up before dawn, commence the day
prakriti, matter; nature
puja, an offering; worship and divine honours offered to the object of devotion
purna swaraj, complete self-rule; total independence
purnavatara, full incarnation of divinity; a perfect *Avatar*
purushartha, the epitome of divine manhood; one of the four ends of human life
purushottama, perfect man, universal man

raj, kingdom, rule, regime
rajas, passion; restlessness; see *gunas*
rajya guru, supreme teacher
ramanama, chanting the name of Rama
ramarajya, the rule of Rama, the Golden Age; the ideal form of government; the kingdom of God on earth
ramdhun, chanting in praise of Rama
rishi, seer
rotli, a flat, round, unleavened bread
ryot (*raiyat*), Indian peasant

sadhana, spiritual discipline
sadhu, ascetic, recluse

saguna, with attributes
samaj, an association, religious or secular
samanvaya, synthesis
samatva, similitude; equanimity in all circumstances
sanatan(a), eternal
sanatana dharma, the eternal verities
sanatani, a faithful follower of Vedic tradition
sandhya, literally 'dawn' or 'sunset'; the *puja* associated with that time of day
sangathan, communalism; loyalty to a sect or group
sangh(a), a voluntary association
sannyasa, renunciation
sannyasi, one who has renounced the world
sardar, a term of respect usually applied to a Sikh
sarvodaya, universal welfare; social good, public interest
sat, abiding, actual, right; self-existent essence
sattva, truth; goodness; purity; see *gunas*
sattvik guna, the qualities of truth, goodness, and purity; see *gunas*
satya, truth; real, existent; valid; sincere, pure; effectual
satyagraha, non-violent resistance; a relentless search for truth; truth-force; holding on to truth
satyagrahi, one who offers *satyagraha*
satyanarayana, Truth as God; God in the form of Truth
savarna, fair in complexion
seva(k), service
shastra(s), Hindu scriptures
shastri, theologian; scholar
shatavadhani, one who can pay simultaneous attention to a hundred things
shraddha, death-anniversary observance
shuddhi, ritual purity; exclusiveness
shudra (sudra), servant or menial caste
siddha, one who has attained spiritual realization
sloka (shloka), verse
smriti, traditional accounts imparted orally, from the word *smriti* or 'memory'; the ceremonial writings of the Hindus, less sacred than the Vedas, which are *sruti* or 'revealed'
swadeshi, self-sufficiency, self-reliance; patriotism

swaraj, freedom; self-rule; political independence

tabligh, ritual purification
takli, spinning-wheel
taluk(a), a district, usually clearly demarcated, in a city or a rural area
tamas, inertia; chaos; darkness; see *gunas*
tamasha, amusement; farcical show
tapas, austerities, penance
tapascharya, meditation and austerities
tapasya, practice of *tapas*
tapovana, hermitage for meditation
tasbih, a Muslim rosary
thugs, predatory groups which used to loot, rob, and kill
tilak(a), auspicious mark on the forehead
tulsi, the basil plant

ulema, a learned man in Islam; an expert in the *Koran*
upas, a poisonous tree

vaid, physician; practitioner of Ayurveda
vairagya, a state of disinterestedness
vaishnava, a devotee of Vishnu; an exemplar of devotion
vaishya, the merchant caste
vakil, lawyer
vanik (vania), see *bania*
varnashram(a), the organization of society into four castes and four stages of life
vedia, literalist
vibhuti, spiritual power
viman(a), flying machine

yajna, sacrifice
yama-niyama, yamas, the chief abstentions, are: *ahimsa* (non-violence), *satya* (truth), *asteya* (non-stealing), *brahmacharya* (chastity), and *aparigraha* (non-possession). The *niyamas*, or main observances, are: *shaucha* (bodily purity), *santosha* (contentment), *tapa* (forbearance), *swadhyaya* (study of scriptures), and *Ishwarapranidhana* (resignation to the will of God)
yoga, spiritual discipline; union with the divine; skill in action
yogabuddhi, spiritual knowledge through a synthesis of *karma yoga*, *bhakti yoga*, and *jnana yoga*
yogi(n), one following a spiritual discipline

Index

Action, 2-4, 10-11, 13, 15, 39, 40-1, 43-4, 66-7, 77, 87, 93-4, 99, 118, 127, 129-30, 135-6, 138, 147, 152, 263, 300, 399, 474, 525, 563, 575
Adharma, 67-8, 71, 435, 552, 581
Advaita, 59, 83
Agnostics, 555, 571
Ahimsa, 4, 7, 39-41, 62, 90, 96, 104, 153, 225, 322-3, 349, 373, 450, 455, 502, 507-8, 511-12, 537, 542, 552, 573
Akbar, 306, 470
Ali Brothers, 392-4, 512, 522
Allah, 322, 515
America, Americans, 26, 30, 37, 48-52, 54-5, 57, 280-2, 293, 359, 362, 445
Anarchist(s), 278, 283, 288
Anarchy, 6, 260, 301, 399
Anasaktiyoga, 93, 514
Andrews, C. F., 113, 388, 517, 536
Antyaja(s), 67-72
Anugita, 82
Aparigraha, 90, 502
Arab(s), Arabia, 150, 468, 472, 475
Arjuna, 21, 57, 82, 85, 88, 93, 96, 129, 183, 513
Arnold, Sir Edwin, 72-3, 78, 476, 480
Art, 458-9, 550
Aryan(s), 465, 471
Ashram(s), 23-5, 32, 63, 68, 103-4, 115, 191-2, 368-9, 556
Asoka, 467, 471
Assassination(s), 11, 35, 133, 236-8, 373
Asteya, 90, 502
Atheism, 72, 74-5, 255, 481, 571, 575, 587
Atheist(s), 9, 116, 129, 463, 503, 555, 570, 572, 574
Atman, 21, 95, 122, 141-2, 146, 151, 309-10, 430, 451, 488, 588
Aurangzeb, 306
Avatar(s), 96, 149, 186
Ayurveda, 96
Azad, Maulana Abul Kalam, 392

Bajaj, Jamnalal, 20, 432-4, 456

Banker, Shankerlal, 159, 163, 438, 441, 443
Belief(s), 8, 11, 27, 182, 579
Bengal, 204-6, 256, 396-7
Besant, Annie, 73-4, 76
Bhagat, Akha, 70, 120, 152, 449, 489-90
Bhagavad Gita, 8, 19, 21, 23, 46, 69, 71-4, 77-9, 81-6, 89-90, 92-100, 114, 120, 123, 128-9, 134, 151-2, 188, 190, 195-6, 225, 273, 299, 339, 379, 462, 484, 488, 510, 512-15, 568
Bhagavata, 151, 182, 450, 489
Bhajans, 22, 196-7
Bhakti, 3, 9, 22-3, 84, 90-1, 95, 99, 136, 339
Bharata, 81, 431, 463
Bhatt, Shamal, 74, 78, 147
Bhave, Vinoba, 21-2, 40, 94, 427-8, 546
Bible, 8, 73, 85, 100, 107, 151, 211, 359, 450, 512, 514, 568
  Genesis, 73
  Mark, 363
  Matthew, 363, 462, 497
  New Testament, 73, 182, 357-8, 503, 510-11, 533
  Numbers, 73
  Old Testament, 73, 182, 329, 503
  Sermon on the Mount, 9, 26, 73-4, 181, 351, 498-501, 506, 517, 539
Blavatsky, H.P., 73, 76, 473, 591
Boehmen, Jacob, 188
Boer War, 528
Bombay, 256, 274, 292-3, 325, 332
Bose, Sir J. C., 312, 315, 428, 584
Boycott, 30, 179, 256, 411-12
Bradlaugh, Charles, 74, 555, 570-1, 573-4
Brahma, 122, 551
Brahmacharya, 13, 80, 87, 90, 285, 339, 502, 546
Brahman, 81, 83, 152, 160, 471, 474, 551
Brahmin, 23, 71, 84-5, 111, 183, 255, 321-2, 471, 481, 487, 504
Brahmo Samaj, 447-9, 473, 475

Bread labour, 119-20, 122-3
Britain, British, 5, 6, 61, 68-70, 78, 110, 133, 201, 277, 293, 295, 306, 308, 325, 331-2, 372, 506
  Empire, 205, 373
  Parliament, 123
  Raj, 16
Brotherhood, 4, 9, 12, 52, 54, 58-9, 76, 455, 515, 527-34
  universal, 4, 391-4
Buddha, Gautama, 9, 12, 16, 18, 41, 44, 122, 289, 329, 359, 390, 467-8, 471, 476-7, 479-83, 491-6, 500-5, 507-8, 512, 550, 559-61, 573
Buddhi, 94
Buddhism, Buddhist(s), 8, 16, 27, 76, 87, 379, 467, 472, 474, 476, 480-3, 494-5, 499, 502, 560, 571, 573-5
  Reformation, 476-7
Burma, 467, 472, 493, 495, 507-8

Caesar, 510-11
Calcutta, 292-3, 325, 332, 505
Canada, Canadian(s), 204, 208
Capital, 51, 350
Capitalism, Capitalist(s), 5, 46-7, 288
Carlyle, Thomas, 74, 210
Caste, 63, 75, 466, 471
Celibacy, 90, 114, 160, 502
Ceylon, 467, 472, 493, 503, 508, 577
Champaran, 43, 45, 364, 367
Charity, 63, 104, 128, 266, 361, 377
Charkha, 315, 353, 428
  see also Spinning-wheel
Charkha Sangh, 425-9, 435, 438, 440, 442
Chastity, 249-51
China, 231, 467, 472, 493, 495, 499, 574
Christians, Christianity, 8-9, 16, 19, 26-7, 61, 69, 71, 76, 111, 144, 153-5, 181-2, 186-7, 210, 217-18, 221-3, 290, 298, 332, 360, 390-1, 455, 468, 472-4, 480, 498-9, 503, 523, 530, 532-3, 544, 550, 553, 569
  esoteric, 329
  see also Jesus Christ
Civil disobedience, 75, 102-3, 442, 510
Civil resistance, *see* Resistance
Civilization, 4-5, 12, 51, 54, 187, 212-15, 218-20, 234, 237, 251, 254, 260-4, 266, 269, 277-9, 283, 288-354, 360-1, 384, 473

  modern, 1, 4-6, 51, 110, 199, 211-12, 217-18, 233, 238, 256, 272, 277, 279, 283, 290-1, 293, 298-9, 312, 325-6, 329-54, 372, 374, 500, 515
  real, 233
  Satanic, 214, 303, 328
Co-operation, 363-9
Communion, 555-7
Communism, Communist, 5, 63-4, 75, 419, 422
Compassion, 12, 14, 78, 80, 107, 112, 117, 150, 266, 339, 474, 489, 499, 519
Conduct, 11, 14-15, 23, 69-70, 85-6, 147, 231-2, 303, 320, 327, 498
Congress, *see* Indian National Congress
Conscience, 7, 11, 13, 19, 122, 127, 211, 245-8, 252, 297, 301, 381, 443, 519, 571-2
Conscientious objector, 327, 571
Constructive programme, 419, 421-43
Conversion, 18, 26, 327, 456-7, 473, 537, 539
Courage, 12, 109, 117, 134, 247-8, 341
Cow-protection, 224, 304, 338, 522, 526, 567
Cowardice, 83, 161, 219, 250, 305, 531, 552-3
Culture, 121, 140, 384, 434, 450-1, 497

Dana, 88
Dastur, 255
Death, 95, 101, 111, 128, 161, 169, 189, 248-50, 563, 585
Deed(s), 9, 107, 117, 182
Deliverance, 21-2, 60, 82, 161
Democracy, 54, 319, 349, 394, 413, 445, 572
Detachment, 10, 15, 17, 66, 94, 437, 446
Devotee(s), 23, 57-8, 89, 552, 581-2
Devotion, 9, 22, 79-81, 90-1, 98, 106, 133, 135-6, 141, 147, 152, 185, 339, 489, 547-65, 581-2
Dharma, 10-11, 23, 56, 68-71, 78, 80, 82-3, 86, 122, 130, 147, 148-51, 183, 284, 303, 321, 435, 438, 552, 581-2
Disciple(s), 25, 58, 103, 136
Discipline, 51, 79, 86, 210, 249, 303, 365, 406, 557
Disease, 212-15, 220, 234, 242-3, 263, 328, 340, 349-50, 361, 464

Index

Divine Mind, 558, 562
Doctor(s), 51, 71, 215, 219-20, 222, 227, 229, 230-2, 258, 263, 294, 304, 340, 426
Dogma(s), 1, 8-10, 26-7, 110, 150, 559
Duragraha, 68-9
Duty, Duties, 2, 10, 12, 19, 22, 25, 39, 57, 68, 70, 77-8, 82, 88, 94, 96, 108, 117, 133, 142, 144, 148-50, 225, 231, 236, 240, 247, 253, 260, 263-4, 286, 373, 394, 488

East India Company, 215-16, 270, 306
East, the, 26, 50-1, 61, 266, 292-3, 318, 328, 324-9
Economics, 355, 357, 361-3
Educated classes, 117, 447-9, 487
Education, 79, 103-5, 129, 208, 232, 251-6, 309, 312-13, 372, 464, 475, 497, 518
End(s), 5, 88, 129, 211, 239, 399
see also Means and ends
England, English, 16-17, 30, 47, 49-50, 54, 58, 63, 72, 78, 90, 102-5, 109, 112, 116, 121, 123, 125, 133-4, 139-40, 147, 153-4, 178, 181, 184, 187, 189, 193, 195, 202-3, 207-9, 211, 214-17, 219-23, 225-30, 233, 235-40, 243-4, 253-4, 259, 261-4, 269, 271-2, 289, 295-6, 304, 323, 334, 337, 344-5, 347-8, 358, 360, 362, 503, 562
Equality, 12, 411, 469, 523-47
Esoteric Christian Union, 154, 288, 329-30
Ethics, 5, 8, 110, 232, 253, 277, 298, 572
Europe, Europeans, 5, 16, 29-32, 47-9, 52, 57, 105-8, 110, 112, 150, 199, 211-15, 230-1, 266, 293, 295, 298, 302-3, 305, 309, 315-17, 325, 345, 349, 353, 372, 489
Evil(s), 5, 7, 10, 17, 47, 62, 70, 74, 90-1, 101, 107-9, 112, 119, 161, 183-4, 212, 220, 243, 279, 289, 298, 310, 344, 346, 418, 452, 531, 538, 553, 575, 578, 585-6
Experiment(s), 1, 3, 36, 39, 42, 50, 53, 55, 62, 159, 280, 289, 350, 355-6, 401, 416, 437, 441
Exploitation, 52, 61, 345, 349
Extremists, 206, 259-60, 331-2

Faith(s), 1-3, 8, 11-12, 21, 23, 26-7, 31, 37, 79, 86, 98, 100, 116, 129, 142, 144, 150-3, 198, 291, 302, 316, 350, 366, 387, 401-2, 430, 448-51, 457-8, 463-4, 489, 497, 499, 534, 536, 554, 563, 569, 577-8, 585-9
Fakir, 111, 126, 232, 470, 473, 513
Family, 7, 47, 53, 152, 245, 267, 420
Fasting, 89, 91, 172, 356
Fear, 21, 24, 43, 56, 83, 117, 119, 138, 210, 219, 238, 241, 249, 251, 308, 309, 335, 551
Fearlessness, 19, 127, 133-4, 185, 249-51, 274, 304, 361, 373, 378, 532, 572
Fellowship, 280, 536
First Cause, 579-80
Fischer, Louis, 75-7, 521
Force, 62, 69, 98-9, 112, 238, 243, 301, 306, 308, 345, 371, 434, 459, 472, 477, 532-3, 592
brute, 109, 238-40, 242-3, 245, 247, 249, 259, 262, 279, 316, 342, 345, 534
parallelogram of, 99
spiritual, 316
Freedom, 4-5, 7, 19, 38, 54, 59, 91, 141-2, 152, 156, 234, 255, 257, 286, 287, 295, 301, 341, 347, 408, 413, 415, 462, 554, 563, 575

Gandhi Seva Sangh, 415-43
Gandhi, Mohandas Karamchand (2 October 1869-30 January 1948)
approach to politics, 381-3
autobiography, 3, 28, 39, 391
books read, 72-4, 154-93, 265
critics, 42-3, 364-5
dharma, 55-8
on disciples, 25
experiments with truth, 39
on his failures, 77
faith in India's economy, 317
followers, 54, 423, 424
on the future, 395
health, 25, 29, 31, 113
as a Hindu, 27
as husband, 28
influences, 66-198
as a labourer, 45-6
as leader, 2, 34-5, 516
on learning to think, 197-8

Gandhi—*cont.*
  on his limitations, 376
  message, 20, 31-2, 35, 37, 50-1, 53, 60, 280, 281
  methods, 50
  mission, 13, 17, 45, 53, 58-9, 381
  on his mistakes, 42, 422
  nationalism, 533
  new order, 62
  not accepting office, 407
  not a perfect man, 17, 19
  patriotism, 59, 113
  as politician, 1, 42
  power over masses, 34
  powers, 33, 39
  public service, 46
  on reading, 196-8, 324
  on his retirement, 24-5, 401-2, 416
  successor, 10, 35
  as a saint, 1, 42
  and Theosophy, 72-3, 76
  as a thinker, 2, 36
  vision, 1, 8, 53, 283, 430, 437
  and the West, 29-32, 48-9, 51, 53-5, 57, 112, 280-1, 521
  on women, 62-3
  writings, 2, 37, 104, 281
Gandhism, 2, 61-2, 64-5, 416, 423-4, 426, 431-2, 437
Ganges, 38, 221, 488, 490, 551-2
Gaol, 3, 157, 178-9, 263
Garibaldi, Giuseppe, 106, 236, 270
Gayatri, 490, 547
Gentleness, 44, 379
Germany, 293, 429
God, 2, 4, 8-10, 13, 16, 19-22, 25-6, 31, 33, 36, 39-41, 45, 48, 53, 57-9, 67, 71, 80, 82-4, 86-9, 91, 95-6, 99-100, 108, 112, 116, 118-19, 122, 137, 150, 155, 160, 170, 181, 185-6, 188, 191, 198, 216-17, 222-3, 225, 234, 239, 246, 259, 266-7, 272, 290, 310, 321-2, 335, 341, 345, 359, 362, 374-5, 379-81, 385-8, 390, 393, 414, 427, 429, 437-9, 449, 451-3, 459, 461-2, 468, 474, 489-90, 494, 501, 515, 518, 520, 535, 538-41, 547-9, 551, 553-4, 558-9, 561, 563-4, 566-92
Godhead, 4, 85, 259, 291
Gods, 302, 359, 568
Gokhale, B.G., 115, 126-39, 201-3, 373, 375-7, 447

Good, 3, 48, 59, 101, 109, 112, 127, 149, 183, 538, 553, 575
Goodness, 100-1, 104, 108, 119, 161, 352
Goodwill, 47-8
Government, 18, 33, 41, 43, 53-4, 56, 71, 120, 126, 301, 391, 345, 400, 413
Grace, 13, 84, 118, 146, 155, 590
Gramodyog Sangh, 425-6, 429, 435, 438
Gratitude, 14, 17, 104
Growth, 10, 27, 37, 182, 280, 357
  moral, 359, 363-9
Gujarat, Gujarati, 17, 55, 68, 72, 92-3, 103, 105, 109-12, 123, 135, 140, 144, 153, 156-7, 232, 270-1, 283, 296, 347
Gunas, 81, 83, 91-2, 152
Gurus, 79, 115, 130, 136, 144, 189, 375, 551-2, 577-8

Happiness, 86, 294, 360, 461, 475, 494, 500-1, 508
Harijans, 63, 76
Harijan Sevak Sangh, 20, 425, 427, 438
Harmony, 23, 47, 71, 117-18, 252, 267
Hatred, 12, 63, 110, 161, 343-5, 347, 379, 590
Health, 53, 120, 232, 296, 453
Heart, 1, 3, 10, 21-2, 31, 41, 47, 55, 58-9, 62-3, 69, 71, 73, 75, 77-8, 82-3, 91, 104-5, 116, 118-19, 126-7, 130, 139-40, 150, 155, 164, 188, 221, 226, 341, 380, 453, 459, 463, 476, 481-2, 497, 531, 539, 541, 547, 556, 562, 585
Heaven, 15, 44, 107, 120, 160, 164, 183, 188
Hell, 164, 183, 188
Hero, 74, 128
Himalayas, 38, 57, 119, 448, 503, 586
Himsa, 7, 90, 323
Hind Swaraj (Gandhi), 5, 17, 199-369
Hindi, 62-3, 154, 156, 179, 184, 261, 425
Hinduism, Hindu(s), 8-9, 11, 19, 26-8, 42, 46, 56, 59, 61, 63, 71, 73, 75-6, 99, 110, 133, 142, 144, 155, 184, 186-7, 190, 214, 216-18, 221-8, 230, 260, 266, 338, 346, 391, 393, 411, 435, 447-8, 454-6, 460, 465-77, 480-1, 484, 487-90, 492-3,

499, 502-3, 505, 507, 513-15, 516, 522, 530, 532, 537, 544, 553, 567
Hindu-Muslim, 63, 76, 133, 227-8, 411, 521, 525-6, 546
Historians, History, 82, 184, 187-8, 215, 225, 235, 244-5, 269, 270, 298, 437, 484-5, 531
Home Rule League, 135-6
Honesty, 149, 211, 283
Honour, 378, 475, 563
Hospitals, 51, 230, 279, 294, 304, 323
Human nature, 9, 42, 549
Humanitarianism, 27
Humanity, 1, 4, 6, 12, 19, 48, 52, 58-9, 109, 128, 153, 288, 291, 296, 330, 531
Hume, A. O., 201, 203, 206, 269
Humility, 10-11, 30-1, 49, 79, 89, 117-18, 120-1, 130, 339, 471, 482, 500, 537, 539-40, 545, 560
Hypocrisy, 67-8, 70, 101, 149, 305, 307-10, 316, 333, 340-1, 354, 384, 552, 564, 578

Ideal(s), 11, 66, 118, 123, 127, 130, 298, 361, 363, 369, 400, 403, 453, 515, 592
Ignorance, 88, 115, 149, 155, 243, 307-9, 572
  spiritual, 149
Illusion, 7, 80, 118
India, 5-6, 9, 15-16, 18, 20, 23, 26, 30-1, 33-5, 38, 43, 46, 48-50, 52-5, 57-9, 61-2, 74-5, 78, 88, 90, 94, 105, 109-13, 116, 120, 123, 125-34, 138-9, 144, 150, 156, 179, 181, 185, 199-200, 202-9, 211, 215-17, 219-23, 227, 229, 231-5, 237-8, 248-9, 261, 266-8, 270, 272, 279, 286, 293, 295, 297, 298, 302-28, 331-3, 344-5, 347, 349, 351-3, 362, 371-3, 375-6, 395, 409, 454, 469-71, 477, 493, 506, 508, 533
  ancient civilization, 233-4, 261, 265 8, 272, 298, 302-6
  Constitution, 403-4, 406-7, 410-11
  future, 365, 454
  independence, 33, 58, 76, 200, 227, 237, 403, 436, 521, 558, 563
  languages, 40, 62, 184, 254-5, 261, 304, 319-20
  peasants, 6, 138, 268, 368, 452

  villages, 269, 285-6, 313-14, 320, 332, 445, 506, 521, 584
  women, 267-9, 305
Indian National Congress, 19-20, 61, 63, 68, 75-6, 124, 129, 134, 136, 137, 139, 201, 204, 206-7, 227, 254, 269-70, 292, 302, 306, 392, 396, 398-402, 404-7, 411, 419-20, 423-5, 431-3, 440-2, 444, 491, 521, 570-2
Indian Opinion, 34, 111, 199-200
Individual(s), 4-5, 7-9, 11, 53-4, 286, 352, 446, 453, 532
Industrialism, 62, 315-18
Injustice, 227, 380
Inner voice, 2-3, 122, 380-1
Institutions, 5, 7, 34, 54, 219, 266, 301, 309, 317, 362, 374-5, 438, 549
Intellect, 77-9, 82, 88, 118, 120, 222, 252, 261, 380, 458, 541, 560, 585, 589
Irreligion, 5, 110, 112, 214, 217-18, 336, 391, 516, 543, 545
Ishopanishad, 444
Ishvara, Ishwara, 322, 520, 573
Islam, 9, 61, 76, 144, 153, 184-7, 391-2, 455, 468-9, 472-5, 478-9, 503, 512, 516, 521-3, 532, 544, 553, 568-9
Isms, 5, 8, 60-1, 64, 423-4
Italy, 105-6, 236-8, 270

Jail, 16, 94, 121, 182, 187, 190, 193
  see also Gaol
Jainism, 8, 27, 117, 143, 151, 153, 223, 467, 480, 571, 573, 575
Jallianwala Bagh, 41, 44
Japan, 208, 217, 231, 298, 325, 418, 467
Jesus Christ, 12, 16, 18, 26, 41, 44, 107, 181-2, 289-90, 299, 319, 329, 357-9, 363, 390, 455, 460-1, 468, 481, 496-8, 500-1, 506, 509-11, 512, 517-21, 533, 535, 538-9, 550, 564
Jews, 16, 69, 76-7, 147
Jinnah, Mohammad Ali, 76-7
Jiva, 87
Jnana, 84, 90, 99
Jnanayoga, 95
Judaism, Jew(s), 16, 69, 76-77, 468, 472, 544
Justice, 7, 69, 86, 104, 112, 134, 203, 227, 233, 360, 478

Kabir, 359, 469, 472-3
Kallenbach, Herman, 132, 273
Karma, 6, 8, 90, 99, 116, 452, 474, 514, 559, 573-5, 579
Karmayoga, 2, 94-5
Khaddar, 315-16, 399
Khadi, 30, 56, 58, 425, 438, 495, 580
Khan, Badshah, 393
Khilafat, 307, 392, 477, 522-3
Kingdom
  of God, 301, 358, 361, 497, 520
  of Heaven, 15-16, 18, 283
Kingsford, Dr. Anna, 154, 531
Knowledge, 22, 37, 69, 71, 78-9, 85, 87-8, 90, 94, 96, 139, 140, 148, 155, 169, 189, 224, 251-2, 314, 337, 416-17, 475, 488-9, 575
  spiritual, 149
Koran, 8, 85, 150-1, 225, 475, 477-8, 512-14, 567-8
Krishna, 21, 23, 57, 82-5, 90-1, 94, 96, 129, 149, 183, 357, 483-5, 513, 550, 567, 591
Kshatriya, 23, 532

Labour union, 425-6, 435, 438
Labour, 45-7, 51, 62, 96, 107, 119-20, 122-3, 130, 132, 164, 188, 213, 228, 286, 288, 339, 396, 546, 548
Law(s), 12, 99, 108, 119, 128, 134, 226-7, 246-7, 279, 344, 356, 358, 362, 366, 379, 389, 462, 557, 564, 575, 584, 589
  divine, 108
  spiritual, 350
Lawyers, 76, 219, 222, 227-9, 232, 258, 262, 294, 339, 432-3
Liberalism, 85, 124, 474
Life, 1, 11, 119, 121, 161, 169, 278, 343, 415, 508, 572
  inner, 122, 266
Light, 2, 10, 20, 49, 58, 86-7, 117, 146, 155, 160, 292, 416, 508, 572, 585-6
Light of Asia, The (Arnold), 73-4, 476, 480, 504
London, 74, 144, 153, 199, 354, 546
Love, 12, 18, 22, 44, 55-7, 67, 78, 104, 109, 122, 125, 128, 131, 136, 139, 161, 242-5, 262, 272, 278-9, 304, 323, 332, 343, 345-6, 372-3, 379, 432, 437, 458, 461, 467, 474, 477, 483, 506, 526, 534-5, 542, 545, 551, 572, 585

Macaulay, Lord, 219, 253, 332
Machinery, 5, 51, 52, 108, 232, 256-9, 279, 322, 332
Mahabharata, 81, 167, 182-4, 187, 303, 305, 484-5, 514-15, 532, 553
Mahavira, 481
Mahomed, Mahomedans, 214, 216, 218, 221-3, 225-6, 230, 260, 217-18, 227-8, 289, 319, 329, 359, 468-70, 500-1, 550, 567
  see also Muslims, Mussalmans
Man, 4, 8, 12, 40-1, 45-6, 72, 334, 459, 532, 553, 591
Manchester, 73, 256-7, 261, 334
Mankind, 1, 7, 48, 50, 54, 185, 231, 286, 334-5, 409
  see also Humanity
Marx, Karl, 64, 283, 403
Mashruwala, Kishorelal G., 40, 95, 418, 421, 430-1, 433, 438, 443
Masses, 31, 34, 43, 62, 127, 281, 315, 348-51, 414, 454, 477, 479, 482, 565
Materialism, 5, 6, 288, 332, 345, 359, 361
Maya, 66-7, 88, 474, 572
Mazzini, Giuseppe, 105-6, 236, 270
Means and ends, 7, 138, 150, 239-42, 416, 445
Meat-eating, 69, 73, 78, 191, 488-9, 505, 509
Medicines, 16-17, 229, 231, 263, 294, 304-6, 323
Meditation, 3, 10, 137, 148, 198
Mehta, Narasinh, 66-8, 71, 137, 321-2, 576
Mehta, Rajchandra Ravjibhai, 115, 139-46, 148-53, 156, 196
Mercy, 337-8
Mind, 5, 11, 13-15, 17, 72, 87, 106, 110, 117, 146, 151, 230, 232, 248-50, 300, 339, 497
Mirabai (Hindu saint), 58, 100
Mirabai, Mirabehn (Madeleine Slade), 31, 38, 192, 194
Missionaries, 26, 73, 319, 448, 456, 467, 471, 517, 539
Moderates, 206, 259-60, 332
Moha, 66
Moksha, 4, 14-15, 18-19, 82-3, 88, 92, 118, 137-8, 140-2, 146, 149, 151-2, 466, 471, 474

Morality, 2, 4, 6, 79, 149, 155, 199, 214, 216, 232, 252, 257, 272, 294, 298-9, 332, 357, 360-1, 363-70, 381, 440, 484, 544, 572
Morals, 27, 100, 107, 305
Morley, Lord, 43, 226, 238, 270
Motive, 148-9, 181, 200, 210, 251, 299, 377, 383
Muhammad, *see* Mahomed
Mukta purusha, 141-2
Mukti, 22, 576
Mullas, 111, 255
Muller, Friedrich Max, 266, 289, 321, 373, 466
Mumukshus, 141, 143, 148
Munis, 16, 137-8, 303
Muslim League, 76, 522-3
Muslims, Mussalmans, 8-9, 11, 19, 56, 61, 63, 69, 75-7, 110, 133, 155, 179, 182, 184-6, 188, 304, 321-2, 338, 392-3, 400, 411, 478, 480, 499, 512, 522, 530
see also Mahomedans
Mystics, 140, 155, 188, 516

Naidu, Sarojini, 11
Nanak, Guru, 359, 469, 473, 484
Naoroji, Dadabhai, 123-6, 202, 269
Narayan, Jaiprakash, 75, 422
Narayana, 321-2
Natal, 17, 130, 290, 343
Nationalism, 220-1, 273, 331, 343-4, 347
Nature, 56, 71, 100, 230, 252, 259, 315-16, 334-5, 360, 451
Nehru, Jawaharlal, 64, 285, 288, 396-7, 409-10, 412
New York, 353-4
Newman, Cardinal, 22, 459, 586
Newspapers, 6, 193, 196, 200, 211, 254, 381
Nirvana, 18, 494, 561
Niyamas, 489-90
Non-attachment, 92-3, 284
Non-co-operation, 7, 32, 47, 55-6, 308, 345-6, 353, 418, 432-3, 487, 510-11, 570
see also Non-violence
Non-possession, 80, 90, 344, 383, 387, 546
Non-resistance, 531-2
Non-retaliation, 108
Non-stealing, 67, 80, 90, 502, 546
Non-violence, 4, 11-13, 19, 23, 27, 34, 38, 45-6, 57, 60, 67, 77-8, 80-3, 87, 90, 96, 101, 103-4, 112-13, 116-17, 122-3, 149, 153, 189, 225, 279, 281, 286, 303-4, 322-3, 349-50, 392-4, 396, 399, 403, 414-20, 422-6, 428-32, 434-5, 437-41, 443, 454, 461, 489, 507, 509, 514, 546, 563

Oneness of all life, 27, 451, 515
Oneness of God, 455
Ormuzd (Hormazd), 160, 544, 591

Pacifist(s), 326-7
Panchamas, 485-7
Pantheism, 466, 468
Paris, 290, 335
Parliament, 13, 208-11, 214, 278, 337, 380
Parsi(s), 69, 76, 110, 221, 223
Partition, Bengal, 205-7
Passion(s), 14, 17, 72, 232, 250, 252
Passive resistance, 43, 102, 134, 243-50, 264, 273, 275-6, 292, 295-6, 299-301, 332, 357, 373
Patanjali, 122, 337
Patel, Sardar Vallabhbhai, 24-5, 192, 396, 405, 415, 417-19, 424, 441, 443, 449
Pathans, 392-3
Patience, 21, 99-100, 112, 134, 141, 209, 219, 266
Patriotism, 19, 41, 48, 105, 125-6, 134, 211, 237-8, 347
Pax Britannica, 218, 332, 348
Peace, 18-19, 23, 25, 31, 42, 48, 54, 56, 90, 144, 146, 219, 267, 295, 326-7, 352-3, 508, 556, 590
Peasants, 6, 107-8, 111, 138, 248, 252
Perfect man, 81, 153
Perfection, 9, 17, 149, 591
Philosophy, Philosophers, 22, 61, 87, 153, 266-7, 269, 340, 353, 450, 482
Phoenix Ashram, 34, 269, 277, 544
Pilgrimage, Pilgrims, 14, 38, 66, 71, 221, 296, 469
Poets, 12, 81-2, 139, 198, 458-9
Political life, 127, 129-30, 134, 374-5
Political power, 396-407, 413-14
Politicians, 33, 41-2

Politics, 1, 4, 6-7, 9, 19, 33, 42, 127, 138, 279, 408, 416-20, 422, 425, 428, 432-3, 435, 439, 441
  true, 415
Politics and religion, 1, 6, 9, 299, 370-446
Politics and society, 408-46
Possessions, 7, 83, 382-6, 388
Poverty, 6-7, 106-8, 119-20, 125, 219, 249-51, 305, 314-15, 322, 352, 359, 362, 365, 390, 444-5, 454
  voluntary, 381-91
  see also Non-possession
Power politics, 7, 415-16, 421, 423-4, 440, 442-4
Power, 2, 7, 11, 12, 21, 39, 52, 66, 121, 141-2, 145, 153, 183, 185, 228, 261, 361, 401-2, 432, 440, 446, 558, 584-5, 591
  occult, 22
  political, 108
  spiritual, 453
  supernatural, 13
Prahlad, 274, 534
Prasad, Rajendra, 415, 424, 441
Prayer, 3, 9, 17, 32, 46, 48, 68, 97, 99, 107, 182, 386-7, 458, 464, 468, 499, 547-65, 590-1
Pride, 32, 56, 66, 89, 106, 338-9, 481
Principle(s), 4, 6, 23, 35, 42, 63, 67, 69, 71, 77-8, 84, 86, 88, 96, 109-10, 117, 137, 150, 277, 283, 299, 303, 324, 345, 370, 402, 406, 423-4, 475, 555
Progress, 118, 138, 152, 328, 344, 356, 359-60, 501
  economic, 355-63
  moral and material, 355-96
  social and moral, 4, 6
  spiritual, 281
Prophet, the, 74, 185-6, 468, 512-13
Prophets, 11, 18, 76, 182, 186, 457, 466, 477
Public, 7, 20, 25, 31-2, 56-7
Public service, 138, 381
Punjab, 201, 205, 304, 307, 381, 469, 473
Purification, 88, 91, 456, 490
Purity, 80, 83, 86, 121, 150, 268, 492, 510

Qaid-e-Azam, 521-2
Quakers, 532, 571, 590

Railways, 51, 219-22, 233, 258, 261, 279, 281, 284, 286, 293-5, 308, 325, 328, 332, 336
  see also Trains
Rajchandra, *see* Rajchandra Ravjibhai Mehta
Rama, 11, 66-7, 69, 80, 118, 149, 161, 322, 338, 380, 431, 463, 482, 490, 544, 547, 551, 567, 582, 591
Ramanama, 322, 489, 556, 569-70, 581-2
Ramarajya, 380
Ramayqna (Tulsidas), 80-1, 126, 157, 568
Ramayana, 182, 305, 340, 514-15
Ranade, Justice Mahadev Govind, 116-17, 129-30, 136, 242, 270, 305
Ravana, 69, 118, 161, 338-9
Reason, 9, 69, 82-3, 85, 87-8, 96, 100, 196, 235, 303, 356, 448, 458, 478, 504, 555, 572, 580, 585-6, 588-9
Reform, 5, 326, 376-7, 400, 408, 413-14, 446, 448, 481
Reformation, Buddhist, 468, 492, 502
Reincarnation, 86, 95, 289
Religion and politics, *see* Politics and religion
Religion(s), 1, 6-10, 19, 26-7, 42, 46-7, 59, 70, 74-5, 85, 99, 107-13, 129, 137-8, 144, 150, 152-3, 155, 182, 190, 214, 217-19, 222-5, 230, 233, 244, 246-7, 260-1, 266, 288, 303, 320-1, 329, 335, 373, 376, 379, 391, 393, 395, 411, 448, 450-1, 454, 459-61, 463-6, 474-5, 481, 488, 502, 519-20, 522-3, 536, 538-43, 546, 548, 553, 555, 569, 577-8
  equality, 542-6
  false, 75
  true, 75, 110, 138, 229, 449
Religious reform, 447-50
Religious study, 503, 516, 518
Renunciation, 4, 72, 74, 278-9, 444-5, 491-2, 500
Resistance, 103, 327, 377-8
Rights, 10, 108, 131, 240, 378, 424-5
Rishi, 16, 45, 124, 137-8, 232, 490
Round Table Conference, 398, 412
Rowlatt Bills, 376-8
Roy, Raja Ram Mohan, 447-8, 473, 475
Rulers, 310, 332, 342-5, 372, 399, 408, 495

Ruskin, John, 103-5, 115, 143-4, 153, 271, 275, 327, 369
Russia, 107-9, 112, 119, 261, 317

Sabarmati Ashram, 436, 544, 557
Sacrifice, 12, 53, 128-9, 181, 183, 238, 243, 307, 415, 453, 490, 492, 502, 514-15, 526, 536, 549-50
Sadhus, 13-14, 129, 133, 137-8, 379
Sage, 16, 80, 116, 118, 122, 294, 303, 320, 372
Saint, 16, 41, 42, 119, 155, 556
Salt tax, 205, 264, 445
Salter, Henry, 573
Salvation, 18, 294, 466, 510, 515
Sanatan dharma, 68, 85
Sannyasa, 138, 514
Sannyasi, 13, 69, 138, 397
Sanskrit, 72, 78, 89, 92-4, 151, 153, 157, 179, 502, 548
Satan, 43, 68, 82, 160, 183, 239, 272, 302, 316, 345, 553
Satya, 4, 90, 502
Satyagraha, 4, 7, 103, 110, 112, 131, 176, 244-5, 278, 280, 327, 338, 341-2, 345, 378, 399-400, 416, 433, 444, 526-7, 564
Satyagrahi, 7, 109, 112, 132, 188, 339, 341, 399, 401
Schools, 67, 69-71, 140
Science, 5, 252, 289, 291, 310-14, 318, 356, 413
  medical, 306
Sect, sectarian, 2, 8, 10, 60-2, 391, 423-4, 442
Seekers, 8, 18, 80, 88, 417, 542
Seer(s), 82, 116, 118, 122, 139
Self, the, 16, 137, 148, 150, 489, 516, 558, 576
Self-control, 4, 14, 15, 66-7, 114, 120, 122, 145-6, 230, 264
  see also Hind Swaraj
Self-denial, 526
Self-discipline, 303
Self-examination, 119, 564-5
Self-government, 54, 204, 349
Self-indulgence, 120-1, 149, 161
Self-knowledge, 576
Self-purification, 49, 83, 299, 416, 508, 576
Self-realization, 84, 122
Self-respect, 46, 587
Self-restraint, 120, 196, 488, 502

Self-rule, 237, 247, 264, 453
  see also Hind Swaraj
Self-sacrifice, 245-6, 279, 307, 309, 500, 510, 531, 563
Self-suffering, 299-300, 346
Self-sufficiency, 425-6
Self-surrender, 90-1, 99, 388, 458, 554
Selflessness, 377
Servants of India Society, 127, 129, 138, 371, 374
Service, 2, 4, 12, 17-18, 24-5, 39, 48, 51, 56, 84-5, 88-90, 105, 107, 120, 123, 128-9, 134-5, 222, 231, 249-50, 287, 321-2, 384, 386-90, 398, 407, 414, 424, 427, 430-1, 444-5, 461-2, 496, 510, 537, 549, 554-7, 575-6, 582, 589
  public, 138, 343
Shah Jehan, 169, 189, 306
Shankaracharya, 16, 84, 88, 122, 150, 183, 359
Shastras, 40, 69-71, 77, 79-80, 88, 95, 100, 127-8, 140, 150-2, 268, 418, 488, 508
Shaw, George Bernard, 412-13
Shibli, Maulana, 185, 512, 546
Sikhs, 77, 188-9, 391-2, 400, 469, 473, 483-5
Silence, 15, 38, 163, 188, 441, 578, 590
Simplicity, 10, 51, 116, 125-6, 153, 185, 229, 278-9, 286, 295, 316, 354, 551
Sin, 12, 25, 67, 70-1, 78, 86, 91, 117, 120, 125-6, 172, 230, 256, 576, 586
Sincerity, 4, 11, 84, 117, 376
Singh, Guru Govind, 391-2, 483-5
Sita, 80-1, 436, 463
Slade, Madeleine, *see* Mirabai (Madeleine Slade)
Slavery, 109-10, 188, 208, 213-14, 230, 234-6, 246-7, 254, 256, 294, 325, 332, 335, 405, 528
Smuts, General Jan, 43, 132, 292, 296, 396
Social service, 460-1
Socialism, 75, 403, 406, 419, 422
Society, 4, 7, 87, 117, 121, 344, 357, 408
Socrates, 5, 100-1, 524
Soldiers, 108, 111-12, 300, 532, 563
Song Celestial, The (Arnold), 72-3
Soul, 2, 5, 8-10, 17-19, 21-2, 36, 42, 52, 54, 86-7, 98, 105, 109, 111, 129-31, 133, 136, 141, 144, 150,

Soul—*cont.*
155, 160, 182, 187, 244-5, 249, 263, 277, 315, 318, 326, 329, 340, 385, 451-2, 458, 494, 542, 571, 590
universal, 87-8
Soul-force, 7, 53-4, 109, 112, 243-8, 259, 264, 277, 279, 299, 332, 364-5, 371
see also Truth-force
South Africa, 2, 21, 43, 46, 102, 113, 125, 131-4, 137, 144, 154, 176, 178-9, 200, 208, 270, 273, 278, 283, 293, 295, 319, 325, 343, 357, 366, 374, 465, 470, 481, 512, 522, 528-9, 591
Speech, 10, 13, 15, 66-7, 70, 87-8, 263, 572, 582
Spinning-wheel, 51-2, 88, 171, 195, 281, 286, 314, 318, 321-2, 426-9, 432, 507
Spirit, 7, 28, 39, 53-4, 80, 117, 118, 127, 155, 183, 283, 300, 468, 519
Spiritual dictionary, 97-8
Spiritual knowledge, 150-1, 588
Spiritual sight, 587
Spiritual transformation, 454-8
Spiritual(ity), 4, 6, 10-11, 16, 51, 91, 134, 136-7, 182, 281, 299, 381, 539
Spiritualize, 127, 129, 134, 138, 374-5
State, 7, 51, 187, 309, 395, 399
ideal, 17, 139, 284, 322, 399
Strength, 1, 23, 49, 57, 67, 88, 109-10, 112, 219, 235, 262, 308, 341, 422
Strikes, 395-6, 446
Sudra, 8, 79, 487
Suffering, 10, 14, 66, 86, 88, 101, 106, 109, 117, 129, 163, 205, 245, 250, 263-4, 266, 326-7, 339, 341, 346, 521, 531, 576
Suicide, 9, 82, 272, 561, 587
Sun, 12, 87, 92, 338, 464
Swadeshi, 205, 258, 264, 399, 525, 546
Swaraj, 15, 47, 50, 57, 67, 120-1, 123, 179, 207-9, 215, 234-5, 264, 278-9, 283, 285, 302-3, 320, 322, 338, 342, 349, 398-400, 404, 426-8, 441, 444, 453, 486, 558
System, 5, 47, 64, 95, 138, 266, 332, 380

Tagore, Devendranath, 448, 473, 475
Tapascharya, 4, 8, 14, 69, 80, 88, 94, 96, 374

Tapasya, 303, 448, 508
Teachers, 301, 359, 390, 450, 466, 483, 489, 492, 496-7, 509, 518, 535, 588, 590
Theology, 9, 61, 516-17
Theosophical Society, 73, 289, 465, 467, 470, 475
Theosophy, 72, 75-6
Thoreau, Henry David, 102-3, 272, 400, 413
Thought, 3, 9-11, 13, 17-18, 49, 66-7, 87, 107, 197-8, 413, 581-3
Tibet, 467, 493, 511-12, 574
Tilak, Bal Gangadhar (Lokamanya), 84, 122, 136, 270, 379, 466, 471
Tolerance, 4, 27, 153, 451, 466-7, 469, 501, 523-47, 534, 542-3, 569
Tolstoy, Count Leo, 5, 45, 103, 107-23, 143-4, 153, 271-2, 274-5, 300, 501, 532, 558, 589
Trains, 71, 82, 213, 284, 363
see also Railways
Transmigration, 27, 573
Transvaal, 43, 109-10, 112, 199, 216, 272, 300-1, 331, 335, 339, 343
Trustee(s), 47, 295, 362, 504
Truth, 3-5, 7-13, 15, 18-19, 23, 26-7, 29-32, 42, 44, 46-9, 57, 60, 67, 72, 78-82, 86, 90, 95, 100, 104, 111, 116-18, 121-2, 126-7, 130, 137, 141-2, 149, 151, 154, 160-1, 182, 188, 200, 249-51, 272, 285-6, 303-4, 309, 316, 320, 322, 326-7, 329, 335-7, 341, 353, 361, 373, 379-81, 414, 417-18, 425, 430-1, 434, 437, 440, 443-4, 450-1, 454, 457-8, 460-2, 464, 477, 483, 488-9, 500-4, 509, 513, 517, 520, 538, 542, 546, 551-2, 555, 569, 572, 585, 587
Truth-force, 244, 250
see also Soul-force
Truthfulness, 134, 403, 475, 536
Tulsidas, 20, 80-1, 126, 157, 244, 338, 453, 463, 489, 568
Tyranny, 109, 237, 247, 260, 295, 337, 583

Unity, 7, 41, 43, 45, 47, 52, 62-3, 105-6, 223, 319, 377, 391, 434-5, 450, 515, 525, 527, 535, 569, 576
Unto This Last (Ruskin), 105, 144, 275, 327
Untouchables, 56, 59, 68, 70-1, 75, 88,

Index 625

346-7, 409, 411, 455-6, 485-7, 489
Untruth, 80, 86, 89, 118-19, 146, 149-50, 160-1, 285, 316, 379, 382, 464
Upanishads, 84, 99, 182, 319, 488, 515
Urdu, 62-3, 179, 184-5

Vaikom, 346-7
Vairagya, 10, 145-6, 152
Vaishnava, 66-72, 322, 487
Varnashrama, 27, 67, 71, 90
Vedanta, 16, 87, 151
Vedas, 67, 69, 79, 84-5, 117, 182, 198, 223, 321, 448, 466, 471, 476-7, 488, 493, 502, 512, 567-8
Vegetarianism, Vegetarian(s), 73, 102, 289, 327
Victor Emmanuel II, King, 106, 236, 270
Violence, 6-7, 10, 75, 80-3, 87, 90, 116, 118-19, 149-50, 239-40, 245, 259, 272-3, 276-9, 285, 295, 297-9, 303, 323, 330-2, 342, 345, 349-50, 378, 392, 429, 435, 454, 553
Virtue(s), 12, 55, 68, 70-1, 86, 104, 111, 114, 128, 134, 140, 148, 269, 316, 489-90, 538
Vivisection, 230, 263, 289, 311, 318
Vow(s), 4, 11, 13, 24-5, 114, 525-7, 546

War, 12, 54, 56, 81-2, 93, 96, 118, 244, 272, 344, 361, 366, 446

Wealth, 5, 107, 111, 119, 185, 213, 228, 257, 263, 306-7, 335, 339, 351, 360-3, 382, 453, 469
Welfare, 7, 38, 47, 144, 172, 356, 363
West, the, 29, 32, 49-51, 53-4, 57, 70, 107, 116, 121-2, 185, 229, 292-3, 302, 318, 324-9, 348
Will, 11, 91, 99, 109, 188, 252
Wisdom, 6, 127, 202, 257, 291, 294, 365, 406
Women, 62-3, 80-1, 214, 256, 305, 338, 582
  see also India, Women, and Gandhi on Women
Words, 2, 9-10, 27, 37, 40, 86, 102, 117, 548
Working Committee, 60, 64, 286-7, 398, 402-4, 406-7, 411
Worship, 56, 128, 154, 221, 547-8, 576-8, 581

Yajna, 88, 120, 122-3, 514
Yamas, 489-90, 502
Yeravda jail, 98, 141, 167, 179, 181, 476, 481, 484, 573
Yoga, Yogi, 21, 122, 137, 148, 340, 470, 473
Yogabuddhi, 94

Zend-Avesta, 151, 568
Zoroaster, Zoroastrianism, 153, 217, 289, 544, 550, 553, 569